Understanding Wine Chemistry

Understanding Wine Chemistry

Andrew L. Waterhouse
University of California
Department of Viticulture and Enology
Davis, CA, USA

Gavin L. Sacks
Cornell University
Department of Food Science
Ithaca, NY, USA

David W. Jeffery
The University of Adelaide
Department of Wine and Food Science
Urrbrae, SA, Australia

WILEY

This edition first published 2016

Registered Office
John Wiley & Sons, Ltd, The Atrium, Southern Gate, Chichester, West Sussex, PO19 8SQ, United Kingdom

For details of our global editorial offices, for customer services and for information about how to apply for permission to reuse the copyright material in this book please see our website at www.wiley.com.

Library of Congress Cataloging-in-Publication Data

Names: Waterhouse, Andrew Leo, 1955– | Sacks, Gavin L., 1978– | Jeffery, David W., 1972–
Title: Understanding wine chemistry / by Andrew L. Waterhouse, University of California, Department of Viticulture and Enology, USA, Gavin L. Sacks, Cornell University, Department of Food Science, USA, David W. Jeffery, The University of Adelaide, Department of Wine and Food Science, Australia.
Description: Chichester, West Sussex : John Wiley & Sons, Inc., 2016. | Includes index.
Identifiers: LCCN 2016008931 | ISBN 9781118627808 (cloth)
Subjects: LCSH: Wine and wine making–Chemistry.
Classification: LCC TP548.5.A5 W38 2016 | DDC 663/.2–dc23
LC record available at http://lccn.loc.gov/2016008931

A catalogue record for this book is available from the British Library.

Set in 10/12pt Times by SPi Global, Pondicherry, India

1 2016

Contents

Foreword

In vino, veritas...a Latin phrase meaning "in wine, truth." Its sentiments, however, are certainly not unique to Roman society. Indeed, numerous civilizations throughout history have similar phrasings given the prominence that wine has played, and continues to play, in diverse religious, cultural, and social events. Yet, despite its nearly seven millenia as part of the human experience, our understanding of this beverage and all of its truths remains woefully incomplete, even in light of the prowess of modern science. The reasons for this state of affairs are complicated, perhaps justifiably so given that it is the complexity of wine that draws many to its taste and to learning the art of its production.

At its most basic level, wine is a mixture of hundreds of different molecules in a constant state of flux, a feature that gives it the quality of being a living, breathing thing. The identity and concentration of these varied compounds at any given time depends on every factor conceivable, from the vine and the soil, to the weather that season, to the full production process, to how a bottle has been stored, to how long a poured glass or opened bottle has had a chance to breathe before being enjoyed. Our perception of its taste is equally fluid, dependent on its temperature, our mood, what else we have recently consumed, and how well our receptors can distinguish those hundreds of molecules in the first place. Thus, if we are even to consider how to unlock the complexity of wine, we must start by understanding this beverage from the standpoint of its chemistry, since it is molecules and what they can do that is at the heart of the matter.

This text by Waterhouse, Sacks, and Jeffery is an excellent starting point for such investigations. Chemistry on its own can seem hopelessly complex, but what these leading scholars have managed to accomplish seemlessly over the course of 33 chapters is the means not only to appreciate, but also to understand the relevant chemistry and chemical phenomena that impact every element of wine from an analytical, organic, and physical perspective. That success results from an approach that first details all of the different compound classes that are found in wine, their reactivities, and how they can contribute to its final taste profile. The text then moves on to the production process and explains, at an appropriately detailed chemical level, not only the fermentation and production process overall, but how each of its steps and certain decisions along the way can impact what molecules, and how much of them, end up in the final product. Finally, by presenting the latest scholarship, the authors highlight the frontiers of wine chemistry research, indicating opportunities for readers to pursue further avenues of discovery if so inclined.

Whether a neophyte, a connoisseur, a production hobbyist, or an aspiring professional, this carefully crafted text offers all readers an opportunity to enhance their knowledge of and appreciation for this wonderful beverage, along with the potential to contribute to enhancements in its future enjoyment. It is certainly hoped that these authors will continue to supplement and refine this already excellent work in the years to come through further editions as the knowledge of wine chemistry continues to grow. However, for now, congratulations are due for their efforts in successfully distilling diverse knowledge drawn from many different disciplines of chemistry into a fully accessible, engaging, and pedagogically powerful approach to the science of wine.

Scott A. Snyder
University of Chicago, Illinois, USA
January 2016

Preface

Having backgrounds in traditional branches of chemistry (organic, analytical, physical), we now feel privileged to be writing about wine chemistry. Not only is wine chemistry a subject that inspires and challenges, it is also a wonderful vehicle for conversation – we often encounter colleagues, visitors and acquaintances who remark on how much they love wine and its complexity. That complexity can be intriguing but it can also be a barrier to further understanding, whether wine is a hobby or an occupation.

Wine chemistry is taught in a growing number of institutions as part of enology and viticulture curricula, as well as in traditional undergraduate chemistry departments as an elective course. Furthermore, there are many individuals in wine production and allied fields (e.g., suppliers) who expect to make science-based decisions or recommendations. With this in mind, we identified the need for a book that could demonstrate to a reader how to utilize a basic knowledge of chemistry to rationally explain – and, better yet, predict – the diversity observed among wines.

Rather than providing only a description of wine constituents, or focusing on sensory characteristics, analytical aspects, or processing issues, we tackle the types of chemical and biochemical reactions that commonly occur in wine – in other words, we interpret winemaking outcomes through the lens of chemical principles. In doing this, we aim to assist students, winemakers, and others in predicting the effects of wine treatments and processes, or interpreting experimental results, based on an understanding of the major reactions that can occur in wine. We assume only a minimal prior knowledge of wine and winemaking, but do expect basic chemistry knowledge including organic chemistry, though we anticipate that our readers may have forgotten a lesson (or three) from these courses. At times we have relied on recent reviews rather than providing extensive citations to the literature, so we encourage the reader to seek out the primary sources of information to enhance their knowledge of any of the topics we have covered.

We approach our objectives by segregating the book into three parts.

- Part A, Wine Components
  To begin, we review the compound classes found in wine, their typical concentrations, their basic chemical reactivities, and their contribution to wine stability or sensory characters. This first part also considers the types of reactions that components can undergo in a wine environment, and is designed to be used as a reference for subsequent sections. Key chemical concepts involving electrophiles and nucleophiles, and electrophilic aromatic substitution are featured in Chapters 10 and 14, respectively.
- Part B, Chemistry of Winemaking
  Following a brief overview of grape composition and wine production practices, we describe the key reactions that occur during and after fermentation. In particular, we highlight how decisions made during winemaking will favor or disfavor certain chemical reactions leading to differences in wine composition.

We expect that this part can be used to generate hypotheses regarding the effect of unfamiliar winemaking processes or changes in juice components on final wine composition. A feature on different bottle closures appears in Chapter 25.

- Part C, Special Topics
  To conclude, we present several case studies that relate the preceding sections to current or emerging areas of wine chemistry. With these examples we aim to demonstrate representative examples of the challenges – and opportunities – facing those who are interested in this amazing natural product called wine.

In preparing such a book we anticipate there will be gaps and errors, and we encourage the reader to send us comments with regard to anything. From simple typographical errors, to missing topics or citations, errors in data or interpretation, and even suggestions for new approaches to explaining wine chemistry, all suggestions are encouraged. We are planning to follow up with a second edition and any comments or ideas for improvement are most welcome. Please send your ideas to: winechem@ucdavis.edu.

In closing, we acknowledge the work of the many researchers in the international wine science community that we have drawn upon in formulating this book, and also appreciate the feedback from our students who helped shape the book by reading draft chapters along the way. We also thank the grapegrowers and winemakers of the world for producing wines in a breathtaking range of styles – and thus chemistries – without which this topic and book would not exist. Lastly, we are eternally grateful to our partners and families who put up with our absence, or absentmindedness, and supported us throughout the process of publishing this book.

Introduction

The chemical diversity of wine

How many choices does a consumer have when they buy a wine? In the United States, all wines sold must have a Certificate of Label Approval (COLA) from the Alcohol and Tobacco Tax and Trade Bureau (TTB), and in 2013 the TTB approved over 93 000 COLA requests.[1] Because many wines are vintage products, that is, a new label will be produced for each harvest year, the true number of wines available in wine stores throughout the United States may be closer to 250 000.[2] In contrast to commodity products where producers strive for homogeneity (e.g., soybeans, milk), variation in specialty products like wine is not only tolerated – it is appreciated and celebrated. Consumers expect that wines with different labels should smell different, taste different, and look different; from a chemist's perspective, consumers expect wines to have different chemical compositions. The study of wine chemistry is the study of these differences – explaining how there can be hundreds of thousands, if not millions, of different wine compositions, and contributing to a winemaker's understanding of how the myriad of choices they are faced with can lead to these differences.

What is wine?

To a first-order approximation, a dry table wine is a mildly acidic (pH 3–4) hydroalcoholic solution. The two major wine components are water and ethanol, typically accounting for about 97% on a weight-for-weight (w/w) basis. The remaining compounds – responsible for most of the flavor and color of wine – are typically present at < 10 g/L (Figure I.1), and many key odorants are found at part-per-trillion (ng/L) concentrations! Notably, none of these compounds appear to be unique to wine – compounds present in wine can also be found in coffee, beer, bread, spices, vegetables, cheese, and other foodstuffs.[3] What distinguishes different wines from other products (and each other) is differences in the relative concentrations of compounds, rather than the presence of unique components.

---

[1] As a caveat, a COLA is a pre-approval process, and not all will become commercial wines.

[2] Furthermore, data based on TTB COLAs only captures those wines approved for sale in the US, which represents only a fraction of wines produced commercially – the wine-searcher.com database reports tracking over 400 000 wines available for sale worldwide as of April 2016.

[3] One compound that is somewhat unique to grape juice and wine is tartaric acid, which is undetectable in most other fruits and vegetables (although it is at a very high concentration in tamarind). Tartaric acid also belongs to the short list of compounds that was first discovered in wine or grapes, and later discovered elsewhere; the monoterpenes wine lactone and Riesling acetal are also on this list, as are some anthocyanin-derived pigments (vitisins and pinotins).

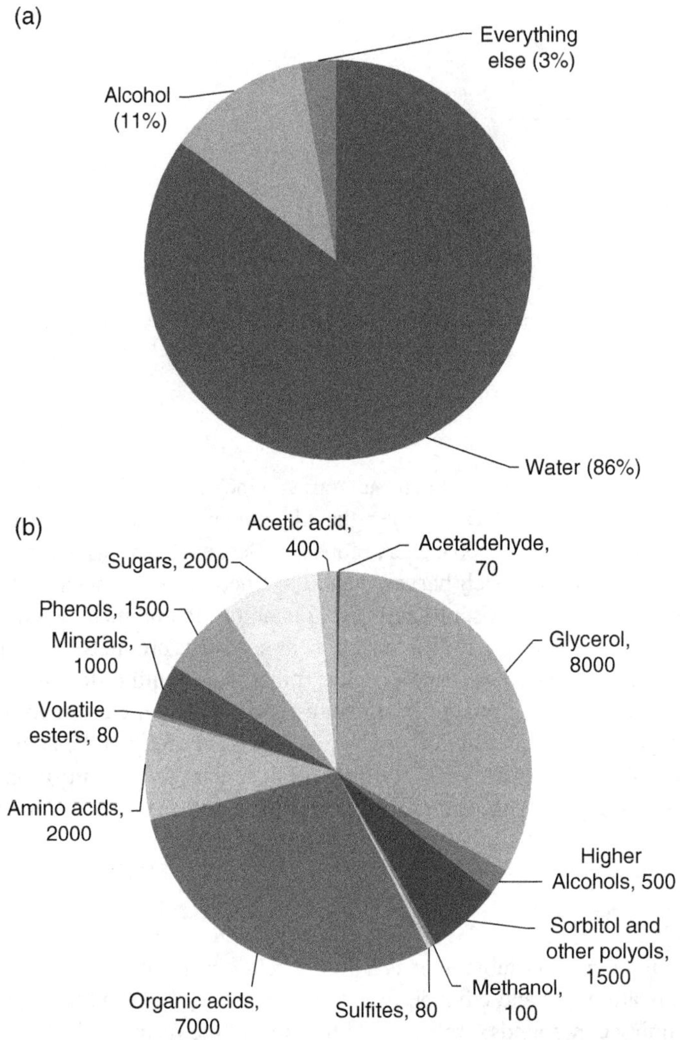

Figure I.1 *Composition of a representative dry red table wine (a) on a % w/w basis and (b) typical concentrations (mg/L) of major wine components excluding water and ethanol, that is, the main contributors to "Everything Else." Key trace components (0.1 ng/L–10 mg/L) would not be visible and are therefore not included*

Wine is produced by the alcoholic fermentation of grape juice or must (juice and solids), which results in the complete or partial transformation of grape sugars to ethanol and CO_2. However, winemaking and wine storage result in many chemical changes beyond simply the consumption of sugars and formation of alcohol. This is readily exemplified by the volatile composition of a wine, which is far more complex than that of grape juice (Figure I.2). These volatile components can contribute to the aroma of wine and such odorants are often classified based on when they are formed; that is, in the grape (*primary*), during fermentation (*secondary*), or during storage (*tertiary*) (Table I.1).

The number of compounds identified in wine follows advances in analytical technology. A survey from 1969 reported that wine and other alcoholic beverages contained 400 volatiles, while a later book from 1983

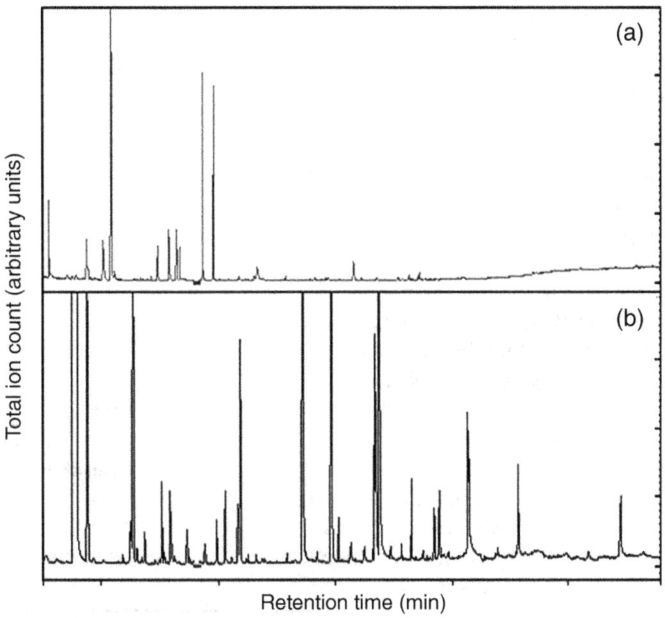

Figure I.2 *Comparison of GC-MS chromatograms for (a) a grape juice and (b) a wine produced from that grape juice. Every peak in the chromatograms represents at least one unique volatile compound*

Table I.1 *Primary, secondary, and tertiary classifications of wine odorants*

Compound classification	Description	Examples (Part and Chapter)
Primary	Compounds present in the grape that persist unchanged into wine	Methoxypyrazines (A.5), rotundone (A.8)
Secondary	Compounds formed as a result of alcoholic or malolactic fermentation due to either i) Normal metabolism of sugars, amino acids, etc. ii) Transformation of grape-specific precursors	i) Ethyl esters (B.22.2), fusel alcohols (B.22.3) ii) Varietal thiols (B.23.2)
Tertiary	Compounds formed during wine storage, for example, as a result of i) Extraction from oak ii) Microbial spoilage or chemical tainting iii) Abiotic transformation of precursor compounds in wine	i) Oak lactones (B.25) ii) Trichloroanisole (A.18) iii) TDN (B.23.1)

reported over 1300 volatiles [1]. A more recent analysis of wines using a state-of-the-art mass spectrometry system (FT-ICR-MS) was able to detect tens of thousands of unique chemical signals across a set of wines, and assign chemical formulae to almost 9000 components [2]. However, the advanced instrumentation in this last report would not distinguish structural isomers – for which there may be billions for a condensed tannin

Table I.2 Summary of major functional classes of interest to wine chemists. Note that compounds may fit into more than one category

Compound functions	Description	Examples (Part and Chapter)
Organoleptic	Compounds that contribute to the taste, odor, or tactile sensations of a wine	Acids (A.3), monoterpenes (A.8), tannins (A.14)
	Compounds that affect wine color or cause a visible haze	Anthocyanins (A.16), proteins (B.26.2)
	Compounds that act as precursors of organoleptically active compounds	Glycosides (B.23.1), *S*-conjugates (B.23.2)
Stability	Compounds that inhibit or promote microbial or abiotic changes during storage	Organic acids (A.3), sulfur dioxide (A.17)
Bioactive	Compounds that may positively or negatively affect human health	Phenolic compounds (A.11), biogenic amines (A.5), ethyl carbamate (A.5)
Matrix	Compounds that affect the speciation or activity of other compounds, usually through non-covalent interactions	Water and ethanol (A.1)
Authenticity	Markers that help distinguish authentic products from fraudulent products	Artificial colors (C.28)

consisting of 30 monomers (Chapter 15). Thus, the number of chemical compounds in wine, like most natural products, is essentially uncountable.

With this in mind, the goal of a wine chemist is not to enumerate every compound, but rather to identify compounds, or in many cases classes of compounds, that will directly or indirectly control key quality aspects of the wine such as organoleptic properties (aroma, flavor, appearance), safety, and stability. Alternatively, compounds may be of interest because they can be used to detect the presence of fraud. These categories, and examples, are summarized in Table I.2.

Chemical reactions in wine

The complexity of its composition would suggest that the range of chemical reactions in wine is limitless. However, as noted above, wine is ~97% ethanol and water, which precludes the large number of reactions in introductory organic chemistry texts that require the absence of protic solvents (e.g., no Grignard reactions). Similarly, the mildly acidic conditions of wines (typically, pH ~3.5) mean that base-catalyzed reactions are usually of low importance (e.g., aldol condensations are unlikely).

As with all chemistry, the key to predicting reactions is to define the components of wine that can react with each other. Many of these reactions will be familiar to students of organic chemistry, and include:

- Reactions between nucleophiles and electrophiles, for example, bisulfite and carbonyls
- Hydrolytic reactions, usually acid-catalyzed, for example, of esters, interflavan bonds, and glycosides
- Addition and elimination reactions, again usually acid-catalyzed.

These reactions and many more are the very essence of this book, and are presented in detail throughout the following chapters. One uniquely challenging aspect of wine chemistry as compared to the organic chemistry lab (and most other foodstuffs) is that reactions are allowed to take place for months, years, or even decades, often at ambient temperatures and in a reductive environment. These conditions can lead to unexpected reaction products – this is especially important since a part-per-trillion of certain compounds may be enough to affect flavor.

Chemistry as a historical record

Many chapters of this text, particularly in Part A, contain tables of "typical concentrations" of various wine components, usually from peer-reviewed reports published since 2000.[4] However, grapegrowing and winemaking practices are not static [3], and typical values may change dramatically with changes in fashion or technology – not to mention climate [4]. In some cases, the analysis of aged wine reveals changes in typical wine composition and lends insight to changes in production practices. For instance, in the nineteenth century, wine drinkers used to prefer much sweeter wines – premium Champagnes would have over 140 g/L of sugars, as compared to < 10 g/L in most modern versions [5]. Control of spoilage organisms like acetic acid bacteria through the use of sulfur dioxide and anaerobic storage during this period was still primitive – a survey of Greek wines from 1872 reported acetic acid concentrations in the range of 1.5 to 3.6 g/L [6], all in excess of modern legal limits. Wine tanks from this era also typically had lead-containing bronze valves (no longer used today), which could be leached under acidic conditions to result in relatively high lead concentrations in wine [5]. Differences in mineral content could also arise from viticulture practices – grapegrowers in the nineteenth and early twentieth century routinely used arsenic to ward off insects and mold [7].[5]

In summary, the values provided in this text should be seen as a snapshot of wine composition circa the early twenty-first century, rather than as fundamental constants. A modern consumer's expectation of a high-quality wine is a reflection of both improved technical capacities as well as accumulated traditions, which in the year 2016 means (for the majority of internationally known premium wines) fermenting with selected strains of *Saccharomyces cerevisiae*, aging in oak barrels, and storing wine in glass bottles with corks. Presumably, the pursuit of different options in the past might have led to alternative perspectives on the idea of wine "perfection" and target chemical compositions. Similar statements can be made about the future of wine, and we expect that the numbers provided here will provide some amusement to the wine chemists of the year 2100.

The chemical senses and wine flavor

The majority of compounds discussed in this text have a role in wine flavor. Because the lexicon used to discuss flavor in scientific publication differs from that used in casual conversation, this introduction will conclude with a brief review of key flavor terminology.

Flavor is defined as the "perception resulting from stimulating a combination of the taste buds, the olfactory organs, and chemesthetic receptors within the oral cavity" [10] – in other words, everything a taster can perceive in the mouth, for example, olfaction, taste, and chemesthesis.

---

[4] These surveys also predominantly consider wines from countries with scholarly activity in enology and viticulture – several European countries, the United States and Canada, South Africa, Australia and New Zealand.

[5] Certain chemical features of wine can also be useful to understanding its use and spread in antiquity. For example, the presence of tartaric acid in a pottery container is considered near-certain evidence that a container held wine [8]. Similarly, the presence of syringic acid was used as evidence that King Tutankhamen drank wine made from red grapes instead of pomegranates. The former contains the pigment malvidin-3-glucoside, which will degrade to the relatively stable syringic acid, while the latter does not [9].

Olfaction, or smell, involves the detection of odorants by olfactory receptors (ORs) located within the nasal cavity. Humans have approximately 700 OR, of which half are usually functioning in any individual [11]. Although each specific OR demonstrates some selectivity towards compound classes, odorants (or mixtures of odorants) typically stimulate combinations of OR, and these combinatorial patterns are associated with particular smells [12]. Olfaction requires that the odorant be volatile to reach the nasal cavity, a process that may occur through two routes:

- Orthonasal olfaction involves the detection of odorants without tasting, for example, by smelling the headspace of the wine. The perceptions arising from orthonasal olfaction are often referred to as *aroma*.
- Retronasal olfaction involves detection of odorants that travel from the oral cavity to the nasal cavity. Most commonly, this occurs following swallowing, after which exhalation drives a small amount of odorants through the nostrils [13].

Although olfaction is selective for volatile compounds, most food volatiles appear to be unimportant to odor. A recent meta-analysis estimated that of the ~10 000 volatiles detected in foodstuffs, <3% were important to food aroma [12]. The same review noted that the aromas of specific foods or beverages (including wine) could be simulated with between 4 and 44 odorants.

Taste involves the detection of small molecules by taste receptors located in the taste buds. Five classes of taste receptors have been established – "sweet," "sour," "bitter," "salty," and "umami" [14], of which only the first three appear to be routinely experienced in wine [15].

Chemesthesis involves the chemical activation of receptors responsible for sensations of pain, temperature, and touch, for example, the "heat" caused by capsaicin in hot chili peppers [16]. There are several critical distinctions between taste and chemesthesis, the most important being that taste is only sensed by the taste receptors of the tongue, while chemesthesis can be detected throughout the oral cavity – and for that matter, throughout the body.[6] The chemesthetic sensations most important to wine are those that cause:

- Pungency and irritation, which can be due to ethanol and CO_2.
- Astringency, or the perceived loss of lubrication in the mouth, which can be triggered by condensed tannins and other phenolic compounds [17].

The perception of "body" is also likely a result of chemesthesis, although the specific compounds responsible for this sensation are still unclear [18].

Classic papers on food analysis (grapes, wine, and otherwise) often focused on identifying or measuring the compounds found in high concentrations, with little emphasis on the sensory relevance of the compounds [19].[7] Since the 1990s, it has been increasingly common to identify organoleptically important compounds through the use of bioassays, for example, using a human sniffer to identify key odorants through GC-olfactometry. Candidate compounds can then be quantified and their relevance evaluated through reconstitution and omission experiments [20].

Perception of flavors. Although the eventual goal of bioassay-based approaches is to recreate the organoleptic property in a model system, a key feature is the use of *activity values* as a rough estimate of a compound's

[6] As an example, the "cooling" sensation of menthol is chemesthetic – the perception of coolness can be felt not only on the tongue but throughout the mouth, in the nose, or on tissue on to which menthol is rubbed. In contrast, a sodium chloride solution would not be perceived as salty anywhere except for the tongue.

[7] As described in Chapter 32, improvements in analytical methodology have renewed interests in using general "non-targeted" approaches to identify potentially important compounds.

Table I.3 Concentrations, odor thresholds, and calculated OAVs for five representative compounds in Sauvignon Blanc wines

Compound	Typical aromas	Concentration range (mg/L)[a]	Odor threshold (mg/L)[b]	Odor activity value, OAV
Water	–	~850 000	–	0
3-Methylbutanol (Isoamyl alcohol)	Solvent, burnt	200–250	30	6–8
1-Hexanol	Grassy	1.5–2.5	8	0.2–0.3
3-Mercaptohexanol	Passionfruit, grapefruit	0.0005–0.0038	0.000060	9–65
3-Isobutyl-2-methoxypyrazine	Bell pepper	0.000008–0.000023	0.000002	4–11

[a] Range of average values observed across 7 wine regions for 2004 and 2005 vintages [23].
[b] From References [24] to [26].

importance. Activity values are calculated as the ratio of a compound's concentration to its sensory threshold in an appropriate matrix:

$$\text{Activity value} = \frac{\text{Concentration}}{\text{Detection threshold}}$$

Typically, compounds with higher activity values have more intense flavors, but the concentration–response function varies among compounds. In simple solutions, the intensity of most taste compounds (sugars, acids) scales as a linear function of their concentration, but the intensity of most odorants increases as roughly the square root of concentration [16].

Use of activity values in evaluating the relevance of odorants to a given foodstuff dates to at least the 1960s [21, 22], and even earlier examples exist for other flavor compounds.[8] As a general rule, compounds with activity values <1 are expected to have a negligible effect on a particular sensory attribute [20]. The utility of the activity value concept can be appreciated from representative data for Sauvignon Blanc wines in Table I.3. Based strictly on concentrations, 1-hexanol appears to be a very important contributor to wine. However, conversion to odor activity values (OAVs) reveals that the "grapefruit" aroma of 3-mercaptohexanol and the "herbaceous" notes of 3-isobutyl-2-methoxypyrazine are far more likely to contribute to Sauvignon Blanc aroma – and, in fact, they do.

Activity values are useful as an initial screen to determine the likely relevance of a given compound. However, simply knowing whether a flavor compound is or is not present at suprathreshold concentrations (activity value > 1) is insufficient to determine if the compound is important to the foodstuff for several reasons:

[8] The first widespread use of activity values in food science is probably the eponymous "Scoville Unit" to describe the pungency of hot chili peppers. Originally, the Scoville value of a pepper was determined by preparing an ethanol extract and determining the dilution necessary before the heat was no longer detectable – a value of 4000 Scovilles, typical of a jalapeño, meant that the extract had to be diluted 4000-fold before the pungency was no longer detectable. Currently, Scovilles are determined indirectly by measuring capsaicin and related compounds using HPLC rather than sensory testing.

- *Masking*. The perceived intensity of a flavorant can be decreased by the presence of other flavor compounds. For example, addition of herbaceous smelling methoxypyrazines to red wine decrease the intensity of fruity aromas [27].
- *Additive or synergistic effects*. Groups of homologous compounds, that is, series of alkyl esters or ketones, can reach sensory threshold through additive effects even if all compounds individually have activity values < 1 [28]. Synergism, that is, an increase in stimulus intensity beyond what is predicted for simple additive effects, can also occur, most often for taste and tactile sensations [16].
- *Matrix effects*. Differences in the matrix (pH, temperature, ethanol concentration, non-covalent interactions with macromolecules) can change the activity of flavor compounds, and particularly the volatility of odorants [29].
- *Synesthesia and confirmation bias*. The different chemosensory modalities (taste, smell, tactile) do not operate in isolation; information from these senses is integrated together. For example, panelists report that increasing the sweetness of a fruity beverage increases the intensity of fruit flavor [30]. A related concept is confirmation bias, in which prior knowledge of a product affects a panelist's perceptions; for example, white wines dyed with tasteless red food coloring are perceived as having fuller body [31].
- *Emergent properties*. Combinations of flavor compounds (particularly odorants) often elicit different percepts than individual compounds. For example, no specific wine compound has a smell exactly like wine, but the combination of odorants at appropriate concentrations allows a sniffer to know that they are smelling wine and not another beverage [32].

Finally, the use of a single value for a sensory threshold obscures the fact that individuals show considerable variation in their sensitivities to different flavor compounds, particularly odorants. One author estimated that a typical 96% confidence interval for odorant thresholds across a population spans a concentration factor of 256 [33], and individuals' thresholds and descriptors may change with repeated experiences [34]. Although this variation does not preclude studies of organoleptic properties, it does necessitate appropriate sensory practices and rigorous statistical analysis of the data (as with other studies involving human subjects). A full discussion of sensory techniques is beyond the scope of this text, but its omission should not be interpreted as trivialization of sensory science. Collecting and interpreting sensory data can be laborious and often represents a limiting step in wine chemistry. We strongly encourage the reader to consult one of the many excellent texts available on sensory science to learn more (e.g., Reference [16]).

References

1. Nykänen, L. and Suomalainen, H. (1983) *Aroma of beer, wine, and distilled alcoholic beverages*, D. Reidel, Dordrecht, Holland.
2. Roullier-Gall, C., Witting, M., Gougeon, R.D., Schmitt-Kopplin, P. (2014) High precision mass measurements for wine metabolomics. *Frontiers in Chemistry*, **2**, 102.
3. McGovern, P.E. (2003) *Ancient wine: the search for the origins of viniculture*. Princeton University Press, Princeton, NJ.
4. Mira de Orduña, R. (2010) Climate change associated effects on grape and wine quality and production. *Food Research International*, **43** (7), 1844–1855.
5. Jeandet, P., Heinzmann, S.S., Roullier-Gall, C., *et al.* (2015) Chemical messages in 170-year-old champagne bottles from the Baltic Sea: revealing tastes from the past. *Proceedings of the National Academy of Sciences of the United States of America*, **112** (19), 5893–5898.

6. Thudichum, J.L.W. and Dupré, A. (1872) *A treatise on the origin, nature, and varieties of wine; being a complete manual of viticulture and oenology*, Macmillan, London.
7. Parascandola, J. (2012) *King of poisons: a history of arsenic*, Potomac Books, Herndon, VA.
8. Michel, R.H., McGovern, P.E., Badler, V.R. (1993) The first wine & beer. *Analytical Chemistry*, **65** (8), 408A–413A.
9. Guasch-Jane, M.R., Andres-Lacueva, C., Jauregui, O., Lamuela-Raventos, R.M. (2006) The origin of the ancient Egyptian drink Shedeh revealed using LC/MS/MS. *Journal of Archaeological Science*, **33** (1), 98–101.
10. Anonymous (2009) *E253-09a, Standard terminology relating to sensory evaluations of materials and products*, ASTM International, West Conshohocken, PA.
11. DeMaria, S. and Ngai, J. (2010) The cell biology of smell. *The Journal of Cell Biology*, **191** (3), 443–452.
12. Dunkel, A., Steinhaus, M., Kotthoff, M., *et al.* (2014) Nature's chemical signatures in human olfaction: a foodborne perspective for future biotechnology. *Angewandte Chemie International Edition*, **53** (28), 7124–7143.
13. Buettner, A., Beer, A., Hannig, C., Settles, M. (2001) Observation of the swallowing process by application of videofluoroscopy and real-time magnetic resonance imaging – consequences for retronasal aroma stimulation. *Chemical Senses*, **26** (9), 1211–1219.
14. Chandrashekar, J., Hoon, M.A., Ryba, N.J.P., Zuker, C.S. (2006) The receptors and cells for mammalian taste. *Nature*, **444** (7117), 288–294.
15. Hufnagel, J.C. and Hofmann, T. (2008) Orosensory-directed identification of astringent mouthfeel and bitter-tasting compounds in red wine. *Journal of Agricultural and Food Chemistry*, **56** (4), 1376–1386.
16. Lawless, H.T. and Heymann, H. (2010) *Sensory evaluation of food principles and practices*. Springer, New York.
17. Schöbel, N., Radtke, D., Kyereme, J., *et al.* (2014) Astringency is a trigeminal sensation that involves the activation of G protein–coupled signaling by phenolic compounds. *Chemical Senses*, **39** (6), 471–487
18. Runnebaum, R.C., Boulton, R.B., Powell, R.L., Heymann, H. (2011) Key constituents affecting wine body – an exploratory study. *Journal of Sensory Studies*, **26** (1), 62–70.
19. Schreier, P., Drawert, F., Junker, A. (1976) Identification of volatile constituents from grapes. *Journal of Agricultural and Food Chemistry*, **24** (2), 331–336.
20. Grosch, W. (2001) Evaluation of the key odorants of foods by dilution experiments, aroma models and omission. *Chemical Senses*, **26** (5), 533–545.
21. Guadagni, D.G., Buttery, R.G., Harris, J. (1966) Odour intensities of hop oil components. *Journal of the Science of Food Agriculture*, **17** (3), 142–144.
22. Rothe, M. and Thomas, B. (1963) Aromastoffe des brotes. *Zeitschrift für Lebensmittel-Untersuchung und Forschung*, **119** (4), 302–310.
23. Benkwitz, F., Tominaga, T., Kilmartin, P.A., *et al.* (2012) Identifying the chemical composition related to the distinct aroma characteristics of New Zealand Sauvignon Blanc wines. *American Journal of Enology and Viticulture*, **63** (1), 62–72.
24. Guth, H. (1997) Quantitation and sensory studies of character impact odorants of different white wine varieties. *Journal of Agricultural and Food Chemistry*, **45** (8), 3027–3032.
25. Tominaga, T., Baltenweck-Guyot, R., Des Gachons, C.P., Dubourdieu, D. (2000) Contribution of volatile thiols to the aromas of white wines made from several *Vitis vinifera* grape varieties. *American Journal of Enology and Viticulture*, **51** (2), 178–181.
26. Buttery, R.G., Seifert, R.M., Guadagni, D.G., Ling, L.C. (1969) Characterization of some volatile constituents of bell peppers. *Journal of Agricultural and Food Chemistry*, **17** (6), 1322–1327.
27. Hein, K., Ebeler, S.E., Heymann, H. (2009) Perception of fruity and vegetative aromas in red wine. *Journal of Sensory Studies*, **24** (3), 441–455.
28. Guadagni, D.G., Buttery, R.G., Okano, S., Burr, H.K. (1963) Additive effect of sub-threshold concentrations of some organic compounds associated with food aromas. *Nature*, **200** (4913), 1288–1289.
29. Pozo-Bayón, M.Á. and Reineccius, G. (2009) Interactions between wine matrix macro-components and aroma compounds, in *Wine chemistry and biochemistry* (eds Moreno-Arribas, M.V. and Polo, M.C.), Springer, New York, pp. 417–436.

30. King, B.M., Duineveld, C.A.A., Arents, P., *et al.* (2007) Retronasal odor dependence on tastants in profiling studies of beverages. *Food Quality and Preference*, **18** (2), 286–295.
31. Delwiche, J. (2004) The impact of perceptual interactions on perceived flavor. *Food Quality and Preference*, **15** (2), 137–146.
32. Ferreira, V., Ortin, N., Escudero, A., *et al.* (2002) Chemical characterization of the aroma of Grenache rosé wines: aroma extract dilution analysis, quantitative determination, and sensory reconstitution studies. *Journal of Agricultural and Food Chemistry*, **50** (14), 4048–4054.
33. Amoore, J.E. (1980) Properties of the olfactory system, in *Odorization* (eds Suchomel, F.H. and Weatherly, J.W.), Institute of Gas Technology, Chicago, IL, pp. 31–35.
34. Stevens, J.C., Cain, W.S., Burke, R.J. (1988) Variability of olfactory thresholds. *Chemical Senses*, **13** (4), 643–653.

Part A

Wine Components and Their Reactions

1

Water and Ethanol

1.1 Introduction

From a macroscopic perspective, wine is a mildly acidic hydroethanolic solution. As shown in Table 1.1, water and ethanol represent ~97% w/w of dry table wines. Ethanol is the major bioactive compound in wine and its presence renders wine and other alcoholic beverages inhospitable to microbial pathogens. Understanding the physiochemical properties of wine will first require a review of the basic properties of water and water–ethanol mixtures. More thorough discussions of the unique properties of water, including those specific to the food chemistry, can be found elsewhere [1].

1.2 Chemical and physical properties of water

Water is a hydride of oxygen, but has unique properties compared to other hydrides of elements nearby on the periodic table, as shown in Table 1.2. For example, the boiling point of water (100 °C) is far above that of hydrides of adjacent elements on the periodic table: HF (19.5 °C), H_2S (–60 °C), and NH_3 (–33 °C). Thus, water exists as a liquid at room temperature, while the other hydrides exist as gases. Similarly, water also has a higher heat of vaporization, heat capacity, and freezing point than would be expected as compared to nearby hydrides.

The unique properties of water are largely due to its ability to engage in intermolecular hydrogen (H) bonding, which results in stronger molecule-to-molecule interactions than in related compounds.

- Oxygen is more electronegative than hydrogen and an O–H bond is more polarized than N–H or S–H.
- The geometry and symmetry of an H_2O molecule allows for four concurrent H bonds per water molecule.

The ability of water to form strong H-bonds explains not only its higher boiling point than homologous hydrides, but also its high *surface tension*. A surface refers to the area in which two phases come into contact (e.g., water–air, water–oil, water–glass), and surface tension refers to the force needed to create an additional surface area between two phases, that is, to spread a water droplet on to a piece of wax paper.

Understanding Wine Chemistry, First Edition. Andrew L. Waterhouse, Gavin L. Sacks, and David W. Jeffery.

Table 1.1 Composition of a typical dry table wine

Compound(s)	Concentration (% w/w)	Major roles in wine
Water	85–89	Tactile (mouthfeel) Major matrix component
Ethanol	9–13%	Tactile (pungency/heat, mouthfeel) Taste (astringency, bitter, sweet) Major matrix component
Glycerol	0.5–1.5%	Negligible, slight contribution to sweetness and body
Acids	0.6–1.0%	Taste (sour), pH buffering
Sugars	0.1–0.5%	Taste (sweet); minor effect on mouthfeel
Polyphenols	0.1–0.2% (red) 0.02–0.05% (white)	Color, mouthfeel (astringency)
Polysaccharides	0.05–0.1%	Mouthfeel
Minerals	0.05–0.2%	pH buffering; minor taste effects
Most odorants	<0.001%	Aroma

Table 1.2 Physical properties of water, ethanol, and their mixture (10% w/w ethanol in water)

Property	Water	Ethanol	10% w/w EtOH
Boiling point (°C) at 100 kPa	100	78	90.85
Density at 20 °C (g/mL)	0.998	0.789	0.983
Surface tension (mN/m)	73	22	48
Viscosity at 20 °C (Pa s) × 1000	1.00	1.14	1.31

Compounds that are polar (or that contain polar functional groups) and are also capable of H-bonding are referred to as *hydrophilic* and tend to be more soluble in water, which in wine would include most sugars and ions like K^+ and SO_4^{2-}. Many compounds of importance to wine flavor, especially odorants, are *hydrophobic* and are characterized by the presence of hydrocarbon groups that are incapable of H-bonding. A snapshot at the molecular level would show water molecules preferably forming H-bonds with each other while interacting minimally with the hydrophobic solute. This imposes order upon the system, and dissolution of hydrophobic compounds in water tends to be entropically unfavorable. Colloquially, the preference of polar solvents to solvate polar compounds rather than non-polar compounds (and vice versa) is referred to as "like dissolves like."

1.3 Properties of ethanol and ethanol–water mixtures

Water and ethanol are completely miscible: that is, they will mix with each other freely at any proportion. The mixing of ethanol and water will have profound effects on the structure of water because ethanol is *amphiphilic* – it has both a hydrophilic alcohol group (–OH) and a hydrophobic hydrocarbon chain ($–CH_2CH_3$). At concentrations <17% v/v, typical for most table wines, ethanol molecules are molecularly dispersed. The –OH group can participate in H-bonding in place of an H_2O molecule, while the $–CH_2CH_3$ group will interact minimally with H_2O. The addition of small amounts of ethanol to water will have several effects on the properties of the matrix:

- Decrease in boiling point. Because it is less capable of H-bonding, ethanol (78 °C) has a lower boiling point than water (100 °C, Table 1.2). Mixtures of ethanol and water have boiling points intermediary to

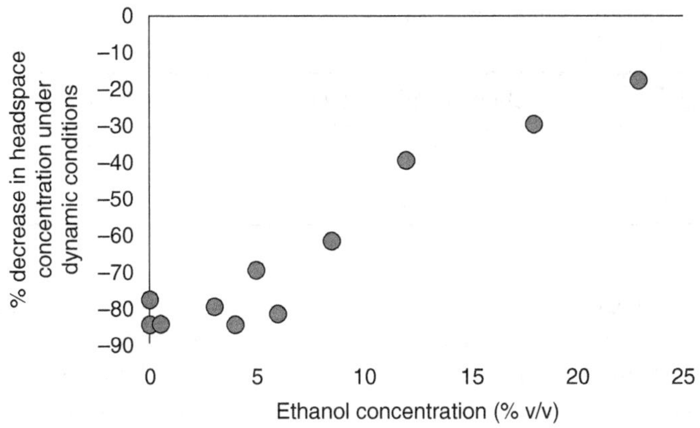

Figure 1.1 *Decrease in ethyl butyrate headspace concentration under dynamic conditions (continuous sparging of the headspace by an inert gas). The smaller decrease observed at higher ethanol concentrations is a result of lower surface tension and faster replenishment of headspace volatiles. Data from Reference [3].*

the pure compounds and show a negative deviation from Raoult's Law (Chapter 26.4). The effect of ethanol on boiling point is exploited in the analytical technique of *ebulliometry*, which uses measurements of the wine boiling point to calculate ethanol concentration [2].

- Decrease in surface tension. Because ethanol is amphiphilic, it will behave like a 'surfactant' – that is, in aqueous solutions it will preferably be found at interfacial surfaces, resulting in a decrease in surface tension (Table 1.2). At the molecular level, the hydrocarbon tail of ethanol will orient itself towards the non-aqueous phase (air, oil, etc.). One consequence of ethanol being both a surfactant and more volatile than water is that wine and other alcoholic beverages will form "tears" or "legs" along the sides of a glass.[1] A second consequence of greater importance to wine sensory properties is that as ethanol migrates to the surface it will bring with it other non-polar volatile compounds. These compounds can then volatilize, resulting in faster equilibration of aroma compounds between headspace and liquid (Figure 1.1 [3]). A practical consequence is that even though the concentration of volatiles in the headspace above water will be higher than over wine under static conditions, this concentration can decrease considerably under dynamic conditions, for example, if a glass is repeatedly sniffed. In contrast, the volatile composition of wine headspace will stay relatively constant.
- Decrease in matrix polarity. Mixing of ethanol and water results in a disruption of water structure and H-bonding. Thus, there will be a decreased entropy loss when hydrophobic volatile compounds dissolve in wine-like solutions as compared to pure water. There may also be a larger increase in enthalpy due to hydrophobic interactions between the hydrocarbon chain of ethanol and the solute. As a result, less polar compounds like vanillin ("vanilla" aroma; Table 1.3) will have greater solubility in ethanol, while polar solutes like sodium chloride will have lower solubility in ethanol because they are less able to participate in H-bonding. The effects of ethanol on solubility will be reconsidered in later chapters, such as on the precipitation of potassium bitartrate (Chapters 4 and 26.1).

---

[1] Wine tears are an example of the *Marangoni effect*, in which liquids move from areas of low surface tension to areas of high surface tension. Evaporation of ethanol from the surface of wine results in an area of higher surface tension, which will be compensated for by migration of ethanol-rich wine from the bulk liquid to the surface. Eventually, this will result in formation of a ring on the walls of the wine glass, which will fall due to gravity.

Table 1.3 Solubility of a non-polar (vanillin) and polar (sodium chloride) compound in water and in ethanol

Compound	Solubility in water at 25 °C	Solubility in ethanol at 25 °C
Vanillin	1.06 g/L	364 g/L
Sodium chloride	365 g/L	0.65 g/L

- Increase in viscosity. Ethanol is only slightly more viscous than water (1.14 versus 1.00 mPa s at room temperature; Table 1.2). Addition of ethanol to water will result in an increase in viscosity greater than the individual components, from 1.31 mPa s at 10% v/v up to a maximum of 2.9 mPa s at 40% v/v.[2] Mixing ethanol and water will also result in a total volume less than the component volumes. For example, mixing of 50 mL of each solvent results in a final volume of 96 mL. These phenomena occur because addition of ethanol will disrupt the more open lattice structure of pure water.
- Formation of ethanol aggregates. The description of ethanol-in-water mixtures as molecular dispersions, e.g., ethanol molecules isolated from each other and completely surrounded by water molecules, is valid at concentrations up to 17% v/v. For solutions with 17–63% v/v ethanol, spectroscopic data indicates that ethanol begins to form molecular aggregates within the solution, as opposed to a true molecular dispersion [4]. These aggregates have been referred to as "micelle-like" and appear to have similar behavior to better known micelle-forming compounds like detergents. At concentrations >63%, water will form a molecular dispersion within ethanol.

1.4 Typical ethanol concentrations in wines

In alcoholic beverages, ethanol is formed by yeast via fermentation of hexose sugars (fructose, glucose). The Gay–Lussac equation describes this reaction:

$$C_6H_{12}O_6\,(\text{hexose sugar}) \rightarrow 2\,CH_3CH_2OH + 2\,CO_2$$

In wines, these sugars are mostly derived from grapes, although sugars can also be legally added prior to fermentation in some regions (*chaptalization*). In principle, one mole of sugar should yield two moles of ethanol, but in practice this value is closer to 1.8 moles of ethanol. Alcoholic fermentation will be described in more detail in later (Chapter 22.1). Wine producers routinely measure ethanol concentrations to track fermentations, for quality control and for legal obligations.[3] Most countries or wine regions place limits on minimum ethanol concentrations for a product to be called a wine, and a tax code may also be based on ethanol concentration.[4]

Unlike most compounds in wine, which are reported in units of w/v (g/L) or w/w (g/kg), it is common in both scientific and commercial settings to report ethanol concentrations in units of % v/v. A wine containing

[2] By comparison, the viscosities of olive oil and honey are usually about 80 and 5000 mPa s, respectively.

[3] Regulatory documents usually refer to "alcohol" rather than "ethanol" because the former term is more widely understood. However, in this book we will preferably use the term ethanol because there are other alcohols in wine.

[4] From a legal perspective, ethanol is the only compound that is required to be in wine, since minimum and maximum ethanol concentrations for dry table wines are regulated in most countries. In the United States, CFR 24.7, a wine labeled "Table Wine" must have between 7 and 14% alcohol. Water is not explicitly required to be present, although it is not clear how one would produce a "water-free" wine!

12% v/v ethanol contains 120 mL of pure ethanol per 1000 mL of wine. Since the density of most wines is close to 1.0 g/mL at 20 °C, % v/v units can be converted to % w/v units by multiplying by the density of ethanol (0.789 g/mL at 20 °C). Thus, a 12% v/v wine with a density of 1.0 g/mL will have (120 mL/L) × (0.789 g/mL) = 94.7 g ethanol per L. A typical ethanol concentration encountered in dry wines is between 11 and 14% v/v. Because the amount of ethanol produced during fermentation is dependent on sugar concentration, wines from warmer regions with longer growing seasons tend to have higher ethanol concentrations than cooler regions. Red winegrapes are usually harvested later than white winegrapes, and as a result red wines typically have a higher ethanol concentration than whites. In recent years, there has been a tendency to pick grapes at higher sugar concentrations [5], such that average ethanol concentrations increased by 0.3–1.0% v/v across different wine regions between 1992 and 2007 [6].

1.5 Sensory effects of ethanol

The organoleptic effects that ethanol can have on wine flavor are diverse and are summarized in Table 1.4.

1.5.1 Major taste/tactile properties of ethanol

Ethanol appears to be a major factor determining bitterness in dry wines. For example, increasing the ethanol content from 8 to 14% results in over a 3-point increase in perceived bitterness on a 10-point scale [13]. By comparison, addition of catechin, a flavan-3-ol associated with bitterness (see Chapter 14) at concentrations well in excess of those found in wine (1500 mg/L) resulted in only a 1-point increase in bitterness. In a

Table 1.4 Summary of the sensory effects of ethanol in wine

Property	Comments
Direct effects	
Bitter taste	Commonly reported as major flavor property of ethanol in wine studies [7] Dominant sensation at 10% v/v ethanol [8]
Pungency ("heat")	Commonly reported as major flavor property of ethanol in wine studies [7] Dominant sensation at 21% v/v ethanol [9] Perception due to activation of the TRPV1 receptor[5]
Sweet taste	Dominant sensation at 4.2% v/v ethanol [9]
Ethereal-sweet smell	Can potentially have additive effects with other odorants
Perceived viscosity	Addition of ethanol to de-alcoholized wine results in maximum perceived viscosity at 10% v/v, but has negligible effects over the range of ethanol found in wines [10]
Indirect effects	
Astringency	Ethanol is reported to be astringent at high concentrations, due to denaturation and precipitation of salivary proteins. In real wines, ethanol results in a decrease in the intensity and duration of perceived astringency, possibly because ethanol disrupts interactions between proteins and tannins [11]
Taste	Ethanol can mask sourness [12]
Aroma	Ethanol can decrease the intensity of other odorants, either by masking or by decreasing their volatility [18]

[5] The TRPV1 channel is responsible for detection of damaging high temperatures, and also capsaicin in hot peppers.

separate study of 13 dry white wines with residual sugars < 10 g/L, differences in bitter intensity were best correlated with ethanol concentration (range = 10.8–14.4%), while no correlation was observed with bitterness and phenolics (range = 169–404 mg/L as gallic acid equivalents) [14].

Beyond bitterness, ethanolic solutions are also frequently described as "pungent" and "sweet" [12]. The dominant sensation will vary with concentration and among individuals, but generally seems to follow the pattern sweet → bitter → pungent with increasing concentration [8, 9]. Despite seeming contradictory, these descriptors can co-exist, e.g., a 10% v/v ethanol solution is reported to be better simulated by a combination of 3% sucrose and 0.005% quinine than either compound in isolation [8].[6]

1.5.2 Ethanol and wine odor

In isolation, ethanol is described as having a "fruity" or "ethereal, solvent-like" odor. Increasing ethanol concentration is usually reported to decrease the intensity and increase the threshold of odorants [15, 16]. For example, reconstitution studies using 7% in place of 10% ethanol resulted in a model wine with greater fruity and floral aromas, and the odor threshold of compounds in model wine are reported to be 10–100-fold higher than in water [15]. These behaviors could be explained by one of two effects:

- Masking. The presence of ethanol odor decreases the perceived intensity of other odors due to cognitive effects
- Matrix effect. Most odorants are hydrophobic and thus will be *more soluble* and *less volatile* in ethanol than in water, for reasons described earlier in this chapter. For example, the gas–liquid partition coefficients ($K_{g,l}$) of two common fermentation metabolites, isoamyl alcohol and ethyl hexanoate, decrease by almost a factor of 2 in a 10% v/v ethanol solution as compared to pure water (Figure 1.2) [17]. The effects of varying ethanol content over the range observed in table wines is more modest, with $K_{g,l}$ changing by less than 10% over an ethanol range of 5–17% v/v [18].

Sensory thresholds of odorants in 10% ethanol can be 10–100 times their threshold in pure water [15], far more than can be explained by the 2-fold decrease in volatility caused by matrix effects. Thus, the major

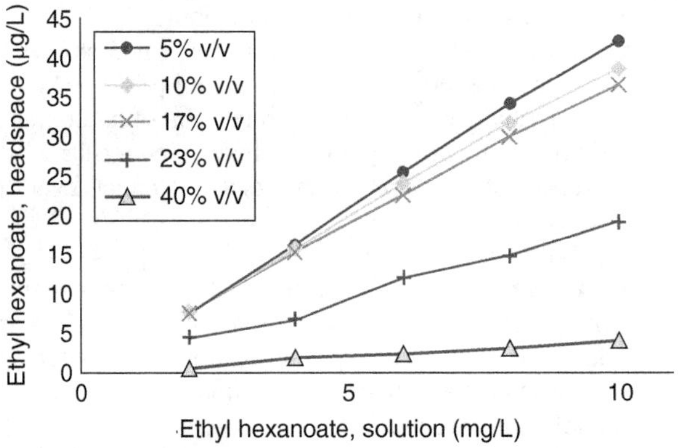

Figure 1.2 Effects of ethanol concentration (5–40% v/v) on ethyl hexanoate volatility. Data from Reference [18]

[6] Bitter and sweet taste receptors are structurally similar, and the property of artificial sweeteners having both bitter and sweet flavor is well known in the flavor industry (e.g., aspartame).

effect of ethanol on wine aroma is probably neurobiological (masking) rather than physiochemical (decreased volatility). However, in distilled spirits, it is possible that decreases in volatility could be of greater significance to sensory attributes. The flavor detection threshold of ethanol in water is reported to be 53 mg/L [19], but this number is not of particular relevance since all table wines will have concentrations well in excess of this value. Of greater importance is the *difference threshold*: that is, the minimum amount of ethanol that must be added to a wine before a sensorially detectable change can be demonstrated. Anecdotally, winemakers often report that differences as small as 0.1% v/v are detectable [20]. However, in formal sensory studies, differences of at least 1%, and sometimes as much as 4%, are necessary to cause detectable changes [21]. To understand this discrepancy it should be noted that many of the studies that have investigated the difference threshold for ethanol relied on addition of pure or near-pure ethanol to low-alcohol wines, which is not a common winemaking practice. In most wineries, differences in ethanol concentration are often realized by less selective approaches, such as allowing grapes to achieve higher initial sugar concentrations or removing ethanol after fermentation by spinning cone, reverse osmosis/distillation, or related techniques. These processes could result in other sensory changes to wine.

References

1. Franks, F. (1972) *Water, a comprehensive treatise*, Plenum Press, New York.
2. Zoecklein, B.W., Fugelsang, K.C., Gump, B.H., Nury, F.S. (1999) *Wine analysis and production*, Kluwer Academic/Plenum Publishers, New York.
3. Tsachaki, M., Linforth, R.S.T., Taylor, A.J. (2005) Dynamic headspace analysis of the release of volatile organic compounds from ethanolic systems by direct APCI-MS. *Journal of Agricultural and Food Chemistry*, **53** (21), 8328–8333.
4. Onori, G. and Santucci, A. (1996) Dynamical and structural properties of water/alcohol mixtures. *Journal of Molecular Liquids*, **69** (1), 161–181.
5. Alston, J.M., Fuller, K.B., Lapsley, J.T., Soleas, G. (2011) Too much of a good thing? Causes and consequences of increases in sugar content of California wine grapes. *Journal of Wine Economics*, **6** (2), 135–159.
6. Alston, J.M., Fuller, K.B., Lapsley, J.T., *et al.* (2011) Splendide mendax: false label claims about high and rising alcohol content of wine. American Association of Wine Economists, AAWE Working Paper No. 82.
7. King, E.S., Dunn, R.L., Heymann, H. (2013) The influence of alcohol on the sensory perception of red wines. *Food Quality and Preference*, **28** (1), 235–243.
8. Scinska, A., Koros, E., Habrat, B., *et al.* (2000) Bitter and sweet components of ethanol taste in humans. *Drug and Alcohol Dependence*, **60** (2), 199–206.
9. Wilson, C.W.M., O'Brien, C., MacAirt, J.G. (1973) The effect of metronidazole on the human taste threshold to alcohol. *British Journal of Addiction to Alcohol and Other Drugs*, **68** (2), 99–110.
10. Pickering, G.J., Heatherbell, D.A., Vanhanen, L.P., Barnes, M.F. (1998) The effect of ethanol concentration on the temporal perception of viscosity and density in white wine. *American Journal of Enology and Viticulture*, **49** (3), 306–318.
11. Siebert, K.J., Carrasco, A., Lynn, P.Y. (1996) Formation of protein-polyphenol haze in beverages. *Journal of Agricultural and Food Chemistry*, **44** (8), 1997–2005.
12. Martin, S. and Pangborn, R.M. (1970) Taste interaction of ethyl alcohol with sweet, salty, sour and bitter compounds. *Journal of the Science of Food and Agriculture*, **21** (12), 653–655.
13. Fischer, U. and Noble, A.C. (1994) The effect of ethanol, catechin concentration, and pH on sourness and bitterness of wine. *American Journal of Enology and Viticulture*, **45** (1), 6–10.
14. Sokolowsky, M. and Fischer, U. (2012) Evaluation of bitterness in white wine applying descriptive analysis, time-intensity analysis, and temporal dominance of sensations analysis. *Analytica Chimica Acta*, **732** (1), 46–52.
15. Grosch, W. (2001) Evaluation of the key odorants of foods by dilution experiments, aroma models and omission. *Chemical Senses*, **26** (5), 533–545.

16. Escudero, A., Campo, E., Farina, L., *et al.* (2007) Analytical characterization of the aroma of five premium red wines. Insights into the role of odor families and the concept of fruitiness of wines. *Journal of Agricultural and Food Chemistry*, **55** (11), 4501–4510.
17. Athes, V., Lillo, M.P.Y., Bernard, C., *et al.* (2004) Comparison of experimental methods for measuring infinite dilution volatilities of aroma compounds in water/ethanol mixtures. *Journal of Agricultural and Food Chemistry*, **52** (7), 2021–2027.
18. Conner, J.M., Birkmyre, L., Paterson, A., Piggott, J.R. (1998) Headspace concentrations of ethyl esters at different alcoholic strengths. *Journal of the Science of Food and Agriculture*, **77** (1), 121–126.
19. Keith, E.S. and Powers, J.J. (1968) Determination of flavor threshold levels and sub-threshold, additive, and concentration effects. *Journal of Food Science*, **33** (2), 213–218.
20. Goode, J. (2005) *The science of wine: from vine to glass*. University of California Press, Berkeley.
21. Meillon, S., Urbano, C., Schlich, P. (2009) Contribution of the temporal dominance of sensations (TDS) method to the sensory description of subtle differences in partially dealcoholized red wines. *Food Quality and Preference*, **20** (7), 490–499.

2

Carbohydrates

2.1 Introduction

Carbohydrates are the most abundant class of biological molecules on Earth. Colloquially, "sugar" is used as either a synonym for carbohydrate or to refer to sucrose (table sugar, a disaccharide composed of glucose and fructose). In food chemistry and in this text, the term *sugar* will refer to low molecular weight carbohydrates, especially those with a sweet taste (e.g. glucose, sucrose).

Sugars are biologically important because they represent a water-soluble energy source. In most organisms glucose is used as a major energy substrate as part of glycolysis, and in humans and most other animals, glucose is used to circulate energy throughout the body. In grapes and many other plants, sugars are transported in the form of sucrose, and will accumulate in the grape berry during ripening in the form of glucose and fructose (Chapter 20). These sugars are the primary substrate for yeast to produce ethanol during alcoholic fermentation (Chapter 22.1) and sugars in wine can contribute to perceived sweetness. Higher molecular weight polymers of sugars (polysaccharides) are major components of the cell walls in grapes and other plants. Finally, covalent bonding between non-sugar compounds (*aglycones*) and sugars to form *glycosides* will result in increased solubility and other chemical changes to the original compound (Chapter 23.1)

2.2 Nomenclature, representation, and occurrence of sugars

The simplest carbohydrates are the monosaccharides, which include the aldoses (polyhydroxylated aldehydes) and ketoses (polyhydroxylated methyl ketones). The name, carbohydrate, refers to the empirical formula $C_x(H_2O)_x$ of aldoses and ketoses, whose general structures are shown in Figure 2.1.

These sugars can be further distinguished by the following properties:

- Carbon number. Pentoses (5 carbons) and hexoses (6 carbons) are the most common in wine.
- Fermentability. Sugars may be distinguished by whether they can be fermented by wine yeast under ordinary circumstances.
- Derivatives. The basic monosaccharide structure may be modified by reduction, oxidation, etc.

Understanding Wine Chemistry, First Edition. Andrew L. Waterhouse, Gavin L. Sacks, and David W. Jeffery.

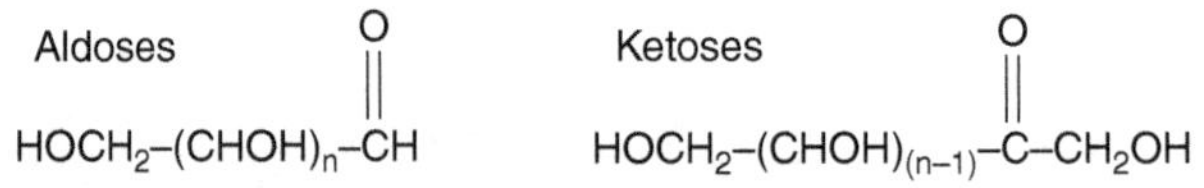

Figure 2.1 General structures for aldose and ketose sugars

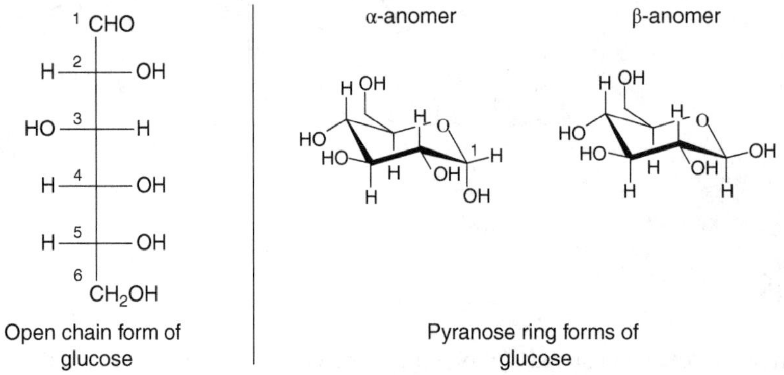

Figure 2.2 Glucose, a hexose sugar, represented as an open-chain Fischer projection (left) and as a chair conformation of its pyranose form (right). The C1–C6 numbering system for hexoses is also shown on the left

In grapes, the most quantitatively important sugars are hexoses: fructose (a ketose) and glucose (an aldose). These compounds are *reducing sugars*, that is, they are capable of reducing Cu(II), Fe(III), and other transition metals under alkaline conditions. The reaction involves oxidation of the aldehyde group of aldoses, but ketoses are also measured because they can isomerize to form aldoses under basic conditions. Importantly, many oligosaccharides (e.g., sucrose) lack a reactive carbonyl group and will not be included in measurements of reducing sugars.

In aqueous solution, hexoses and pentoses predominantly exist as hemiacetal rings, formed by the reversible intramolecular reaction of the carbonyl group with an –OH group. The resulting forms are referred to as either furanoses (5-member rings) or pyranoses (6-membered rings). For each ring form, the hemiacetal group can exist as one of two stereoisomers, referred to as anomers (α- or β-; Figure 2.2). The different forms exist in equilibrium, and at room temperature the most abundant isomers of glucose and fructose are the β-pyranose forms. The hemiacetal form of sugars can react with alcohols to form acetals (see Chapter 9). The resulting compounds are called *glycosides*. Glycosidic bonds between the hemiacetal of one sugar ring and the –OH group of another sugar result in the formation of disaccharides, trisaccharides, and larger polysaccharides. Sucrose, for example, is a disaccharide of glucose and fructose. A more complete description of sugar nomenclature can be found in an introductory biochemisty or food chemistry textbook [1].

Typical concentrations of sugars in wine are reported in Table 2.1, and their structures are shown in Figure 2.3. In grapes, fructose and glucose increase during ripening from near-undetectable before veraison to a total concentration in the range of 180–250 g/kg at harvest (Chapter 20). In the wine industry, it is more common to report total soluble solids (TSS) rather than individual sugars in juices. Typically, TSS measurements are performed by density or refractometry, and are calibrated against sucrose standards. Because sugars represent only 90–95% of soluble solids, these measurements are more accurately described as "apparent sugars". TSS can be reported in units of Brix, where 1 Brix = 1% w/w soluble solids as sucrose, but may also be reported as Baumé, calculated as Brix ÷ 1.8. Baumé is a rough measure of the "potential alcohol" of a wine: that is, the % v/v ethanol concentration that would be achieved if the juice was fermented to dryness.

Table 2.1 *Concentrations and taste properties of major sugars in wine*

Sugar or sugar derivatives	Typical concentration in dry wine (g/L)[a]	Taste threshold in H_2O (g/L) [4–6]	Sweet intensity of a 10% w/w solution [5][b]	Notes
Sugars				
Fructose	0.2–4	1.8–2.4	114	Major grape sugar, hexose
Glucose	0.5–1	3.6–12	69	Major grape sugar, hexose
Sucrose	0–0.2	3.6	100	Disaccharide of glucose and fructose
Arabinose	0.5–1	2.5		Non-fermentable pentose, component of glycosides and pectin
Galactose	0.1	9.0		Isomer of glucose, component of pectin
Rhamnose	0.2–0.4			Deoxy sugar Forms glycosides
Sugar alcohols				
Glycerol	7–10	5.2–7.7		Fermentation metabolite
Mannitol	0.01–0.05	7.3	69	Indicative of grape rot or wine spoilage
Arabitol		6.5		
Sorbitol (glucitol)	0–0.05	6.2	51	
Inositols, total	0.2–0.7	3.2		
Sugar acids				
Gluconic acid	Up to 2[c]		Sour	Indicative of grape rot
Galacturonic acid	0.1–1		Sour/astringent	Major component of pectin
2-Oxogluconic acid	Up to 0.1[c]			Indicative of grape rot

[a] Sugars and sugar alcohol concentrations compiled from multiple sources from Reference [7]. Data on sugar acids compiled from Reference [8].
[b] Reference is sucrose = 100.
[c] High values in grapes afflicted by rot.

During alcoholic fermentation, hexose sugars are largely converted to ethanol and CO_2, but some *residual sugar (RS)* will still be detectable in wines. Potential sources of residual sugars in finished wines include:

- Incomplete fermentation, either because fermentation is stopped or because sugars are non-fermentable (e.g. arabinose, xylose)
- Back-sweetening of wine with sucrose, grape must, or other sources after fermentation is complete
- Hydrolysis of glycosides during storage (Chapter 23.1)
- Extraction from oak [2]

The species included in RS measurements will vary with methodology. For example, enzymatic methods will generally only measure fructose, glucose, and possibly sucrose. Methods based on copper reduction assays will include other reducing sugars (e.g., arabinose), but not sucrose. Typical dry table wines will have residual sugar concentrations between 1 and 4 g/L, but sweet wines may have greater than 100 g/L (Chapter 19). Because yeast are glucophilic, fructose is usually at higher concentrations than glucose [3], although this difference will be obscured by back-sweetening. Several other monosaccharides are present in wine at concentrations ranging from 0.1 g/L to 1 g/L, including arabinose, galactose, and rhamnose (Table 2.1). Most of these sugars cannot be fermented by wine yeast strains.

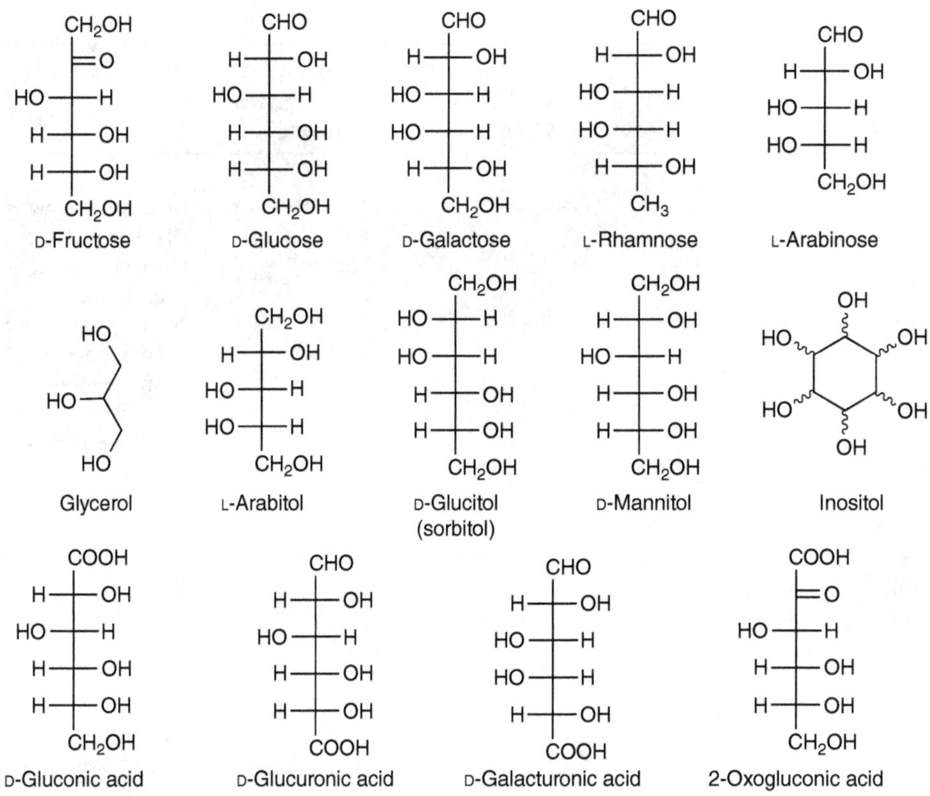

Figure 2.3 Structures of major sugars and sugar derivatives found in wine

2.3 Physical, chemical, and sensory properties of sugars

Sugars are hydrophilic due to their large number of –OH groups (Figure 2.2). Monosaccharides are soluble in both ethanol and water, but solubility (particularly in ethanol) decreases with increasing chain length [1]. Sugars can participate in a range of reactions in aqueous solutions, including wine.

Enzymatic reactions of sugars in grape must or non-stabilized wine include:

- Glycolysis – the metabolism of sugars (Chapter 22.1)
- Enzymatic oxidation to form sugar acids or enzymatic reduction to form sugar alcohols
- Enzymatic hydrolysis of glycoside bonds, for example, hydrolysis of sucrose to yield glucose and fructose by *invertase*.

Important non-enzymatic reactions in finished wine include:

- Acid hydrolysis of glycoside bonds – for example, sucrose added to wine will be ~50% hydrolyzed to fructose and glucose when stored at room temperature for two months at pH 3.0 [9].
- Sugar degradation – under acidic conditions, sugars can be degraded to furanic carbonyl compounds (furfural, 5-(hydroxymethyl)furfural (HMF), sotolon) and related products, particularly at higher temperatures (Chapter 25).

- Carbonyl chemistry – the carbonyl group of reducing sugars exists primarily as a hemiacetal and thus is less available to react than analogous carbonyls discussed in Chapter 9. Nonetheless, the carbonyl group can react with bisulfite to yield an adduct (Chapter 17). Reactions with amine groups to yield browning products, as occurs in the thermal processing of many foods (Maillard reactions) is expected to be negligible due to the low pH and low amino acid concentrations of wine (Chapter 5).

The notable flavor property of most sugars is sweetness (Table 2.1). Most monosaccharides have sweet detection thresholds in the range of 10–50 mM (0.2–1.0% w/w). Of the sugars found in wine, the most potent is fructose, which is almost twice as sweet as glucose and 15% sweeter than sucrose at a concentration of 10% w/w [5]. The sweet taste of sugars will increase perception of body and mask other taste and tactile sensations like sourness, bitterness, astringency, and pungency, and vice versa [10]. For example, increasing the sugar concentration of a citric acid solution results in a decrease in perceived acidity [11]. Similar observations have been made in wine reconstitution and omission studies near the sensory thresholds of sugars, in which increasing the concentration of sugars decreases astringency and sourness [6]. The sweetness threshold of sugars also provides a rationale for labeling or classifying wines as "dry" versus "sweet" based on quantitative analysis of residual sugars. For example, in EU regulations, "dry" (or its translation) indicates <4 g/L of residual sugar, corresponding roughly to the detection threshold for sugars in wines.[1]

Sugars are non-volatile and have no aroma – the sweet caramel aroma of cooked sugar is due to degradation products like sotolon, discussed later (Chapter 9). However, the presence of sugar has well-known cognitive effects on non-taste attributes of food flavor. For example, in coffee, addition of sweeteners increases the retronasal intensity of "caramel," while decreasing "roasty" and "coffee" flavors [12]. The reverse can also happen, where sweet aromas can increase perception of sweetness by taste [13]. This phenomenon of combining information from two different sensory modalities is referred to as *synesthesia* [14]. High sugar concentrations can also have quantifiable effects on the volatility of odorant compounds, by decreasing the availability of water for solvation ("salting-out").[2] The magnitude of these effects are usually small, with <20% increase in volatility observed in a 15% w/v sucrose solution versus water for three representative odorants (isoamyl acetate, ethyl hexanoate, eugenol) [15].

2.3.1 Sugar alcohols

Polyalcohols (polyols) describe aliphatic compounds with multiple –OH groups and no other functional groups (Figure 2.2). The term is used synonymously with *sugar alcohols* in some texts, since they are often derived from sugars and usually have a sweet taste. The major class of sugar alcohols in wine are the *alditols*, which are formed by reduction of the sugar carbonyl group to an alcohol during fermentation. They are named by replacing the *–ose* suffix of the corresponding aldose sugar with *–itol*.

The major sugar alcohol in found in wine is glycerol, which is usually the most abundant compound in dry wines after water and ethanol. Glycerol is common to all alcoholic fermentations, where it has several important physiological functions as an osmoregulant and in redox balancing (Chapter 22.1). Glycerol concentrations are usually in the range of 7–10 g/L [16], but may be over 15 g/L in high sugar fermentations like ice

[1] Allowances may be made for the effects of acidity on perceived sweetness, such that a high acid wine may have as much as 9 g/L of residual sugar in some EU regions. Also, in most New World countries, use of the term "dry" is not regulated.

[2] The salting-out effect of sugars is of minimal significance to wine aroma, but will impact ethanol measurement by the ebulliometry technique, which relies on differences in boiling point in ethanol–water mixtures versus water. The presence of sugars will decrease the boiling point of wine and result in incorrectly high apparent ethanol contents, such that a correction factor must be applied for RS >20 g/L.

wines [17]. Other alditols, such as mannitol, arabitol, and sorbitol, are present in the range of 20–300 mg/L (total < 1 g/L, Table 2.1). Higher concentrations of sugar alcohols may be a sign of microbial spoilage; for example, high concentrations of mannitol formed by enzymatic reduction of fructose are indicative of lactic spoilage [18]. Other quantitatively important polyalcohols are 1,2-propanediol (propylene glycol) and 2,3-butanediol, formed by reduction of lactic acid and acetoin, respectively. Another class of polyalcohols found in wines is the *inositols* (cyclitols), which have perhydroxylated cyclohexane rings. The inositols are components of phosphoinositols, used by yeast and other eukaryotes in cell membranes and signal transduction. One major (*myo-*) and two minor (*scyllo-* and *chiro-*) inositols are observed, and typical total concentrations are ~300 mg/L [19].

Sugar alcohols are chemically and microbiologically stable under reductive wine conditions.[3] Sugar alcohols lack a reactive carbonyl group, and thus will not participate in electrophilic reactions as is observed with sugars. Sugar alcohols do not appear to be metabolized by commercial yeast or LAB strains under normal wine conditions, although some rare spoilage LAB can degrade glycerol [18].

Sugar alcohols have about 50% of the sweet intensity of sucrose on a w/w basis (Table 2.1). In wine, only glycerol is usually present at concentrations above its taste threshold (5.2 g/L) [4]. Historically, glycerol was implicated as an important contributor to the mouthfeel of a wine, but the actual impact on mouthfeel appears to be minor. For example, addition of >25 g/L of glycerol to a model wine was necessary to cause a detectable change in mouthfeel [4], and in a non-targeted study on white wines, no correlation was observed between glycerol and wine body [20]. Although these results cast doubt on the impact of glycerol in isolation, a reconstitution study on dry red wines observed that elimination of all alditols (glycerol, sorbitol, mannitol, and others) results in a significant decrease in wine mouthfulness/body [6].

2.3.2 Sugar acids

Aldonic acids have a carboxylic acid group in place of the aldehyde group and are named by replacing the *–ose* suffix of the corresponding aldose sugar with *–onic acid,* for example, glucose and gluconic acid. *Uronic acids* have a carboxyl acid group in place of the terminal hydroxyl group. Uronic acid derivatives of aldoses are named by replacing the *–ose* suffix of the corresponding sugar with *–uronic acid*, e.g. glucose and glucuronic acid. Confusingly, uronic acid derivatives of ketoses are named as *oxo* derivatives of aldonic acids, for example, 2-oxogluconic acid. Uronic acids possess an electrophilic aldehyde group and at high concentrations can contribute to SO_2 binding (Chapter 17). These sugar acids (Figure 2.2, Table 2.1) are generally formed by enzymatic oxidation and high concentrations, particularly of gluconic acid, are associated with grape spoilage organisms such as *Gluconobacter* [21]. One exception is galacturonic acid, which is typically present at several hundred mg/L in dry wines, and is likely formed by pectin hydrolysis during or after fermentation.

The impact of uronic and aldonic acids in wine is poorly studied. They will contribute to titratable acidity ($pK_a = \sim 3.5$), but at typical concentrations found in wine (<0.5 g/L as tartaric acid equivalents) they are likely to be of negligible sensory importance. Although higher concentrations are found in wines produced from grapes compromised by *Acetobacter* and *Gluconobacter*, it is likely that other off-flavors (e.g. volatile acidity) will have a more noticeable role.

[3] In other foods, sugar alcohols like sorbitol and mannitol are often used as sweeteners in chewing gums, since they are not fermentable by oral microflora and thus do not promote dental caries. They can also be used as part of low-calorie products because they are poorly absorbed during digestion, although this has the unfortunate side-effect of serving as a laxative.

2.4 Polysaccharides

Polysaccharides are polymers of carbohydrates. The major polysaccharides in grapes are structural components:

- Cellulose, a water-insoluble β-(1 → 4) polymer of D-glucopyranose
- Pectin, a water-soluble heteropolymer that consists primarily of α-(1 → 4) D-galacturonic acid linkages
- Hemicellulose, a diverse class of heteropolymers related to pectin that consists of xylose, arabinose, galactose, glucose, and several other sugars.

In wine, polysaccharides may be derived from both grape and yeast cell walls (Table 2.2). Because of their polymeric nature, polysaccharides are not ordinarily characterized as individual molecules. Instead, it is common to isolate and quantify fractions based on size (e.g., by gel permeation chromatography) or charge (ion exchange chromatography). The fractions may then be hydrolyzed to determine the constituent sugars [22].

Grape polysaccharides can decrease yield and extraction during winemaking (Chapters 19 and 21) and decrease the efficiency of filtration (Chapter 26.3), and winemakers will often use carbohydrase enzymes to decrease these impacts. Grape polysaccharides will be further degraded by microbial enzymes during fermentation and are not efficiently transferred to wine during fermentation either due to degradation or poor solubility.

Two exceptions are rhamnogalacturonans I and II (RG-I and RG-II), pectin fragments that contain a high degree of branching from several classes of sugar residues. Arabinogalactan proteins (AGPs) consist primarily of arabinose, galactose, and glucuronic acid, and also contain about 10% of hydroxyproline-rich protein. Both appear to be less susceptible to hydrolysis than other pectic or hemicellulosic substances. The major yeast-derived polysaccharides are mannoproteins, the primary constituents of yeast cell walls. They are released during yeast autolysis and thus are expected to be at higher concentrations in wines aged on lees. They consist primarily of mannose, with smaller amounts of protein. More detailed information on polysaccharide composition can be found elsewhere [22, 23].

Polysaccharides are widely used in the food industry to increase perceived viscosity and mouthfeel in beverages, although typically at higher concentrations than those in wine [1]. The effects of wine polysaccharides on mouthfeel and other flavor properties are an active area of investigation. Addition of isolated polysaccharide fractions to model wine are reported to slightly increase body, and the acidic RG-II fraction also decreased perceived astringency [24]. Polysaccharides may also have a modest indirect effect on wine aroma by decreasing volatility of some odorants [25], although the effect appears to be small. Mannoproteins can inhibit potassium bitartrate crystallization (Chapters 26.1 and 27) and stabilize the foam of sparkling wines [26]. Finally, some polysaccharides are associated with wine faults. For example, β-glucans produced by *Pediococcus* (and other bacteria) can cause a visible fault called "ropiness" (viscous and oily texture) [18] which can also affect mouthfeel and filterability.

Table 2.2 Typical concentrations and sources of major polysaccharides in wine. Adapted from Reference [23]

		Concentration (mg/L)			
Class	Source	Red wine	White wine	Typical MW (kDa)	Charge
Rhamnogalacturonans (RG-I and -II)	Grape	50–250	10–50	10	Acidic
Arabinogalactan proteins (AGPs)	Grape	100–150	50–150	100–250	Neutral
Mannoproteins	Yeast	100–150	100–150	50–500	Neutral

References

1. Brady, J.W. (2013) *Introductory food chemistry*, Comstock Pub. Associates, Ithaca, NY.
2. del Alamo, M., Bernal, J.L., del Nozal, M.J., Gomez-Cordoves, C. (2000) Red wine aging in oak barrels: evolution of the monosaccharides content. *Food Chemistry*, **71** (2), 189–193.
3. Fugelsang, K.C. and Edwards, C.G. (2007) *Wine microbiology practical applications and procedures*, Springer, New York.
4. Noble, A.C. and Bursick, G.F. (1984) The contribution of glycerol to perceived viscosity and sweetness in white wine. *American Journal of Enology and Viticulture*, **35** (2), 110112.
5. Belitz, H.D., Grosch, W., Schieberle, P. (2009) *Food chemistry*, Springer-Verlag, Berlin.
6. Hufnagel, J.C. and Hofmann, T. (2008) Quantitative reconstruction of the nonvolatile sensometabolome of a red wine. *Journal of Agricultural and Food Chemistry*, **56** (19), 9190–9199.
7. Liu, S.-Q. and Davis, C.R. (1994) Analysis of wine carbohydrates using capillary gas liquid chromatography. *American Journal of Enology and Viticulture*, **45** (2), 229–234.
8. Luz Sanz, M. and Martinez-Castro, I. (2009) Carbohydrates, in *Wine chemistry and biochemistry* (eds Moreno-Arribas M.V. and Polo M.C.), Springer, New York, pp. xv, 735 pp.
9. Wilker, K.L. (1992) Hydrolysis of sucrose in Eastern US table wines. *American Journal of Enology and Viticulture*, **43** (4), 381–383.
10. Lawless, H.T. and Heymann, H. (2010) *Sensory evaluation of food principles and practices*. Springer, New York.
11. McBride, R.L. and Johnson, R.L. (1987) Perception of sugar–acid mixtures in lemon juice drink. *International Journal of Food Science and Technology*, **22** (4), 399–408.
12. Chiralertpong, A., Acree, T.E., Barnard, J., Siebert, K.J. (2008) Taste–odor integration in espresso coffee. *Chemosensory Perception*, **1** (2), 147–152.
13. Stevenson, R.J., Prescott, J., Boakes, R.A. (1999) Confusing tastes and smells: how odours can influence the perception of sweet and sour tastes. *Chemical Senses*, **24** (6), 627–635.
14. Auvray, M. and Spence, C. (2008) The multisensory perception of flavor. *Consciousness and Cognition*, **17** (3), 1016–1031.
15. Friel, E.N., Linforth, R.S.T., Taylor, A.J. (2000) An empirical model to predict the headspace concentration of volatile compounds above solutions containing sucrose. *Food Chemistry*, **71** (3), 309–317.
16. Mattick, L.R. and Rice, A.C. (1970) Survey of the glycerol content of New York State wines. *American Journal of Enology and Viticulture*, **21** (4), 213–215.
17. Pigeau, G.M., Bozza, E., Kaiser, K., Inglis, D.L. (2007) Concentration effect of Riesling Icewine juice on yeast performance and wine acidity. *Journal of Applied Microbiology*, **103** (5), 1691–1698.
18. Bartowsky, E.J. (2009) Bacterial spoilage of wine and approaches to minimize it. *Letters in Applied Microbiology*, **48** (2), 149–156.
19. Carlavilla, D., Villamiel, M., Martínez-Castro, I., Moreno-Arribas, M.V. (2006) Occurrence and significance of quercitol and other inositols in wines during oak wood aging. *American Journal of Enology and Viticulture*, **57** (4), 468–473.
20. Skogerson, K., Runnebaum, R., Wohlgemuth, G., *et al.* (2009) Comparison of gas chromatography-coupled time-of-flight mass spectrometry and 1H nuclear magnetic resonance spectroscopy metabolite identification in white wines from a sensory study investigating wine body. *Journal of Agricultural and Food Chemistry*, **57** (15), 6899–6907.
21. Sponholz, W.R. and Dittrich, H.H. (1985) Origin of gluconic, 2-oxogluconic, and 5-oxogluconic, glucuronic and galacturonic acids in musts and wines. *Vitis*, **24** (1), 51–58.
22. Vidal, S., Williams, P., Doco, T., *et al.* (2003) The polysaccharides of red wine: total fractionation and characterization. *Carbohydrate Polymers*, **54** (4), 439–447.
23. Cheynier, V. and Sarni-Manchado, P. (2010) Wine taste and mouthfeel, in *Managing wine quality*, Vol. **1**, *Viticulture and wine quality* (ed. Reynolds, A.G.), Woodhead Publishing and CRC Press, Oxford and Boca Raton, FL.
24. Vidal, S., Francis, L., Williams, P., *et al.* (2004) The mouth-feel properties of polysaccharides and anthocyanins in a wine like medium. *Food Chemistry*, **85** (4), 519–525.
25. Villamor, R.R. and Ross, C.F. (2013) Wine matrix compounds affect perception of wine aromas, in *Annual review of food science and technology*, Vol. **4** (eds Doyle, M.P. and Klaenhammer, T.R.), Annual Reviews, Palo Alto, CA, pp. 1–20.
26. Blasco, L., Vinas, M., Villa, T.G. (2011) Proteins influencing foam formation in wine and beer: the role of yeast. *International Microbiology*, **14** (2), 61–71.

3

Acids

3.1 Introduction

Organic acids are weak acids bearing a carbon chain and at least one acidic carboxyl group, –COOH (Figure 3.1, HA). They may contain other functional groups such as alcohols, ketones, or double bonds. The lower molecular weight organic acids (C_1–C_4) are highly water-soluble but as the carbon chain length increases water solubility is reduced. Organic acids are widespread in the plant kingdom, where they are involved in primary metabolic pathways such as energy production and amino acid biosynthesis, as well as non-fundamental roles like response to osmotic stress and discouraging predation of fruit. In wine, organic acids have two critical roles: (i) they are a major determinant of wine pH, which affects the appearance, microbial stability, and chemical stability of wine, and (ii) they have direct impacts on taste, particularly sourness.

3.2 Organic acids in wine

3.2.1 Major organic acids

Six acids represent >95% of the total organic acids found in wine (Table 3.1). These compounds can be extracted from the grape, formed by microbial metabolism, or may be added exogenously by the winemaker [1,2]. With the exception of acetic acid, they are non-volatile, and may cumulatively be referred to as "fixed acids" in older literature. Also, many of these acids are *polyprotic*; that is, they contain more than one –COOH group and thus have more than one H^+ equivalent per mole.

Tartaric and malic acids are generally at the highest concentration in wines following alcoholic fermentation. Both of these compounds are formed in grapes early in the growing season, but their behavior during grape maturation and winemaking is dissimilar:

- Malic acid is present at very high concentrations (>20 g/kg) prior to veraison, but will be actively metabolized during berry ripening. Concentrations are generally lower in warmer regions and more mature grapes. Malic acid may also be metabolized during fermentation, with conversion to lactic acid being of particular importance to winemaking (Chapter 22.1).

Understanding Wine Chemistry, First Edition. Andrew L. Waterhouse, Gavin L. Sacks, and David W. Jeffery.

Table 3.1 *Structures, indicative concentration ranges, and sensory properties, of major organic acids important to wine [1–11]*

Acid[a] (pK_a in water)	Structure	Concentration (g/L, typical)	Sensory notes[b]	Common sources[c]
Acetic (4.76)	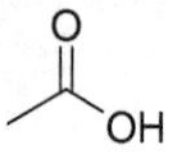	0.1–0.5 Legal limit in US: 1.4 g/L (red), 1.2 g/L (white)	Volatile – vinegar, pungent aroma (200 mg/L threshold)	Y, LAB, AAB
Citric (3.13, 4.76, 6.40)	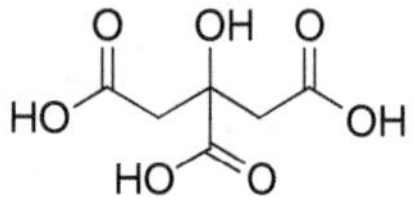	0.1–0.7	Sour, astringent	G, Y
L-Lactic[d] (3.86)	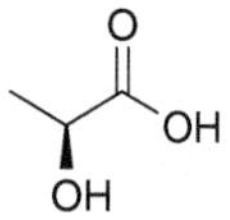	0–3	Sour, astringent	LAB
L-Malic[d] (3.40, 5.11)	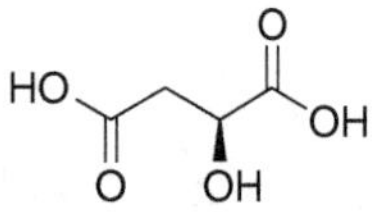	2–7 (higher possible when under-ripe fruit is used)	Sour, astringent	G
Succinic (4.21, 5.64)	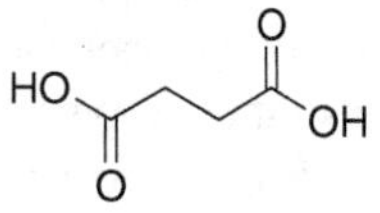	0.5–1.0	Sour, salty, bitter	Y
L-Tartaric[d] (2.98, 4.34)	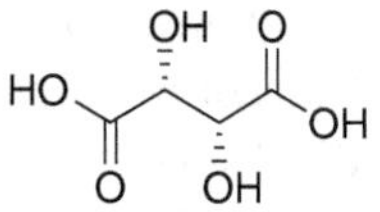	2–6	Sour, astringent	G

[a] Common names are presented rather than systematic (IUPAC) names. Multiple pK_a values are reported for polyprotic acids with multiple –COOH groups (i.e., pK_{a1}, pK_{a2}, etc.).
[b] Sensory descriptors refer to perception of acid in isolation in simple matrices (water or hydroalcoholic solution).
[c] G = grape, Y = yeast, LAB = lactic acid bacteria, AAB = acetic acid bacteria. In many wine regions, tartaric, malic, and citric acid may be legally added by the winemaker (Chapter 27).
[d] Stereochemistry of the dominant stereoisomer is shown for chiral compounds. For lactic acid, D-lactic acid may be formed by lactic acid bacterial metabolism of sugars and can serve as a marker for LAB spoilage.

- Tartaric acid is formed during initial berry cell division and is stable throughout berry ripening: its concentration is generally constant on a per berry basis. Tartaric acid is not metabolized during winemaking, but it can be lost through physiochemical mechanisms like precipitation (Chapter 26.1).

Citric, succinic, and acetic acid will be formed as typical products of alcoholic fermentation (Chapter 22.1). Acetic acid production will be higher under conditions of high osmotic stress (Chapter 22.1) and can also increase during malolactic fermentation (Chapter 22.5), and very high concentrations of acetic acid may indicate either lactic acid or acetic acid bacterial spoilage (Chapter 22.5).

3.2.2 Minor volatile fermentation-derived acids and other organic acids

Several other volatile aliphatic organic acids have been identified in wine (Table 3.2), the most plentiful (mg/L concentrations) being even numbered, straight-chain fatty acids. These are produced as byproducts of fatty acid metabolism, and are common to all alcoholic fermentations. Short branched-chain fatty acids

Table 3.2 Structures, indicative concentration ranges, odor thresholds, and odor descriptors of minor microbially-derived volatile fatty acids important to wine [1–11]

Acid (pK_a)	Structure	Typical range (mg/L)	Detection threshold (mg/L)[a]	Odor descriptors[a]
Butanoic (butyric) (4.83)	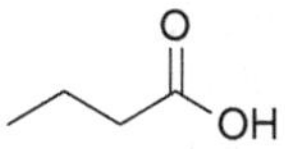	0.4–5	0.2	Rancid, sweat, cheese
Decanoic (4.90)	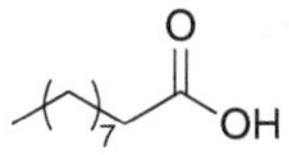	0.06–0.8	1	Fatty, rancid
Hexanoic (caproic) (4.85)	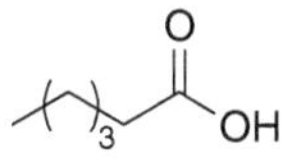	0.8–4	0.4	Fatty, rancid, cheese
3-Methylbutanoic (isovaleric) (4.77)	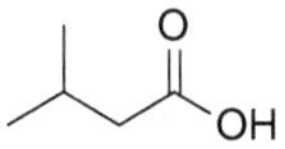	0.3–1	0.03	Rancid, sweat, cheese
2-Methylpropanoic (isobutyric) (4.84)	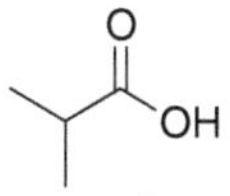	0.4–2	2	Rancid, butter, cheese
Octanoic (caprylic) (4.89)	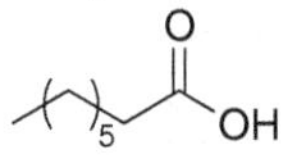	0.6–5	0.5	Fatty, rancid
Propanoic (propionic) (4.87)	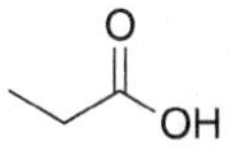	Up to 100	8	Pungent, rancid

[a]Threshold and descriptor data determined in multiple matrices, including water, hydroalcoholic solution, beer, white wine, and red wine.

(isovaleric, isobutyric) are also found at slightly lower concentrations. Yeast production of these volatile acids is sensitive to several physiological factors (nutrient status, oxygen availability, temperature), and these are discussed in detail in Chapter 22.2.

Several other acids will be discussed in later chapters:

- Hydroxycinnamic acids and their derivatives (Chapter 13, 23.3)
- Pyruvic acid, α-ketoglutaric acid, and other α-keto acids (Chapter 9, 22.1, 22.3)
- Ascorbic acid (Chapter 24)
- Sorbic acid (Chapter 18).

3.3 Organic acids, pH, and wine acidity

The major importance of organic acids to wine is, unsurprisingly, wine acidity. There are several metrics associated with acidity that are used to describe wine, as summarized in Table 3.3.

3.3.1 pH, pK_a, and simple solutions of acids

In aqueous solutions like wine, weak organic acids will be partially dissociated, as shown in Figure 3.1. pH is a measure of the free proton concentration of a *solution*, calculated as $pH = -\log [H^+]$. The degree of dissociation for a given acid is described by the acid dissociation constant (K_a), where a larger K_a is associated

Table 3.3 Summary of acidity-related metrics important to wine and juice

Measurement	Calculated as	Typical values in dry wine	Relevance
pH	$-\log [H^+]$[a]	3.0-3.7[b]	Higher pH results in decreased microbial stability and effectiveness of SO_2 (Chapter 17), decreased anthocyanin pigment color (Chapter 16), and decreased rate of acid-catalyzed reactions (Chapter 25)
Titratable acidity (TA)	$[H^+] + [COOH]$[c]	67–107 mEq/L (5–8 g/L as tartaric equivalents)	Correlated with perceived sourness
Total acidity	$[COOH] + [COO^-]$	100–150 mEq/L	Correlated with buffer capacity
Buffer capacity	OH^- concentration necessary to raise pH by 1 unit	25–75 mEq/pH unit	Lower value predicts a larger change to pH following a given change to TA

[a] $[H^+]$ is the concentration of free protons.
[b] Typically, a higher pH range (3.3–3.7) and lower TA values (5–6.5 g/L) are observed in red wines.
[c] In practice, the TA is determined by titration of a wine sample to endpoint (pH 8.2 in the US and Australia, pH 7.0 in the European Union).

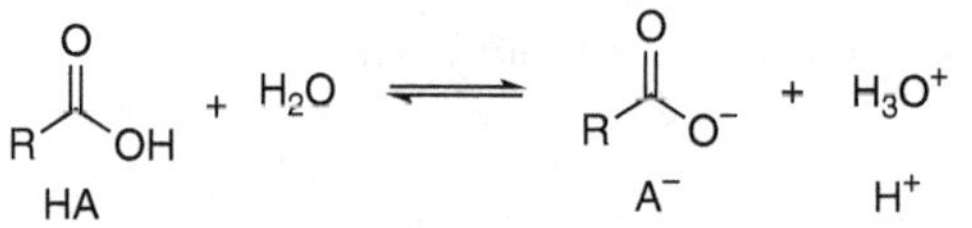

Figure 3.1 Dissociation reaction involving a weak monoprotic organic acid (HA) forming its conjugate base (A^-)

with greater dissociation and a stronger acid (Equation (3.1)).[1] For the weak organic acids in wine, K_a is $<< 1$, and values are reported as pK_a (i.e., $-\log K_a$).

$$K_a = \frac{[H^+][A^-]}{[HA]} \tag{3.1}$$

pK_a values determined in water for major wine acids are shown in Tables 3.1 and 3.2. Polyprotic acids, with multiple –COOH groups, have one pK_a for each group. The pK_a values in these tables and elsewhere in the literature are usually for dilute solutions in pure water at 20 °C, but these values will be slightly different in real wines. The major factors affecting pK_a values are [12]:

- Ethanol – less polar solvents like ethanol result in less ionization and higher pK_a values. As a rule of thumb, pK_a typically increases ~0.25 units for model solutions containing 12% v/v ethanol, e.g. the pK_{a1} of tartaric in 12% v/v ethanol increases from 2.98 to 3.23 [13].

[1] To summarize, pH describes the acidity of a solution, and pK_a is a property of a compound in that solution. One would not say "the pH of that compound is…," but rather "the pH of solution containing 10 g/L of that compound is…"

- Ionic strength – ionic strength (I) is a measure of the concentration of ions in solution, and higher ionic strength will lead to a decrease in pK_a values. Ionic strength is defined as

$$I = \frac{1}{2}\sum_{i=1}^{n} c_i z_i^2 \tag{3.2}$$

  where i is the ionic species number, n is the total number of different ions, c is the molar concentration of a species, and z is the species charge state (e.g., $z=1$ for K^+, $z=2$ for Ca^{2+}). In low ionic strength solutions ($I<0.1$ M) the most common approach for estimating the ionic strength corrected pK_a' is derived from the Debye–Huckel equation:[2]

$$\mathrm{p}K_a' = \mathrm{p}K_a - \frac{0.51z^2\sqrt{I}}{1+\sqrt{I}} \tag{3.3}$$

  The ionic strength of most wines is estimated to be 0.075–0.085 M [14]. From the Debye–Huckel theory, this will lead to a ~0.1 unit decrease in pK_a for monoprotic acids and pK_{a1} of polyprotic acids, and even larger decreases for the formation of multiply charged anions [12].
- Temperature – the degree of dissociation of weak acids will usually (but not invariably) increase with increasing temperature. For weak organic acids these effects are usually small over a typical range of room temperatures (<0.1 unit decrease in pK_a).

The effects of ethanol and ionic strength will partially offset each other, and a good rule is that the first pK_a value of an organic acid will be 0.10–0.15 units higher in wine than in water, and 0.10–0.15 units lower for second dissociation constants.

The pH of a simple aqueous solution containing only a single monoprotic weak acid can be calculated readily assuming the acid's concentration ([HA]) and K_a value are known. Because identical amounts of H^+ and A^- will be formed following dissociation, $\boldsymbol{x}$ can be substituted for [H^+] and [A^-], and **[HA]-*x*** substituted into the denominator of the acid equilibrium expression (Equation (3.1)). This yields Equation (3.4), which can then be solved by the quadratic formula of Equation (3.5) (only the positive solution from the quadratic formula is used):

$$x^2 + K_a x - K_a\mathrm{HA} = 0 \tag{3.4}$$

$$\left[\mathrm{H}^+\right] = x = \frac{-K_a + \sqrt{K_a^{\,2} - 4K_a\mathrm{HA}}}{2} \tag{3.5}$$

A simplified version to calculate [H^+] is based on the assumption that dissociation of HA is small, that is, K_a is around 1×10^{-4} or less and ([HA]-x) ≈ [HA]:

$$\left[\mathrm{H}^+\right] = \sqrt{K_a\left[\mathrm{HA}\right]} \tag{3.6}$$

Calculating the pH of solutions containing multiple organic acids and/or conjugate bases (M^+COO^-) is more complex. Typically in these situations, the pH calculations are performed numerically by computer. Free software programs are available for these calculations [15].

[2] More complicated empirical approaches to correcting for ionic strength appropriate for high ionic strength systems may be found in physical chemistry texts.

3.3.2 Buffers and pK_a

Wines are *buffer* solutions – that is, they contain a mixture of a weak acid(s) and conjugate base(s), which limits (i.e., buffers) the change in pH following addition of a strong acid or base. It is this resistance to pH change that makes a weak acid solution sour when tasted, as unbuffered solutions would have the H_3O^+ ions neutralized by salts in saliva. As a strong acid (e.g., HCl) is added to a buffer, it dissociates to form H_3O^+ and Cl^-. To re-establish equilibrium, the concentration of added H_3O^+ must decrease. This occurs by the reaction of H_3O^+ with A^- to form HA, resulting in only a marginal decrease in pH as compared to the original solution. Analogous reactions occur as base is added.

At any pH, the relative concentrations of a weak acid HA and its conjugate base A^- can be calculated from the Henderson–Hasselbalch (H-H) equation:

$$pH = \mathrm{p}K_a + \log\left(\frac{[\mathrm{A}^-]}{[\mathrm{HA}]}\right) \tag{3.7}$$

From the H-H equation it follows for a monoprotic acid that:

- When pH is 1 unit greater than pK_a, 90% of the weak acid is ionized.
- When pH is equal to pK_a, 50% of the weak acid is ionized.
- When pH is 1 unit less than pK_a, 10% of the weak acid is ionized.

The distribution of acid species in wine is further complicated because the two major acids in juice or following alcoholic fermentation (malic and tartaric) each contain two carboxylic acids groups (H_2A, *diprotic*). The first dissociation constant (pK_{a1}) for both acids falls within the range of normal wine pH (i.e., 3.0–4.0) and the second dissociation constant (pK_{a2}) falls between pH 4.5 and 5. In a typical wine, these acids will exist primarily in their fully protonated (i.e. H_2A) and associated conjugate base (i.e., HA^-) forms, and to a smaller extent as the fully ionized acid (i.e., A_2^-, Figure 3.2).

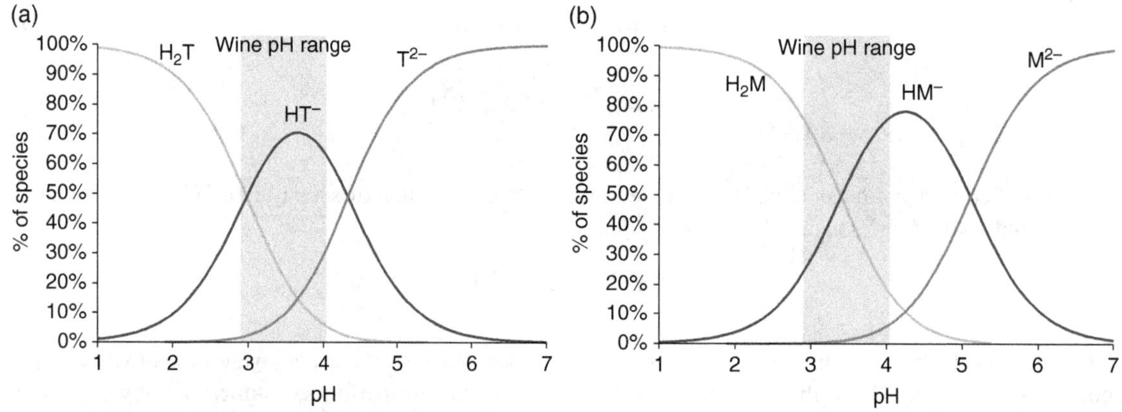

Figure 3.2 *Relative concentrations of neutral organic acid and its ionized forms in aqueous solution as a function of pH for (a) tartaric acid (H_2T), bitartrate anion (HT^-), and tartrate anion (T^{2-}), and (b) malic acid (H_2M), bimalate anion (HM^-), and malate anion (M^{2-})*

3.3.3 Titratable acidity

The titratable acidity (TA) is the concentration of titratable protons in a sample.

- The TA is determined by measuring the concentration of base (usually NaOH) that must be titrated to bring the sample pH to a particular value (endpoint pH) near neutrality.
- The H-H equation shows that acids with pK_a values more than 1 pH unit above the endpoint pH will dissociate minimally, and thus contribute negligibly to TA.
- Because the endpoint pH for wine titrations is generally either 7.0 or 8.2, carboxylic acid groups (typically, $pK_a = 3–5$) will contribute to TA, while polyphenols will not ($pK_a = 9–11$, Chapter 11). Thus, TA measures the sum of free H^+ and undissociated weak organic acid groups (i.e., $[H^+] + [COOH]$).
- TA will include both volatile and non-volatile acids. Volatile acidity (VA) can be determined separately by including a distillation step prior to titration.

Wine TA is generally in the range of 0.05–0.15 moles of titratrable proton equivalents per liter (0.05–0.15 Eq/L). By comparison, at a typical wine pH of 3–4, the concentration of free H^+ is negligible (0.001–0.0001 Eq/L) and the majority of TA is thus due to undissociated weak organic acids. In practice, it is more common to report wine in g/L equivalents of an acid, often as tartaric acid equivalents (Table 3.4).[3] For example, if a wine contains 0.10 $[H^+]$ Eq/L, the concentration as tartaric acid (molecular weight = 150 g/mol, 2 $[H^+]$ Eq/mol) is shown as

$$TA = \frac{0.10\left[H^+\right]Eq}{L} \times \frac{mol\,H_2T}{2\left[H^+\right]Eq} \times \frac{150\,g}{mol\,H_2T} = 7.5\,g/L\,as\,H_2T \qquad (3.8)$$

Typical values for white wine are pH 3.0–3.4 and TA of 6–9 g/L as tartaric acid, while those for red wine are pH 3.3–3.7 and TA of 5–8 g/L as tartaric acid.[4] In certain countries sulfuric acid (98 g/mol, 2 $[H^+]$ Eq/mol) is often used as a reporting acid in European countries.

Table 3.4 Summary of acidity-related metrics important to wine and juice

Organic acid	MW (g/mol)	Number of titratable protons (Eq/mol)	g/Eq	Conversion factor to g/L of tartaric equivalents[a]	Comments
Tartaric	150	2	75	1.0	Common for New World wines
Malic	134	2	67	1.12	Common for ciders
Citric	192	3	64	1.17	Common for citrus wines
Sulfuric	98	2	49	1.53	Common for wines produced in the EU

[a] Multiplying factor to convert from reporting as one acid to tartaric equivalents, for example, a solution with a TA of 1 g/L as citric acid has a TA of 1.17 g/L as tartaric acid.

[3] Reporting the acid concentration "as tartaric" is convenient for winemakers, since acid additions are frequently performed with tartaric acid.
[4] Note that a solution does not need to contain the reporting acid: writing "this wine contains 7.5 g/L as tartaric" does not mean "this wine contains 7.5 g/L *of* tartaric acid." Rather, it means "this wine has the same titratable acid concentration as a 7.5 g/L tartaric acid solution."

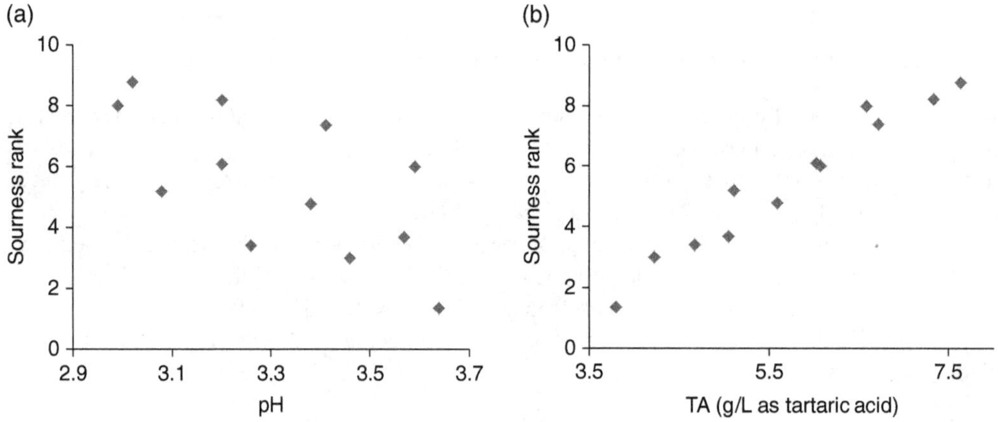

Figure 3.3 *Sourness rankings from a sensory panel (higher rank has greater "acid taste") plotted against (a) pH and (b) titratable acidity (TA) for 12 model wine solutions composed of varying amounts of tartaric acid, malic acid, and potassium hydroxide. Data from Plane 1980 [16]*

The importance of TA to winemakers is that it is well correlated with perceived sourness over the range of values typically observed in wines, as shown in Figure 3.3. The pH has a slight effect independent of TA, and Plane and colleagues [16] proposed that sourness was proportional to TA (g/L) – pH. Since the range of TA far exceeds the range of pH in wine, TA dominates this expression. Although TA is an excellent proxy for the sourness of a wine, other components may mask sourness (e.g., ethanol, sugar).

3.3.4 Total acidity

As defined by Boulton and colleagues [17], the total acidity of a wine refers to the sum concentration of all carboxylic acid and carboxylate groups [COOH] + [COO^-], and is expressed in units of molar or tartaric acid equivalents per L.[5] In practice, total acidity can be determined (i) by quantifying and summing individual organic acids using a technique like HPLC or (ii) by treating a wine with strong cation exchange resin prior to titration. The importance of total acidity is that it is well correlated with the buffer capacity of a wine or juice, as described in the next section.

The total acidity is invariably larger than the TA, and the difference between the two (total acidity – titratable acidity) is equal to the fraction of the acid that has been neutralized at a given pH, or equivalent to the concentration of metal cations in the wine [18]. As shown in Figure 3.4, the slope of a plot of TA versus total acidity (dark line) is only 0.74, indicating that on average approximately one-quarter of protons are missing from organic acids. Adding the molar equivalents of two of the major metal cation species (K^+, Na^+) results in both a better correlation and a slope closer to 1 (Figure 3.4, dashed line).

Viticultural or winemaking factors that result in K^+/H^+ exchange, e.g. K^+ uptake by the grape during ripening or hard pressing, can decrease TA while leaving the total acid nearly constant (Table 3.5). The difference between TA and total acidity (0.029 and 0.037 Eq/L for free run and hard pressed, respectively) is due primarily to an exchange of K^+ for H^+ (0.019 and 0.28 Eq/L). The remaining difference arises from other metal cations, for example, Ca^{2+}, Na^+, Mg^{2+} (Chapter 4).

[5] Although the concepts are distinguished here, in some texts total acidity is used as a synonym for titratable acidity (TA).

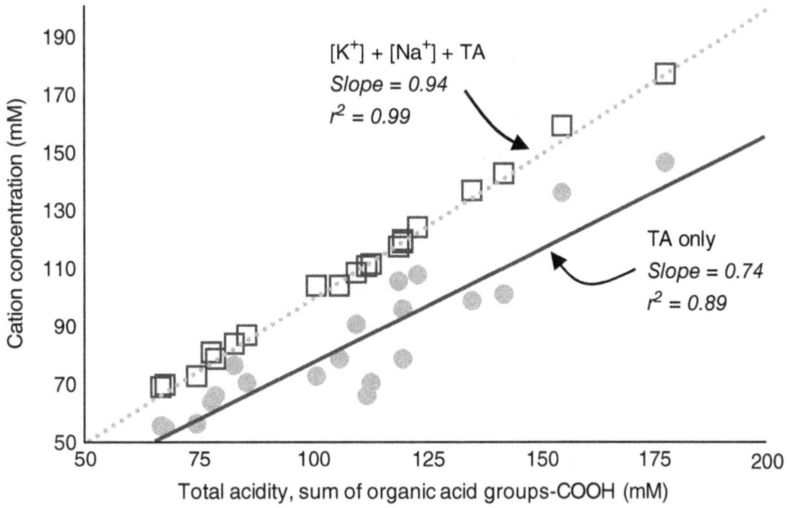

Figure 3.4 Relationship between total acidity, titratable acidity (TA), and other metal cations in wines. The correlation between TA and total acidity is modest and has a non-unity slope (solid line) because a portion of titratable protons are replaced with K^+ and other metal cations. The correlation is improved by summing TA with major metal cations (dashed line). Data from Boulton 1980 [18]

Table 3.5 Relationship of TA, K^+, and total acidity in a free-run and hard-pressed Baco noir juice

Sample	pH	TA in g/L as tartaric (Eq/L)	K^+ in g/L (Eq/L)	Total acidity in g/L as tartaric (Eq/L)
Free run	3.50	5.01 (0.066)	0.76 (0.019)	7.10 (0.095)
Hard-pressed	3.63	4.52 (0.059)	1.10 (0.028)	7.20 (0.096)

3.3.5 Buffer capacity

The buffer capacity (BC) is the ability of a solution to resist changes in pH following addition of a strong acid or base, and is defined as follows:

$$BC = \frac{\Delta TA}{\Delta pH} \quad (3.9)$$

The buffer capacity in wine is correlated with the concentration of buffers in wine. Since the major buffering compounds in wine are organic acids and their conjugate bases, higher buffering capacity is correlated with higher total acidity. Typical values for buffering capacity in wine are in the order of 25–75 mEq/L (1.9–5.6 g/L as tartaric) per pH unit.

The *buffer range* is the range of pH values over which buffering can be observed. For a simple system containing HA and A^-, the buffer range is generally defined as the pH range within ±1 unit of the pK_a, and the maximum buffer capacity will be observed at $pH = pK_a$. However, a complex system like wine containing a

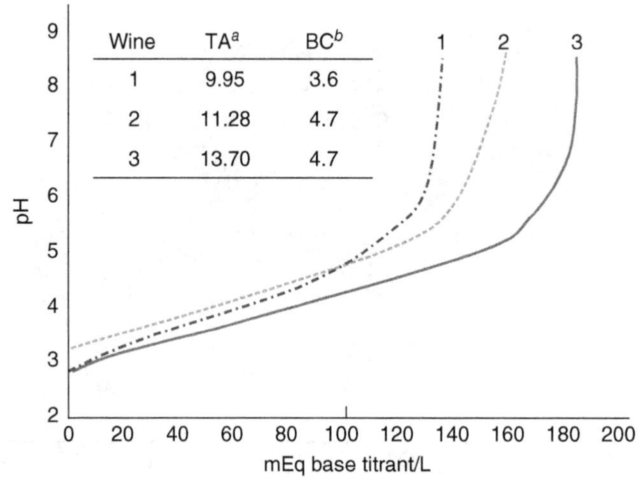

Figure 3.5 *Titration curves for three wines with addition of a strong base. The endpoint for titration is defined as pH = 8.2, but note that a near identical endpoint would be achieved for pH 7.0. [a]Titratable acidity, g/L as tartaric acid. [b]Buffer capacity, g/L as tartaric acid per pH unit, calculated between pH 3 and 5. Data from Mattick 1980 [19]*

number of conjugate acid–base pairs with multiple pK_a values will have an extended buffer range, so the point of maximum buffering may not be obvious. As seen in Figure 3.5, BC will vary among wines, but is generally constant for a given wine (i.e., linear relationship between pH and TA) over a normal wine pH range of 3–5 [19]. Buffer capacity is of importance to winemakers when acid adjustments are performed: wines with larger buffer capacities will require greater removal or addition of acid to effect a pH change. High buffer capacity can present a problem in certain cases, for example, if the aim of a winemaker is to reduce the pH to decrease the microbial spoilage risk.

3.4 Acid adjustments

3.4.1 Acid additions

In many regions, organic acids such as tartaric, citric, and malic or their mixtures can be added to either juice or wine, while the addition of mineral acids such as sulfuric is universally prohibited. All of these acids will result in an increase in TA. Beyond this, addition of tartaric acid may result in precipitation of potassium bitartrate, as described below, such that the final change to TA is lower than the amount of acid added.

3.4.2 Neutralization and/or precipitation of organic acids with carbonate salts

Titratable acidity can be decreased and pH increased by addition of carbonate salts. In most winemaking regions, the legally allowable forms of carbonate salts (Chapter 27) are:

- Potassium salts – $KHCO_3$ or K_2CO_3 are more commonly used.
- Calcium salts – $CaCO_3$ is less commonly used because of problems associated with calcium tartrate instabilities (Chapter 26.1).

Beyond neutralization, addition of potassium and calcium salts to wine may also result in precipitation of potassium bitartrate (KHT) or calcium tartrate (CaT), both of which have poor solubility at low temperatures in ethanolic solutions (Chapter 26.1). Precipitation can result in further decreases in titratable and/or total acidity. KHT precipitation is more commonly observed than CaT because (i) K^+ is at higher concentrations in grapes, (ii) K^+ salts are more commonly used for deacidification, and (iii) the bitartrate (HT^-) species is usually at higher concentrations than tartrate (T^{2-}, Figure 3.2).

KHT precipitation will always be accompanied by a decrease in TA because KHT bears a titratable proton that contributes to TA, but the pH of a wine can increase or decrease depending on the pH prior to precipitation. Wines below pH 3.65 will experience a pH decrease as the dominant H_2T/HT^- equilibrium shifts right to increase [HT^-], thereby increasing [H_3O^+], whereas above pH 3.65 there is a pH increase due to the dominant HT^-/T^{2-} equilibrium shifting left, to increase [HT^-] (Figure 3.6).[6]

Unlike KHT precipitation, the precipitation of CaT will not affect TA, since the compound has no titratable proton. However, it will decrease pH because CaT is a base: total acidity and buffer capacity will also decrease.

Winemakers typically wish to prevent precipitation of KHT, CaT, and other less common salts of organic acids in the bottle. The factors that affect stability of these salts – as well as strategies to test for and prevent instabilities – are discussed in more detail in Chapter 26.1.

A less common precipitation used for deacidification during winemaking is of calcium malate (CaM). The pK_{a2} of malic acid is relatively high (5.11), and malate will be at negligible concentrations at wine pH. However, in a technique called "double salt acid reduction," a portion of wine (20–30%) is near-completely neutralized with $CaCO_3$. This results in a pH >5 and precipitation of CaM and CaT [17]. The resulting wine can be racked and added back to the unadjusted wine. This process will increase pH and decrease TA.

When calculating the effects of carbonate salt addition on TA, it is useful to consider the neutralization and precipitation steps individually. The change in TA resulting from neutralization will occur quickly and stoichiometrically. For example, a 1 g/L solution of $KHCO_3$ (0.01 Eq/L) added to wine will neutralize 0.75 g/L (0.01 Eq/L) of TA as tartaric acid. Similar calculations can be used for K_2CO_3 or $CaCO_3$, requiring 0.69 g/L or 0.5 g/L, respectively, to achieve a 0.75 g/L change. For potassium salts, precipitation of KHT will be

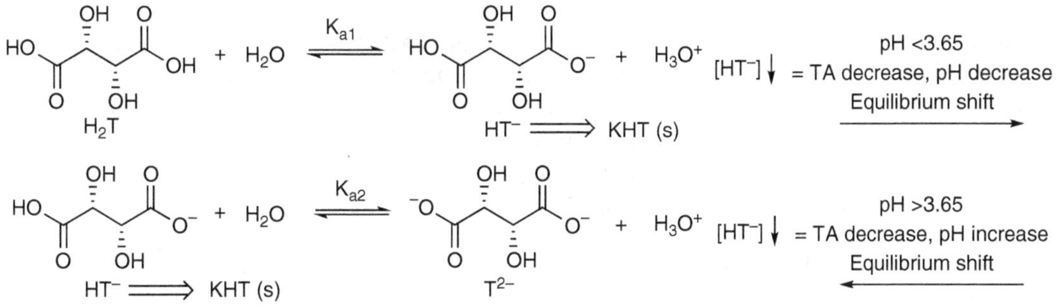

Figure 3.6 Tartaric acid equilibrium and effect of KHT precipitation on wine TA and pH depending on whether the wine is initially above or below pH 3.65

[6] An alternate approach to understanding why pH can increase or decrease following KHT precipitation is to consider what would happen if KHT was added instead of removed. Because KHT is amphoteric (can act as both an acid and base), a KHT solution acts like a buffer and at saturation will form a solution with pH ~ 3.65. What would be the effect if you added this KHT solution to a wine with pH >3.65, assuming nothing precipitates out? The answer is that the pH decreases to a value between 3.65 and the original wine pH, since the KHT solution buffers things to a pH of 3.65. Now, what if you reversed the previous experiment, and magically removed the added KHT? The pH would increase back to the original pH. These effects would be reversed if the original wine has a pH < 3.65.

non-quantitative and occur more slowly. Each mol/L decrease in KHT will result in a 0.75 g/L decrease in TA. Thus, if both neutralization and precipitation go to completion, every 1 g/L addition of $KHCO_3$ will eventually result in a 1.5 g/L decrease in TA.

3.4.3 Biological deacidification

Conversion of malic acid to lactic acid by lactic acid bacteria ("malolactic fermentation," MLF) is a common approach to decreasing titratable acidity (Chapter 22.5). Lactic acid has only one –COOH group as compared to two for malic, and a complete MLF will result in a TA decrease equal to the initial molar concentration of malic acid; pH will also increase. Even when a wine starts at an appropriate TA, MLF may still be advantageous because a solution containing equal $[H^+]$ equivalents of lactic and tartaric acid will have a lower pH than a solution containing the same total $[H^+]$ equivalents of malic acid (Table 3.6). Maloethanolic conversion – in which yeast cells metabolize malic acid to ethanol and CO_2 – can also occur to a limited extent during alcoholic fermentations (Chapter 22.1). Tartaric acid is generally stable throughout wine production,[7] although in rare cases it can be degraded by spoilage lactic strains, a phenomenon called *tourné*. Citric acid will also be partially or fully metabolized by lactic acid bacteria (Chapter 22.5), but because of the low concentrations of citric acid in wine and because one of the end products is acetic acid, the net effect on pH and TA is low.

3.4.4 Physiochemical approaches to altering pH and TA

Electrodialysis can be used to decrease TA while minimally affecting pH, and cation exchange resins can be used to increase TA and decrease pH. These approaches are discussed in more detail in Chapter 26.1

3.4.5 The balance of pH and TA

Winemakers often have competing goals of trying to keep pH low (to avoid microbial spoilage, among other reasons), while also preventing TA from getting too high (to avoid excessive sourness). TA and pH are inversely correlated in wines, but their correlation is not perfect for two reasons:

- Organic acid groups vary in pK_a. Stronger acids (lower pK_a) will result in a greater pH decrease for the same contribution to TA. For example, based on the distribution shown in Figure 3.2, addition of one equivalent of tartaric acid to a solution will cause a greater pH decrease (more ionized so increases $[H_3O^+]$) than one equivalent of malic acid, although both will cause a similar increase in TA.
- At a given TA, a higher buffering capacity generally corresponds to a higher pH. This buffering capacity is due to the presence of metal cations (e.g., K^+) and conjugate bases.

The goals of low TA and low pH are most readily achieved when the wine contains proportionally higher concentrations of stronger acids (e.g., tartaric) and lower concentrations of weaker acids (malic, lactic, succinic, acetic) and salts of conjugate bases. As shown in Table 3.6, adjustment using $KHCO_3$ to a typical target of pH 3.5 will yield different final TA values depending on initial acid composition. A high malic acid red wine will end up with a higher TA value because malic is a weaker acid (higher pK_a, decreases pH less) than tartaric.

In practice, it is common to use potassium salts for deacidifying high-acid wines when small adjustments are necessary (1–3 g/L as tartaric acid) and when the pH starts relatively low (<3.4). Larger TA corrections by

[7] As mentioned in the Introduction chapter, tartaric acid is so stable that it can be found on centuries old archeological samples. Its presence on shards from pottery vessels is considered evidence that the vessel contained wine, as there are no other common sources of liquids that possess tartaric acid.

Table 3.6 Titratable acidities of model solutions following adjustment to pH 3.5 with either $KHCO_3$ or tartaric acid, and assuming no precipitation. TA values were calculated using CurTiPot speciation software [15]

Matrix	Concentrations of acids in model wine[a]			TA after adjusting to pH 3.5
	Tartaric acid	Malic acid	Lactic acid	
Typical red, pre-MLF	0.03 M (4.5 g/L)	0.02 M (2.7 g/L)	0	6.4 g/L
Typical red, post-MLF	0.03 M	0	0.02 M (1.8 g/L)	5.3 g/L
High malic red, pre-MLF	0.03 M	0.06 M (8 g/L)	0	10.6 g/L
High malic red, post-MLF	0.03 M	0	0.06 M (5.4 g/L)	7.2 g/L

[a] Model wine containing 1.3 g/L K^+ and 10% EtOH.

this approach would result in excessively high pH. Since high TA is usually a result of high malic acid, partial or complete MLF is also commonly used for deacidification, although this approach will change other sensory properties and is thus not appropriate for all wine styles (Chapter 22.5). For wines with very high malic acid, a combination of techniques including calcium malate precipitation ("double salt"), standard carbonate addition, and MLF may be employed.

3.5 General roles of organic acids and pH in wine reactions

All organic acids can by esterified by ethanol and other alcohols during fermentation or wine storage, and some of these esters have important aroma qualities (Chapters 7, 22.2, and 25). Conversely, esters can be hydrolyzed to form organic acids. These equilibria are considered in Chapter 7. Lower pH values will also accelerate acid-catalyzed reactions in wine as follows (note that the effect is not dependent on TA):

- Hydrolysis and/or rearrangements of glycosidic aroma precursors will occur faster (Chapters 8, 23.1, and 25).
- Ester hydrolysis and formation will be faster (Chapters 7 and 25).
- Several reactions involving red wine polyphenolics will increase, including hydrolysis of polymeric polyphenols and activation of carbonyls to react with flavonoids (Chapters 16 and 25).
- pH will affect the equilibria, reactivity, and sensory effects of acidic compounds like anthocyanins (Chapter 16), SO_2 (Chapter 17), and CO_2.

3.6 Sensory effects of acids

The major sensory effect of acids in wine is to increase perceived sourness. As discussed in Section 3.3, sourness is well correlated with TA and only weakly correlated with pH. Low pH solutions can also induce astringent perceptions [20], likely due to precipitation and functional loss of lubricating salivary proteins [21].[8] Additionally, some acids are described as having characteristics in addition to sourness, for example,

[8] This astringency is readily appreciated by tasting lemon juice (pH 2.2, ~5% citric acid), which causes a gritty feeling in the mouth in addition to intense sourness.

"bitter-salty" taste of succinic acid (Table 3.1). However, the relevance of these extra-sour sensory attributes in wine-like matrices (rather than water–ethanol solutions) has not been well studied.

Volatile acidity (VA) is comprised primarily (>90%) of acetic acid. While acetic acid concentrations approaching the sensory threshold (400 mg/L) are commonly found in wine, higher concentrations are typically a result of bacterial spoilage (Chapter 22.5) [22]. Specific regulations vary among regions and wine types, but typical limits are around 1.0–1.5 g/L. A complication of evaluating the effects of volatile acids on wine aroma is that their concentrations are usually correlated with their corresponding ethyl esters, and the latter are generally more potent. For example, acetic acid has a pungent, vinegar-like aroma in isolation, but the off-aromas associated with VA appear to be due primarily to the more potent ethyl acetate, formed by esterification of acetic acid (Chapter 7). Furthermore, although most other volatile acids have cheesy or rancid odors in isolation, their sensory contribution in real wine seems to be a general contribution to vinous character.

References

1. Fowles, G.W.A. (1992) Acids in grapes and wines: a review. *Journal of Wine Research*, **3** (1), 25–41.
2. Swiegers, J.H., Bartowsky, E.J., Henschke, P.A., Pretorius, I.S. (2005) Yeast and bacterial modulation of wine aroma and flavour. *Australian Journal of Grape and Wine Research*, **11** (2), 139–173.
3. Gardner, R.J. (1980) Lipid solubility and the sourness of acids: implications for models of the acid taste receptor. *Chemical Senses*, **5** (3), 185–194.
4. Sowalsky, R.A. and Noble, A.C. (1998) Comparison of the effects of concentration, pH and anion species on astringency and sourness of organic acids. *Chemical Senses*, **23** (3), 343–349.
5. Siebert, K.J. (1999) Modeling the flavor thresholds of organic acids in beer as a function of their molecular properties. *Food Quality and Preference*, **10** (2), 129–137.
6. Lambrechts, M.G. and Pretorius, I.S. (2000) Yeast and its importance to wine aroma – a review. *South African Journal for Enology and Viticulture*, **21**, 97–129.
7. Francis, I.L. and Newton, J.L. (2005) Determining wine aroma from compositional data. *Australian Journal of Grape and Wine Research*, **11** (2), 114–126.
8. Da Conceicao Neta, E.R., Johanningsmeier, S.D., McFeeters, R.F. (2007) The chemistry and physiology of sour taste – a review. *Journal of Food Science*, **72** (2), R33–R38.
9. Burdock, G.A. (2010) *Fenaroli's handbook of flavor ingredients*, 6th edn, CRC Press, Boca Raton, FL.
10. (2012) Dissociation constants of organic acids and bases, in *CRC Handbook of Chemistry and Physics*, 93rd edn (ed. Haynes, W.M.), CRC Press, Boca Raton, FL, pp. 5-94–5-103.
11. Ferreira, V., Lopez, R., Cacho, J.F. (2000) Quantitative determination of the odorants of young red wines from different grape varieties. *Journal of the Science of Food and Agriculture*, **80** (11), 1659–1667.
12. Reijenga, J., van Hoof, A., van Loon, A., Teunissen, B. (2013) Development of methods for the determination of p*K(a)* values. *Analytical Chemistry Insights*, **8**, 53–71.
13. Usseglio-Tomasset, L. and Bosia, P. (1978) Determinazione delle constanti di dissociazione dei principali acidi del vino in soluzioni idroalcoliche di interesse enologico. *Rivista Vitic Enol*, **31** (9), 380–403.
14. Abguéguen, O. and Boulton, R.B. (1993) The crystallization kinetics of calcium tartrate from model solutions and wines. *American Journal of Enology and Viticulture*, **44** (1), 65–75.
15. Gutz, I.G.R. CurTiPot. Available from: http://www2.iq.usp.br/docente/gutz/Curtipot.html.
16. Plane, R.A., Mattick, L.R., Weirs, L.D. (1980) An acidity index for the taste of wines. *American Journal of Enology and Viticulture*, **31** (3), 265–268.
17. Boulton, R.B., Singleton, V.L., Bisson, L.F., Kunkee, R.E. (1999) *Principles and practices of winemaking*, Kluwer Academic/Plenum Publishers, New York.
18. Boulton, R. (1980) The relationships between total acidity, titratable acidity and pH in wine. *American Journal of Enology and Viticulture*, **31** (1), 76–80.

19. Mattick, L.R., Plane, R.A., Weirs, L.D. (1980) Lowering wine acidity with carbonates. *American Journal of Enology and Viticulture*, **31** (4), 350–355.
20. Gawel, R. (1998) Red wine astringency; a review. *Australian Journal of Grape and Wine Research*, **4** (2), 74–95.
21. Siebert, K.J. and Chassy, A.W. (2004) An alternate mechanism for the astringent sensation of acids. *Food Quality and Preference*, **15** (1), 13–18.
22. Bartowsky, E.J. (2009) Bacterial spoilage of wine and approaches to minimize it. *Letters in Applied Microbiology*, **48** (2), 149–156.

4

Minerals

4.1 Introduction

A mineral is a naturally occurring solid substance with a defined chemical composition and ordered atomic structure. Minerals are major constituents of rocks and can be classified based on their abundance, that is, major rock-forming elements (e.g., oxygen, silicon, iron, magnesium, aluminum, calcium, sodium, and potassium) and trace elements (e.g. fluorine, phosphorus and precious metals such as silver and gold). The term "mineral" also refers to mineral substances solubilized or otherwise dispersed in matrices (e.g., soil or foodstuffs), excluding components comprised of the organic elements carbon, hydrogen, oxygen, and nitrogen.[1] For example, a solution of copper sulfate is said to have a mineral content even though it lacks an ordered structure. Many of these minerals are a familiar sight on food labels (e.g., potassium, iron, sodium), and are essential nutrients for living organisms, including plants. Due to their widespread natural presence, minerals in wine arise from a number of sources, including processing aids or additives used during winemaking, but are primarily extracted from the grape berry.

The main dissolved mineral in wine arises from salts of the metal potassium (K), with smaller contributions from salts of iron (Fe), sodium (Na), copper (Cu), calcium (Ca), magnesium (Mg), aluminum (Al), manganese (Mn), and zinc (Zn) (Table 4.1). These metals have a positive charge in solution (i.e., are present as *cations*), and their negatively charged inorganic counterparts include phosphate, sulfate/sulfite, nitrate, bromide, and chloride anions [1], yielding a mineral content of wine in the order of 1.5–3 g/L.[2] Trace heavy metals such as lead (Pb), cadmium (Cd), and nickel (Ni) have also been identified in wine; concentrations of these and other metals may be regulated internationally (including limits on impurities in additives/processing aids) due to wine quality and health (toxicity) concerns.[3]

[1] Using the term "mineral" as a synonym for components other than C, H, N, and O is somewhat archaic but can still be found in the literature when describing inorganic components present in different matrices, particularly foods.

[2] Organic acids discussed in Chapter 3 can also serve as quantitatively important anions, although they would not be considered "mineral."

[3] Example agencies are usually the same ones responsible for determining allowable additives (Chapter 27), for example, Food Standards in Australia and New Zealand, the European Commission, and the US Alcohol and Tobacco Tax and Trade Bureau. Typical maximum limits for lead are 100–300 μg/L across a range of countries, for instance, while those for copper are 1–2 mg/L, but as high as 5–10 mg/L in some cases. Where limits are not specified in regulations for a given country, the default limits are based on impurity specifications imposed by the International Oenological Codex or Food Chemicals Codex.

Understanding Wine Chemistry, First Edition. Andrew L. Waterhouse, Gavin L. Sacks, and David W. Jeffery.

Table 4.1 Oxidation states, indicative concentration ranges, and potential impacts on wine properties of the predominant metals found in wine [2–9]

Metal	Oxidation state	Concentration range (mg/L)	Effect on wine properties
Al	+3	ND[a]–14	Haze formation, polyphenol complexation, taste
Ca	+2	7–310	Precipitation with resultant pH and buffering capacity decrease
Cu	+1, +2	ND–3	Oxidation, haze formation, binding of sulfur compounds, polyphenol complexation, taste
Fe	+2, +3	0.06–55	Oxidation, haze formation, polyphenol complexation
K	+1	125–3060	Precipitation with resultant pH change and titratable acidity decrease
Mg	+2	8–720	Polyphenol complexation
Mn	+2, +3, +4, +6, +7	0.1–6	Oxidation
Na	+1	ND–320	Taste
Pb	+2, +4	ND–1.25	Potential health effects
Zn	+2	ND–9	Haze formation, taste

[a] ND, not detected.

4.2 Origins of metals in wine

Sources of metals in wine are diverse and include vineyard soil, sprays, and exogenous contaminants (e.g., Pb from atmospheric emissions, although this has decreased with phasing out of leaded fuel), fining agents, winery equipment, and winemaking treatments (Table 4.2).

- The alkali and alkaline earth metals (i.e., Na, K, Mg, and Ca) present in wine accumulate during berry ripening, and the grape is the major source of these metals.
- Carbonate salts of K and Ca may also be intentionally added in the winery during deacidification treatments, and these same metals may be lost through precipitation as their tartrate salts (Chapters 3 and 26.1).
- Winemaking equipment was historically a major source of transition and heavy metals (e.g. Cu and Pb from brass fittings, Fe from iron–enamel tanks), but stainless steel equipment and plastic materials in modern wineries have greatly diminished this occurrence. However, impurities in additives and processing aids (e.g., bentonite, Chapter 26.2) may still contribute to the concentrations of these metals in wine. These metals may also decrease during fermentation due to adsorption on to yeast cells.

4.3 Reactions involving metals

The most important reactions involving transition metals are those associated with redox phenomena (Chapters 11 and 24) due to the catalytic action of these metals in wine, particularly copper and iron, but also manganese [12–15]. This reactivity relates to the reduction potentials of the different metals, but also those of oxygen, phenolic compounds, and other wine components (i.e., antioxidants such as SO_2 and ascorbic acid, ligands such as carboxylate ions and water) at wine pH [12]. Overall, redox reactions are

Table 4.2 Potential origins of metals that affect wine chemistry and sensory properties [5, 6, 8, 10, 11]

Metal	Origin[a]
Al	Soil (V), bentonite (W), metal alloys (W), filter media (W)
Ca	Soil (V), fertilizers (V), fungicides (V), bentonite (W), deacidification (W), fining agents (W), filter media (W), concrete tanks (W)
Cu	Soil (V), fungicides (V), fertilizers (V), pesticides (V), metal alloys (W), fining agents (W), stabilizers (W), filter media (W)
Fe	Soil (V), metal alloys (W), bentonite (W), filter media (W), fining agents (W), yeast supplements (W), stabilizers (W)
K	Soil (V), fertilizers (V), fining agents (W), stabilizers (W), potassium metabisulfite (W)
Mg	Soil (V), bentonite (W), concrete tanks (W), fining agents (W), yeast supplements (W)
Mn	Soil (V), fungicides (V), fertilizers (V), pesticides (V), filter media (W)
Na	Irrigation (V), coastal environment (V)[b], soil (V), bentonite (W), sodium metabisulfite (W), stabilizers (W), fining agents (W), cation exchange (W)
Zn	Soil (V), fungicides (V), fertilizers (V), pesticides (V), metal alloys (W), fining agents (W), stabilizers (W), filter media (W), yeast supplements (W)

[a]Vineyard, (V); winemaking, (W).

[b]Enhanced concentrations, presumably from increased content of Na in soil but with a potential effect from direct contact of berries with saline aerosol (derived from sea spray).

facilitated by metals (as redox couples, e.g. Fe^{3+}/Fe^{2+}) such that phenolics are oxidized and oxygen is reduced to form other powerful oxidants, including radicals (via a one-electron process) of varying reactivity. These aspects are described in more detail in Chapter 24 but the following provides a brief summary of the impact of transition metals:

- Transition metals in their reduced states (e.g., Fe^{2+} or Cu^{+}) can accelerate the reaction of oxygen and wine components by catalyzing the formation of reactive oxygen species (ROS).
- These metals promote formation of ROS through activation of oxygen, yielding the hydroperoxyl radical ($HOO^{\bullet}$), and via the Fenton reaction of hydrogen peroxide (H_2O_2) to produce the highly reactive hydroxyl radical ($^{\bullet}OH$).
- The requisite oxidation states of metal catalysts are regenerated through redox cycling [14], which effectively promotes phenolic oxidation and formation of ROS.
- The presence of multiple transition metals can have a synergistic effect in redox cycling [14].
- Phenolic compounds may act as ligands, which bind transition metal ions, thereby influencing (i.e., promoting or retarding) oxidative processes [16].

Transition metals are also implicated in redox reactions of sulfur compounds, with sulfur oxidation potentially leading to formation of di- and trisulfides (Chapter 10), and various combinations of metals (particularly Cu) leading to increases in hydrogen sulfide and methanethiol in wine stored under anaerobic (reductive) conditions (Chapter 30) [9]. However, Cu can also form insoluble precipitates with hydrogen sulfide and thiols (e.g., methanethiol and ethanethiol). The deliberate addition of small amounts of Cu^{2+} as $CuSO_4$ (ca. 0.1–0.5 mg/L) is used in this way to remove undesirable sulfur compounds that negatively impact on wine aroma (Chapter 26.2). The treatment is not specific, however, and can affect concentrations of desirable thiols, either through binding or as a result of redox reactions due to residual copper.

4.4 Sensory effects of metals

Effects on wine properties due to metals are summarized in Table 4.1. Transition metals can form hazes (i.e., metal casse) through interaction with other wine components (Chapter 26.2) [5, 6]. At around 10 mg/L, Fe^{3+} formed under oxidative conditions can form insoluble complexes with phenolic compounds or phosphates and proteins, although such concentrations of Fe are now rare when using modern equipment. In contrast, under reductive conditions Cu concentrations of about 1 mg/L can lead to insoluble complexes with proteins. Haze formation has also been attributed to aluminum and zinc at similar concentrations to that of Fe^{3+}.

The presence of these metals is thought to contribute to undesirable astringency and metallic,[4] bitter, or sour tastes at similar concentrations to those that cause haze problems, whereas sodium may increase saltiness while decreasing bitter and sour tastes [5, 6, 18, 19]. Precipitation of Ca^{2+} and K^{+} salts of tartaric acid also influences the taste and mouthfeel properties of wine by altering wine pH and, in the case of potassium bitartrate, lowering titratable acidity (Chapter 3). Additionally, organoleptic defects can arise as an indirect result of transition metals in wine by increasing the rate of oxidation and formation of volatile and non-volatile wine components (e.g., ethanol, aroma compounds, polyphenols, added antioxidants), resulting in changes to wine aroma and color. Another possibility is that metals may affect the color of red wine by complexing with anthocyanins (and other co-pigments) and shifting the absorbance maxima. While not well studied in wine, metalloanthocyanins formed from natural or synthetic pigments are well known in other systems [6, 7, 20] (Figure 4.1).

4.5 Metals and wine authenticity

Metal concentrations are dependent on soil type and geology (e.g. aluminum, barium, lithium, manganese), both soil and winemaking (e.g., sodium, potassium, calcium, magnesium, copper, iron, zinc), and exogenous contamination (e.g., lead, chromium, and nickel from pollution or metallic components). As a result, metal concentrations have been investigated as markers for authenticity (Chapter 28) – particularly for geographical

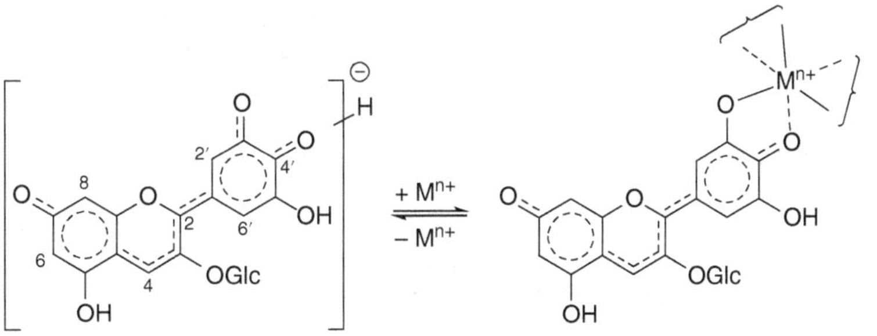

Figure 4.1 *Representation of metalloanthocyanin formation through complexation of a metal with the quinoidal anion (i.e., deprotonated form of quinoidal base, Chapter 16) of an anthocyanin glucoside (Glc). Dashed lines represent delocalized electrons and –H indicates the proton can be attached to an oxygen at position 7, 3′ or 4′. M^{n+} refers to a di- or trivalent metal cation such as Mg^{2+}, Al^{3+}, or Fe^{3+}*

[4] Metallic "taste" is likely the result of retronasal perception of volatile carbonyl compounds arising in the mouth through oxidative degradation of lipids induced by the presence of metals. Such metal-catalyzed lipid oxidation and generation of aroma compounds labeled as "metallic" is well known in the food industry. A further possibility is that metallic taste may be used by some as a descriptor for other tactile or taste sensations where there is no retronasal component, but the semantics of the terms used to describe metallic taste needs further exploration [17].

origin – using spectroscopic techniques or inductively coupled plasma-mass spectrometry (ICP-MS) [5, 8, 21–25]. Multivariate data analysis (chemometrics, Chapter 32) then allows the development of models and classification of wines based on their metal content. Applications of these approaches are discussed further in Chapter 28.

References

1. Sirén, H., Sirén, K., Sirén, J. (2015) Evaluation of organic and inorganic compounds levels of red wines processed from Pinot Noir grapes. *Analytical Chemistry Research*, **3**, 26–36.
2. Cox, R.J., Eitenmiller, R.R., Powers, J.J. (1977) Mineral content of some California wines. *Journal of Food Science*, **42** (3), 849–850.
3. Ough, C.S., Crowell, E.A., Benz, J. (1982) Metal content of California wines. *Journal of Food Science*, **47** (3), 825–828.
4. Aceto, M., Abollino, O., Bruzzoniti, M.C., *et al.* (2002) Determination of metals in wine with atomic spectroscopy (flame-AAS, GF-AAS and ICP-AES); a review. *Food Additives and Contaminants*, **19** (2), 126–133.
5. Aceto, M. (2003) Metals in wine, in *Reviews in food and nutrition toxicity* (eds Preedy V.R. and Watson R.R.), Taylor & Francis, London, pp. 169–203.
6. Pohl, P. (2007) What do metals tell us about wine? *Trends in Analytical Chemistry*, **26** (9), 941–949.
7. Ibanez, J.G., Carreon-Alvarez, A., Barcena-Soto, M., Casillas, N. (2008) Metals in alcoholic beverages: a review of sources, effects, concentrations, removal, speciation, and analysis. *Journal of Food Composition and Analysis*, **21** (8), 672–683.
8. Tariba, B. (2011) Metals in wine – impact on wine quality and health outcomes. *Biological Trace Element Research*, **144** (1–3), 143–156.
9. Viviers, M.Z., Smith, M.E., Wilkes, E., Smith, P. (2013) Effects of five metals on the evolution of hydrogen sulfide, methanethiol, and dimethyl sulfide during anaerobic storage of Chardonnay and Shiraz wines. *Journal of Agricultural and Food Chemistry*, **61** (50), 12385–12396.
10. Zoecklein, B.W., Fugelsang, K.C., Gump, B.H., Nury, F.S. (1995) Metals, cations and anions, in *Wine analysis and production*, Chapman and Hall, New York, pp. 199–208.
11. Bimpilas, A., Tsimogiannis, D., Balta-Brouma, K., *et al.* (2015) Evolution of phenolic compounds and metal content of wine during alcoholic fermentation and storage. *Food Chemistry*, **178**, 164–171.
12. Danilewicz, J.C. (2003) Review of reaction mechanisms of oxygen and proposed intermediate reduction products in wine: central role of iron and copper. *American Journal of Enology and Viticulture*, **54** (2), 73–85.
13. Waterhouse, A.L. and Laurie, V.F. (2006) Oxidation of wine phenolics: a critical evaluation and hypotheses. *American Journal of Enology and Viticulture*, **57** (3), 306–313.
14. Danilewicz, J.C. (2007) Interaction of sulfur dioxide, polyphenols, and oxygen in a wine-model system: central role of iron and copper. *American Journal of Enology and Viticulture*, **58** (1), 53–60.
15. Danilewicz, J.C. (2012) Review of oxidative processes in wine and value of reduction potentials in enology. *American Journal of Enology and Viticulture*, **63** (1), 1–10.
16. Dangles, O. (2012) Antioxidant activity of plant phenols: chemical mechanisms and biological significance. *Current Organic Chemistry*, **16** (6), 692–714.
17. Lawless, H.T., Schlake, S., Smythe, J., *et al.* (2004) Metallic taste and retronasal smell. *Chemical Senses*, **29** (1), 25–33.
18. Breslin, P.A.S. (1996) Interactions among salty, sour and bitter compounds. *Trends in Food Science and Technology*, **7** (12), 390–399.
19. Keast, R.S.J. and Breslin, P.A.S. (2002) An overview of binary taste–taste interactions. *Food Quality and Preference*, **14** (2), 111–124.
20. Brouillard, R., Chassaing, S., Isorez, G., *et al.* (2010) The visible flavonoids or anthocyanins: from research to applications, in *Recent advances in polyphenol research*, Wiley-Blackwell, pp. 1–22.
21. Sauvage, L., Frank, D., Stearne, J., Millikan, M.B. (2002) Trace metal studies of selected white wines: an alternative approach. *Analytica Chimica Acta*, **458** (1), 223–230.

22. Cozzolino, D., Kwiatkowski, M.J., Dambergs, R.G., *et al.* (2008) Analysis of elements in wine using near infrared spectroscopy and partial least squares regression. *Talanta*, **74** (4), 711–716.
23. Martin, A.E., Watling, R.J., Lee, G.S. (2012) The multi-element determination and regional discrimination of Australian wines. *Food Chemistry*, **133** (3), 1081–1089.
24. Geana, I., Iordache, A., Ionete, R., *et al.* (2013) Geographical origin identification of Romanian wines by ICP-MS elemental analysis. *Food Chemistry*, **138** (2–3), 1125–1134.
25. Versari, A., Laurie, V.F., Ricci, A., *et al.* (2014) Progress in authentication, typification and traceability of grapes and wines by chemometric approaches. *Food Research International*, **60**, 2–18.

5

Amines, Amino Acids, and Proteins

5.1 Introduction

Characterized by the presence of a nitrogen atom with a lone pair of electrons, amines are a class of compounds derived from ammonia (NH_3) by replacement of one, two, or three hydrogens to yield primary, secondary, and tertiary amines, respectively. Additional substitution on the amino nitrogen with a hydrocarbyl group yields a quaternary ammonium cation whereas protonation of the amino nitrogen yields an aminium cation. Amines can also be present in molecules containing additional functional groups (e.g. carboxyl, carbonyl, thiol, etc.), and in grapes, the major amine-group-containing species are amino acids and their polymers (e.g., proteins).

5.2 Chemistry of amines

The presence of the lone electron pair results in amines and other amine-containing compounds behaving as weak bases. Analogous to acids (Chapter 3), the predilection of bases to be protonated can be described by an equilibrium constant. It is common to report the pK_a value of the conjugate acid, where a higher pK_a value indicates a weaker conjugate acid and a stronger amine base. Values of pK_a for different amine classes are summarized in Table 5.1.

The nitrogen lone pair of amino acids and many other amines is protonated at wine pH, and thus many well-known food chemistry reactions that involve amine groups acting as nucleophiles are expected to occur very slowly, for example, the reaction of amino acids with sugars as occurs in Maillard reactions. Amines with low pK_a values will also have minimal volatility and thus no odor in wine.

Amino acids and polypeptides have multiple ionizable groups at wine and juice pH. As an example, simple amino acids with neutral hydrocarbon side chains (e.g., leucine, glycine) have two ionizable groups: the carboxylic acid (pK_{a1} = ~2) and the amine (pK_{a2} = 9–10 for the $-NH_3^+$ conjugate acid). Application of the Henderson–Hasselbalch equation (Chapter 3) reveals that the majority of amino acids and proteins will exist as *zwitterions* (both positive and negative charges on the same molecule) in juice or wine.

Understanding Wine Chemistry, First Edition. Andrew L. Waterhouse, Gavin L. Sacks, and David W. Jeffery.

Table 5.1 *Representative structures and acidity constants for weakly basic nitrogenous compounds*

Compound class	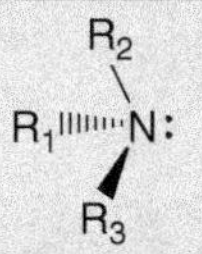 Amine	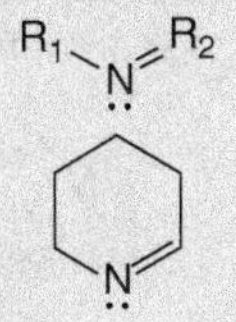 Imine	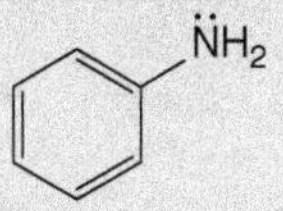 Arylamine	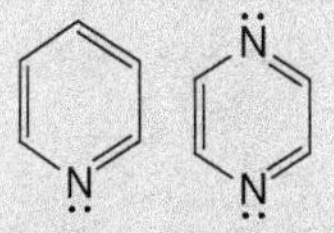 Heteroaromatic amine
Typical pK_a of corresponding weak acid	9–11	5–7	2–6	Pyridine (left): 5 Pyrazine (right): 0.6
Examples in wine	Amine group of α-amino acids	2-Acetyl-3,4,5,6-tetrahydropyridine	Methyl anthranilate	3-Isobutyl-2-methoxypyrazine

The charge of a zwitterionic species at a given pH can be determined from its isoelectric point (pI), giving the following possibilities:

- pH<pI – the protein (or amino acid) has a net *positive* charge.
- pH=pI – the protein has *no* net charge.
- pH>pI – the protein has a net *negative* charge.

The importance of pI to wine and other food systems is that it defines the pH at which a protein will be at minimum solubility.[1] Most wine proteins are reported to have pI values just above wine pH, in the range of 4–6 [1], and will thus have a net positive charge in wine or juice. As a result, proteins can non-covalently bond to species that have formal negative charges or are H-bond acceptors, such as bentonite and polyphenols. These properties will be discussed more in Chapter 26.2.

5.3 Amino acids and related major nitrogenous compounds in wines

The major nitrogenous constituents of grapes and wines are summarized in Table 5.2, and include [2]:

- Ammonium (NH_4^+), which represents a major source of usable nitrogen during fermentation.
- Amino acids, characterized by the presence of both a carboxylic acid group (R–COOH) and an amine group (R–NH_2 or R–NH–R′).
- Proteins, which are polymers of amino acids (i.e., large polypeptides) linked by amide groups (R–CONH–R′), and usually containing >100 amino acid monomers.
- Oligopeptides, typically defined as polymers containing 2 to 20 amino acid monomers, again joined through amide bonds.

5.3.1 Amino acids and ammonia

In must, the major soluble nitrogen forms exist as ammonium and free amino acids. Yeast require nitrogen for several roles – primarily as a component of proteins, cell walls, and nucleic acids[2] – and ammonium and free

[1] The first step of most cheesemaking processes is to form a casein protein precipitate ("curds") in milk either by decreasing the milk pH through acidification or by increasing the pI of the casein by the action of a protease. Both approaches bring the pI of casein closer to the pH of the milk.

[2] Inspection of the nutritional data on a bread yeast packet reveals that yeast is about 50% protein by dry weight. Approximately one-sixth of the weight of protein is due to nitrogen.

Table 5.2 *Concentrations of major nitrogenous species in must and wine*

Nitrogenous compounds	Concentration in grape must [3–5]			Concentration in wine (mg/L)
	mg/L	mg/L as N	Contribute to yeast assimilable nitrogen (YAN)?	
Ammonium	100 ± 45	79 ± 35	Yes	Lower than in must
α-Amino acids (non-proline)	843 ± 51	135 ± 51	Yes	
Proline	Up to 4000	Up to 500	No	Similar to must concentrations
Glutathione (major oligopeptide)	15–100	3–15	Yes	ND – 27 [5]
Proteins	20–250	3–15	No	30-275 [4,6]

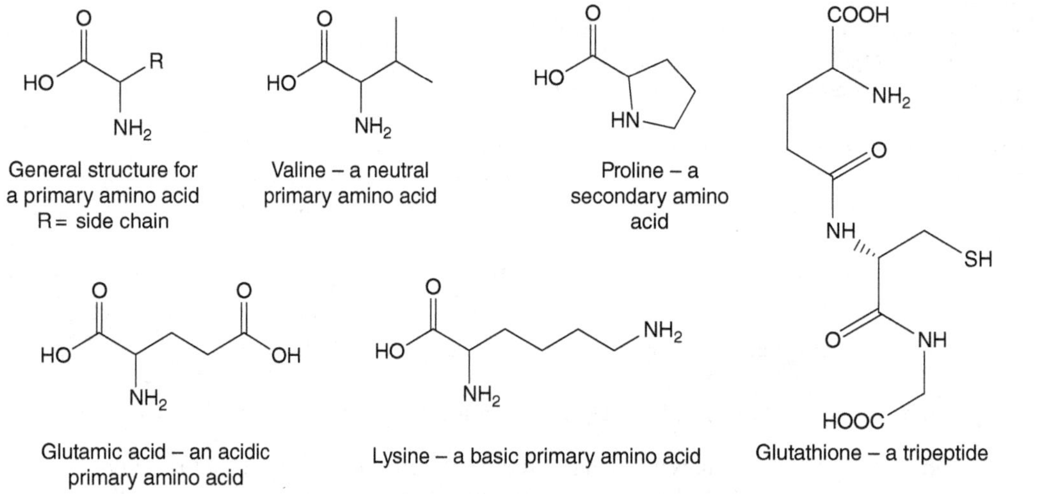

Figure 5.1 *Structures of major nitrogenous species in must: proline, primary α-amino acids, and glutathione (tripeptide)*

amino acids serve as the main nitrogen source during alcoholic fermentation. The structures of several representative amino acids in grapes are shown in Figure 5.1. The majority of amino acids in grapes and in wines are primary α-amino acids; that is, the amine group is bonded to only one carbon ($R–NH_2$), and the acid and amine groups form bonds to the same carbon. Also, most of the amino acids in grapes and wines are proteinogenic, that is, they have an associated codon that utilizes them in the synthesis of proteins during transcription. A few non-proteinogenic amino acids can also be found in must, especially γ-aminobutyric acid (GABA), which can exist at concentrations up to 580 mg/L [7].[3]

The predominant amino acid in musts is proline, which can exist at concentrations up to 4000 mg/L, followed by arginine, valine, and alanine [2, 8]. Not all forms of amino acids are equally useful to yeast as a

[3] GABA can be metabolized to γ-butyrolactone (GBL), discussed later in a Chapter 7 footnote for its role as a "date-rape" drug. GABA concentrations naturally found in wine are typically much less than concentrations used to adulterate drinks (2000 mg/L or more).

nitrogen source, and in particular the secondary amino acids (proline and hydroxyproline) are not well used under anaerobic conditions. The "yeast assimilable nitrogen" (YAN) fraction is defined as [3]

$$\text{YAN}(\text{mg/L as N}) = \text{Ammonium}(\text{mg/L as N}) + \text{Primary amino acids}(\text{mg/L as N}) \quad (5.1)$$

The importance of YAN to nitrogen metabolism is described in more detail later (Chapter 22.3). The major factors affecting must ammonium, amino acid, and YAN concentrations have been reviewed [2] and include cultivar and growing conditions, as well as pre-fermentation additions of diammonium hydrogen phosphate (DAP) or other supplements.

Concentrations of ammonium and α-amino acids in finished wines are typically lower than in must due to utilization by yeast [9]. A survey of 128 commercial wines found YAN concentrations ranging from 11–586 mg/L as N. Higher concentrations of YAN in finished wines may result from excessive must nitrogen supplementation [10] and are generally discouraged since they can promote microbial instability [11]. In wine, most amino acids exist at concentrations 1–2 orders of magnitude below their taste thresholds [12]. The only amino acids that approach their thresholds in model wine are proline ("sweet") and glutamate ("umami"), but reconstitution studies with model wines showed that omitting all amino acids at usual wine concentrations had no effect on wine flavor [12]. Interestingly, proline in wines is reported to correlate with perception of "body" in dry white wines [13], although this may be because it serves as a maturity marker.

5.3.2 Oligopeptides

The best-studied oligopeptide in grapes and wines is glutathione (GSH, Figure 5.1). GSH is a tripeptide formed from glycine, cysteine, and glutamine. GSH is produced by a wide range of microorganisms, plants, and animals, where it has important roles in preventing oxidative damage and in the metabolism of toxic compounds.

Factors affecting GSH in grapes and wines have been reviewed thoroughly [14]. GSH present in grapes (Table 5.2) can be utilized as a nitrogen source by yeast during fermentation. However, yeast will also synthesize considerable amounts of GSH during fermentation – the intracellular GSH content of *S. cerevisiae* can be up to 1% dry weight [15] – and some will be excreted. Although it has minor importance as a nitrogen source, GSH has a key role as a nucleophile in reaction with *o*-quinones to inhibit browning and other oxidative reactions (Chapters 13 and 24), and with (*E*)-2-hexenal or other unsaturated aldehydes to form *S*-glutathione conjugates (Chapters 10 and 23.2).

Yeast autolysis is known to release other oligopeptides into wine, but, historically, their contribution has not been well studied. Recently, an oligopeptide ($MW < 3\,kDa$) formed by degradation of a heat-shock protein during yeast autolysis was shown to have a sweet taste [16]. This observation may help explain why wines aged on lees are often described as being less astringent or acidic ("softer") even though they have subthreshold concentrations of sugars.

5.3.3 Proteins

The protein concentration of white musts and wines is reportedly 20–250 mg/L [4] and 30–275 mg/L [4, 6], respectively, although these values are presented cautiously because common methods for protein analysis in wine can suffer from interferences [1]. Wine peptides with $MW > 3\,kDa$, which includes all proteins, are reported to have no flavor at concentrations found in wine [16]. However, the major soluble proteins are heat-unstable and can denature to cause haziness. The two major classes of grape proteins with molecular weights in the range of 21–32 kDa have been implicated in white wine haze [17], and are discussed in more detail elsewhere (Chapter 26.2).

- Chitinases, glucanases, and other pathogenesis-related (PR) proteins, which are increased in response to disease pressure.
- Thaumatin-like proteins, which are associated with grape ripening and berry softening.

S. cerevisiae lacks strong proteolytic activity and thus grape proteins are not used as a nitrogen source. However, most proteins are poorly extracted during fermentation, and likely are bound to the pomace and lees. Protein concentrations in white wines will be further decreased by the use of fining agents, particularly bentonite (Chapter 26.2). Yeast-derived proteins are not major contributors to the protein concentration of white wines [4]. However, mannoproteins (typically at concentrations of 100–150 mg/L, 30% protein) may be minor contributors to protein content, especially when added exogenously to inhibit potassium bitartrate crystal formation (Chapter 26.1).

Grape and yeast-derived proteins in red wines are not as well studied as in white wines, in part because of the greater economic importance of protein haze in white wine and in part because common colorimetric methods for protein determination require modification for use in red wines. The protein concentrations of red wines are reported to range from 50 to 100 mg/L [18]. Proteins are well known to bind to grape or wine tannins (Chapter 14), a property exploited by winemakers to decrease tannins during cellar operations (Chapter 26.2). The lower concentration of protein in red wines likely arises from binding to tannin during maceration. Conversely, very high concentrations of proteins in grapes have been correlated with lower concentrations of tannins in finished wines [19].

5.4 Nitrogenous compounds with health effects

5.4.1 Biogenic amines

In fermented products, biogenic amines describe those compounds produced by decarboxylation of amino acids. Commercial and wild yeasts have low ability to form biogenic amines, and their primary source in wines appears to be lactic acid bacteria [20], particularly some strains of the spoilage species *Pedioccocus* (Chapter 22.5). The major biogenic amines (histamine, tyramine, putrescine) in red wines are shown in Table 5.3. Total concentrations in red wines are usually <50 mg/L, about an order of magnitude less than concentrations observed in sauerkraut and many other lactic fermented foods. Because malolactic fermentation is less common in white wines, concentrations of biogenic amines in whites are typically lower than in reds (<4 mg/L total) [21]. The primary strategies to prevent biogenic amine formation have been reviewed elsewhere [22], and include:

- Using starter bacterial cultures without amino acid decarboxylases
- Preventing the growth of spoilage bacteria
- Decreasing the concentrations of amino acid precursors, that is, by avoiding excess nitrogen additions during fermentation (Chapter 22.3).

Biogenic amines, particularly histamine, have well-known adverse health effects and can lead to headaches, heart palpitations, diarrhea, and other ill-effects at high concentrations [21]. They have been implicated as a potential cause of the "Red Wine Headache" phenomenon, and the European Union has recommended histamine limits of 10 mg/L, although the correlation between histamine and headaches is still not well established [23].

Finally, indole is not a classic biogenic amine, but does appear to be formed as a result of microbial metabolism of an amino acid (tryptophan). Unlike other biogenic amines, indole is a heterocyclic amine and will be predominantly in its volatile (and odorous) form at wine pH. Indole is detectable in most wines at concentrations of 1–10 μg/L [24]. Higher concentrations, up to 350 μg/L [24], appear to arise during sluggish fermentations, resulting in an off-flavor described as "plastic" at concentrations in excess of 30 μg/L [25].

Table 5.3 Major biogenic amines in wines

Biogenic amine[a] (*corresponding amino acid*)	Structure	Representative concentrations [24, 26] Mean (*range*), mg/L
Putrescine (*Ornithine*)	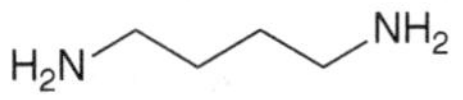	19.4 (*2.9–122*)
Tyramine (*Tyrosine*)	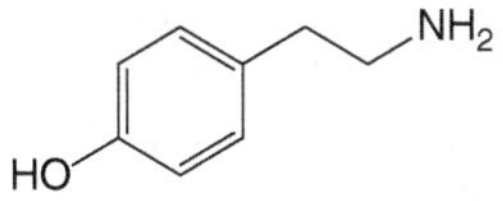	3.5 (*1.1–10.7*)
Histamine (*Histidine*)	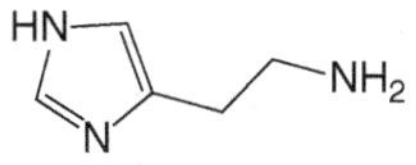	7.2 (*0.5–26.9*)
Indole[b] (*Tryptophan*)	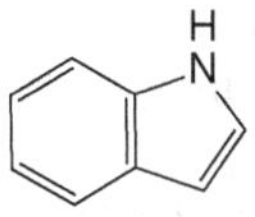	Sound wine: 1–10 µg/L Faulty wine: up to 350 µg/L

[a] Common name.
[b] Indole is not a classic biogenic amine in that it is not formed by decarboxylation of an amino acid. However, it does appear to be produced from tryptophan by microorganisms during fermentation through an unknown pathway.

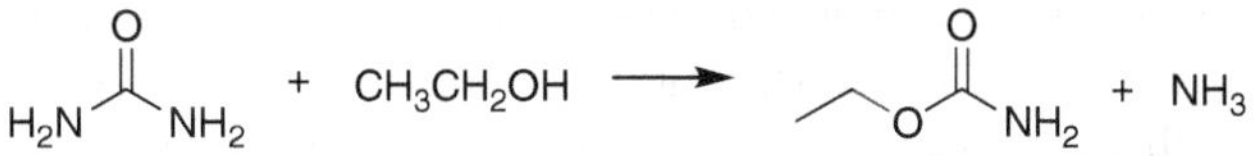

Figure 5.2 Reaction of urea with ethanol (left) to yield ethyl carbamate and ammonia (right)

5.4.2 Ethyl carbamate

Ethyl carbamate (EC) is a known carcinogen and is present in detectable amounts in many fermented products [27]. In Canada, EC must be below 30 µg/L in table wines and 100 µg/L in dessert wines, while in the US a voluntary industry target of 15 µg/L has been established. Two key pathways have been identified for EC formation in wine: (i) a major pathway involving yeast catabolism of arginine to release urea and (ii) a minor amount formed from lactic acid bacteria degradation of arginine to citrulline [2]. Both urea and citrulline can undergo non-enzymatic ethanolysis to yield EC (Figure 5.2). This conversion of EC from these precursors is non-quantitative (<10%) and highly temperature dependent (Chapter 25). In a survey of 27 wines, increasing storage temperature from 20 °C to 40 °C resulted in a 20–40-fold increase in the EC formation rate [28]. Additionally, factors that increase urea, citrulline, or their precursor (arginine) such as excessive vineyard fertilization will potentially increase EC in wine [2]. Several strategies for decreasing EC have been proposed, including enzymatic degradation of EC or urea, and the use of sorbents [29].

5.5 Odor-active amines

With the exception of indole, the amines listed above contribute negligibly to wine aroma because their nitrogen lone pair will be protonated at wine pH, rendering the amine non-volatile. Two classes of grape-derived amines have pK_a values < 3 either due to inductive effects (methoxypyrazines) or resonance (aniline derivatives), and are thus volatile at wine pH. Both of these compound classes are *primary odorants* (see Introduction chapter), and will contribute to *varietal aromas*. Another compound class, the microbially produced cyclic imines, are not volatile at wine pH (pK_a = 5–7), but can deprotonate at mouth pH to become odor-active.

5.5.1 Varietal amines – methoxypyrazines and "foxy" aniline derivatives

The 3-alkyl-2-methoxypyrazines, also called methoxypyrazines, pyrazines, or MPs, possess odors generally described as "vegetal" or "earthy" (Table 5.4). These compounds have some of the lowest odor thresholds (~1 ng/L) of any compound found in wine. The MPs most often reported in grapes above their sensory thresholds are 3-isobutyl-2-methoxypyrazine (IBMP, "bell pepper") and 3-isopropyl-2-methoxypyrazine (IPMP, "peas"). Other MPs, including 3-*sec*-butyl-2-methoxypyrazine (sBMP), 3-ethyl-2-methoxypyrazine (EMP), and 2,5-dimethyl-3-methoxypyrazine (DMMP) are also reported in grapes and wines, but usually at concentrations below their respective odor thresholds [30,31]. In addition to grapes, IPMP and several other MPs in wines can also derive from contamination of ferments by insects like the multicolored Asian ladybeetle (MALB, *Harmonia axyridis*) or the 7-spot ladybeetle (*Coccinella septempunctata*) [32]. Finally, 2-methoxy-3,5-dimethylpyrazine ("musty, fungal"), has been identified in tainted corks, and will be discussed in more detail in Chapter 18.

While MPs may add complexity or typicity to some wine styles, for example, Sauvignon Blanc, winemakers are often interested in avoiding excess MP concentrations to avoid masking fruity aromas, especially in red wines [39].

The concentration of MPs in grape berries is well-known to be dependent on genotype, with suprathreshold MP concentrations most often observed in the so-called "Bordeaux" varieties like Cabernet Sauvignon and Sauvignon Blanc. MPs reach a maximum concentration about 1–2 weeks prior to veraison and degrade during maturation. Generally, lower MP concentrations at harvest are correlated with warmer, drier, and longer growing seasons, well-exposed clusters pre-veraison, and less vigorous sites [40]. Within grapes, MPs are located mostly in the skins and stems, and are readily extracted during skin fermentation [41]. MPs appear to be stable post-fermentation, as wine age is not correlated with MP concentration [42], and widely used fining agents cannot selectively remove MPs [43]. MPs are expected to be less reactive than the major aromatic ring-containing compounds in wines – the polyphenols – because the ring nitrogens of MPs are more electronegative and will make the ring less susceptible to electrophilic addition (Chapter 11).

Two aniline derivatives, methyl anthranilate (MA, "artificial grape") and *o*-aminoacetophenone (*o*-AAP, "acacia, mothball") are found in supra-threshold concentrations in large-berried American grape species and their offspring, for example, *V. labruscana* cultivars like Concord and Niagara (Table 5.4). MA and *o*-AAP are critical compounds for the so-called "foxy" aroma of *labruscana* and related American grape species,[4] MA can be below the sensory threshold in certain foxy-smelling grapes (e.g., Catawba, *V. rotundifolia*) [47], with more recent work implicating *o*-AAP as the key foxy-smelling compound [45].[5] Other non-nitrogenous compounds such as furaneol and methylfuraneol ("strawberry, caramel") may contribute to the native character of American grapes [49], although these compounds are more often observed as sugar degradation products of toasted oak (Chapter 25). MA and *o*-AAP are detectable in *vinifera* wines but usually at

[4] The "foxy" descriptor was applied to large-berried wild American grapes like *V. labrusca* and *V. rotundifolia* (Muscadine) as early as the 1620s by European settlers [44]. The most probable explanation is that Europeans associated the aroma of these grape species with the musky aromas of the European foxes. This hypothesis is given more credibility by the fact that *o*-AAP is present in the anal sac of carnivores like the Japanese weasel [45], and MA and *o*-AAP are both bird deterrents [46]. Several other characteristics (large berries, thick skins, downward growth, thick pulp, dull color) support the hypothesis that large-berried grapes were not selected for by birds, but instead by small mammals like raccoons.

[5] In 1964, the discovery of *o*-AAP as a key off-flavor in stale dry milk was sufficiently exciting to justify publication in the prestigious journal *Nature* [48].

Table 5.4 *Summary of properties for key varietal amine odorants in wines*

Name	Structure	Typical range in Bordeaux-grape varietals [33–36]	Threshold in wine [33, 37, 38]	Odor descriptor
3-Isopropyl-2-methoxypyrazine (IPMP)	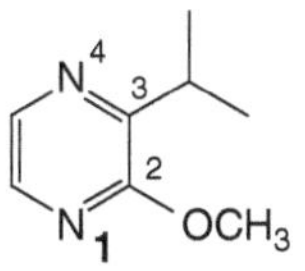	<0.5–5.6 ng/L	0.3–2 ng/L	Asparagus, earth, peas
3-Isobutyl-2-methoxypyrazine (IBMP)	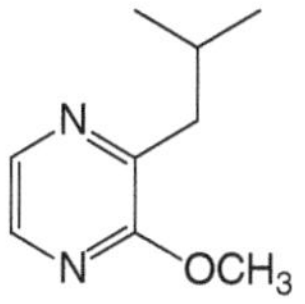	4–30 ng/L	2 ng/L (detection) 8–16 ng/L (recognition)	Bell pepper, vegetal
Methyl anthranilate (MA)	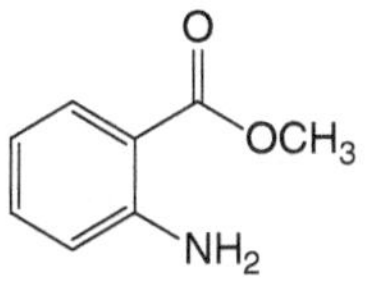	600–3000 μg/L (*labruscana*) 0.06–0.6 μg/L (*vinifera*)	300 μg/L	Artificial grape
ortho-Aminoacetophenone (*o*-AAP)	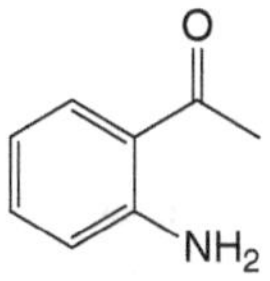	8–12 μg/L (*labruscana*) <0.5 μg/L (*vinifera*) 0.8–13 μg/L (ATA)	0.5 μg/L	Corn tortilla, mothball, acacia

subthreshold concentrations. One exception to this statement is wines afflicted by so-called atypical aging (ATA), referred to as UTA in German (*Untypische Alterungsnote*) [50]. Early work on ATA/UTA suggested that drought stress and nitrogen deficiency induced formation of indole-3-acetic acid (IAA) in grapes, and studies on model solutions suggested an oxidative pathway to form *o*-AAP from IAA during storage, although more recent investigations have been more equivocal about these hypotheses [50].

MA is highly stable – supratheshold concentrations (2–3 mg/L) can be observed in 5 year old Concord and Niagara wines [47], comparable to concentrations observed in Concord grapes and young Concord wines. Based on thermodynamic predictions, methyl anthranilate should be lost and anthranilic acid and ethyl anthranilate formed during storage via acid-catalyzed hydrolysis and ethanolysis (Chapter 25). The slow kinetics of MA solvolysis as compared to aliphatic esters may be due either to resonance stabilization of the carbonyl group (so the carbon has less partial positive charge) or because of steric interference (from the *ortho*-substituent blocking access to the carboxyl group) [51].

5.5.2 "Mousy" imines

Several cyclic imines, shown in Table 5.5, have been detected in wines: 2-ethyl-3,4,5,6-tetrahydropyridine (ETHP), 2-acetyl-3,4,5,6-tetrahydropyridine (ATHP), and 2-acetylpyrroline (APY). Although non-volatile in wine, these compounds can deprotonate and become volatile in the mouth, resulting in "mousy," "cardboard," or "cracker" retronasal aromas. These compounds had been previously identified in baked and roasted products like bread crust, where they can contribute positively. They are formed from polynitrogenous amino acids (e.g., ornithine, lysine) by spoilage organisms, and will be discussed in more detail later (Chapter 22.5).

Table 5.5 "Mousy" cyclic imines in wine

Mousy-smelling compound	Structure	Retronasal threshold (μg/L) [52]	Maximum reported concentration (μg/L) [52]
2-Ethyl-3,4,5,6-tetrahydropyridine (ETHP)	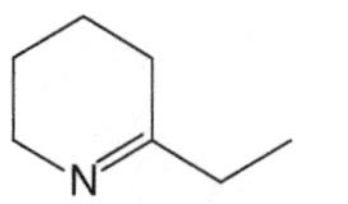	150	162
2-Acetyl-3,4,5,6-tetrahydropyridine (ATHP)	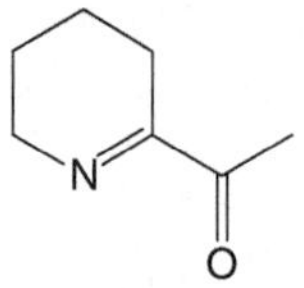	16	108
2-Acetylpyrroline (APY)	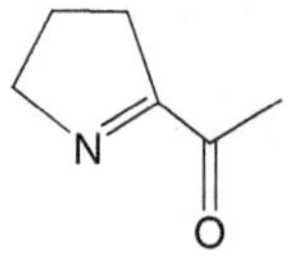	0.1	7.8

References

1. Marchal, R. and Waters, E.J. (2011) New directions in stabilization, clarification, and fining of white wines, in *Managing wine quality*, Vol. **2**, Oenology and wine quality (ed. Reynolds, A.G.), Woodhead Publishing and CRC Press, Oxford and Boca Raton, FL.
2. Bell, S.J. and Henschke, P.A. (2005) Implications of nitrogen nutrition for grapes, fermentation and wine. *Australian Journal of Grape and Wine Research*, **11** (3), 242–295.
3. Dukes, B.C. and Butzke, C.E. (1998) Rapid determination of primary amino acids in grape juice using an *o*-phthaldialdehyde/*n*-acetyl-l-cysteine spectrophotometric assay. *American Journal of Enology and Viticulture*, **49** (2), 125–134.
4. Bayly, F.C. and Berg, H.W. (1967) Grape and wine proteins of white wine varietals. *American Journal of Enology and Viticulture*, **18** (1), 18–32.
5. Fracassetti, D., Lawrence, N., Tredoux, A.G.J., *et al.* (2011) Quantification of glutathione, catechin and caffeic acid in grape juice and wine by a novel ultra-performance liquid chromatography method. *Food Chemistry*, **128** (4), 1136–1142.
6. Santoro, M. (1995) Fractionation and characterization of must and wine proteins. *American Journal of Enology and Viticulture*, **46** (2), 250–254.
7. Wurz, R.E.M., Kepner, R.E., Webb, A.D. (1988) The biosynthesis of certain gamma-lactones from glutamic acid by film yeast activity on the surface of flor sherry. *American Journal of Enology and Viticulture*, **39** (3), 234–238.
8. Spayd, S.E. and Andersen-Bagge, J. (1996) Free amino acid composition of grape juice from 12 *Vitis vinifera* cultivars in Washington. *American Journal of Enology and Viticulture*, **47** (4), 389–402.
9. Huang, Z. and Ough, C.S. (1991) Amino-acid profiles of commercial grape juices and wines. *American Journal of Enology and Viticulture*, **42** (3), 261–267.
10. Jiranek, V., Langridge, P., Henschke, P.A. (1995) Amino acid and ammonium utilization by *Saccharomyces cerevisiae* wine yeasts from a chemically defined medium. *American Journal of Enology and Viticulture*, **46** (1), 75–83.
11. Fugelsang, K.C. and Edwards, C.G. (2007) *Wine microbiology practical applications and procedures*, Springer, New York.
12. Hufnagel, J.C. and Hofmann, T. (2008) Quantitative reconstruction of the nonvolatile sensometabolome of a red wine. *Journal of Agricultural and Food Chemistry*, **56** (19), 9190–9199.

13. Skogerson, K., Runnebaum, R., Wohlgemuth, G., *et al.* (2009) Comparison of gas chromatography-coupled time-of-flight mass spectrometry and ^{1}H nuclear magnetic resonance spectroscopy metabolite identification in white wines from a sensory study investigating wine body. *Journal of Agricultural and Food Chemistry*, **57** (15), 6899–6907.
14. Kritzinger, E.C., Bauer, F.F., du Toit, W.J. (2012) Role of glutathione in winemaking: a review. *Journal of Agricultural and Food Chemistry*, **61** (2), 269–277.
15. Bachhawat, A., Ganguli, D., Kaur, J., *et al.* (2009) Glutathione production in yeast, in *Yeast biotechnology: diversity and applications* (eds Satyanarayana, T. and Kunze, G.), Springer, Netherlands, pp. 259–280.
16. Marchal, A., Marullo, P., Moine, V., Dubourdieu, D. (2011) Influence of yeast macromolecules on sweetness in dry wines: role of the *Saccharomyces cerevisiae* protein Hsp12. *Journal of Agricultural and Food Chemistry*, **59** (5), 2004–2010.
17. Waters, E.J., Shirley, N.J., Williams, P.J. (1996) Nuisance proteins of wine are grape pathogenesis-related proteins. *Journal of Agricultural and Food Chemistry*, **44** (1), 3–5.
18. Smith, M.R., Penner, M.H., Bennett, S.E., Bakalinsky, A.T. (2011) Quantitative colorimetric assay for total protein applied to the red wine Pinot Noir. *Journal of Agricultural and Food Chemistry*, **59** (13), 6871–6876.
19. Springer, L.F. and Sacks, G.L. (2014) Limits on red wine tannin extraction and additions: the role of pathogenesis-related proteins. *65th ASEV National Conference*, **62** (3), 390A–391A.
20. Landete, J.M., Ferrer, S., Pardo, I. (2007) Biogenic amine production by lactic acid bacteria, acetic bacteria and yeast isolated from wine. *Food Control*, **18** (12), 1569–1574.
21. Peña-Gallego, A., Hernández-Orte, P., Cacho, J., Ferreira, V. (2011) High-performance liquid chromatography analysis of amines in must and wine: a review. *Food Reviews International*, **28** (1), 71–96.
22. Guo, Y.-Y., Yang, Y.-P., Peng, Q., Han, Y. (2015) Biogenic amines in wine: a review. *International Journal of Food Science and Technology*, **50** (7), 1523–1532.
23. Kanny, G., Gerbaux, V., Olszewski, A., *et al.* (2001) No correlation between wine intolerance and histamine content of wine. *Journal of Allergy and Clinical Immunology*, **107** (2), 375–378.
24. Capone, D.L., Van Leeuwen, K.A., Pardon, K.H., *et al.* (2010) Identification and analysis of 2-chloro-6-methylphenol, 2,6-dichlorophenol and indole: causes of taints and off-flavours in wines. *Australian Journal of Grape and Wine Research*, **16** (1), 210–217.
25. Arevalo-Villena, M., Bartowsky, E.J., Capone, D., Sefton, M.A. (2010) Production of indole by wine-associated microorganisms under oenological conditions. *Food Microbiology*, **27** (5), 685–690.
26. Konakovsky, V., Focke, M., Hoffmann-Sommergruber, K., *et al.* (2011) Levels of histamine and other biogenic amines in high-quality red wines. *Food Additives and Contaminants: Part A*, **28** (4), 408–416.
27. Ough, C.S. (1976) Ethyl carbamate in fermented beverages and foods. 1. Naturally occuring ethyl carbamate. *Journal of Agricultural and Food Chemistry*, **24** (2), 323–328.
28. Kodama, S., Suzuki, T., Fujinawa, S., *et al.* (1994) Urea contribution to ethyl carbamate formation in commercial wines during storage. *American Journal of Enology and Viticulture*, **45** (1), 17–24.
29. Zhao, X., Du, G., Zou, H., *et al.* (2013) Progress in preventing the accumulation of ethyl carbamate in alcoholic beverages. *Trends in Food Science and Technology*, **32** (2), 97–107.
30. Lacey, M.J., Allen, M.S., Harris, R.L.N., Brown, W.V. (1991) Methoxypyrazines in Sauvignon Blanc grapes and wines. *American Journal of Enology and Viticulture*, **42** (2), 103–108.
31. Botezatu, A. and Pickering, G.J. (2012) Determination of ortho- and retronasal detection thresholds and odor impact of 2,5-dimethyl-3-methoxypyrazine in wine. *Journal of Food Science*, **77** (11), S394–S398.
32. Botezatu, A.I., Kotseridis, Y., Inglis, D., Pickering, G.J. (2013) Occurrence and contribution of alkyl methoxypyrazines in wine tainted by *Harmonia axyridis* and *Coccinella septempunctata*. *Journal of the Science of Food and Agriculture*, **93** (4), 803–810.
33. Roujou de Boubee, D., Van Leeuwen, C., Dubourdieu, D. (2000) Organoleptic impact of 2-methoxy-3-isobutylpyrazine on red Bordeaux and Loire wines. Effect of environmental conditions on concentrations in grapes during ripening. *Journal of Agricultural and Food Chemistry*, **48** (10), 4830–4834.
34. Aubry, V., Etievant, P.X., Ginies, C., Henry, R. (1997) Quantitative determination of potent flavor compounds in Burgundy Pinot noir wines using a stable isotope dilution assay. *Journal of Agricultural and Food Chemistry*, **45** (6), 2120–2123.

35. Panighel, A., Dalla Vedova, A., De Rosso, M., *et al.* (2010) A solid-phase microextraction gas chromatography/ion trap tandem mass spectrometry method for simultaneous determination of "foxy smelling compounds" and 3-alkyl-2-methoxypyrazines in grape juice. *Rapid Communications in Mass Spectrometry*, **24** (14), 2023–2029.
36. Dollmann, B., Wichmann, D., Schmitt, A., *et al.* (1996) Quantitative analysis of 2-aminoacetophenone in off-flavored wines by stable isotope dilution assay. *Journal of AOAC International*, **79** (2), 583–586.
37. Allen, M.S., Lacey, M.J., Harris, R.L.N., Brown, W.V. (1991) Contribution of methoxypyrazines to Sauvignon Blanc wine aroma. *American Journal of Enology and Viticulture*, **42** (2), 109–112.
38. Pickering, G.J., Ker, K., Soleas, G.J. (2007) Determination of the critical stages of processing and tolerance limits for *Harmonia axyridis* for "ladybug taint" in wine. *Vitis*, **46** (2), 85–90.
39. Hein, K., Ebeler, S.E., Heymann, H. (2009) Perception of fruity and vegetative aromas in red wine. *Journal of Sensory Studies*, **24** (3), 441–455.
40. Scheiner, J.J., Vanden Heuvel, J.E., Sacks, G.L. (2009) How viticultural factors affect methoxypyrazines. *Wines and Vines*, **Nov**, 113–117.
41. Ryona, I., Pan, B.S., Sacks, G.L. (2009) Rapid measurement of 3-alkyl-2-methoxypyrazine content of winegrapes to predict levels in resultant wines. *Journal of Agricultural and Food Chemistry*, **57** (18), 8250–8257.
42. Alberts, P., Stander, M.A., Paul, S.O., de Villiers, A. (2009) Survey of 3-alkyl-2-methoxypyrazine content of South African Sauvignon blanc wines using a novel LC-APCI-MS/MS method. *Journal of Agricultural and Food Chemistry*, **57** (20), 9347–9355.
43. Pickering, G., Lin, J., Reynolds, A., *et al.* (2006) The evaluation of remedial treatments for wine affected by *Harmonia axyridis*. *International Journal of Food Science and Technology*, **41** (1), 77–86.
44. Pinney, T. (1989) *A history of wine in America: from the beginnings to prohibition*, University of California Press, Berkeley, CA.
45. Acree, T.E., Lavin, E.H., Nishida, R., Watanabe, S. (1990) *O*-amino acetophenone as the foxy smelling component of *labruscana* grapes, in *Flavor science and technology, 6th Weurman Symposium*, Geneva, Switzerland, pp. 49–52.
46. Mason, J.R., Clark, L., Shah, P.S. (1991) Ortho-aminoacetophenone repellency to birds: similarities to methyl anthranilate. *The Journal of Wildlife Management*, **55** (2), 334–340.
47. Nelson, R.R., Acree, T.E., Lee, C.Y., Butts, R.M. (1977) Methyl anthranilate as an aroma constituent of American wine. *Journal of Food Science*, **42** (1), 57–59.
48. Parks, O.W., Schwartz, D.P., Keeney, M. (1964) Identification of *o*-aminoacetophenone as a flavour compound in stale dry milk. *Nature*, **202** (4928), 185–187.
49. Rapp, A. (1998) Volatile flavour of wine: correlation between instrumental analysis and sensory perception. *Nahrung-Food*, **42** (6), 351–363.
50. Linsenmeier, A., Rauhut, D., Sponholz, W.R. (2010) Ageing and flavour deterioration in wine, in *Managing wine quality*, Vol. **2**, Oenology and wine quality (ed. Reynolds, A.G.), Woodhead Publishing and CRC Press, Oxford and Boca Raton, FL.
51. Chapman, N.B., Shorter, J., Utley, J.H.P. (1963) The separation of polar and steric effects. Part II. The basic and the acidic hydrolysis of substituted methyl benzoates. *Journal of the Chemical Society*, **241**, 1291–1299.
52. Snowdon, E.M., Bowyer, M.C., Grbin, P.R., Bowyer, P.K. (2006) Mousy off-flavor: a review. *Journal of Agricultural and Food Chemistry*, **54** (18), 6465–6474.

6
Higher Alcohols

6.1 Introduction

The term "higher alcohols" refers to volatile alcohols with more than two carbon atoms.[1] These compounds are produced as a byproduct of yeast amino acid metabolism and are common to all products of alcoholic fermentation (e.g., beer, wine, cider, etc.). The higher alcohol category excludes polyols like glycerol and other sugar alcohols (Chapter 2), volatile phenols (Chapter 12), and grape-derived varietal aroma compounds like monoterpene alcohols (Chapter 8). In addition to fermentation-derived higher alcohols, this chapter will also discuss grape-derived six carbon alcohols (e.g., 1-hexanol) and methanol since they share some chemical properties with the fermentation-derived higher alcohols.

6.2 Properties of higher alcohols

The higher alcohols are amphiphilic, possessing a non-polar hydrocarbon group and a polar (and H-bonding) –OH group. Most of the higher alcohols discussed in this chapter are moderately non-polar, have low volatility, and, except for the fully water-miscible *n*-propanol, are only somewhat soluble in water.

Higher alcohols are not strongly reactive under wine conditions, although the alcohol group is a weak nucleophile at wine pH. They represent the most reduced form of oxygen-containing compounds and are generally stable in the reducing environment of wine. The most important reactions that the higher alcohols participate in are as follows:

- Esterification. Higher alcohols can combine with carboxylic acids to form esters (Chapter 7). During fermentation, this can result in suprathreshold concentrations of acetate esters as a result of acetyltransferase

[1] Higher alcohols are sometimes referred to as "fusel alcohols." Fusel is German for "bad liquor," likely derived from *fuseln* v. "to bungle." Historically, the "fusel oil" or "tails" fractions – that is, the volatiles recovered after distilling the majority of ethanol – received considerable attention from chemists dating back to the nineteenth century. This fraction is ~50% fusel alcohols, and was of economic importance because their presence decreased the price of spirits. In 1864 Henry Watts stated [1] that over half of the section on fusel oils is dedicated to a discussion of "defuselization" of spirits, not only with the familiar charcoal but also milk, soap, bleach, and other oddities.

Understanding Wine Chemistry, First Edition. Andrew L. Waterhouse, Gavin L. Sacks, and David W. Jeffery.

enzymes (Chapter 22.3). Esterification of higher alcohols can occur non-enzymatically during storage through an acid-catalyzed mechanism but the concentrations of resulting esters formed are typically well below threshold and not of clear sensory importance.

- Substrate for oxidation. Higher alcohols can be oxidized to form their corresponding aldehydes, many of which are contributors to "oxidized aromas." The mechanism for this reaction is discussed later (Chapter 24).

6.3 Origins and concentrations of higher alcohols

The majority of the higher alcohols are formed as a byproduct of yeast amino acid metabolism. Amino acids are formed and degraded via α-keto acid carbon skeletons. Under certain conditions, such as the case of nitrogen limitation, these skeletons will be decarboxylated and reduced to form higher alcohols. This process may be anabolic, in which the carbon skeleton is biosynthesized by yeast from sugars, or catabolic, in which an existing amino acid is broken down. This will be discussed in more detail later (Chapter 22.3).

Representative concentrations of the several higher alcohols, which are largely absent from grapes and are almost exclusively formed during fermentation, appear in Table 6.1. Most of the higher alcohols have a clear structural relationship with a specific amino acid.

Table 6.1 Summary of characteristics of major fermentation-derived higher alcohols

Name/structure	Related amino acid	Typical range (mg/L)	Threshold in model wine (mg/L)[2]	Odor descriptor
2-Methyl-1-propanol (isobutanol) 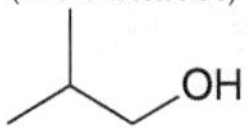	Valine	25–87[a]	40	Solvent
2-Methyl-1-butanol (active amyl alcohol)[c,d] 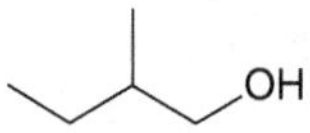	Isoleucine	16–31[b]	1.2 [3][e]	Solvent, fusel
3-Methyl-1-butanol (isoamyl alcohol)[c] 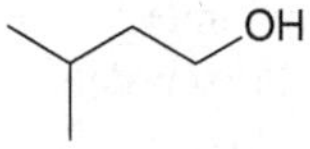	Leucine	84–333[b]	30	Solvent, fusel
3-Methylsulfanyl-1-propanol (methionol) 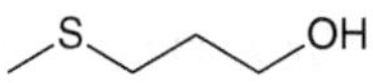	Methionine	0.16–2.4[a]	1	Boiled potato
2-Phenylethanol (β-phenylethanol) 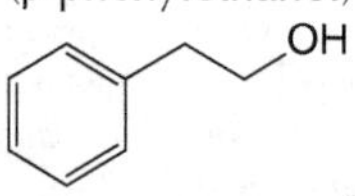	Phenylalanine	40–153[a]	14	Rose, honey

[a] Spanish red wines [2].
[b] New York State white wines [4].
[c] "Amyl" is the common name for a pentyl (five-carbon) hydrocarbon chain.
[d] Exists as a mixture of R/S enantiomers. Threshold reported for racemic mixture.
[e] Threshold determined in water.

Because the higher alcohols are formed during fermentation as part of basic yeast nitrogen metabolism, their production is dependent on both fermentation conditions and nutrient availability. Factors that increase production of higher alcohols relate to metabolism of amino acids and include [5]:

- Low yeast assimilable nitrogen or other nutrient deficiencies related to amino acid biosynthesis
- Higher concentrations of suspended solids
- Higher temperatures
- Certain yeast strains.

In addition, the relative concentrations of these higher alcohols will be dependent on the initial distribution of amino acids. All of these factors are discussed in more detail later (Chapter 22.3). Small amounts of 2-phenylethanol may be formed from glycosylated precursors (Chapter 23.1), although it is not a major source in wine [6]. After fermentation, the higher alcohols in Table 6.1 appear to be stable during storage. For example, there is no significant change in isoamyl alcohol, isobutyl alcohol, and 2-phenylethanol after 1 year of bottle storage at temperatures ranging from 5 to 18 °C [7].

With the exception of 2-phenylethanol ("rose, honey") the higher alcohols do not typically have desirable odors in isolation. However, reconstitution studies indicate that the individual higher alcohols have only minor effects. For example, removal of select higher alcohols resulted in a detectable but indescribable difference in reconstituted model Grenache rosé [8] and had no significant effect on a model Gewurztraminer [9]. Furthermore, no correlation was observed between concentrations of isoamyl alcohol, 2-phenylethanol, or methionol and red wine quality scores [10]. In all likelihood, higher alcohols make a modest contribution to the vinous nature of all wines (and fermented beverages more generally), but are unlikely to serve as impact odorants. Their more important role may be as substrates in the formation of more potent odorants, e.g. acetate esters and aldehydes (Chapters 7 and 9).

6.4 Six-carbon (C_6) alcohols

Several C_6 alcohols and aldehydes are formed enzymatically by grapes and other plants following mechanical damage, for example, crushing grapes, mowing the lawn, or chewing on fresh parsley. These compounds are largely absent from the intact grape berry (or other plant tissues), but are instead formed by enzymatic oxidation of polyunsaturated fatty acids. Because they typically have "herbaceous" and "green grass" aromas, these compounds are also referred to as green leaf volatiles (GLVs). Formation of C_6 compounds appears to be a ubiquitous response of plants to mechanical wounding, possibly for inter- or intraplant signaling to result in an upregulation of antiherbivory compounds like tannins [11].

The enzymatic lipid oxidation pathway responsible for formation of C_6 compounds in grapes and many other plants is discussed in more detail later in the book (Chapter 23.3). This pathway yields C_6 aldehydes and alcohols as primary products, but the C_6 aldehydes are mostly lost during fermentation (Chapters 22.1 and 23.3). Typical C_6 compound concentrations are shown in Table 6.2. The primary C_6 compounds that persist at concentrations around or above their sensory thresholds following fermentation are 1-hexanol and *cis*-3-hexenol. Addition of these compounds at typical red wine concentrations of (1.48 mg/L of 1-hexanol and 234 µg/L of *cis*-3-hexenol) to both dearomatized and neutral wines had no significant effect on aroma. Addition of the same compounds to neutral wines containing spikes of perithreshold concentrations of IBMP (Chapter 5) changed perception of the wines from "earthy" to "green pepper" [13], suggesting that they may contribute additively. Multivariate analyses also show a correlation between C_6 compounds and "green" or "leafy" aromas [12], although this may be an indirect relationship resulting from the fact that fewer C_6 compounds are formed during crushing with increased grape maturity.

***Table 6.2** Typical concentrations of C_6 alcohols in wine*

Name	Wine source	Typical range (μg/L)[a]	Threshold[b] (μg/L) [9]
1-Hexanol (*n*-hexanol)[c]	Spanish red wines [2] Marlborough (NZ) Sauvignon Blanc [12]	2100–13800 1327–3739	8000
cis-3-Hexenol ((*Z*)-3-hexenol)	Spanish red wines Marlborough (NZ) Sauvignon Blanc	8–651 331–711	400
trans-3-Hexenol ((*E*)-3-hexenol)	Marlborough (NZ) Sauvignon Blanc	66–130	1550[d]
cis-2-Hexenol ((*Z*)-2-hexenol)	Marlborough (NZ) Sauvignon Blanc	6–18	Unknown
trans-2-Hexenol ((*E*)-2-hexenol)	Marlborough (NZ) Sauvignon Blanc	ND–8	400[d]

[a] 95% confidence interval reported for NZ Sauvignon Blanc.
[b] Threshold determined in 10% ethanol.
[c] All C_6 alcohols in the table have the –OH group at the 1-position, for example, *cis*-3-hexenol is *cis*-3-hexen-1-ol.
[d] Threshold in water (Leffingwell)

6.5 Methanol

In terms of origin and wine chemistry, methanol does not fit neatly into any other chapter, and is included in the discussion of higher alcohols because of its shared –OH functional group and volatility. Unlike higher alcohols, methanol is not a yeast fermentation metabolite, but is instead formed by acidic or enzymatic hydrolysis of methylated galacturonic acid residues of pectin, an important grape polysaccharide (Chapter 2). Enzymes capable of hydrolyzing methanol from esterified pectin are referred to as *pectin esterases* (PE) or *pectin methylesterases* (PME). These are present in fruits like grapes [14],[2] which can be released by *S. cerevesiae* and other yeasts during fermentation [15], or may be a part of exogenous enzymes added by winemakers to increase extraction of components. Methanol can thus be found at mg/L concentrations in a wide range of fruit juices and wines, particularly those with high pectin or that are processed with pectinases.

In humans, methanol is metabolized to formaldehyde by alcohol dehydrogenase, which can lead to neurotoxicological effects (e.g., blindness) and death. The maximum safe amount of methanol that can be ingested is estimated to be 2 grams, and for acute toxicity is estimated to be 8 grams [16].[3] Although most alcohol-producing nations impose limits on methanol, in many countries (including the US) these limits are only explicitly stated for distilled spirits. The European Union limits methanol in red wine to 400 mg/L and in rosé and white wine to 250 mg/L. This would equate to consumption of 20 L of a red wine at the legal limit to reach toxic levels of methanol, and it appears that commercial wines rarely approach this limit. A review of ~20 surveys of methanol content in commercial wines was published in 1976 (Figure 6.1) [17].[4] The survey reported that

[2] In other areas of fruit juice processing, native *pectin methylesterases* will be deactivated thermally because they can be detrimental to product quality and not just because of methanol production. For example, PME enzymes can result in a loss of cloudiness in orange juice, since de-esterification increases the solubility of pectin.

[3] The metabolism of methanol to formaldehyde in humans utilizes the same dehydrogenase as for ethanol. Because ethanol has a 20-fold higher affinity for the dehydrogenase enzyme, it can be used to competitively inhibit methanol metabolism and formation of toxic formaldehyde. However, methanol is excreted slowly through alternate pathways (1–2% decrease per hour through urination and respiration), and therapeutic use of ethanol in treating methanol poisoning requires continual and long-term exposure until methanol drops to safe concentrations in the blood [16].

[4] More recent international meta-analyses of methanol in table wines are not available, presumably because of a lack of concern about the danger posed by the typically low levels of methanol observed in grape wines.

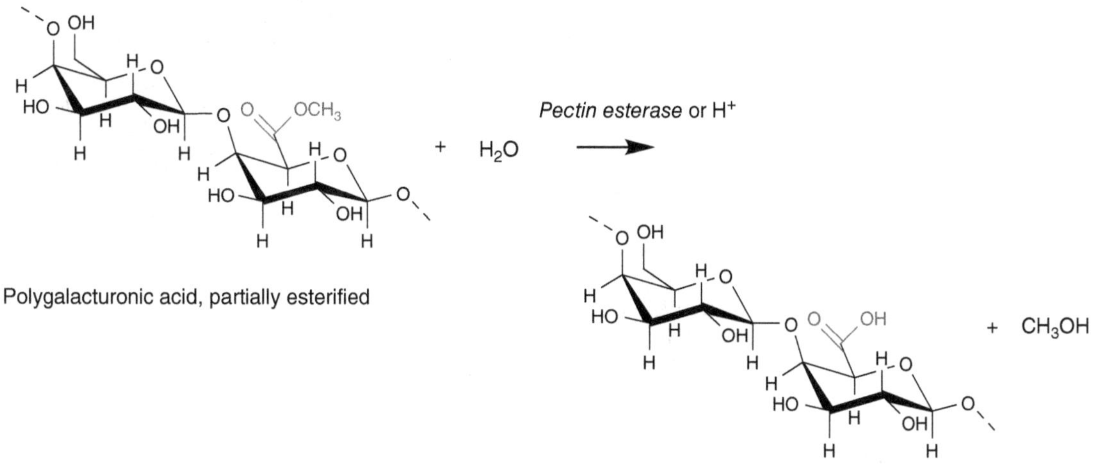

Figure 6.1 *Hydrolytic reaction of esterified polygalacturonic acid to yield methanol [17]*

average methanol values ranged from 21 to 194 mg/L in red table wines and from 16 to 80 mg/L in white table wines, with a high concentration of 635 mg/L in an Italian red table wine.

Because it is derived from pectin hydrolysis, methanol content will start at negligible concentrations and increase during fermentation. Factors that affect methanol concentration in finished wines have been explored in the literature [17–19]:

- Grape variety. Grapes with high concentrations of pectin, for example, *V. labruscana* grapes such as Concord (Chapter 20) will result in wines with higher methanol concentrations (up to 500 mg/L) [18]. Analogously, wines produced from high pectin fruits like plums can yield wines with 0.15% v/v methanol [20].
- Pectolytic enzymes. While these enzymes are typically selected for their ability to cleave bonds between sugars (e.g., galacturonase, Chapter 2), most commercial pectinase preparations also possess PME activity, and use of pectinases will increase methanol content by >50% [19].
- Maceration time and fermentation temperature. Most pectin will be bound to the pomace and longer contact with grape cell wall material during fermentation increases methanol content. In one report, an increase of 50–60 mg/L was seen when pressing occurred after 8 hours (presumably, before fermentation commenced) as opposed to 64 hours [17]. Generally, higher temperatures will also increase hydrolysis, although very high temperature, for example, thermovinification, can inactivate PME [18].

The effects of maceration time and fermentation temperature likely explain the differences in average methanol content observed between red and white table wines, described above. Finally, methanol will be further concentrated during distillation, as described later (Chapter 26.4).

References

1. Watts, H. (1864) *A dictionary of chemistry and the allied branches of other sciences*, Vol. **2**, William Wood & Co., New York.
2. Ferreira, V., Lopez, R., Cacho, J.F. (2000) Quantitative determination of the odorants of young red wines from different grape varieties. *Journal of the Science of Food and Agriculture*, **80** (11), 1659–1667.

3. Czerny, M., Christlbauer, M., Christlbauer, M., *et al.* (2008) Re-investigation on odour thresholds of key food aroma compounds and development of an aroma language based on odour qualities of defined aqueous odorant solutions. *European Food Research and Technology*, **228** (2), 265–273.
4. Lee, C.Y. and Cooley, H.J. (1981) Higher-alcohol contents in New York wines. *American Journal of Enology and Viticulture*, **32** (3), 244–246.
5. Bell, S.J. and Henschke, P.A. (2005) Implications of nitrogen nutrition for grapes, fermentation and wine. *Australian Journal of Grape and Wine Research*, **11** (3), 242–295.
6. Ugliano, M., Bartowsky, E.J., McCarthy, J., *et al.* (2006) Hydrolysis and transformation of grape glycosidically bound volatile compounds during fermentation with three *Saccharomyces* yeast strains. *Journal of Agricultural and Food Chemistry*, **54** (17), 6322–6331.
7. Makhotkina, O. and Kilmartin, P.A. (2012) Hydrolysis and formation of volatile esters in New Zealand Sauvignon blanc wine. *Food Chemistry*, **135** (2), 486–493.
8. Ferreira, V., Ortin, N., Escudero, A., *et al.* (2002) Chemical characterization of the aroma of Grenache rose wines: aroma extract dilution analysis, quantitative determination, and sensory reconstitution studies. *Journal of Agricultural and Food Chemistry*, **50** (14), 4048–4054.
9. Guth, H. (1997) Quantitation and sensory studies of character impact odorants of different white wine varieties. *Journal of Agricultural and Food Chemistry*, **45** (8), 3027–3032.
10. Ferreira, V., San Juan, F., Escudero, A., *et al.* (2009) Modeling quality of premium Spanish red wines from gas chromatography–olfactometry data. *Journal of Agricultural and Food Chemistry*, **57** (16), 7490–7498.
11. Matsui, K. (2006) Green leaf volatiles: hydroperoxide lyase pathway of oxylipin metabolism. *Current Opinion in Plant Biology*, **9** (3), 274–280.
12. Benkwitz, F., Tominaga, T., Kilmartin, P.A., *et al.* (2012) Identifying the chemical composition related to the distinct aroma characteristics of New Zealand Sauvignon blanc wines. *American Journal of Enology and Viticulture*, **63** (1), 62–72.
13. Escudero, A., Campo, E., Farina, L., *et al.* (2007) Analytical characterization of the aroma of five premium red wines. Insights into the role of odor families and the concept of fruitiness of wines. *Journal of Agricultural and Food Chemistry*, **55** (11), 4501–4510.
14. Barnavon, L., Doco, T., Terrier, N., *et al.* (2001) Involvement of pectin methyl-esterase during the ripening of grape berries: partial cDNA isolation, transcript expression and changes in the degree of methyl-esterification of cell wall pectins. *Phytochemistry*, **58** (5), 693–701.
15. Jayani, R.S., Saxena, S., Gupta, R. (2005) Microbial pectinolytic enzymes: a review. *Process Biochemistry*, **40** (9), 2931–2944.
16. Paine, A. and Davan, A.D. (2001) Defining a tolerable concentration of methanol in alcoholic drinks. *Human and Experimental Toxicology*, **20** (11), 563–568.
17. Gnekow, B. and Ough, C.S. (1976) Methanol in wines – source and amounts. *American Journal of Enology and Viticulture*, **27** (1), 1–6.
18. Lee, C.Y., Robinson, W.B., Van Buren, J.P., *et al.* (1975) Methanol in wines in relation to processing and variety. *American Journal of Enology and Viticulture*, **26** (4), 184–187.
19. Cabaroglu, T. (2005) Methanol contents of Turkish varietal wines and effect of processing. *Food Control*, **16** (2), 177–181.
20. Zhang, H., Woodams, E.E., Hang, Y.D. (2012) Factors affecting the methanol content and yield of plum brandy. *Journal of Food Science*, **77** (4), T79–T82.

7

Esters

7.1 Introduction

In nature, esters are well-known contributors to the aroma of flowers and ripe fruits. Despite their low total concentration (<0.1% w/w), esters are also characteristic products of alcoholic fermentation and are critical to the aroma of most alcoholic beverages, and wine is no exception.[1] Some esters may contribute to bitterness, serve as flavor precursors, or participate in reactions with phenolics. In wines, the majority of esters are *secondary* or *tertiary flavor compounds* (see Introduction chapter) that are largely absent from grapes but are instead formed during fermentation and storage, respectively. The two main classes of esters important to wine are ethyl esters and acetate esters, although cyclic esters with 5-membered rings (γ-lactones) may also contribute to wine aroma (Figure 7.1).

7.2 Chemistry of esters

The ester functional group is characterized by the presence of a carbonyl group (C=O) in which the carbonyl carbon is covalently bound to an alkoxy group (–OR, Figure 7.1). The carbonyl group of an ester is relatively stable due to resonance stabilization through the –OR group. As a result, reactions of the ester carbonyl group with nucleophiles like bisulfite (HSO_3^-) are less favorable than for aldehyde and ketone carbonyl groups.

In wine and related beverages, esters are formed through reaction of a carboxylic acid (R–COOH) and an alcohol (R′–OH) (Figure 7.2).[2] The ester-forming reaction is referred to as *esterification* and the reverse reaction is referred to as *ester hydrolysis*. In wine, these reactions can occur during fermentation through enzymatic processes (Chapter 22.2 and 22.3), and can also occur post-fermentation through non-enzymatic, acid-catalyzed reactions (Chapter 25). The reaction mechanism for acid-catalyzed esterification is shown in Figure 7.3.

[1] Based on GC-MS/O studies, esters are reported to contribute to the aroma of table wines, fortified wines, rum, pisco, tequila, beer, bourbon whisky, sake, apple cider, Cognac and other brandies, and likely all other alcoholic beverages other than vodka.

[2] statement is also true for botanicals that are stored in alcohol, e.g. vanilla pods. Several important volatile compounds in vanilla extracts are formed by esterification of vanilla-derived carboxylic acids.

Understanding Wine Chemistry, First Edition. Andrew L. Waterhouse, Gavin L. Sacks, and David W. Jeffery.

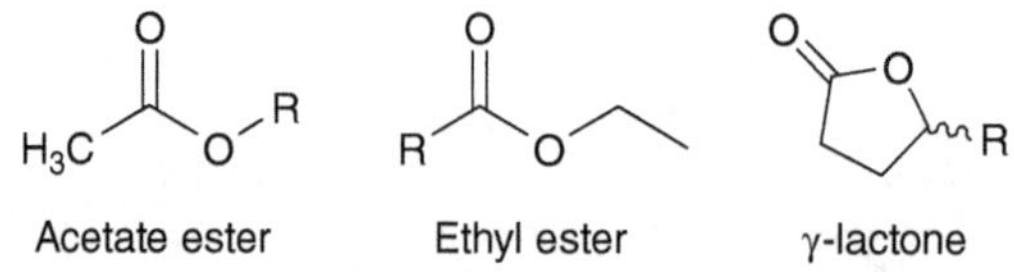

Figure 7.1 General structures for major classes of esters that contribute to wine flavor

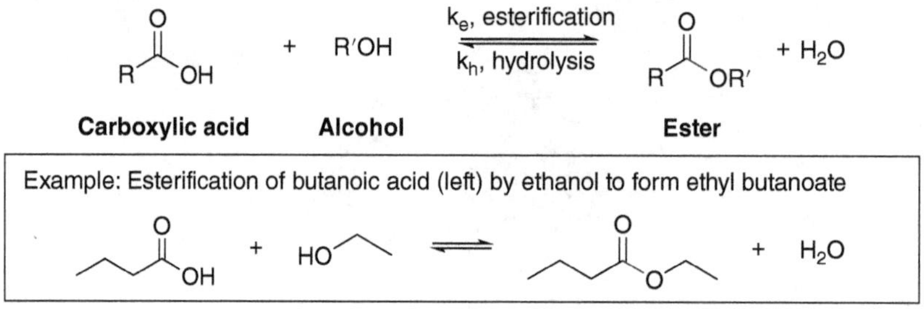

Figure 7.2 General reaction for the formation and hydrolysis of esters (top) and esterification reaction to form ethyl butanoate (bottom)

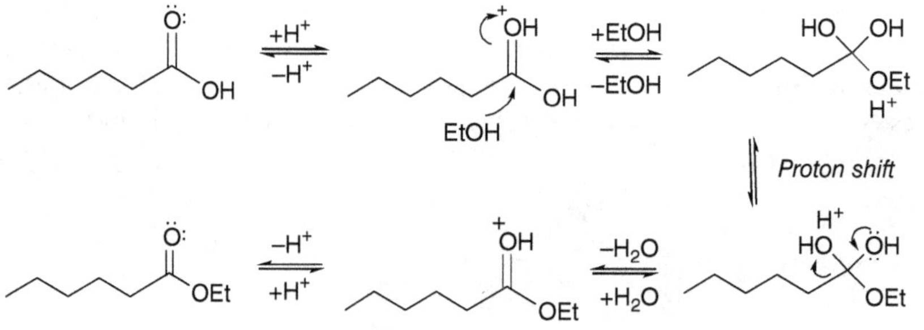

Figure 7.3 Reaction mechanism for acid-catalyzed esterification of an alcohol (ethanol) and a carboxylic acid (hexanoic acid) to yield an ester (ethyl hexanoate). All steps are reversible, and the reverse reaction represents ester hydrolysis

Esterification reactions are typically reversible – that is, both hydrolysis and esterification are always occurring at measurable rates. The relative proportions of carboxylic acid, alcohol, and ester forms will move towards equilibrium during wine storage. The equilibrium expression for ester formation is determined as the ratio of the esterification reaction rate constant (k_e) to the reverse hydrolysis reaction (k_h), and can also be defined in terms of the ratios of the reaction constituents:

$$K_{eq} = \frac{k_e}{k_h} = \frac{[\text{Ester}][\text{H}_2\text{O}]}{[\text{Alcohol}][\text{Carboxylic acid}]} \tag{7.1}$$

For most acyclic esters, $K_{eq} = 4$ is a commonly accepted value [1]. Because systems will move towards equilibrium (the "Law of Mass Action"), the concentration of an individual ester can increase or decrease during storage, depending on the initial concentrations of acid, ester, and alcohol. Water is generally constant, around 50 M. Equation (7.1) can be used to predict equilibrium concentrations of esters in wine and changes that will occur during storage. For example:

- Aliphatic ethyl esters, for example, ethyl hexanoate, tend to increase or decrease only slightly during storage. Using ethyl hexanoate as an example: in a typical wine, [Ethanol] = 100 g/L or 2 M. From Equation (7.1) and assuming $K = 4$, the ratio of [Ester]/[Acid] should be approximately 1:6. After fermentation, a typical wine may have 0.01 mM (1.44 mg/L) ethyl hexanoate and 0.05 mM (5.8 mg/L) hexanoic acid, or a ratio of 1:5. Thus, ethyl hexanoate would decrease only slightly during storage.
- Acetate esters, for example, isoamyl acetate, tend to be at much greater concentrations than equilibrium following fermentation, and thus will decrease during storage. Using isoamyl acetate as an example: in a typical wine, [Acetic acid] = 0.5 g/L or 8.3 mM, and thus the ratio of [Ester]/[Alcohol] should be approximately 1:1600. However, after fermentation, a typical wine may have 0.01 mM (1.3 mg/L) isoamyl acetate and 1 mM (90 mg/L) isoamyl alcohol, or a ratio of 1:100. Thus, isoamyl acetate is expected to decrease by over an order of magnitude during storage.
- Esters of organic acids and other more complex acids can increase by over an order of magnitude during storage, for the reason that these esters tend to be at negligible concentrations immediately after fermentation.
- Lactones, particularly oak lactones, can increase during storage because the closed-ring form is strongly favored thermodynamically (see Section 7.4).

In wine-like solutions, ester hydrolysis will obey pseudo-first-order reaction kinetics:[3]

$$\ln[C] = \ln[C_0] - kt \quad \text{or} \quad \log[C] = \log[C_0] - \frac{kt}{2.303} \tag{7.2}$$

where C represents the concentration at a given time t and C_0 is the initial concentration of the ester. The time necessary for the ester to decrease by 50% (the half-life, $t_{½}$) is inversely related to k:

$$t_{1/2} = \frac{0.693}{k} \tag{7.3}$$

Assuming that the ester is in large excess of its corresponding acid, that is esterification is negligible, it is possible to calculate k and $t_{½}$ from the slope of a plot of log [C] versus t (see Figure 7.4). These values have been tabulated for a wide range of esters at different pH values and ethanol concentrations [2].

At a normal wine pH of 3.4–3.7 and room temperature, the apparent half-life of most aliphatic esters will be several months, but many factors are known to affect the kinetics of ester formation and hydrolysis [1]. Importantly, these factors will not dramatically alter the equilibrium constant. This is reviewed in further detail in Chapter 25.

- pH. Esterification in wine is an acid-catalyzed reaction (Figure 7.3), so decreasing pH will increase the rate of esterification or hydrolysis at a factor roughly proportional to the change in [H^+]. For example, increasing the pH from 2.9 to 4.1 in a model wine solution increases the half-life for isoamyl acetate hydrolysis from 60 days to 570 days [1]. A review of predicted and calculated pH-dependent ester hydrolysis values can be found in Reference [2].

[3] The reaction is pseudo-first-order because even though the rate constant is dependent on the concentration of water, H^+, and the ester, only the last one of this list changes significantly during hydrolysis. Pseudo-first-order rates can also describe esterification reactions involving ethanol, for similar reasons.

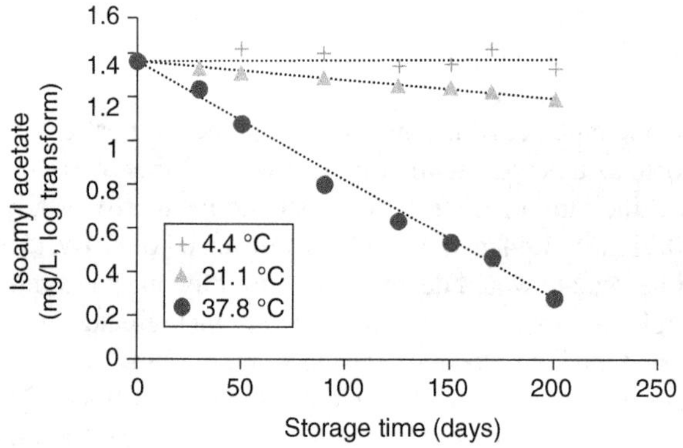

Figure 7.4 Isoamyl acetate hydrolysis over time in a model wine as a function of temperature. Data were adapted from Reference [1]

- Temperature. Higher temperatures will also increase reaction rates (Figure 7.4), with a 10 °C increase in temperature roughly doubling the rate of hydrolysis (E_a = 45–65 kJ/mol; see Chapter 25 for more on activation energy). Similar increases in reaction rates are expected for esterification reactions.
- Substrate. Esters with longer carbon chains tend to approach equilibrium faster. For example, the hydrolysis rate of ethyl decanoate is about 3-fold faster than ethyl octanoate and 15-fold faster than ethyl hexanoate under similar model wine conditions. Conjugated acids, for example, sorbic acid, have been observed to have very slow esterification kinetics [3], presumably due to resonance stabilization of the carbonyl group.

7.3 Esters in grapes

Many fruits accumulate suprathreshold concentrations of volatile esters during ripening, presumably to attract frugivores.[4] By comparison, most grapes accumulate negligible concentrations of volatile esters: in one study of Cabernet Sauvignon and Riesling berries, only methyl hexanoate and (*Z*)-3-hexenyl butanoate were detectable during ripening, and the total ester concentration was < 1 mg/kg [5]. These grape-derived esters would be expected to be hydrolyzed during fermentation and storage, and to subsequently have a negligible impact on wine flavor. One notable exception is methyl anthranilate (MA), responsible for the "foxy, Concord" aromas of mature *V. labruscana* and *V. rotundifolia* grapes (Chapter 5).

Hydroxycinnamic acids (HCA, e.g., coumaric acid) exist in grapes primarily as their tartrate esters. These esters have no aroma and their role is discussed in Chapter 13. Some flavan-3-ols are found as gallate esters, as described in Chapter 14.

7.4 Esters formed during winemaking and storage

The majority of odorous wine esters are formed during fermentation and storage via enzymatic or non-enzymatic esterification of carboxylic acids. The factors affecting these reactions are discussed later in this chapter.

---

[4] For example, the ester concentration in ripe bananas (*Musa*) can range from 57 to 89 mg/kg, consisting mostly of isoamyl acetate, isoamyl butanoate, and butyl butanoate [4].

7.4.1 Ethyl and acetate esters

All carboxylic acids and alcohols in wine can potentially esterify. Since there are dozens (if not hundreds) of both of these compound classes in wines, one could expect to find thousands of different esters in wine, if only at vanishingly small concentrations. Because ethanol is the dominant alcohol in wine, the majority of esters formed in wine are ethyl esters, with the most potent typically being fatty acid ethyl esters (FAEE). Fatty acids are formed as part of yeast lipid metabolism (Chapter 22.2), and their concentrations and sensory properties are discussed in Chapter 3. A list of FAEE with high OAV in wine are reported in Table 7.1.

Beyond FAEE, several other ethyl esters can form during storage and have been detected in wine:

- Ethyl lactate, ethyl hydrogen malate, diethyl malate, and other esters of the major wine organic acids
- Ethyl cinnamate, ethyl vanillate, and related ethyl esters of cinnamic and phenolic acids
- Ethyl sorbate and ethyl phenylacetate, implicated in some wine faults, as described below
- Ethyl leucinate, ethyl valinate, and other ethyl esters of amino acids.

In wine, volatile esters other than ethyl esters are generally at concentrations well below their sensory thresholds. One exception to this rule is the acetate esters, which are distinguished by the presence of an acetyl moiety (Table 7.2). Acetate esters are formed by enzymatic acetylation of alcohols during fermentation [11], discussed in more detail in Chapter 22.3. The alcohols that participate in these reactions are often byproducts of amino acid biosynthesis (e.g., isoamyl alcohol, 2-phenylethanol) or can be grape-derived (e.g., 1-hexanol, *cis*-3-hexen-1-ol), described in Chapter 6. 3-Mercaptohexanol, a varietal thiol released during fermentation, can also be acetylated to form its corresponding *O*-acetate ester (Chapter 10).

Formation of the FAEE and acetate esters during fermentation is governed by several factors, which will be discussed in more detail in later chapters. Briefly, factors include:

- Concentrations of FAEE are affected by the free fatty acid concentration formed during fermentation, which is influenced by must composition, insoluble solids concentration, oxygen availability, temperature, and yeast strain, as well as the degree of enzymatic esterification (Chapter 22.2).
- Formation of the acetate esters is governed by the availability of alcohol substrates and the activity of acetyltransferase enzymes capable of esterifying higher alcohols to acetate esters (Chapter 22.3).

As mentioned above in Section 7.1 and Chapter 25, acetate esters tend to decrease during storage, FAEE stay roughly constant, and esters of other wine carboxylic acids like organic acids tend to increase. The rate of increase or decrease (i.e., the rate at which equilibrium is approached) will be faster at lower pH and higher temperature, for reasons previously discussed.

7.4.2 Aliphatic γ-lactones

Cyclic esters formed by intramolecular condensation of carboxylic acid and alcohol groups (hydroxycarboxylic acids) are referred to as *lactones*. In principle, the lactone ring can range from three atoms to infinitely large, but in wine the most important lactones are those that are the most thermodynamically stable rings, specifically those containing five atoms (γ-lactone) and to a lesser extent those containing six atoms (δ-lactones).

One important category is the unsubstituted γ-lactones, shown in Table 7.3. Except for γ-butyrolactone (where R = H), these lactones exist as enantiomers with slightly different sensory properties. In most wines, the distribution of enantiomers is nearly racemic (60:40) [12], and stereochemical distinctions can often be ignored. Relatively little work has gone into understanding factors that affect production of these unsubstituted lactones. Butyrolactone appears to be formed by yeast catabolism of glutamic acid [13]. The source of

Table 7.1 Odor descriptors, structures, detection thresholds, and concentrations for key fatty acid ethyl esters (FAEE) in wines. The ethoxy group is highlighted

Examples	Structure	Odor descriptor[a]	Odor threshold (μg/L)[b]	Typical range (μg/L)[c]
Ethyl acetate	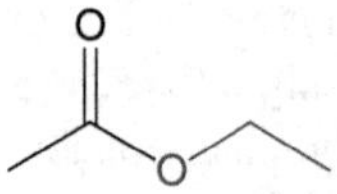	Nail polish remover, fruity	12 000	5000–63 000[d]
Ethyl butanoate	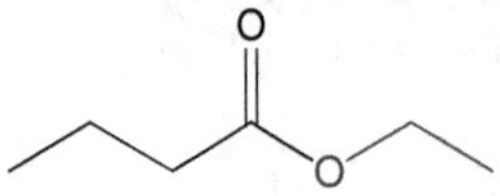	Apple, fruity	20	70–540
Ethyl hexanoate (*ethyl caproate*)	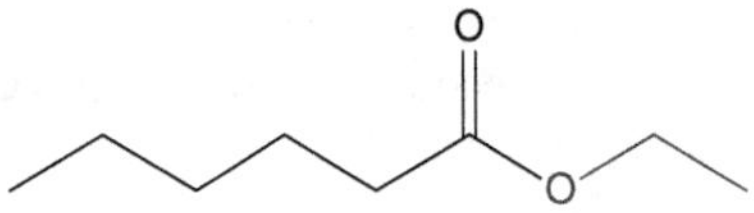	Green apple	14	150–1500
Ethyl octanoate (*ethyl caprylate*)	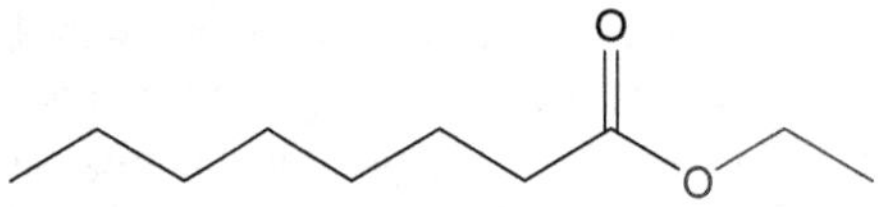	Fruity, peach	5	140–2500
Ethyl decanoate	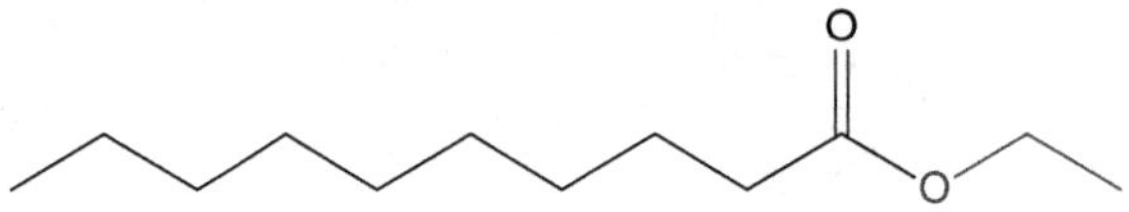	Fruity	200	14–910
Ethyl 2-methylpropanoate (*ethyl isobutyrate*)	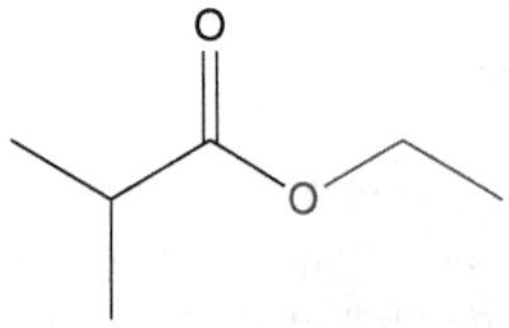	Sweet	15	10–120
Ethyl 2-methylbutanoate	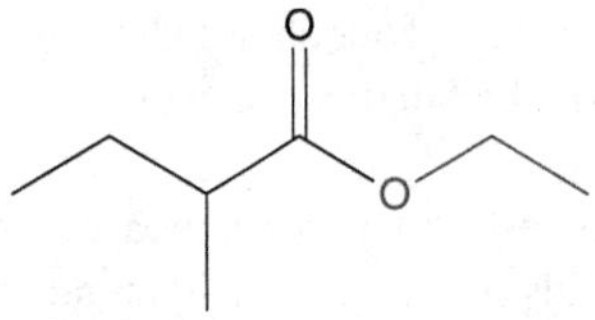	Apple	18	n/a
Ethyl 3-methylbutanoate (*ethyl isovalerate*)	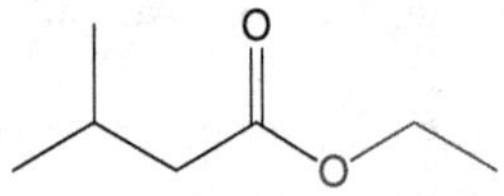	Fruity	3	2–30
Ethyl 3-methylpentanoate	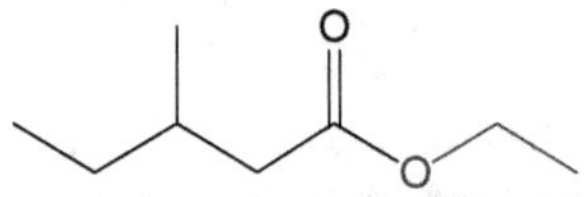	Strawberry	0.5[c]	30 ng/L[e]
Ethyl 4-methylpentanoate	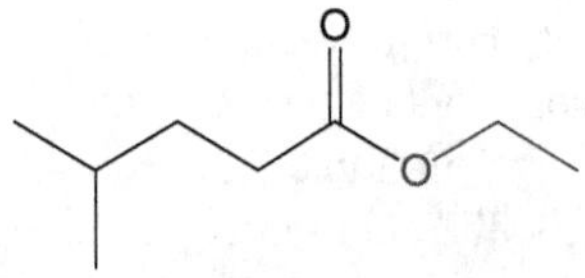	Strawberry	0.75[c]	230 ng/L[e]

[a] Descriptors from Flavornet (www.flavornet.org) and Reference [6].

[b] Thresholds in model wine from References [7] and [8].

[c] Ranges summarized from Spanish red wines [9] and New Zealand Sauvignon Blanc white wines [10]. Mean values from Reference [10].

[d] Higher concentrations possible in wines that are microbially spoiled, for example, by *Acetobacter*.

[e] Concentrations in two aged red wines (>5 years old). Concentrations of all but ethyl 4-methylpentanoate were undetectable in young wines.

Table 7.2 Odor descriptors, structures, detection thresholds, and concentrations for key acetate esters in wines. The acetate moiety is highlighted

Examples	Structure	Odor descriptor[a]	Odor threshold in wine (μg/L) [11]	Typical range (μg/L) [11]
2-Methylpropyl acetate (*isobutyl acetate*)	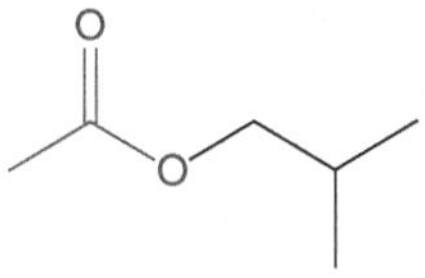	Banana, cherry	1600	ND–170[b]
3-Methylbutyl acetate (*isoamyl acetate*)	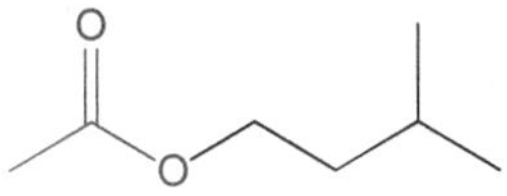	Banana	160	30–5500
Hexyl acetate	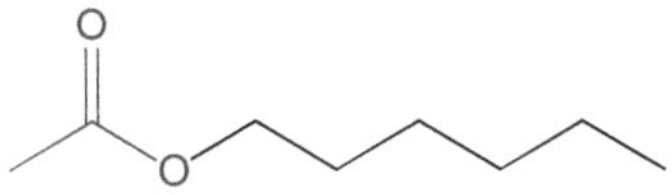	Green apple, sweet	1800	ND–260
2-Phenylethyl acetate	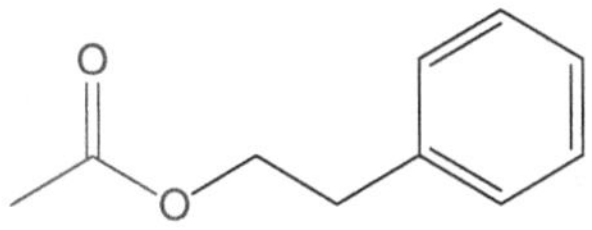	Honey, rose	2400	ND–260

[a] Descriptors from Flavornet (www.flavornet.org).
[b] ND = Not detectable.

the γ-hydroxycarboxylic acid precursors for the other lactones have not been well studied. Several of the other lactones (e.g., octalactone, nonalactone, decalactone) are also found in beer, so it is possible that they are formed *de novo* during fermentation. Red wines generally have higher concentrations of longer-chain lactones than white wines and botrytized wines can have up to 59 μg/L of nonalactone, but explanations for these effects are unclear [12].

A second important class of aliphatic γ-lactones are the oak-derived 3-methyl-γ-octalactones, also referred to as oak lactones or whiskey lactones (Table 7.3). These compounds can exist as four stereoisomers, of which two are the dominant forms detected in wines: the (*4S,5S*)-isomer, referred to as *cis*-oak lactone, and the (*4S,5R*)-isomer, referred to as *trans*-oak lactone. These compounds are usually at undetectable concentrations in wines made without oak contact, but can be extracted during contact of wine with oak barrels or barrel alternatives (e.g., oak staves, oak chips).

The factors affecting *cis*-oak lactone variation in oak and oaked wines is discussed in more detail in Chapter 25. Briefly, oak lactones are present in untoasted oak at total concentrations ranging from 10 to 100 mg/kg, with the majority (>75%) existing as the *cis*-isomer [17]. In wine, the *cis*-isomer can be detected at concentrations close to 700 μg/L, and is usually present at two-fold higher concentrations (and 10-fold greater odor activity) than the *trans*-isomer.

Similar to some monoterpenoids (Chapter 23.1), oak lactones can be derived from non-volatile precursors, one of which is the lactone open-ring form, 3-(*S*)-hydroxy-4-methylcarboxylic acid (Figure 7.5). Formation of the oak lactone from the precursor is acid-catalyzed, similar to that for other esters, and also appears to be irreversible. The rate of lactonization is over 10-fold faster for the *trans*-isomer formation than for the *cis*-isomer (half-life = 3 h versus 40 h at pH 2.9 and room temperature), likely due to greater thermodynamic stability of the *trans*-isomer [18]. The lactonization rate is such that oak lactones can continue to form and oak aromas continue to intensify even once the wine is no longer in contact with oak, and the faster rate of *trans*-isomer formation explains the decrease in the *cis/trans* rate during storage. These carboxylic acid precursors appear

Table 7.3 *Odor descriptors, structures, detection thresholds, and concentrations for key lactones in wines*

Examples	Structure	Odor descriptor[a]	Odor threshold (μg/L) [14,15][b]	Typical range (μg/L) [8,12,14,16]
γ-Butyrolactone (C_4)	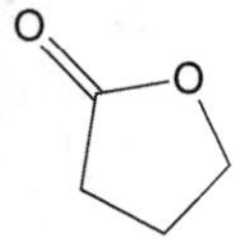	Caramel, sweet	Unknown (high)	4100–21 400
γ-Octalactone (C_8)	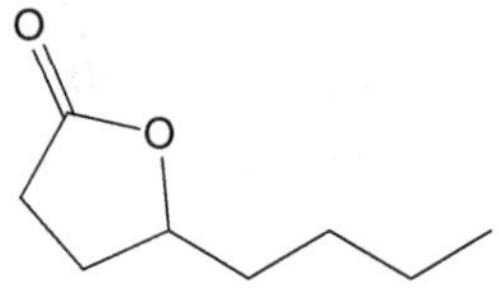	Coconut	7	<0.1–5
γ-Nonalactone (C_9)	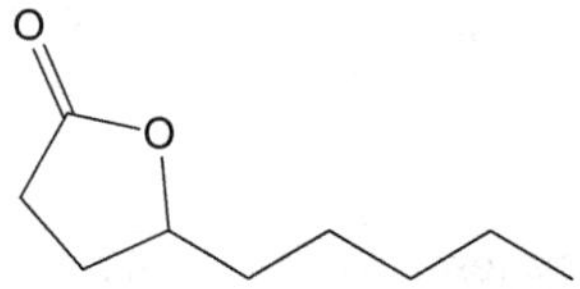	Peach	25	2–30
γ-Decalactone (C_{10})	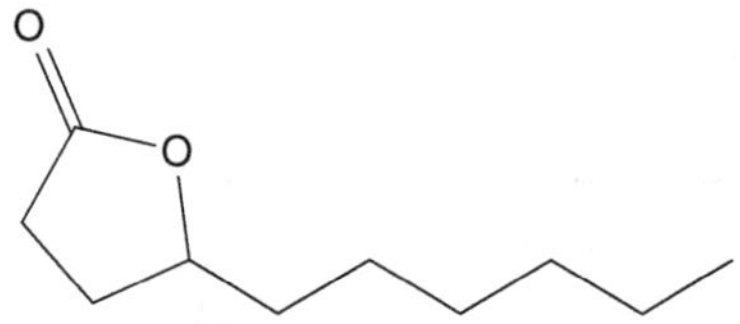	Fatty, peach	0.7	<0.1–1.5
γ-Dodecalactone (C_{12})	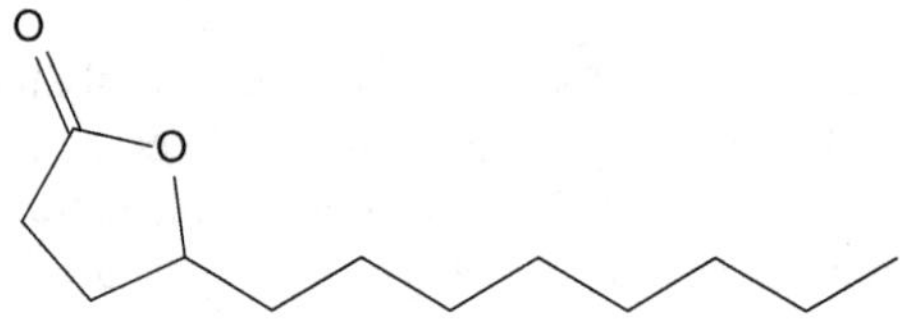	Coconut	7	<0.1–20
(4*S*,5*S*)-3-Methyl-γ-octalactone (*cis*-oak lactone)	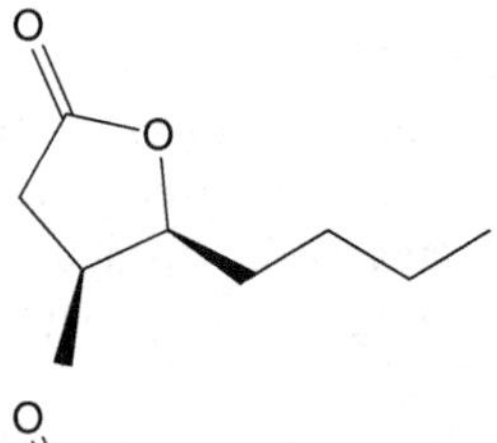	Coconut	25	<0.1–700
(4*S*,5*R*)-3-Methyl-γ-octalactone (*trans*-oak lactone)	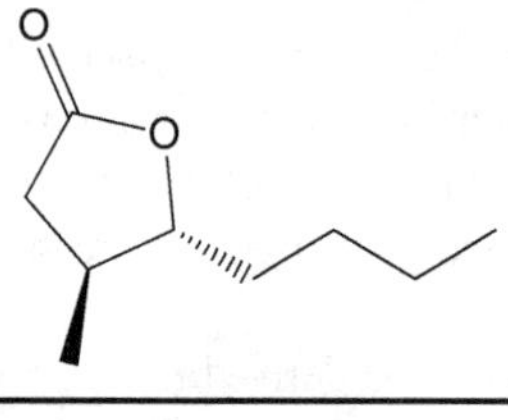	Coconut	110	<0.02–350

[a] From www.flavornet.org and Reference [15].
[b] In model wine, for racemic mixture if not specified.

likely to form either through pyrolysis (during toasting) or hydrolysis of glycoconjugates of 3-hydroxy-4-methylcarboxylic acid in oak, notably the galloylglucoside (present at up to 500 mg/kg in oak), rutinoside, and glucoside forms (Figure 7.5) [19]. The glucoside is more stable than the free 3-hydroxy-4-methylcarboxylic acid, and conversion to oak lactones can take over 1 year under model conditions [19]. Other odorous lactones including wine lactone (Chapter 8 and Chapter 25) and sotolon (Chapter 9) are discussed later.

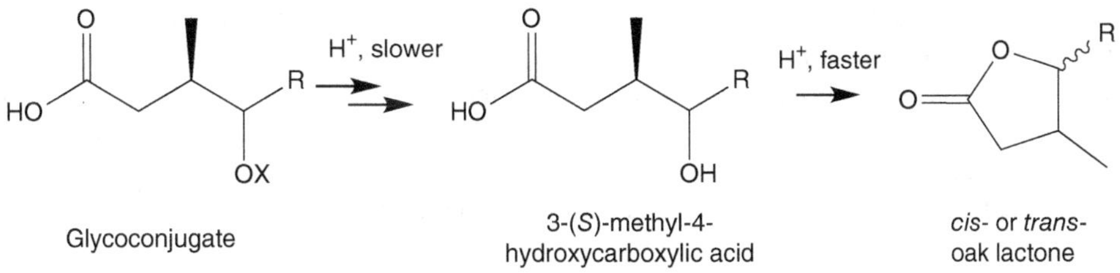

Figure 7.5 Formation of cis- and trans-oak lactones from lactonization of 3-(S)-hydroxy-4-methylcarboxylic acid in aqueous acidic environments (adapted from Reference [19]). The carboxylic acid precursor is detectable in oak or may be formed during storage from glycoconjugate precursors (X = gallolylglucosyl, rutinosyl, or glucosyl group)

7.5 Sensory effects

In most gas chromatography–olfatometry studies, the straight chain and branched-chain FAEE are invariably included in the list of the most potent compounds (highest OAV), in particular ethyl butanoate, ethyl hexanoate, ethyl octanoate, and ethyl 2- and 3-methylbutanoate. Elimination of any individual FAEE generally has minor effects on the overall aroma, but in combination these appear to be responsible for red- and dark-fruit aromas in wines [20]. The esters of most other organic acids, e.g., diethyl tartrate or diethyl succinate, have low volatility and sensory thresholds well above their concentrations in wine, although an additive contribution cannot be ruled out.

A handful of ethyl esters and their corresponding acids may be responsible for off-aromas (also see Chapter 18) in some wines, such as:

- High concentrations of ethyl acetate and acetic acid are well known to contribute to perception of "volatile acidity," a wine fault characterized by pungent "nail polish remover" and "vinegar" aromas [21].
- Ethyl phenylacetate and phenylacetic acid are reported to be characteristic odorants in wines produced from "sour rot" afflicted grapes [22].
- Ethyl sorbate, formed by esterification of sorbic acid (a preservative) is reported to yield a "candy, pineapple" aroma during storage [3].

In contrast to ethyl esters, which appear to act in combination, acetate esters (particularly isoamyl acetate) may exert an effect on the fruity aromas of young wines in isolation. Removal of isoamyl acetate from a reconstituted Grenache rosé wine is reported to diminish fruitiness [7], and spiking a Maccabeo white wine to yield a 200% increase is reported to increase "banana" aroma [23]. However, since acetate ester concentrations are often well correlated, these experiments are not very realistic. Reconstitution studies in which all acetate esters are selectively omitted have not been reported. One acetate ester that deserves special consideration is the polyfunctional thiol 3-mercaptohexyl acetate (3-MHA), whose concentration is strongly correlated with "passionfruit" aromas (Chapter 10). As described previously, acetate ester concentrations will decrease during wine storage through acid hydrolysis to well below threshold, and thus will have a minimal effect on the aroma of aged wines.

The C_8–C_{12} lactones have similar "peach, fruity, coconut" aromas, and may have a minor effect on wine aroma even at subthreshold concentrations through additive effects [24]. Since they are present in most wines and arise from fermentation, they may contribute to the typical vinous character of wine. Despite its relatively high concentration, γ-butyrolactone likely does not affect wine aroma.[5]

[5] γ-Butyrolactone (GBL) and its open ring form, γ-hydroxybutyrate (GHB), have much more notorious roles in wines and other alcoholic beverages as so-called "date-rape" drugs added to the drinks of unsuspecting victims [16]. The total concentrations of GBL and GHB, which can introconvert in the body, naturally found in wine are typically much less than concentrations used to adulterate drinks (2000 mg/L or more).

Because of its lower threshold and higher concentrations in most wines, the *cis*-isomer of oak lactone has a more profound effect on the aroma of wine and other oak-aged beverages. The *cis*-isomer is reported to correlate positively with "coconut" aroma in Chardonnay, and with "coconut", "vanilla", "berry," and "dark chocolate" aromas in Cabernet Sauvignon wine [17]. Oak lactone concentrations correlate with consumer liking of oaked wines, but only to a point – a racemic mixture of *cis*- and *trans*-isomers in red wine was reported to decrease consumer preference at concentrations above 235 μg/L [25].

References

1. Ramey, D.D. and Ough, C.S. (1980) Volatile ester hydrolysis or formation during storage of model solutions and wines. *Journal of Agricultural and Food Chemistry*, **28** (5), 928–934.
2. Rayne, S. and Forest, K. (2011) Estimated carboxylic acid ester hydrolysis rate constants for food and beverage aroma compounds. *Nature Proceedings*. doi: 10.1038/npre.2011.6471.1.
3. Derosa, T., Margheri, G., Moret, I., *et al.* (1983) Sorbic acid as a preservative in sparkling wine – its efficacy and adverse flavor effect associated with ethyl sorbate formation. *American Journal of Enology and Viticulture*, **34** (2), 98–102.
4. Nogueira, J.M.F., Fernandes, P.J.P., Nascimento, A.M.D. (2003) Composition of volatiles of banana cultivars from Madeira Island. *Phytochemical Analysis*, **14** (2), 87–90.
5. Kalua, C.M. and Boss, P.K. (2009) Evolution of volatile compounds during the development of Cabernet Sauvignon grapes (*Vitis vinifera L.*). *Journal of Agricultural and Food Chemistry*, **57** (9), 3818–3830.
6. Campo, E., Ferreira, V., Escudero, A., Cacho, J. (2005) Prediction of the wine sensory properties related to grape variety from dynamic-headspace gas chromatography–olfactometry data. *Journal of Agricultural and Food Chemistry*, **53** (14), 5682–5690.
7. Ferreira, V., Ortin, N., Escudero, A., *et al.* (2002) Chemical characterization of the aroma of Grenache rose wines: aroma extract dilution analysis, quantitative determination, and sensory reconstitution studies. *Journal of Agricultural and Food Chemistry*, **50** (14), 4048–4054.
8. San Juan, F., Cacho, J., Ferreira, V., Escudero, A. (2012) Aroma chemical composition of red wines from different price categories and its relationship to quality. *Journal of Agricultural and Food Chemistry*, **60** (20), 5045–5056.
9. Ferreira, V., Lopez, R., Cacho, J.F. (2000) Quantitative determination of the odorants of young red wines from different grape varieties. *Journal of the Science of Food and Agriculture*, **80** (11), 1659–1667.
10. Benkwitz, F., Tominaga, T., Kilmartin, P.A., *et al.* (2012) Identifying the chemical composition related to the distinct aroma characteristics of New Zealand Sauvignon Blanc wines. *American Journal of Enology and Viticulture*, **63** (1), 62–72.
11. Sumby, K.M., Grbin, P.R., Jiranek, V. (2010) Microbial modulation of aromatic esters in wine: current knowledge and future prospects. *Food Chemistry*, **121** (1), 1–16.
12. Cooke, R.C., Capone, D.L., van Leeuwen, K.A., *et al.* (2009) Quantification of several 4-alkyl substituted gamma-lactones in Australian wines. *Journal of Agricultural and Food Chemistry*, **57** (2), 348–352.
13. Wurz, R.E.M., Kepner, R.E., Webb, A.D. (1988) The biosynthesis of certain gamma-lactones from glutamic acid by film yeast activity on the surface of flor sherry. *American Journal of Enology and Viticulture*, **39** (3), 234–238.
14. Ferreira, V., Jarauta, I., Ortega, L., Cacho, J. (2004) Simple strategy for the optimization of solid-phase extraction procedures through the use of solid–liquid distribution coefficients. Application to the determination of aliphatic lactones in wine. *Journal of Chromatography A*, **1025** (2), 147–156.
15. Spillman, P.J., Sefton, M.A., Gawel, R. (2004) The contribution of volatile compounds derived during oak barrel maturation to the aroma of a Chardonnay and Cabernet Sauvignon wine. *Australian Journal of Grape and Wine Research*, **10** (3), 227–235.
16. Elliott, S. and Burgess, V. (2005) The presence of gamma-hydroxybutyric acid (GHB) and gamma-butyrolactone (GBL) in alcoholic and non-alcoholic beverages. *Forensic Science International*, **151** (2–3), 289–292.
17. Pollnitz, A.P., Jones, G.P., Sefton, M.A. (1999) Determination of oak lactones in barrel-aged wines and in oak extracts by stable isotope dilution analysis. *Journal of Chromatography A*, **857** (1–2), 239–246.

18. Wilkinson, K.L., Elsey, G.M., Prager, R.H., *et al.* (2004) Rates of formation of *cis*- and *trans*-oak lactone from 3-methyl-4-hydroxyoctanoic acid. *Journal of Agricultural and Food Chemistry*, **52** (13), 4213–4218.
19. Wilkinson, K.L., Prida, A., Hayasaka, Y. (2013) Role of glycoconjugates of 3-methyl-4-hydroxyoctanoic acid in the evolution of oak lactone in wine during oak maturation. *Journal of Agricultural and Food Chemistry*, **61** (18), 4411–4416.
20. Lytra, G., Tempere, S., de Revel, G., Barbe, J.C. (2012) Impact of perceptive interactions on red wine fruity aroma. *Journal of Agricultural and Food Chemistry*, **60** (50), 12260–12269.
21. Fugelsang, K.C. and Edwards, C.G. (2007) *Wine microbiology practical applications and procedures*, Springer, New York.
22. Campo, E., Saenz-Navajas, M.P., Cacho, J., Ferreira, V. (2012) Consumer rejection threshold of ethyl phenylacetate and phenylacetic acid, compounds responsible for the sweet-like off odour in wines made from sour rotten grapes. *Australian Journal of Grape and Wine Research*, **18** (3), 280–286.
23. Escudero, A., Gogorza, B., Melus, M.A., *et al.* (2004) Characterization of the aroma of a wine from Maccabeo. Key role played by compounds with low odor activity values. *Journal of Agricultural and Food Chemistry*, **52** (11), 3516–3524.
24. Ferreira, V. (2010) Volatile aroma compounds and wine sensory attributes, in *Managing wine quality*, Vol. **1**, *Viticulture and wine quality* (ed. Reynolds, A.G.), Woodhead Publishing and CRC Press, Oxford and Boca Raton, FL.
25. Chatonnet, P., Boidron, J.N., Pons, M. (1990) Maturation of red wines in oak barrels – evolution of some volatile compounds and their aromatic impact. *Sciences Des Aliments*, **10** (3), 565–587.

8

Isoprenoids

8.1 Introduction

The term isoprenoid encompasses a diverse and complex range of hydrocarbons and their oxygenated derivatives (cf. terpenoids, which applies to oxygenated terpenes, although the terms are often used synonymously) based on the repeated presence of the C_5 isoprene unit (2-methylbuta-1,3-diene). The carbon skeleton may consist of simple inclusion of isoprene units, giving rise to compounds that are multiples of C_5 (Figure 8.1), or there may be deviation from the basic addition of C_5 units through loss or shifting of a fragment, such as a methyl group. Isoprenoids can exist as saturated/unsaturated and cyclic/acyclic hydrocarbons, and can contain alcohol, aldehyde, ketone, ester, ether, and acetal functionalities. Volatile isoprenoids are widespread aroma and flavor compounds produced enzymatically by different plants as secondary metabolites via the terpenoid pathway [1, 2].

8.2 General chemical and sensory properties of isoprenoids

Of importance to wine aroma are the isoprenoids classified as monoterpenoids (C_{10} compounds), sesquiterpenoids (C_{15} compounds), and C_{13}-norisoprenoids (C_{13} compounds derived from tetraterpenoids[1]) (Table 8.1). Many of these impart desirable aromas and show dependence on grape variety. Isoprenoids tend to be very non-polar compounds, although oxygenation increases water solubility compared to the hydrocarbon variants (e.g., the monoterpene myrcene has log $P=4.3$ and water solubility = 0.004 g/L, whereas its alcohol derivative geraniol has log $P=3.6$ and water solubility = 0.4 g/L). While aliphatic hydrocarbon forms are usually not highly reactive, the presence of different functional groups (i.e., alkenes, alcohols, ketones) can increase reactivity and allow for the formation of new aroma compounds (e.g., see Figure 8.2).

[1] More specifically, C_{40} isoprenoid pigments known as carotenoids are cleaved to yield a range of structurally diverse compounds collectively termed apocarotenoids (i.e., norisoprenoids) differing in carbon chain length and functionality. Such compounds include the phytohormone abscisic acid (C_{15}), vitamin A (C_{20} retinoids), and volatile compounds with C_9, C_{10}, C_{11}, and C_{13} carbon skeletons. Volatile norisoprenoids constitute an important group of natural aroma and flavor molecules, of which the C_{13}-norisoprenoids are the most widely abundant compounds derived from carotenoids.

Understanding Wine Chemistry, First Edition. Andrew L. Waterhouse, Gavin L. Sacks, and David W. Jeffery.

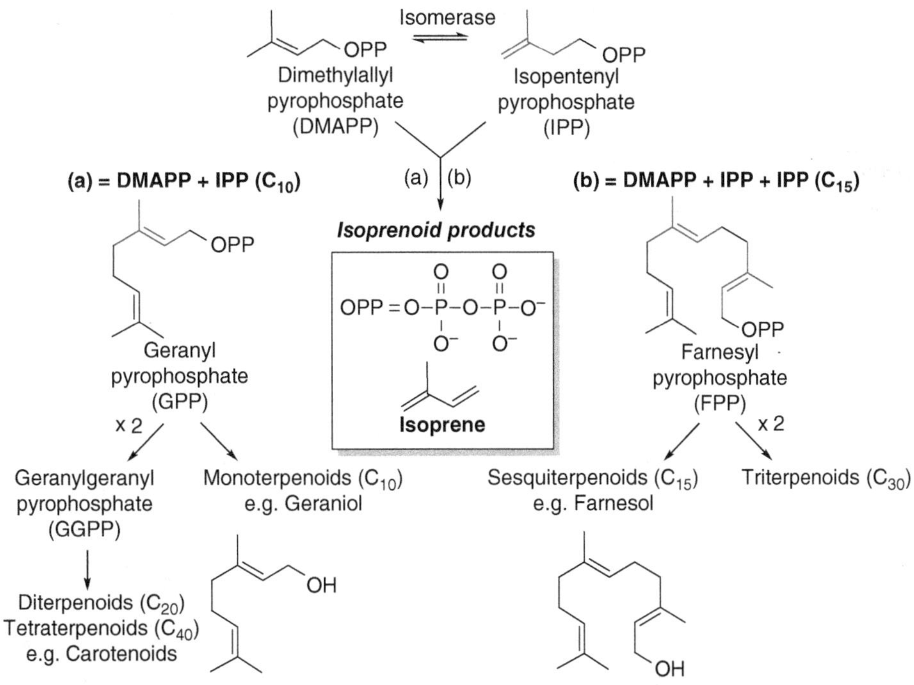

Figure 8.1 Representation of isoprenoid biosynthetic pathway from activated isoprene units (DMAPP and IPP), giving rise to a diverse range of natural products based on this C_5 structural unit

Isoprenoids are often present in water-soluble bound forms (i.e., non-volatile glycosides) in the grape berry, many of which can be hydrolyzed under mildly acidic wine conditions to release the aglycone (Chapter 23.1). Some glycosides are not hydrolyzed effectively at wine pH (e.g. unactivated alcohols such as citronellol), and harsher conditions, including lower pH and higher temperature, are required. In contrast, enzymatic hydrolysis of glycosides (i.e., glycosidases from yeast or commercial preparations) under normal winemaking conditions yields the corresponding volatile compounds (Chapter 23.1), which may then undergo acid-catalyzed rearrangement during aging at wine pH (Chapter 25) to afford different aroma compounds, potentially altering the aroma profile of a wine as a result.

8.3 Monoterpenoids

Characteristics of key monoterpenoids in wines are shown in Table 8.1. This class of isoprenoids is typically associated with white wines produced from aromatic Muscat grape varieties, where concentrations may exceed threshold by 100-fold [8]. Monoterpenoids are also found at suprathreshold concentrations in aromatic non-Muscat cultivars such as Traminer and Riesling, and at subthreshold concentrations in neutral varieties including Cabernet Sauvignon, Merlot, Chardonnay, and Sauvignon Blanc [8]. The grape-derived monoterpenoids of greatest significance to wine aroma include linalool and geraniol ("floral" character of most Muscat varieties) and (–)-*cis*-rose oxide (which contributes to a "lychee" aroma in Gewurztraminer [19]); the impact of selected grape varieties on wine concentrations of these three aroma compounds is highlighted in Table 8.2. Wine lactone, found in wine as a single stereoisomer out of eight possibilities [20], may also be an important contributor to the aroma of varietal wines [21, 22], although surveys of typical wine

Table 8.1 *Indicative odor descriptors, detection thresholds, and concentration ranges for classes of isoprenoids found in wine [3–18]*

Examples[a]	Structure[b]	Odor descriptor[c]	Threshold (μg/L)[c]	Range[d,e] (μg/L)
Monoterpenoid				
Wine lactone	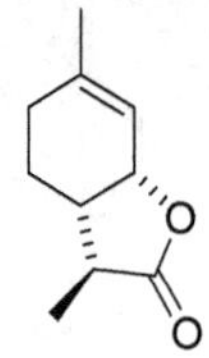	Coconut, woody, sweet	0.01	0.1[f]
(–)-*cis*-Rose oxide	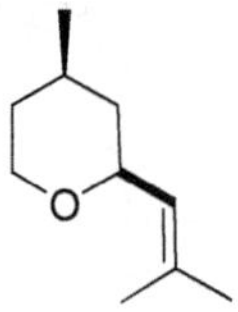	Floral, green, rose	0.2	Trace–22
1,8-Cineole (eucalyptol)	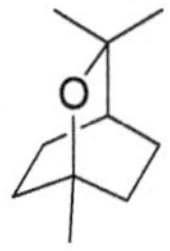	Eucalyptus, fresh	1.1	ND–33
Linalool	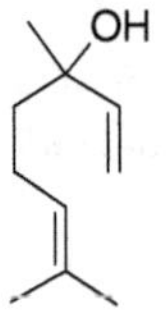	Floral, citrus	25	ND–370
Geraniol	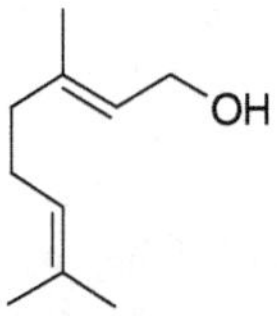	Floral, citrus	30	ND–290
Citronellol	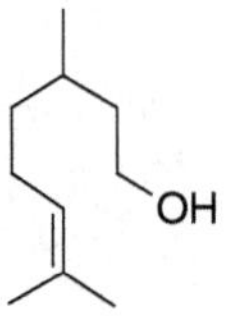	Rose, citrus	100	1–50
Hotrienol	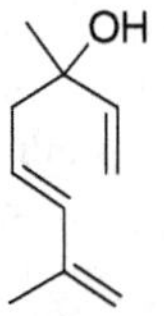	Floral, citrus	110	3–240
α-Terpineol	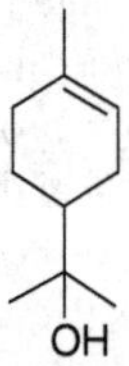	Lilac	250	ND–400

(continued overleaf)

Table 8.1 *(continued)*

Examples[a]	Structure[b]	Odor descriptor[c]	Threshold (µg/L)[c]	Range[d,e] (µg/L)
Nerol	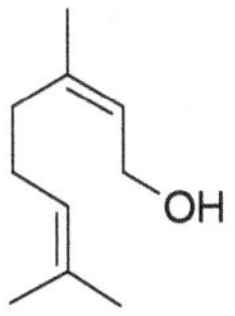	Floral, green	400	ND–360
Sesquiterpenoid				
(–)-Rotundone	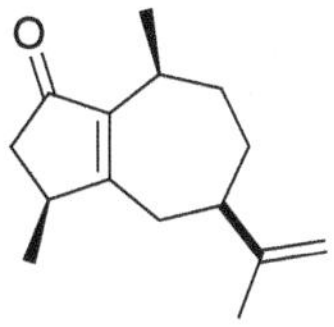	Black pepper	0.016	Trace–0.56
Nerolidol	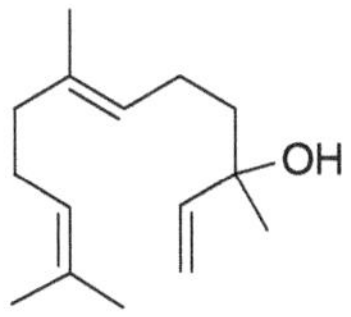	Floral, apple, green	10	ND–29
Farnesol	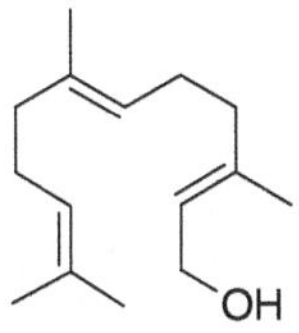	Floral, rose	20	ND–180
C_{13}-Norisoprenoid				
β-Damascenone	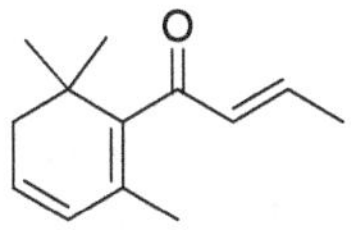	Cooked apple, quince, floral	0.050	0.3–45
β-Ionone	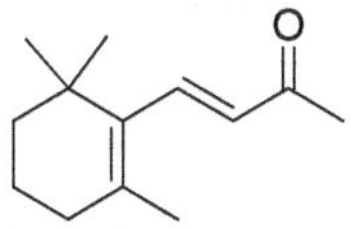	Violet, wood, raspberry	0.090	ND–18
1,1,6-Trimethyl-1,2-dihydronapthalene (TDN)	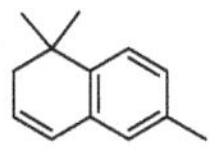	Kerosene, petrol	2	ND–54

[a]Common names are widely used in the literature since the systematic names are more complex; for example, geraniol is (*E*)-3,7-dimethyl-2,6-octadien-1-ol and β-damascenone is (*E*)-1-(2,6,6-trimethyl-1-cyclohexa-1,3-dienyl)but-2-en-1-one.
[b]Stereochemistry is shown for compounds that have data reported for individual stereoisomers. Note that other compounds with chiral centers also exist as stereoisomers but are not always studied individually and are therefore reported together as a single component.
[c]Descriptor and threshold data may refer to matrices such as air, water, hydroalcoholic solution, and white or red wine.
[d]Trace, present but not quantified.
[e]ND, not detected.
[f]Determined in a study of two wines.

Table 8.2 Effect of grape variety on typical concentrations of linalool, geraniol, and (–)-cis-rose oxide in wines [6, 14, 18, 19, 24–26]

Grape varietal	Concentration range in wine (μg/L)		
	Linaool	Geraniol	*cis*-Rose oxide
Muscat	78–462	44–256	ND
Gewurztraminer	60–225	45–221	8–21
Riesling	ND[a]–230	ND–109	ND
Sauvignon Blanc	10–58	ND–6	NR[b]

[a] ND, not detected.
[b] NR, not reported.

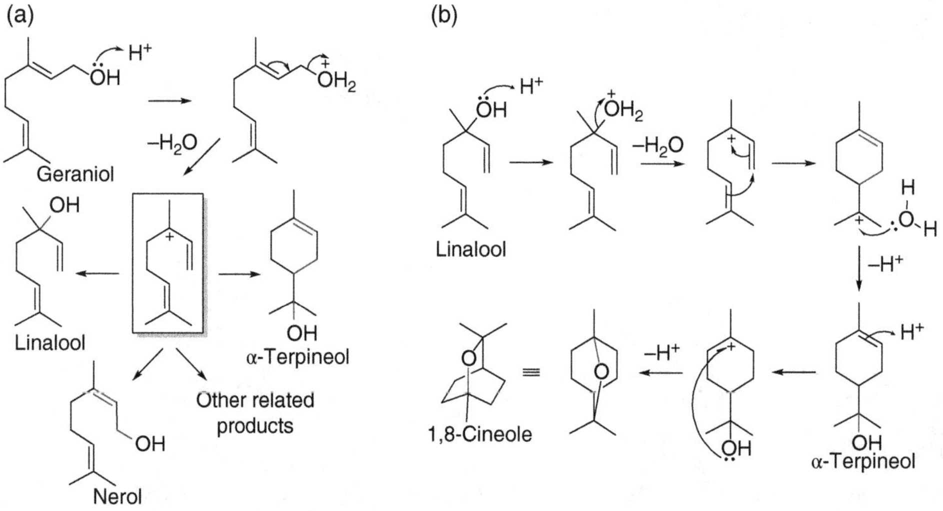

Figure 8.2 Acid-catalyzed reactions of monterpenoids through carbocation intermediates, showing (a) hydrolysis of geraniol and subsequent rearrangement to form other monoterpenoids, including linalool, α-terpineol, and nerol, and (b) formation of 1,8-cineole from grape-derived linalool through hydrolysis and intramolecular cyclization

concentrations are absent in the literature. With descriptors such as "coconut" and "woody", wine lactone was reported at suprathreshold concentrations in Scheurebe and Gewurztraminer wines [6], and potentially arises from acid-catalyzed cyclization of grape-derived acid or glucose ester forms [23] (Chapter 25). Citronellol, nerol, α-terpineol, and dozens of other monoterpenoids (and some monoterpenes such as limonene, *p*-cymene, γ-terpinene, and myrcene[2]) have also been identified in grape musts and wines [5], although often at concentrations well below their sensory thresholds. Note that monoterpenoids can be found as free volatile compounds in grapes, but largely exist as glycosylated (up to 95% of the total) or polyhydroxylated precursors (Chapter 23.1), both of which are odorless.

[2]The non-polar nature of these monoterpene hydrocarbons would limit extraction during winemaking and there may be losses due to volatilization or binding to solids.

Apart from the major impact of cultivar, growing conditions may have a slight effect on monoterpenoids in grapes. Beyond that, absolute or relative concentrations in wine will be influenced by extraction from grapes and transformations of precursors or free volatiles during fermentation and storage [5, 8]. The extent of precursor hydrolysis and rearrangement is affected by temperature, ethanol content, and pH, as described further in Chapters 23.1 and 25. Briefly, factors that can influence wine aroma through changes to monoterpenoid profile include:

- Extraction. Extended skin-contact, harder pressing of grape musts, increased temperature and use of pectolytic enzymes typically increase extraction of free and bound monoterpenoids.
- Enzymatic transformations. Microorganisms are capable of hydrolyzing monoterpenoid glycosides, as can exogenous glycosidases added during winemaking. Additionally, the fermentation process may result in enzymatic reduction of alkenes and acetylation of alcohols.
- Chemical transformations. Acid-catalyzed hydrolysis or rearrangement of precursors yields a range of monoterpenoids,[3] which may undergo further reactions at ordinary winemaking pH and temperature (Figure 8.2a). Lower pH and elevated temperature (e.g., from flash pasteurization or during storage) will accelerate these reactions.[4]

Most monoterpenoids in wine are derived from grape secondary metabolites, but 1,8-cineole (eucalyptol, Table 8.1) can have an alternate origin. Predominantly limited to red wines, 1,8-cineole also happens to be one of the principal essential oils in *Eucalyptus* leaves. It could be envisaged to form through acid-catalyzed cyclization of α-terpineol, which arises from linalool (or the monoterpene limonene) (Figure 8.2b); transformation of such compounds can occur under acidic model wine conditions to yield small amounts of 1,8-cineole [13]. A slightly more important pathway appears to be airborne transfer of volatiles from *Eucalyptus* trees growing in the vicinity of grapevines [16], as previously suggested [27]. However, the major source of 1,8-cineole in red wines appears to be from *Eucalyptus* leaves and other matter (i.e., material other than grapes (MOG), such as grape vine leaves and stems) being present during red wine fermentations [16] (Figure 8.3). Studies such as these highlight the need to consider alternative (i.e., environmental) sources of wine aromas that are not definitively linked to grape and yeast metabolites or maturation and storage phenomena (beyond the known exogenous contaminants, Chapter 18).

8.4 Sesquiterpenoids

A number of sesquiterpenoids have been identified in wine as free volatile compounds, including farnesol, nerolidol, and rotundone (Table 8.1). Presumably hydroxylated variants exist in glycosidic forms, based on their presence in other plants [28], an assumption supported by the increase in concentration of farnesol during malolactic fermentation or hydrolysis of grape glycosidic extracts [29]. However, in contrast to other isoprenoid secondary metabolites, no sesquiterpenoid glycosides have been identified to date. Sesquiterpenes such as α-cedrene, α-farnesene, α-ylangene, and α-guaiene have also been found in wine but are not studied in depth due to their low concentrations with respect to sensory thresholds. Nonetheless, as is common in the plant kingdom, a wide range of functional sesquiterpenes and related compounds are biosynthesized in grapes

[3] Products of acid hydrolysis can be different than those derived from hydrolysis by glycosidases, with the latter tending to better preserve the natural monoterpenoid profile of the grapes.

[4] In a similar way, isolation conditions employed during analysis can contribute to reactions that alter the apparent distribution of monoterpenoids determined in wine.

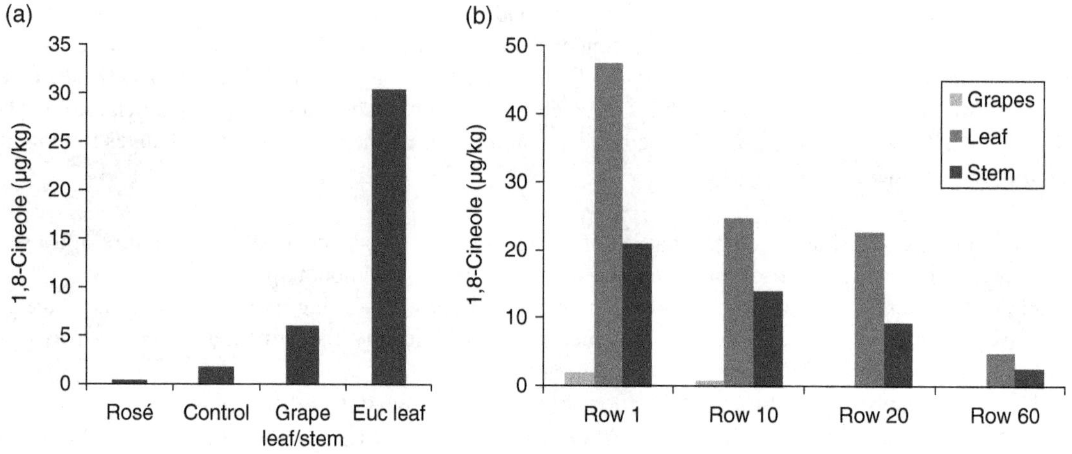

Figure 8.3 *1,8-Cineole concentrations in (a) Shiraz wines made either without skin contact (i.e., rosé style), with maceration on skins (control), and maceration with the addition of small proportions of grapevine leaf and stem (grape leaf/stem) or Eucalyptus leaves and bark (Euc leaf), and (b) extracts from Shiraz grapes, grapevine leaves, and grapevine stems sampled from a vineyard at increasing distances from a stand of Eucalyptus trees, with Row 60 being furthest away. Data from Reference [16]*

(and by yeast to some extent) in cultivars such as Cabernet Sauvignon, Shiraz, Riesling, Gewurztraminer, and some Portuguese varieties (e.g., References [30] to [32]).

From a flavor perspective, the primary aroma compound, rotundone, is the most notable sesquiterpenoid found in wine. Rotundone possesses a characteristic "black pepper" aroma and a low odor detection threshold (16 ng/L) and was first identified in the essential oil of nut grass (*Cyperus rotundus*, hence the name given to this bicyclic ketone). It is also found in high concentrations in black and white pepper and at lower levels in various herbs [33]. Rotundone was first identified in Shiraz wines (up to 145 ng/L) [33] and has since been detected in many other red wines, such as Cabernet Sauvignon, Durif, Mourvedre, Vespolina, and Schioppettino (up to ca. 560 ng/L in the latter two varieties). It may also be found in white wines, particularly Gruner Veltliner (up to 266 ng/L) [34].

Beyond grape variety, rotundone concentrations increase with grape maturation, and higher concentrations are correlated with cooler viticultural regions and cooler vintages; analogously, shading and cooler berry temperatures are reported to lead to higher rotundone, even within the same cluster [35, 36]. Similar to flavonoids (Chapter 11), rotundone is located primarily in the grape skins, which explains the higher concentrations in red wines. Rotundone is highly hydrophobic, and ~90% is bound to lees and marc, with further losses occurring during filtration operations [37]. Inclusion of grape leaves and stems can increase rotundone by up to 6-fold as compared to fermentation of grapes alone [16]. This reinforces the notion presented above (for 1,8-cineole) that MOG (and non-grapevine material) included in a fermentation can have important influences on wine aroma.

8.5 C_{13}-Norisoprenoids

An array of C_{13}-norisoprenoids have been found in all of the internationally important varietal wines, for example, Chardonnay, Riesling, Semillon, Sauvignon Blanc, Cabernet Sauvignon, Pinot Noir, and Shiraz [11]. From a sensory standpoint, the more important contributions to wine aroma appear to be β-damascenone, β-ionone, and

TDN (Table 8.1), although several others (vitispirane, Riesling acetal, actinidols, and (*E*)-1-(2,3,6-trimethylphenyl) buta-1,3-diene (TPB)) have also been detected. More generally, norisoprenoids (especially C_{13} derivatives; C_9–C_{11} also exist) are ubiquitous flavor and fragrance compounds, and arise from the enzymatic or chemical breakdown of carotenoid pigments (C_{40}) such as neoxanthin and β-carotene [38], either directly or through intermediate glycosides.

As is a common theme for the isoprenoid class of compounds, the acidic nature of grape and wine matrices (and the presence of enzymes/act of fermentation) means that significant transformations of non-volatile carotenoid-derived precursors take place to yield compounds that then impact on wine aroma. Formation of β-damascenone, for example, involves oxidative cleavage of neoxanthin in grapes to yield grasshopper ketone, its subsequent enzymatic reduction to an allenic triol intermediate, and finally acid-catalyzed hydrolysis steps during winemaking (potentially of glycosylated intermediates as well, Chapter 23.1) to produce β-damascenone (Figure 8.4a). In contrast, β-ionone can arise directly in grapes after oxidative cleavage of β-carotene (Figure 8.4b). Carotenoid cleavage dioxygenases (CCDs) are the enzymes responsible for the regiospecific oxidative degradation of a range of carotenoids, leading to different norisoprenoid and polyene primary cleavage products [1, 11, 39].

Of the C_{13}-norisoprenoids, the best studied are TDN and β-damascenone. TDN has an aroma reminiscent of "kerosene" and a sensory threshold of 2 μg/L (Table 8.1). It has been detected in several varietal wines, such as Chardonnay, Sauvignon Blanc, Pinot Noir, and Cabernet Sauvignon at near-threshold concentrations (e.g., mean = 1.3 μg/L in non-Riesling wines, maximum = 6.4 μg/L in a Cabernet Franc) [14]. However, TDN is best associated with Riesling wines and particularly aged Rieslings, where it can be present at concentrations over 50 μg/L and dominate the aroma of the wine. The sensory effects of lower concentrations of suprathreshold TDN, as are found in younger Riesling wines (e.g., mean = 6.4 μg/L, maximum = 17.1 μg/L)

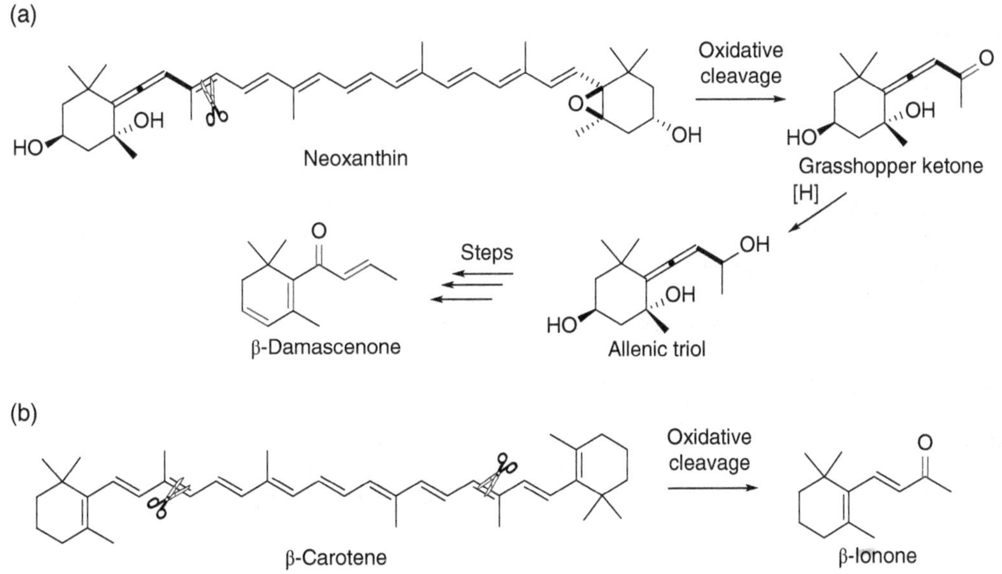

Figure 8.4 Formation of C_{13}-norisoprenoids through oxidative degradation of carotenoid (a) neoxanthin to form β-damascenone after reduction (designated as [H]) and acid-catalyzed hydrolysis (Chapter 23.1) and (b) β-carotene to give two molecules of β-ionone directly (whereas α-carotene gives one molecule of α-ionone and one of β-ionone)

are not clear [14], but these concentrations do not appear to lead to consumer rejection [40]. Apart from the grape variety, the factors that affect TDN are:

- Wine age, pH, and storage temperature. TDN is at negligible concentrations in grapes, but can instead be formed by acid-catalyzed hydrolysis of compounds such as Riesling acetal and glycosidic precursors during storage [11, 14]. Older wines, wines with lower pH, and those stored at elevated temperatures exhibit higher concentrations of TDN (Chapter 25).
- Growing conditions. Cluster light exposure and warmer growing conditions can result in 2–4-fold increases in TDN precursors. The biological mechanism for this effect is still unresolved, but may be related to altered carotenoid profiles or carotenoid degradation.
- Absorption. Once formed, TDN is stable and may accumulate to concentrations above 50 μg/L over time. However, being a highly non-polar compound, it may be lost due to scalping by non-polar packaging materials[5] (i.e., absorption of aroma volatiles by cork or synthetic closures [41]).

β-Damascenone ("cooked apple") is widely detected in wines, and unlike TDN its concentration does not appear to be dependent on variety [42]. Although β-damascenone has a very low sensory threshold in water and is often reported to have the highest odor activity of any compound in wine, it is not an impact odorant and its aroma is rarely if ever the dominant sensation perceived in a wine [42]. Instead, it appears to modify the perception of other odorants – low concentrations of β-damascenone can enhance fruitiness associated with esters while suppressing green characters from methoxypyrazines [43]. β-Damascenone also possesses a highly matrix-dependent sensory threshold – its detection threshold in a red wine is reported to be over 3 orders of magnitude higher than its threshold in water [44]. While true of all compounds, the statement that OAVs merely act as a guide to the impact of an odorant (see the Introduction chapter) is of particular relevance to β-damascenone.

References

1. Schwab, W., Davidovich-Rikanati, R., Lewinsohn, E. (2008) Biosynthesis of plant-derived flavor compounds. *The Plant Journal*, **54** (4), 712–732.
2. Martin, D.M., Chiang, A., Lund, S.T., Bohlmann, J. (2012) Biosynthesis of wine aroma: transcript profiles of hydroxymethylbutenyl diphosphate reductase, geranyl diphosphate synthase, and linalool/nerolidol synthase parallel monoterpenol glycoside accumulation in Gewürztraminer grapes. *Planta*, **236** (3), 919–929.
3. Ribereau-Gayon, P., Boidron, J.N., Terrier, A. (1975) Aroma of Muscat grape varieties. *Journal of Agricultural and Food Chemistry*, **23** (6), 1042–1047.
4. Ohloff, G. (1978) Importance of minor components in flavors and fragrances. *Perfumer and Flavorist*, **3**, 11–22.
5. Marais, J. (1983) Terpenes in the aroma of grapes and wines: a review. *South African Journal for Enology and Viticulture*, **4** (2), 49–58.
6. Guth, H. (1997) Quantitation and sensory studies of character impact odorants of different white wine varieties. *Journal of Agricultural and Food Chemistry*, **45** (8), 3027–3032.
7. Ferreira, V., López, R., Cacho, J.F. (2000) Quantitative determination of the odorants of young red wines from different grape varieties. *Journal of the Science of Food and Agriculture*, **80** (11), 1659–1667.
8. Mateo, J.J. and Jimenez, M. (2000) Monoterpenes in grape juice and wines. *Journal of Chromatography A*, **881** (1–2), 557–567.

[5] This is also the case for other non-polar compounds, such as naphthalene and long-chain ethyl esters, which are absorbed by cork and especially synthetic closures, but not screw caps. More polar compounds, such as oak volatiles, short-chain ethyl esters, and methoxypyrazines, tend not to be affected by closure type. The absorption of non-polar aroma compounds on hydrophobic surfaces would appear to be a generally applicable phenomenon in winemaking.

9. Castro Vázquez, L., Pérez-Coello, M.S., Cabezudo, M.D. (2002) Effects of enzyme treatment and skin extraction on varietal volatiles in Spanish wines made from Chardonnay, Muscat, Airén, and Macabeo grapes. *Analytica Chimica Acta*, **458** (1), 39–44.
10. Coelho, E., Perestrelo, R., Neng, N.R., *et al.* (2008) Optimisation of stir bar sorptive extraction and liquid desorption combined with large volume injection-gas chromatography-quadrupole mass spectrometry for the determination of volatile compounds in wines. *Analytica Chimica Acta*, **624** (1), 79–89.
11. Mendes-Pinto, M.M. (2009) Carotenoid breakdown products the-norisoprenoids-in wine aroma. *Archives of Biochemistry and Biophysics*, **483** (2), 236–245.
12. Burdock, G.A. (2010) *Fenaroli's handbook of flavor ingredients*, 6th edn, CRC Press, Boca Raton, FL.
13. Capone, D.L., Van Leeuwen, K., Taylor, D.K., *et al.* (2011) Evolution and occurrence of 1,8-cineole (eucalyptol) in Australian wine. *Journal of Agricultural and Food Chemistry*, **59**, 953–959.
14. Sacks, G.L., Gates, M.J., Ferry, F.X., *et al.* (2012) Sensory threshold of 1,1,6-trimethyl-1,2-dihydronaphthalene (TDN) and concentrations in young Riesling and non-Riesling wines. *Journal of Agricultural and Food Chemistry*, **60** (12), 2998–3004.
15. Bowen, A.J. and Reynolds, A.G. (2012) Odor potency of aroma compounds in Riesling and Vidal blanc table wines and icewines by Gas Chromatography–Olfactometry–Mass Spectrometry. *Journal of Agricultural and Food Chemistry*, **60** (11), 2874–2883.
16. Capone, D.L., Jeffery, D.W., Sefton, M.A. (2012) Vineyard and fermentation studies to elucidate the origin of 1,8-cineole in Australian red wine. *Journal of Agricultural and Food Chemistry*, **60** (9), 2281–2287.
17. Coetzee, C., Lisjak, K., Nicolau, L., *et al.* (2013) Oxygen and sulfur dioxide additions to Sauvignon blanc must: effect on must and wine composition. *Flavour and Fragrance Journal*, **28** (3), 155-–167.
18. Paula Barros, E., Moreira, N., Elias Pereira, G., *et al.* (2012) Development and validation of automatic HS-SPME with a gas chromatography-ion trap/mass spectrometry method for analysis of volatiles in wines. *Talanta*, **101**, 177–186.
19. Ong, P.K.C. and Acree, T.E. (1999) Similarities in the aroma chemistry of Gewürztraminer variety wines and lychee (*Litchi chinesis* Sonn.) fruit. *Journal of Agricultural and Food Chemistry*, **47** (2), 665–670.
20. Guth, H. (1996) Determination of the configuration of wine lactone. *Helvetica Chimica Acta*, **79** (6), 1559–1571.
21. Guth, H. (1997) Identification of character impact odorants of different white wine varieties. *Journal of Agricultural and Food Chemistry*, **45** (8), 3022–3026.
22. López, R., Ferreira, V., Hernández, P., Cacho, J.F. (1999) Identification of impact odorants of young red wines made with Merlot, Cabernet Sauvignon and Grenache grape varieties: a comparative study. *Journal of the Science of Food and Agriculture*, **79** (11), 1461–1467.
23. Giaccio, J., Capone, D.L., Håkansson, A.E., *et al.* (2011) The formation of wine lactone from grape-derived secondary metabolites. *Journal of Agricultural and Food Chemistry*, **59** (2), 660–664.
24. Selli, S., Canbas, A., Cabaroglu, T., *et al.* (2006) Aroma components of cv. Muscat of Bornova wines and influence of skin contact treatment. *Food Chemistry*, **94** (3), 319–326.
25. Kozina, B., Karoglan, M., Herjavec, S., *et al.* (2008) Influence of basal leaf removal on the chemical composition of Sauvignon Blanc and Riesling wines. *Journal of Food Agriculture and Environment*, **6** (1), 28–33.
26. Jouanneau, S., Weaver, R.J., Nicolau, L., *et al.* (2012) Subregional survey of aroma compounds in Marlborough Sauvignon Blanc wines. *Australian Journal of Grape and Wine Research*, **18** (3), 329–343.
27. Herve, E., Price, S., Burns, G. (eds) (2003) Eucalyptol in wines showing a "Eucalyptus" aroma. VIIème Symposium International d'Œnologie, Actualités Œnologiques, 19–21 June 2003, Bordeaux, France, Tec & Doc Lavoisier, Paris.
28. Stahl-Biskup, E., Intert, F., Holthuijzen, J., *et al.* (1993) Glycosidically bound volatiles – a review 1986–1991. *Flavour and Fragrance Journal*, **8** (2), 61–80.
29. Ugliano, M. and Moio, L. (2006) The influence of malolactic fermentation and *Oenococcus oeni* strain on glycosidic aroma precursors and related volatile compounds of red wine. *Journal of the Science of Food and Agriculture*, **86** (14), 2468–2476.
30. Parker, M., Pollnitz, A.P., Cozzolino, D., *et al.* (2007) Identification and quantification of a marker compound for "Pepper" aroma and flavor in Shiraz grape berries by combination of chemometrics and gas chromatography–mass spectrometry. *Journal of Agricultural and Food Chemistry*, **55** (15), 5948–5955.

31. Coelho, E., Coimbra, M.A., Nogueira, J.M.F., Rocha, S.M. (2009) Quantification approach for assessment of sparkling wine volatiles from different soils, ripening stages, and varieties by stir bar sorptive extraction with liquid desorption. *Analytica Chimica Acta*, **635** (2), 214–221.
32. May, B. and Wüst, M. (2012) Temporal development of sesquiterpene hydrocarbon profiles of different grape varieties during ripening. *Flavour and Fragrance Journal*, **27** (4), 280–285.
33. Wood, C., Siebert, T.E., Parker, M., *et al.* (2008) From wine to pepper: Rotundone, an obscure sesquiterpene, is a potent spicy aroma compound. *Journal of Agricultural and Food Chemistry*, **56** (10), 3738–3744.
34. Mattivi, F., Caputi, L., Carlin, S., *et al.* (2011) Effective analysis of rotundone at below-threshold levels in red and white wines using solid-phase microextraction gas chromatography/tandem mass spectrometry. *Rapid Communications in Mass Spectrometry*, **25** (4), 483–488.
35. Scarlett, N.J., Bramley, R.G.V., Siebert, T.E. (2014) Within-vineyard variation in the 'pepper' compound rotundone is spatially structured and related to variation in the land underlying the vineyard. *Australian Journal of Grape and Wine Research*, **20**, 214–222.
36. Zhang, P., Barlow, S., Krstic, M., *et al.* (2015) Within-vineyard, within-vine, and within-bunch variability of the rotundone concentration in berries of *Vitis vinifera* L. cv. Shiraz. *Journal of Agricultural and Food Chemistry*, **63** (17), 4276–4283.
37. Caputi, L., Carlin, S., Ghiglieno, I., *et al.* (2011) Relationship of changes in rotundone content during grape ripening and winemaking to manipulation of the "peppery" character of wine. *Journal of Agricultural and Food Chemistry*, **59** (10), 5565–5571.
38. Winterhalter, P. and Rouseff, R. (2001) Carotenoid-derived aroma compounds: an introduction, in *Carotenoid-derived aroma compounds*, American Chemical Society, pp. 1–17.
39. Walter, M., Floss, D., Strack, D. (2010) Apocarotenoids: hormones, mycorrhizal metabolites and aroma volatiles. *Planta*, **232** (1), 1–17.
40. Ross, C.F., Zwink, A.C., Castro, L., Harrison, R. (2014) Odour detection threshold and consumer rejection of 1,1,6-trimethyl-1,2-dihydronaphthalene in 1-year-old Riesling wines. *Australian Journal of Grape and Wine Research*, **20** (3), 335–339.
41. Capone, D., Sefton, M., Pretorius, I., Høj, P. (2003) Flavour "scalping" by wine bottle closures – the "winemaking" continues post vineyard and winery. *Australia and New Zealand Wine Industry Journal*, **18** (5), 16, 18–20.
42. Sefton, M.A., Skouroumounis, G.K., Elsey, G.M., Taylor, D.K. (2011) Occurrence, sensory impact, formation, and fate of damascenone in grapes, wines, and other foods and beverages. *Journal of Agricultural and Food Chemistry*, **59** (18), 9717–9746.
43. Escudero, A., Campo, E., Farina, L., *et al.* (2007) Analytical characterization of the aroma of five premium red wines. Insights into the role of odor families and the concept of fruitiness of wines. *Journal of Agricultural and Food Chemistry*, **55** (11), 4501–4510.
44. Pineau, B., Barbe, J.C., Van Leeuwen, C., Dubourdieu, D. (2007) Which impact for beta-damascenone on red wines aroma? *Journal of Agricultural and Food Chemistry*, **55** (10), 4103–4108.

9

Aldehydes, Ketones, and Related Compounds

9.1 Introduction

Aldehydes and ketones originate as fermentation metabolites as well as oxidation products and in many situations the source is difficult to discern. However, once a wine is *microbially stable*, additional aldehydes or ketones arise from non-enzymatic oxidation pathways, generally from oxidation of the analogous alcohol, but also from less direct routes such as reaction of amino acids with α-dicarbonyls [1] (Figure 9.1). Grape-derived aldehydes are also detectable in juice following crushing, and particularly six-carbon aldehydes formed by enzymatic oxidation of grape lipids (Chapters 6 and 23.3). However, these compounds will be reduced to their corresponding alcohols during winemaking.

The critical phenomena associated with these compounds include:

- The carbonyl carbon of aldehydes, α-diketones, and, to a lesser extent, ketones is electropositive, especially in acid, and these compounds are fairly reactive *electrophiles* (Chapter 10). Because wine is rich in nucleophiles (e.g., phenolics, sulfites), the most reactive carbonyl compounds will not accumulate in wine, but instead react with other compounds as a part of wine aging; significant accumulation of these carbonyls is a mark of a highly oxidized wine (Chapter 24).[1]
- Volatile aldehydes and ketones often have 100–10000-fold lower sensory thresholds than their corresponding alcohols and thus small degrees of oxidation can affect wine aroma.
- Aldehydes and ketones can form reversible adducts with bisulfite (yielding sulfonates, Chapter 17), alcohols (yielding hemiacetals and acetals), or create adducts or bridges between two phenolic structures through addition to the A-ring of flavonoids (Chapter 24).[2]
- The sensory perception of odorous carbonyl compounds can be attenuated by the reversible formation of sulfonates, so sensory studies should control for the concentration of sulfites.

[1] Some wine styles are typified by an "oxidized" aroma; for instance Sherry or Madeira are the result of air exposure with or without yeast.
[2] Proper color development in red wine is dependent on the reactions between oxidation products and the tannins and anthocyanins, as described in Chapter 25.

Understanding Wine Chemistry, First Edition. Andrew L. Waterhouse, Gavin L. Sacks, and David W. Jeffery.

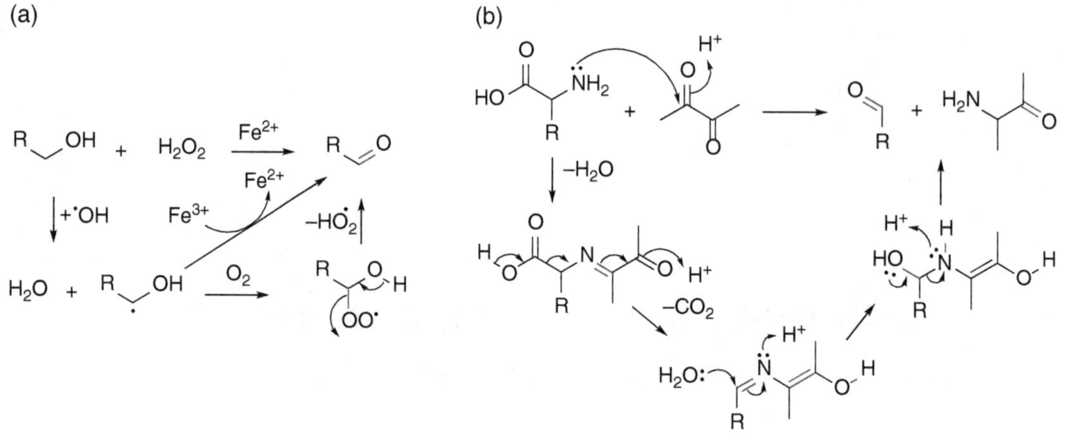

Figure 9.1 Examples of two pathways to form aldehydes and ketones via (a) Fenton oxidation of alcohols and (b) Strecker reaction of an amino acid with diacetyl, a common wine component (note that the enol tautomerization step to give the α-aminoketone as the other product is not shown)

9.2 Acetaldehyde

Acetaldehyde, the best documented of the wine aldehydes and certainly the most abundant, is chemically related to ethanol, from which much is derived by oxidation. Acetaldehyde concentrations in standard table wine are in the 20–100 mg/L range, and often at levels described as the sensory threshold (40–100 mg/L). However, it is also a very strong SO_2 binder ($K_d = 1\times10^{-6}$, Chapter 17) such that <1% of acetaldehyde typically exists in the free, volatile form in wines with free SO_2 present. Consequently, acetaldehyde may be of negligible importance to table wine aroma but could contribute to wines with low SO_2 such as Sherries [2]. The sensory descriptors of acetaldehyde vary with concentration; at low levels, it enhances fruitiness, but with higher concentrations, it is reminiscent of nuts, and at still higher levels smells like rotten apple (Table 9.1).

Acetaldehyde is produced by yeast during alcoholic fermentation as the penultimate step before ethanol formation (Chapter 22.1). As a result, the total amount of acetaldehyde produced during fermentation is nearly equivalent to the molar amount of ethanol produced (~2 M or 90 g/L). However, since most acetaldehyde is immediately reduced to ethanol by yeast, the final wine concentration is reported to average 25 mg/L in reds and 40 mg/L in whites [3], with variations due to yeast strain and initial must SO_2 concentration (Chapter 17). After fermentation, additional acetaldehyde will appear due to the oxidation of ethanol via the Fenton reaction (Figure 9.1).

9.3 Short and medium chain aldehydes

9.3.1 Higher alcohol-derived aldehydes

Higher alcohols are readily oxidized to aldehydes, and due to their relatively high concentrations (i.e., hundreds of mg/L, Chapter 6), oxidation can yield suprathreshold concentrations of some odorous aldehydes. Short branched-chain aldehydes (C_4 and C_5), including 2-methylpropanal and 2- and 3-methylbutanal, are associated with dried fruit, woody, and sweet and fusel type aromas (Table 9.1). These compounds are found in many wines and are particularly high in aged ports and Sherries, although they are found in any aged wine. The amounts of these aldehydes in young wines are between 5 to 10 μg/L, but in older oxidized wines, the

Table 9.1 *Aldehydes and ketones in wines. Table derived from References [3] to [14]*

Name	Structure	Odor descriptor	Threshold (μg/L)	Range (μg/L)
Acetaldehyde (Ethanal)	CH_3CHO	Fruity to rotting apple	100000	ND–211 000
2-Methylpropanal (Isobutyraldehyde)	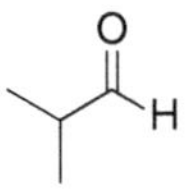	Banana, melon, varnish, cheese	6	1–200
2-Methylbutanal	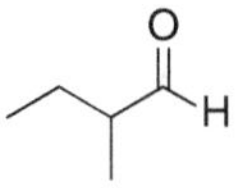	Green grass, fruity	16	3–100
3-Methylbutanal (Isovaleraldehyde)	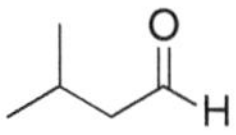	Unripe banana, apple, cheese, amylic	4	40–250
Octanal	$CH_3(CH_2)_6CHO$	Citrus	2.5	0.04–6
Nonanal	$CH_3(CH_2)_7CHO$	Citrus	2.5	0.05–9
Decanal	$CH_3(CH_2)_8CHO$	Citrus	1.25	0.07–1.6
(*E*)-2-octenal	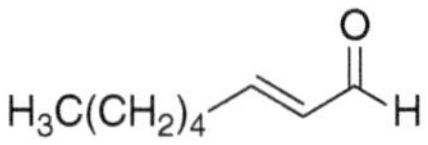		3	0.04–4
(*E*)-2-nonenal	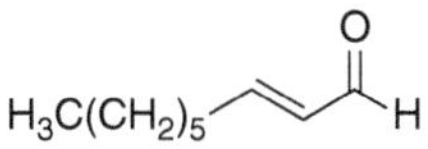	Greasy	0.6	0.1–9
Methional [9]	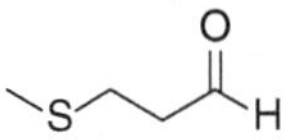	Cooked potato, cabbage	0.5	0.5–80
Phenylacetaldehyde	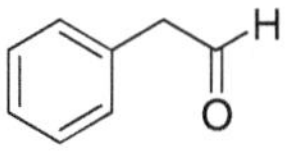	Floral	1	2.5–130
Diacetyl	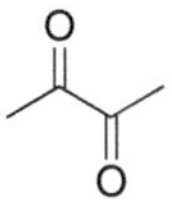	Butter	100	5–7500
Acetoin	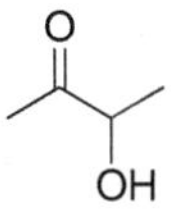	Butter, cream	150 000	100–60000

(continued overleaf)

Table 9.1 *(continued)*

Name	Structure	Odor descriptor	Threshold (μg/L)	Range (μg/L)
Furfural	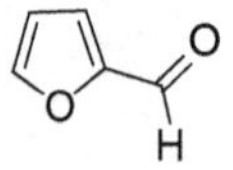	Caramel	15,000	0–5000
Sotolon	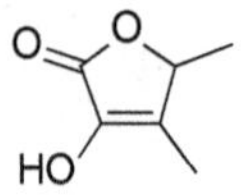	Maple syrup	8	1–6
3-Methyl-2,4-nonanedione	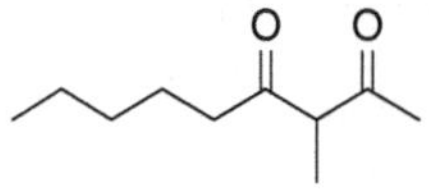	Prune, minty, anise	0.016	0.004–0.3
1,1-Diethoxyethane	$CH_3CH(OCH_2CH_3)_2$	Licorice, green fruit	1400	500–70 000
2-Methyl-4-hydroxymethyl-dioxolane and dioxane isomers	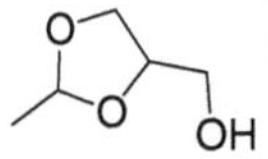	Sweet, port-like	100 000 (mix of 4 isomers)	200–1400
Non-volatile				*Average level*
Pyruvic acid	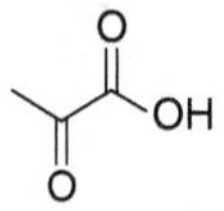			14 000 reds, 25 000 whites
Glyoxylic acid	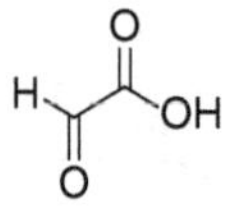			NR[a]
Glyoxal	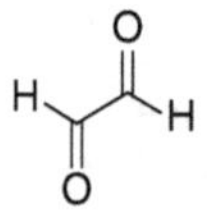			100–2000
Methylglyoxal	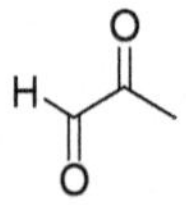			100–1000

[a] Not reported, presence inferred by reaction products.

levels are much higher, at 20–80 μg/L [4]. These aldehydes have been shown to be involved in reactions with flavonoids, analogous to acetaldehyde [15].

Methional (cooked potato aroma) and phenylacetaldehyde (floral aroma) are very potent aldehydes arising from the oxidation of corresponding alcohols, which are derived from amino acids methionine and phenylalanine, respectively (Chapter 6). These have the aroma of cooked vegetables, honey, and other fermented foods. Higher concentrations of these compounds are more often observed in older wines, and are strongly associated with quality degradation [4] and oxidized character [9].

9.3.2 Medium chain aldehydes (C_8–C_{10})

These are found in many foods, and impart citrus fruit type aromas to wine (Table 9.1). These compounds are added to many foods, cosmetics, and cleaning agents, so they are ubiquitous in the modern environment.[3] Their concentrations are usually correlated in a given wine and their sensory effect appears to be additive. However, due to their low levels they are thought to only rarely impact sensory attributes, and even in their absence wines can still present similar citrus fruit nuances [7].

9.3.3 (*E*)-2-Alkenals

These are α,β-unsaturated aldehydes with an (*E*)- (*trans*-) double bond at the 2 position. The most potent (*E*)-2-alkenals in wine have medium chain length (C_6–C_9), of which (*E*)-2-nonenal is the most potent. The C_6 form ((*E*)-2-hexenal, "grassy") is formed at high concentrations in must by enzymatic oxidation of linolenic acid (Chapter 23.3), but is largely metabolized during fermentation. The other (*E*)-2-alkenals have aromas of mushroom, mold, earth, and dust (Table 9.1). In finished wines, these compounds are likely formed through non-enzymatic degradation of unsaturated fatty acids. The (*E*)-2-alkenals are generally correlated with negative sensory attributes and low price points, and most wines do not have suprathreshold levels unless they are oxidized or long-aged [5].

9.3.4 1,2- or α-dicarbonyl compounds

These compounds contain two adjacent carbonyl groups. Many of these species are produced by microbial metabolism. For example, diacetyl ("butter") primarily arises from lactic acid bacterial activity (Chapter 22.5) and is associated with the buttery aromas of wine (Table 9.1). It is closely related to the less potent α-hydroxy-carbonyl compound, acetoin, via reduction of one of the carbonyl moieties, and 2,3-butanediol via further reduction. The proportion of diacetyl versus acetoin is diagnostic for the status of lactic acid oxidation [16] and these compounds are thought to react with thiols to yield potent heterocyclic products with roasted and herbaceous aromas [17]. Diacetyl is typically present at 0.2–2.5 mg/L in commercial wine but can be 2–3 times higher during malolactic fermentations. As noted above in Figure 9.1, diacetyl and other 1,2-dicarbonyls can react in a Strecker reaction with amino acids to produce other odor-active aldehydes.

Glyoxal and methylglyoxal can be produced both by microbial metabolism and by abiotic oxidative processes such as the Maillard reaction [18]. These are present in wine at concentrations ranging from 0.1 to 2 mg/L. As noted for other aldehydes, levels are expected to decrease with time due to the reaction of glyoxal with flavonoids. Glyoxal is near odorless, but it appears that some of its secondary reaction products may form odorous heterocyclic reaction products, such as thiazoles [19]. In addition to the more abundant α-dicarbonyl compounds such as diacetyl and glyoxal, there is also the homologous pentane-2,3-dione as well as phenylglyoxal. All of these compounds increase during lactic acid fermentations, diminish if wine is subsequently aged on lees due to enzymatic reduction (Chapter 22.5), but remain relatively static if the wine is racked [8].

The medium-chain β-diketone, 3-methyl-2,4-nonanedione has recently been identified in red wines. It was found in a search for the prune character of oxidatively aged red wines, although its descriptors in model wine are concentration dependent and also include mint and anise [20]. It has an exceptionally low aroma detection threshold of 16 ng/L in model wine, and appears to be present in aged red Bordeaux wines well above threshold. This diketone may be a general contributor to aged wine character Its origin is unclear, although its formation in soybean oil is linked to unsaturated fatty acids.

---

[3] Synthetic aldehydes ranging in length from eight and thirteen carbons were a critical innovation in the famed perfume Chanel No. 5, created by Ernest Beaux in 1921.

9.4 Complex carbonyls

A significant number of carbonyls originate from oxidation of complex precursor molecules such as organic acids, which result in highly functionalized products. Many of these precursors are themselves produced by oxidation, but also exist as microbial metabolites from fermentation, so their source in a particular wine is difficult to discern. In general, these compounds have no aroma due to the multiple oxygenated groups, but their importance lies in their reactivity and ability to modify wine color, and possibly astringency, through reactions with anthocyanins and/or tannins (Chapters 14, 16, and 25).

Several α-keto-carboxylic acids are reported in wine (i.e. α-keto acids, Chapter 22.3) and may contribute to SO_2 binding due to their relatively high concentrations (Chapter 17). Pyruvic acid is produced by yeast metabolism (Chapter 22.1) as well as by Fenton oxidation of lactic or malic acid. Pyruvic acid is of particular interest for its ability to react with anthocyanins to yield stabilized pigments (Chapter 16). Published values for typical pyruvic acid levels cover a wide range, with concentrations after fermentation reportedly about 100 mg/L, and mean values of 14 mg/L in commercial reds and 25 mg/l in whites [21]. The higher concentrations in white wines likely reflect lower concentrations of phenolic species that would react with the pyruvic acid. Glyoxylic acid has rarely been directly measured in wine [22], but many products of its reaction with flavonoids have been observed [23] (Chapter 24). It is presumed to arise from the Fenton oxidation of tartaric acid, as reported by Fenton himself [24]. α-Ketoglutaric acid is a key intermediate in nitrogen metabolism (Chapter 22.3), and typical wine concentrations are 22 mg/L in whites and 74 mg/L in reds [21].

Sotolon is a potent aroma compound that is reminiscent of maple syrup and is an important aroma component of oxidized wine styles, such as port, vin jaune or sherry, whereas in white wine it appears to diminish varietal character.[4] Studies have shown that sotolon arises in dry wines from the reaction between 2-keto-butyric acid, formed by either yeast metabolism or ascorbic acid degradation, and acetaldehyde [25]. Levels in table wines are usually below the threshold of 8 μg/L in dry white wines, but much higher in oxidized styles [26]. Concentrations in sweet, thermally oxidized wines like Madeira can approach 2000 μg/L. The mechanism for sotolon formation in wine under high sugar and high temperature conditions is not established.

Furfural and related compounds are found in styles of wines that are heated, such as Madeira,[5] or in oak aged wines, where the aldehydes are formed during oak toasting (Chapter 25). In cooked wines like Madeira, these compounds can be formed through acid-catalyzed degradation of sugars (Chapter 25 for the mechanism). Such wines contain related aldehydes such as maltol and cyclotene, which can also arise from sugar degradation [27]. Extraction of furfural into Chardonnay wine from barrels was observed at 2–6 mg/L in the first cycle of use (6 months aging) and, in the same barrels, only 0.2–1.1 mg/L on the second use [28]. 5-Methylfurfural was also observed at 0.4–1 mg/L in the first year of use and 0.1–0.4 mg/L in the second year of use (Chapter 25 for more on oak extractives). In Madeira, levels of furfural between 1 and 24 mg/L and between 1 and 74 mg/L of hydroxymethylfurfural have been reported [29].

Several other carbonyls have been historically implicated as important to wine aroma, but occur at concentrations well below threshold, for example, solerone in sherry. Benzaldehyde ("marzipan", "cherry") may exceed its threshold in wines contaminated by epoxy-resin tanks, but is usually found far below its detection threshold of 2000 μg/L [30], although it may contribute to aromas that are characteristic of wines produced with carbonic maceration.

---

[4] Sotolon is also at high concentrations in fenugreek, a spice common to Mediterranean, Middle Eastern, and Indian cuisines. A maple syrup smell in Manhattan in 2009, suspected by some to be a terrorist chemical weapon attack on the city, was eventually traced to fenugreek being processed at a plant in Bergen County, New Jersey.

[5] Furfural and hydroxymethylfurfural (HMF) are also classic markers of excessive thermal processing in the fruit juice industry, for example, in the production of orange juice.

9.5 Carbonyl reactivity

Due to the electrophilic nature of aldehydes and ketones, one common reaction of these substances is the addition of water and/or alcohols to create hydrates (i.e., $R_2C(OH)_2$), hemiacetals (i.e., $R_2C(OH)OR'$, $R' \neq H$) or acetals (i.e., $R_2C(OR')_2$, $R' \neq H$, and R' groups may be different, affording a mixed acetal) (Chapter 2). Hydrates and hemiacetals rapidly equilibrate with their parent carbonyl compounds in the presence of aqueous acid, but acetals can have slow equilibration times with their respective hemiacetal. Common examples of mixed acetals in wine include the glycosidic forms of alcohols and phenols, for instance geraniol or quercetin glucosides (Chapters 8 and 15), but here the focus is on the reactions of the simpler, volatile compounds.

The simplest of the acetals is 1,1-diethoxyethane, derived from the reaction of ethanol and acetaldehyde. (Figure 9.2) It has been reported at above threshold concentrations in some wines, and has an aroma of apple and licorice in model wine [31] (Table 9.1). Its initial concentration will be dependent on the amount of acetaldehyde produced, for example, by ethanol oxidation, and will be depleted as oxidation slows and the acetaldehyde is consumed by other reactions, for example, with flavanoids (Figure 9.3) or by microbial metabolism.

Acetaldehyde also reacts with other alcohols such as glycerol, an abundant polyol in wine (Chapters 2 and 22.1). With alcohols at three positions, glycerol can form two different cyclic acetals, a dioxolane and a dioxane (Figure 9.2) [32]. These cyclic structures have *cis*- and *trans*- (i.e., (*Z*)- and (*E*)-) isomers, each with two stereoisomers, for a total of four compounds with aromas described as "sweet" and "old port" (Table 9.1). These products are relatively polar and have weak aroma impact unless the wine has been oxidized quite a bit, but they have been proposed as markers of wine oxidation or poor storage conditions [11]. Another related cyclic acetal is the one formed by acetaldehyde and 2,3-butanediol, the reduced form of diacetyl (and acetoin), yielding 2,4,5-trimethyl-1,3-dioxolane. This compound has been reported a few times in studies of

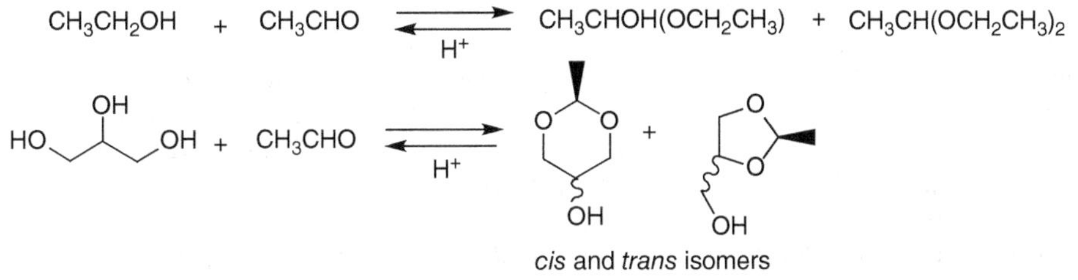

Figure 9.2 Acetal equilibria in the presence of acid in hydroalcoholic media

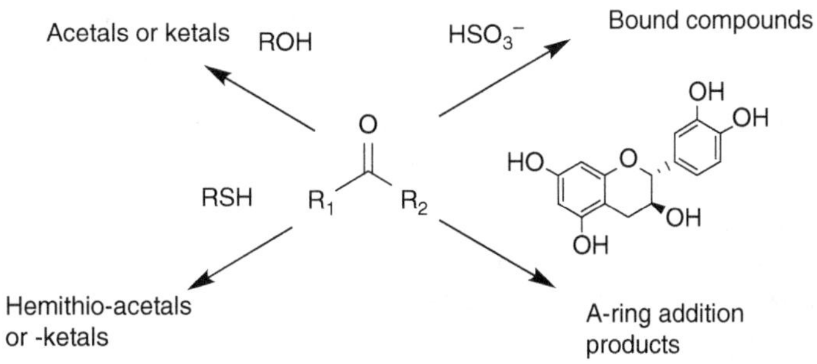

Figure 9.3 Possible reactions of carbonyls in wine

oxidized wine [33], where it appears to contribute green and aldehydic characters. Similar aromas were attributed to acetals arising from acetaldehyde and other diols found in wine.

In addition to these acetals, aldehydes and ketones generally exist in wine partially as their hydrated forms. As there is a rapid equilibrium between the forms the hydrates are often not observed analytically, for example, by gas chromatography, as dehydration occurs during injection.

Carbonyl compounds also react reversibly with other wine nucleophiles, including bisulfite, the phloroglucinol ring of flavonoids, and thiols (Figure 9.3), since carbonyls represent a major class of SO_2 binders, and these bound forms of SO_2 have diminished antimcrobial and antioxidant activity (Chapter 17) [34]. Reactions with flavonoids are also well documented [35] (Chapters 14 and 16), while reactions with thiols are not as well studied [36] (Chapter 24). Any analysis of aldehydes or ketones should account for the possibility of these various addition products, releasing the carbonyls in the course of the analysis.

References

1. Pripis-Nicolau, L., de Revel, G., Bertrand, A., Maujean, A. (2000) Formation of flavor components by the reaction of amino acid and carbonyl compounds in mild conditions. *Journal of Agricultural and Food Chemistry*, **48** (9), 3761–3766.
2. Escudero, A., Asensio, E., Cacho, J., Ferreira, V. (2002) Sensory and chemical changes of young white wines stored under oxygen. An assessment of the role played by aldehydes and some other important odorants. *Food Chemistry*, **77** (3), 325–331.
3. Jackowetz, J.N. and Mira de Orduña, R. (2013) Survey of SO_2 binding carbonyls in 237 red and white table wines. *Food Control*, **32** (2), 687–692.
4. Cullere, L., Cacho, J., Ferreira, V. (2007) An assessment of the role played by some oxidation-related aldehydes in wine aroma. *Journal of Agricultural and Food Chemistry*, **55** (3), 876–881.
5. San Juan, F., Cacho, J., Ferreira, V., Escudero, A. (2012) Aroma chemical composition of red wines from different price categories and its relationship to quality. *Journal of Agricultural and Food Chemistry*, **60** (20), 5045–5056.
6. Bartowsky, E.J., Francis, I.L., Bellon, J.R., Henschke, P.A. (2002) Is buttery aroma perception in wines predictable from the diacetyl concentration? *Australian Journal of Grape and Wine Research*, **8** (3), 180–185.
7. Cullere, L., Ferreira, V., Cacho, J. (2011) Analysis, occurrence and potential sensory significance of aliphatic aldehydes in white wines. *Food Chemistry*, **127** (3), 1397–1403.
8. de Revel, G., Pripis-Nicolau, L., Barbe, J.C., Bertrand, A. (2000) The detection of alpha-dicarbonyl compounds in wine by formation of quinoxaline derivatives. *Journal of the Science of Food and Agriculture*, **80** (1), 102–108.
9. San-Juan, F., Ferreira, V., Cacho, J., Escudero, A. (2011) Quality and aromatic sensory descriptors (mainly fresh and dry fruit character) of Spanish red wines can be predicted from their aroma-active chemical composition. *Journal of Agricultural and Food Chemistry*, **59** (14), 7916–7924.
10. Lavigne, V., Pons, A., Darriet, P., Dubourdieu, D. (2008) Changes in the sotolon content of dry white wines during barrel and bottle aging. *Journal of Agricultural and Food Chemistry*, **56** (8), 2688–2693.
11. Ferreira, A.C.D., Barbe, J.C., Bertrand, A. (2002) Heterocyclic acetals from glycerol and acetaldehyde in port wines: evolution with aging. *Journal of Agricultural and Food Chemistry*, **50** (9), 2560–2564.
12. Pons, A., Lavigne, V., Darriet, P., Dubourdieu, D. (2011) Determination of 3-methyl-2,4-nonanedione in red wines using methanol chemical ionization ion trap mass spectrometry. *Journal of Chromatography A*, **1218** (39), 7023–7030.
13. Guth, H. (1997) Quantitation and sensory studies of character impact odorants of different white wine varieties. *Journal of Agricultural and Food Chemistry*, **45** (8), 3027–3032.
14. Cutzach, I., Chatonnet, P., Dubourdieu, D. (2000) Influence of storage conditions on the formation of some volatile compounds in white fortified wines (vins doux naturels) during the aging process. *Journal of Agricultural and Food Chemistry*, **48** (6), 2340–2345.
15. Pissarra, J., Lourenco, S., Gonzalez-Paramas, A.M., et al. (2004) Structural characterization of new malvidin 3-glucoside-catechin aryl/alkyl-linked pigments. *Journal of Agricultural and Food Chemistry*, **52** (17), 5519–5526.

16. Nielsen, J.C. and Richelieu, M. (1999) Control of flavor development in wine during and after malolactic fermentation by *Oenococcus oeni. Applied and Environmental Microbiology*, **65** (2), 740–745.
17. Marchand, S., de Revel, G., Bertrand, A. (2000) Approaches to wine aroma: release of aroma compounds from reactions between cysteine and carbonyl compounds in wine. *Journal of Agricultural and Food Chemistry*, **48** (10), 4890–4895.
18. Wang, Y. and Ho, C.T. (2012) Flavour chemistry of methylglyoxal and glyoxal. *Chemical Society Reviews*, **41** (11), 4140–4149.
19. de Revel, G., Marchand, S., Bertrand, A. (2004) Identification of Maillard-type aroma compounds in winelike model systems of cysteine–carbonyls: cccurrence in wine, in *Nutraceutical beverages: chemistry, nutrition, and health effects* (eds Shahidi, F. and Weerasinghe, D.K.), American Chemical Society, Washington, pp. 353–364.
20. Pons, A., Lavigne, V., Eric, F., *et al.* (2008) Identification of volatile compounds responsible for prune aroma in prematurely aged red wines. *Journal of Agricultural and Food Chemistry*, **56** (13), 5285–5290.
21. Jackowetz, J.N. and de Orduna, R.M. (2013) Improved sample preparation and rapid UHPLC analysis of SO_2 binding carbonyls in wine by derivatisation to 2,4-dinitrophenylhydrazine. *Food Chemistry*, **139** (1–4), 100–104.
22. Blouin, J. (1966) Contribution to study of binding of sulphur dioxide in musts and wines. *I. Annales de Technologie Agricole*, **15** (3), 223–287.
23. Es-Safi, N.E., Le Guerneve, C., Cheynier, V., Moutounet, M. (2000) New phenolic compounds formed by evolution of (+)-catechin and glyoxylic acid in hydroalcoholic solution and their implication in color changes of grape-derived foods. *Journal of Agricultural and Food Chemistry*, **48** (9), 4233–4240.
24. Fenton, H. (1894) Oxidation of tartaric acid in the presence of iron. *Journal of the Chemical Society*, **75**, 1–11.
25. Pons, A., Lavigne, V., Landais, Y., *et al.* (2010) Identification of a Sotolon Pathway in dry white wines. *Journal of Agricultural and Food Chemistry*, **58** (12), 7273–7279.
26. Moyano, L., Zea, L., Moreno, J.A., Medina, M. (2010) Evaluation of the active odorants in Amontillado sherry wines during the aging process. *Journal of Agricultural and Food Chemistry*, **58** (11), 6900–6904.
27. Belitz, H.D., Grosch, W., Schieberle, P. (2009) *Food Chemistry*, Springer-Verlag, Berlin.
28. Towey, J.P. and Waterhouse, A.L. (1996) The extraction of volatile compounds from French and American oak barrels in Chardonnay during three successive vintages. *American Journal of Enology and Viticulture*, **47**, 163–172.
29. Camara, J.S., Alves, M.A., Marques, J.C. (2006) Changes in volatile composition of Madeira wines during their oxidative ageing. *Analytica Chimica Acta*, **563** (1–2), 188–197.
30. Peinado, R.A., Moreno, J., Bueno, J.E., *et al.* (2004) Comparative study of aromatic compounds in two young white wines subjected to pre-fermentative cryomaceration. *Food Chemistry*, **84** (4), 585–590.
31. Moyano, L., Zea, L., Moreno, J., Medina, M. (2002) Analytical study of aromatic series in sherry wines subjected to biological aging. *Journal of Agricultural and Food Chemistry*, **50** (25), 7356–7361.
32. Peterson, A.L., Gambuti, A., Waterhouse, A.L. (2015) Rapid analysis of heterocyclic acetals in wine by stable isotope dilution gas chromatography–mass spectrometry. *Tetrahedron*, **71** (20), 3032–3038.
33. Escudero, A., Cacho, J., Ferreira, V. (2000) Isolation and identification of odorants generated in wine during its oxidation: a gas chromatography–olfactometric study. *European Food Research and Technology*, **211** (2), 105–110.
34. Han, G.M., Wang, H., Webb, M.R., Waterhouse, A.L. (2015) A rapid, one step preparation for measuring selected free plus SO_2-bound wine carbonyls by HPLC-DAD/MS. *Talanta*, **134**, 596–602.
35. Drinkine, J., Lopes, P., Kennedy, J.A., *et al.* (2007) Analysis of ethylidene-bridged flavan-3-ols in wine. *Journal of Agricultural and Food Chemistry*, **55** (4), 1109–1116.
36. Sonni, F., Clark, A.C., Prenzler, P.D., *et al.* (2011) Antioxidant action of glutathione and the ascorbic acid/glutathione pair in a model white wine. *Journal of Agricultural and Food Chemistry*, **59** (8), 3940–3949.

10

Thiols and Related Sulfur Compounds

10.1 Introduction

Along with oxygen and nitrogen, sulfur is a common heteroatom found in organic molecules derived from natural sources. Sulfur sits below oxygen on the periodic table so has a similar electronic configuration and forms analogous compounds, including many organic ones (cf. alcohol, ROH, and thiol, RSH), but displays dissimilar properties due to the presence of *d*-orbitals and ability to form multiple bonds as a result (i.e., sulfur can expand its octet). In addition, sulfur is larger and less electronegative than oxygen, which influences the relative reactivity of sulfur compounds as compared to oxygen analogs. This chapter will discuss sulfur-containing species with redox states ≤0, also called "reduced" forms of sulfur, which have relevance to both wine aroma and wine redox chemistry. The chemistry of sulfur dioxide, SO_2, will be discussed in Chapter 17.

10.1.1 Key chemical properties and reactions

The chemistry of sulfur compounds is most conveniently discussed in contrast to their oxygen analogs. Like oxygen, the prevalent oxidation state for sulfur is –2, but a range of other oxidation states are possible (up to +6, Table 10.1). As a result, many of the reactions that sulfur compounds partake in are redox reactions that involve a change in oxidation state. The key chemistry of reduced sulfur compounds in wine is summarized below:

- Thiol nucleophilicity. Thiols are better nucleophiles than alcohols, as sulfur is more polarizable than oxygen due to the greater distance of the valence electrons from the nucleus, and the lower electronegativity of sulfur means its lone pair electrons are more available for bond formation. As *nucleophiles*, thiols can participate in reactions with *electrophiles* like aldehydes and quinones.
- Thiols and hydrogen bonding. S–H bonds are less polar than O–H bonds, which results in a poorer hydrogen-bonding capability, as evident in the respective properties of H_2O (liquid) and H_2S (gas) at room temperature (see Chapter 1 for more on this).
- Acidity of thiols. S–H bonds are weaker than O–H bonds and there is greater delocalization of negative charge on the larger sulfur atom of the thiolate, such that H_2S and thiols are stronger acids than H_2O and alcohols (cf. ethanethiol, $pK_a = 8.5$ versus ethanol, $pK_a = 15.9$).

Understanding Wine Chemistry, First Edition. Andrew L. Waterhouse, Gavin L. Sacks, and David W. Jeffery.

Table 10.1 *Reactions of sulfur compounds particularly relevant to wine chemistry*

Oxidation state[a]	Compound type	Example reaction[b]
	Sulfide	$2H_2S + CO_2 \xrightarrow{H^+} CS_2 + 2H_2O$
		$H_2S + ROH \text{ or } RCHO \xrightarrow{H^+} RSH + H_2O$
		$2RSH \xrightarrow{[O]} RSSR$
		$RSH + \text{Electrophile} \longrightarrow \text{RS-Electrophile}$
−2	Thiol	$RSH \xrightarrow{Cu^+/Cu^{2+}} CuSR$
	Alkyl sulfide	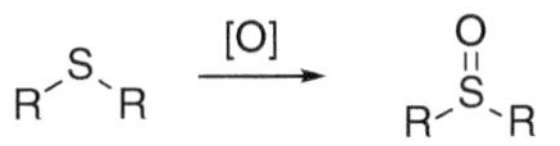
	Thioacetate	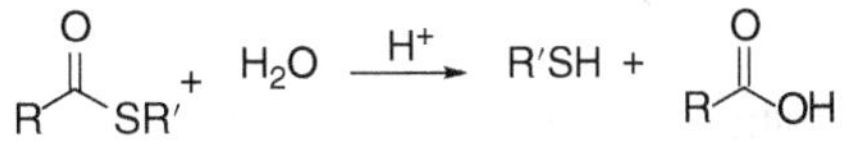
		$RSSR \xrightarrow{[H]} 2RSH$
−1	Alkyl disulfide	$RSSR + R'SH \longrightarrow RSSR' + RSH$
0	Sulfoxide	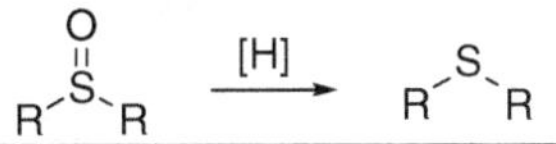

[a] Oxidation states of +4 (e.g., sulfur dioxide, bisulfite, and sulfite) and +6 (e.g., sulfate) are also possible; reactions involving these compounds are described elsewhere.
[b] May involve chemical and/or enzymatic transformations, some of which are putative; [O] = oxidation, [H] = reduction.

- Thiol/disulfide redox chemistry. C–S bonds are similar in strength to C–O bonds, and C=S bonds are weaker than C=O bonds; however, S–S bonds are much stronger than O–O bonds (250 versus 150 kJ/mol).[1] S–S bonds can be formed by oxidation of thiols to yield disulfides (RSSR),[2] with disulfide bridges being important to the structure and function of proteins.
- Reactions with transition metals. Metal ions (e.g., copper and iron) contribute to redox reactions such as formation of disulfides, and are involved in the formation of metal sulfides and sulfur–metal complexes.

10.1.2 Sulfur compounds and wine aroma

Sulfur-containing compounds relevant to wine aroma are mainly ascribed to microbial metabolic activities acting on grape constituents or other substrates. There are effects of viticulture, winemaking, and storage on the sulfur compounds present in different wine styles and their resulting diversity contributes significantly to wine sensory properties and quality. Various classes of volatile sulfur compounds (primarily those with reduced forms of sulfur, that is, an oxidation state of −1 or −2, Table 10.1) are especially important to aroma,

[1] This additional bond strength leads to a striking difference compared to oxygen, with the propensity for *catenation* of sulfur atoms (linking with other sulfur atoms into chains), as in the cyclic S_8 allotrope of elemental sulfur. Among non-metallic elements, this ability to form chains or rings between like atoms is only rivaled by carbon.
[2] Note that S-S bonds are susceptible to attack by nucleophiles including other thiols, in that case yielding mixed disulfides (i.e. RSSR′) and another thiol via thiol-disulfide exchange.

Table 10.2 Indicative odor descriptors, detection thresholds, and odor activity values for classes of volatile sulfur compounds found in wine [2–8]

Compound type	Examples[a]	Structure	Odor descriptor	Threshold (ng/L)	OAV (max.)
Sulfide	Hydrogen sulfide (H_2S)	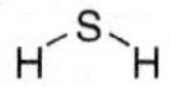	Rotten egg	1000	35
Alkyl thiol	Methanethiol (MeSH)	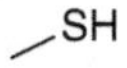	Putrefaction	2000	8
	Ethanethiol (EtSH)	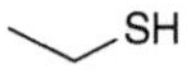	Onion, rubber	1000	19
Alkyl sulfide	Dimethyl sulfide (DMS)	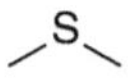	Cabbage, asparagus, truffle	25 000	30
	Diethyl sulfide (DES)	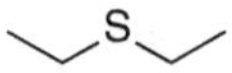	Garlic, rubber	1000	30
	3-(Methylsulfanyl)propan-1-ol (methionol)		Potato, cauliflower	1 000 000	6
Alkyl disulfide	Dimethyl disulfide (DMDS)	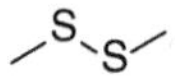	Onion, cabbage, asparagus	29 000	<1
	Diethyl disulfide (DEDS)	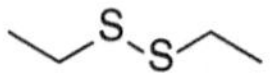	Onion	4000	22
Thioacetate	Methyl thioacetate (MeSAc)	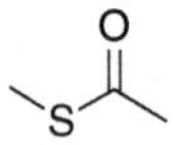	Cheese, egg	50 000	2
	Ethyl thioacetate (EtSAc)	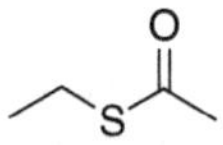	Garlic, onion	10 000	18
Polyfunctional thiol	3-Mercaptohexan-1-ol (3-MH)	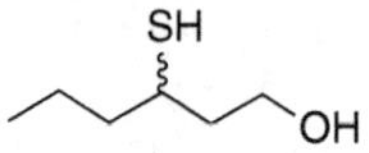	Grapefruit, passionfruit	60	310
	3-Mercaptohexyl acetate (3-MHA)	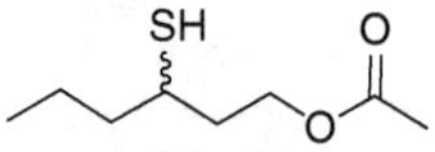	Passionfruit, box tree	4	625
	4-Mercapto-4-methylpentan-2-one (4-MMP)	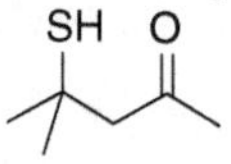	Box tree, guava	3	30
	4-Mercapto-4-methylpentan-2-ol (4-MMPOH)	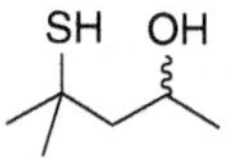	Citrus	55	2
S-Heterocycle	Benzothiazole (BT)	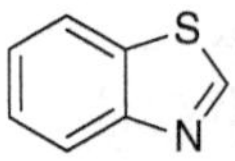	Rubber	50	<1
	2-Methyltetrahydrothiophen-3-one (MTHT)	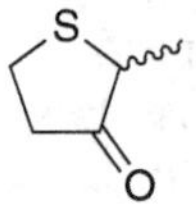	Natural gas, sulfurous	90	5

Table 10.2 (*continued*)

Compound type	Examples[a]	Structure	Odor descriptor	Threshold (ng/L)	OAV (max.)
Aryl thiol	Benzenemethanethiol (BMT)	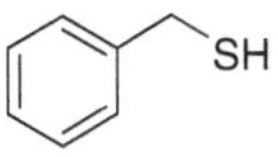	Smoke, struck flint	0.3	440 (1330)[b]
	2-Furanmethanethiol (2-furfurylthiol (FFT))	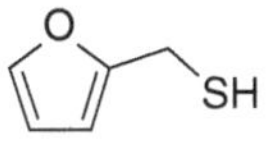	Roasted coffee	0.4	560 (13 750)[b]
	2-Methyl-3-furanthiol (MFT)	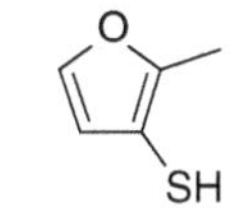	Cooked meat	3	66

[a]The IUPAC terms "sulfanyl" (i.e., RS–, R≠H) and "thiol" (i.e., RSH, R≠H) are preferably used in place of obsolete "mercapto" and "mercaptan" in the systematic naming of these compounds, although the older terms are still used interchangeably due to their familiarity; for example, 3-mercaptohexan-1-ol can be found in the literature as 3-sulfanylhexan-1-ol (3-SH) and methanethiol as methyl mercaptan. The IUPAC prefix "thio" denotes the replacement of oxygen with sulfur (as in *thioacetate*, for example), so "sulfanyl" is used in the systematic name 3-(methylsulfanyl)propan-1-ol (methionol) rather than "thio"; both prefixes are found in the literature and 3-(methylthio)propan-1-ol is also used to refer to methionol. Chiral compounds are presented as their racemates.

[b]From one study of aged Champagne wines (up to 27 years in bottle) [9].

as well as in other foods (e.g., garlic, cabbage, roasted meat) and beverages (e.g., coffee, beer) [1]. Volatile sulfur compounds possess a wide range of aroma descriptors and detection thresholds, and contribute aromas to wine that can be perceived as positive or negative, depending on the nature of the compound and its concentration. Table 10.2 outlines some aspects of the types of volatile sulfur compounds found in wine.

The odorous sulfur compounds listed in Table 10.2 are typically undetectable in grapes, but may be classified based on their origins during the winemaking process. Some compounds have precursors that can be traced to the grape (varietal), whereas others are yeast metabolites (fermentative) or arise during storage of wine in bulk vessels (e.g., oak barrels) or bottles. In many cases, the precursors are non-volatile organic sulfur compounds in the form of amino acids, peptides, and their conjugates, although inorganic sulfur compounds (such as sulfite, sulfate, or elemental sulfur) can strongly impact wine aroma when metabolized by yeast. The non-volatile precursors are introduced below where relevant to description of the associated sulfur aroma compounds, although bisulfite/sulfur dioxide is treated separately (Chapter 17) due to its overall significance in winemaking. Aspects related to conversion of sulfur compounds during fermentation are presented in Chapter 22.4.

10.2 Varietal sulfur aroma compounds – polyfunctional thiols

Some thiols that contribute to wine aroma have their origins in the grape berry. *Varietal thiols* such as 3-MH, 3-MHA, and 4-MMP, which impart desirable citrus and tropical fruit notes (Table 10.2), are of particular importance to certain grape varieties. Such thiols are denoted *polyfunctional* due to the presence of additional functional groups containing oxygen, and are notable for their very low sensory thresholds (and generally high OAVs) and major impact on wine aroma. The range of polyfunctional thiols continues to grow as new compounds are identified and assessed (e.g., 3-sulfanyl forms of 2-methylbutan-1-ol, pentan-1-ol, and heptan-1-ol), but most of the understanding has come from studies on key odorants 4-MMP and 3-MH/3-MHA, which are particularly abundant in Sauvignon Blanc [8,10,11]. These and similar thiols have also been identified as important contributors to the aroma of other fruits and

Table 10.3 Effect of grape variety on typical concentrations of 3-MH, 3-MHA, and 4-MMP in wines [3,5,17–19]

Grape varietal	Concentration range in wine (ng/L)		
	3-MH	3-MHA	4-MMP
Sauvignon Blanc	25.8–18 700	ND[a]–2510	<0.6–87.9
Red blends[b]	678–11 500	4.62–154	4.83–54.2
Botrytized wine[c]	2330–9650	ND	8.5–40
Gewurztraminer	1340–3280	0.5–5.7	ND–15
Riesling	407–562	ND–6.4	ND–7.6

[a] ND, not detected.
[b] Different proportions of Syrah, Grenache, Mourvedre, Cinsault, and Carignan.
[c] Sauvignon Blanc or Semillon.

plants including grapefruit, passionfruit, guava, box tree and broom. With the exception of 3-MHA and other *O*-acetylated derivatives arising from acetylation of 3-sulfanyl alcohols, varietal thiols can be released from non-volatile, grape-derived precursors (amino acid *S*-conjugates) during fermentation [8,11–13] (Chapter 23.2).

4-MMP and the more widespread 3-MH and 3-MHA have low ng/L odor detection thresholds and are frequently present at concentrations in excess of those thresholds (Table 10.2). These three aroma impact compounds, which are particularly important to the varietal aroma of young Sauvignon Blanc wines, have been identified in a range of white (including botrytized), red, and rosé wines from grape varietals such as Scheurebe, Colombard, Riesling, Semillon, Petit and Gros Manseng, Gewurztraminer, Muscat, Grenache, Merlot, Syrah, and Cabernet Sauvignon [8,10,11] (e.g., Table 10.3). Note that 3-MH and 3-MHA (and other chiral thiols) exist as pairs of enantiomers whose profiles are not necessarily 50:50 (i.e., a racemic mixture). Studies on the individual enantiomers reveal slightly different odor detection thresholds and aroma qualities in hydroalcoholic media [14], as could be expected of these chiral aroma molecules. (*R*)-3-MH has a reported threshold of 50 ng/L and an aroma of "citrus peel" and "passionfruit," whereas (*S*)-3-MH has a threshold of 60 ng/L and a "grapefruit" aroma. For the acetates, (*R*)-3-MHA has a slightly higher reported threshold of 9 ng/L and a "passionfruit" aroma when compared to (*S*)-3-MHA with its threshold of 2.5 ng/L and aroma of "box wood". The profile of 3-MH and 3-MHA enantiomers can therefore influence the overall fruit, vegetal, and tropical aromas of wine [15]. Furthermore, there is good correlation between the concentration of 3-MH/3-MHA and tropical/passionfruit characters for Sauvignon Blanc [3,16].

Varietal thiols will participate in many of the reactions described in Table 10.1, of which nucleophilic addition to *o*-quinones is particularly well studied [e.g., 11, 20, 21]. Varietal thiols may therefore be lost due to reaction with *o*-quinones formed through polyphenol oxidation, in an analogous manner to the reaction of glutathione (GSH) in the formation of GRP (Chapters 11 and 24). In fact, as described in Chapter 24, GSH may help preserve varietal thiols in wine [22].

Thiols can also participate in some unique thiol-disulfide chemistry following oxidation [8]; for instance, 3,3′-dithiobis(hexan-1-ol) and related chiral oxidation products (3-propyl-1,2-oxathiolane and its 2-oxide, 3-propyl-γ-sultine, Figure 10.1), identified in botrytized Sauternes wines, can arise from 3-MH oxidation [23]. Such disulfides and oxidation products (which can also be artifacts from aroma compound extraction and analysis [24]) are expected to have different detection thresholds and altered aroma qualities compared to the varietal thiols themselves.

Beyond oxidative reactions, hydrolysis of thiol-containing acetate esters (e.g. 3-MHA to 3-MH and acetic acid) will occur during storage, with an important effect of higher temperature increasing the rate of 3-MHA

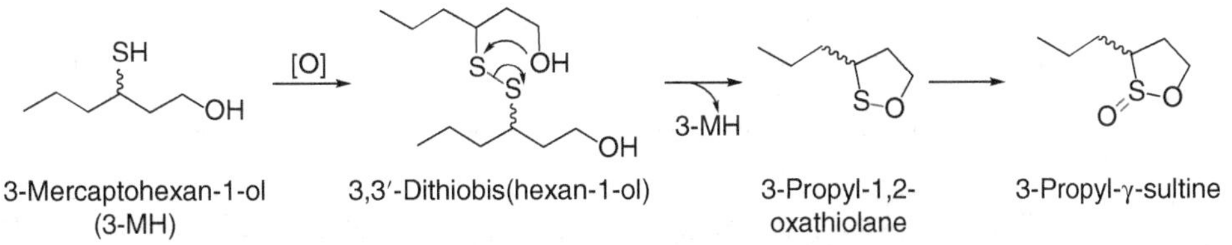

Figure 10.1 Sulfur compounds resulting from oxidation of 3-MH and thermal disproportionation of its disulfide

hydrolysis (as with other acetate esters) [25]. Since the acetate esters are more potent than their corresponding hydrolysis products, this will result in a loss of odor intensity. Factors governing ester hydrolysis are discussed in more detail in Chapter 7.

10.3 Fermentative sulfur aroma compounds

A range of sulfur-containing aroma compounds are produced as general products of fermentation, including H_2S, MeSH, DMS, DMDS, thioacetates, methionol, and heterocycles (Table 10.2); polysulfides such as dimethyl tri- and tetrasulfide have also been identified. These compounds typically have a negative impact on wine aroma and are often linked to H_2S production by yeast. Yeast may utilize a variety of grape-derived sources of sulfur (organic or inorganic, including sulfur fungicides, GSH and SO_2, but primarily SO_4^{2-}) to produce H_2S for incorporation into sulfur-containing amino acids [26]. Fermentation conditions, nutrients, and yeast strain can therefore have a large impact on the biosynthesis of these sulfur compounds (Chapter 22.4), with suboptimal conditions leading to greater production. Such sulfur aroma compounds are important to the aromas of other foods and beverages (particularly those involving fermentation) such as beer, cider, cheese, heated foods, and vegetables [1].

Of particular relevance to wine aroma are the low molecular weight, low boiling point fermentative sulfur compounds, which have thresholds in the low μg/L range and may commonly be encountered at these concentrations or higher in wine (Table 10.2). At low levels, relatively more abundant compounds such as H_2S and DMS may contribute positively to wine aroma; H_2S forms part of a young wine's fermentation bouquet whereas DMS has been shown to increase perceived fruitiness of wines, up to around 100 μg/L [4,26,27]. When concentrations of fermentative sulfur compounds in wine are excessive (i.e., OAV >1), especially where they occur in combination, it leads to perceptions that the wine is "reduced," due to the "reductive" aromas associated with rotten egg, putrefaction (sewage), onion, garlic, and cooked/canned vegetables from compounds containing reduced forms of sulfur (i.e., –1 or –2 oxidation states, Table 10.1). The presence of particular sulfur compounds (e.g., H_2S, MeSH, and DMS) can provide some idea of whether a wine will be perceived as "reduced," as indicated in Table 10.4 for a range of commercial bottled wines with evident reductive sensory characteristics. Concentrations of other fermentative sulfur compounds are also elevated under conditions that favor H_2S formation, such that CS_2, EtSH, DEDS, and thioacetates are also implicated in "reduced" aromas. As indicated below, there are aroma defects that may only become evident after bottling and storage [28] (e.g., increases in H_2S or DMS) but more research on fermentative sulfur compounds is required to properly understand the influences of bottle closure, redox status of wine, and additive or masking sensory effects, among other factors.

Fermentative sulfur compounds such as thiols and thioacetates can enter into similar reactions [26, 29] to those described previously for varietal thiols. Methyl or ethyl thioacetate can hydrolyze over time [30] to release more potent thiols. Thiols can also potentially be oxidized to their corresponding disulfides, trisulfides, and presumably mixed forms (remembering that artifact formation is possible upon analysis). These compounds may constitute a latent "reservoir" of malodorous thiols that could be formed during anaerobic wine storage – this hypothesis is discussed in more detail in Chapter 30. Similar to varietal thiols, low molecular weight thiols and H_2S can be lost through reaction with *o*-quinones [31], and these losses will be limited in the presence of

Table 10.4 Volatile sulfur compounds determined in commercial bottled white and red wines that had evident reductive sensory characteristics. DMDS, EtSH, EtSAc, and DEDS have been omitted due to not being detected in the majority of these wines. Data from Reference [4]

Variety (no. assessed)	Concentration range (μg/L)					
	H_2S	MeSH	MeSAc	DMS	CS_2	DES
White wine						
Chardonnay (4)	1.5–5.0	3.0–8.0	ND[a]–7.0	20.0–185.0	0.5–5.0	ND
Pinot Gris (1)	2.0	3.0	ND	11.0	0.5	ND
Riesling (10)	0.5–35.0	ND–3.0	ND	11.0–37.1	ND–21.1	ND–0.4
Sauvignon Blanc (6)	0.8–4.0	1.7–6.0	ND	25.0–118.2	1.0–13.5	ND–0.4
Sauvignon Blanc Semillon (4)	2.0–13.0	1.0–4.0	ND–2.1	25.0–76.0	0.5–14.8	ND–0.4
Verdelho (1)	1.0	1.6	ND	47.7	18.6	0.4
Viognier (1)	0.5	3.0	ND	78.0	6.0	ND
Red wine						
Cabernet Merlot (2)	0.5–0.8	0.4–1.0	ND	102.5–106.0	3.5–15.6	ND–0.4
Cabernet Sauvignon (5)	ND–1.6	ND–1.5	ND–10.0	88.0–379.5	3.0–20.0	ND–0.4
Durif (1)	2.0	2.0	18.0	61.0	1.0	ND
Grenache Shiraz Merlot (1)	0.7	0.7	ND	111.0	18.0	0.4
Merlot (3)	0.5–1.2	ND–1.6	3.0–8.0	48.0–235.0	8.0–17.0	ND–0.4
Sangiovese (1)	ND	ND	ND	68.0	4.0	ND
Shiraz (22)	ND–8.7	ND–5.0	ND–12.5	28.0–765.0	2.0–45.1	ND–0.5
Shiraz blends (4)	ND–1.0	1.0–1.2	ND	57.0–228.4	2.0–17.4	ND–0.4

[a] ND, not detected.

other thiols like GSH [32]. As an alkyl sulfide the chemistry of DMS is distinct from the other species discussed in this chapter. DMS does not appear to be susceptible to oxidation, and its concentration will rise during wine storage as a result of hydrolysis of *S*-methyl methionine (Chapter 23.3) [33].

Because most low-molecular weight sulfur compounds have undesirable "reductive" aromas, winemakers employ several tools to remove them from wine when present. Most commonly, Cu^{2+} (in the form of copper sulfate) is used to remove H_2S and thiols[3] from wine as their insoluble sulfides, but not dialkyl sulfides, di- (or poly-)sulfides, or thioacetates (Chapter 26.2). Storage of wine on yeast lees may decrease concentrations of fermentative sulfur compounds through adsorption, although care must be taken with additions of SO_2 if storing on lees, since sulfite reductases can produce H_2S for some time after fermentation has ceased (Chapter 22.4).

10.4 Other sulfur-containing aroma compounds

Apart from fermentative and varietal sulfur compounds, a number of other sulfur aroma compounds can form during storage. The aryl thiols BMT, FFT, and MFT are particularly important odorants with low ng/L sensory thresholds (and high OAV in some wines) and are responsible for smoke/struck flint, roasted coffee and meat (or "toasty") aromas, respectively (Table 10.2). Naturally, BMT, MFT, and FFT are significant contributors to the aroma of other foods and beverages such as garden cress, meat and seafood, fruit juice, beer, roasted hazelnuts, and coffee [1]. Aged wines are commonly associated with higher concentrations of these thiols,

[3] Cu(II) thiolates (cupric mercaptides, i.e., $Cu(SR)_2$) are not expected to be stable complexes. Cu^{2+} could conceivably decrease the concentration of thiols either by forming $Cu(SR)_2$, which disproportionates to yield insoluble CuSR (along with RSSR) or by being reduced to Cu^+ in the presence of SO_2 (or ascorbate), with subsequent formation of CuSR.

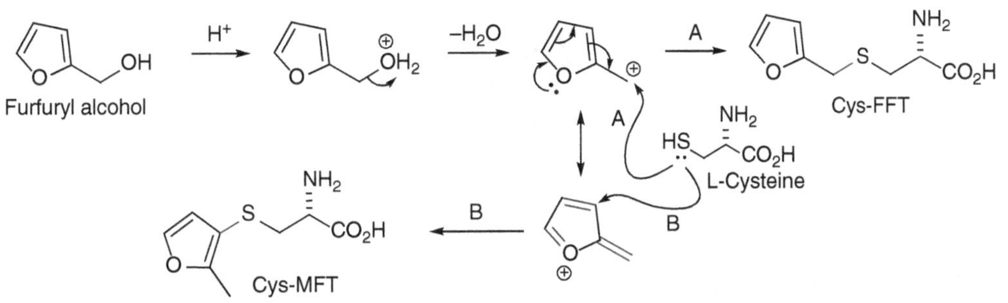

Figure 10.2 Proposed mechanisms for formation of odorless cysteine-S-conjugates of FFT (Cys-FFT, pathway A) and MFT (Cys-MFT, pathway B) isolated and identified from the reaction of xylose (or furfuryl alcohol) with L-cysteine

where they have been detected in Champagne and white (e.g., Chardonnay, Sauvignon Blanc, Semillon) and red (e.g. Cabernet Sauvignon, Merlot) wines [10].

FFT is believed to be formed primarily by reaction of H_2S (from fermentation) with furfural extracted from oak barrels (Chapter 25), and is thus at particularly high concentrations in barrel fermented wines [34]. In support of this, furfural and FFT are typically well correlated with each other and with the extent of wood toasting and age of oak barrels [34]. In addition, toasted oak furanyl precursors may also be formed through acid-catalyzed sugar degradation [35]. In contrast, BMT and benzaldehyde (from benzyl alcohol) are not well correlated, and BMT shows strong varietal dependence, suggesting that a different (unknown) pathway exists for its formation [36]. MFT has no formation pathway proposed for wine, but it is thermally generated in other foodstuffs, along with FFT, from Maillard reactions involving cysteine (or H_2S) and pentoses [37]. Interestingly, cysteine *S*-conjugates of FFT and MFT (analogous to those that act as varietal thiol precursors, Chapter 23.2) have been identified in Maillard reactions between xylose (which gives furfuryl alcohol) and cysteine [38] (Figure 10.2). Whether such conjugates have a role in the formation of FFT and MFT in wine is yet to be determined, but this would likely require an enzymatic thiolysis process, which is harder to rationalize considering the effect of storage.

Although there are no specific studies to draw on, the reactivity of BMT, MFT, and FFT in wine is envisaged to be comparable to other thiols, with similar guiding principles about subsequent contributions to wine aroma. Oxidation is expected to lead to disulfide formation (again, analytical artifacts should be considered), and BMT, FFT, and MFT could conceivably react with *o*-quinones in the same manner as 3-MH, H_2S, and GSH.

Chemical principles: nucleophiles and electrophiles

Nucelophilicity relates to the relative reactivity of nucleophiles (i.e., nucleus-loving, electron-rich) and their ability to donate an electron pair to form new bonds with electrophiles. The greater availability of electron pairs on a nucleophilic center (e.g., S, O, N, C=C) relates to the increased reactivity of that species. Nucleophiles are also Lewis bases and their reactivity is influenced by the following factors:

- Increased electronegativity across a period decreases availability of lone pair electrons and decreases nucleophilicity (e.g., N is more nucleophilic than O and nucleophilicity parallels basicity).
- Higher electron density increases electron pair availability, meaning an anion is a better nucleophile than its neutral form (e.g., RS^- is more nucleophilic than RSH and nucleophilicity parallels basicity).
- Increased polarizability down a row (i.e., ease of electron density distortion of larger atoms) enhances nucleophilicity (e.g., RSH is a better nucleophile than ROH and nucleophilicity is opposite to basicity).

Electrophilicity relates to the relative reactivity of electrophiles (i.e., electron-loving, electron-poor) and their ability to accept an electron pair to form new bonds with nucleophiles. The greater electron deficiency of an electrophilic center (e.g., H^+, and especially C in C=O, flavylium cation, quinone) explains the increased reactivity of that species. Electrophiles are also Lewis acids and their reactivity is influenced by the following factors:

- Increased degree of positive charge for a given atom increases electrophilicity (e.g., a carbocation is a better electrophile than a polar, neutral molecule).
- Increased bond polarity (or polarizability, e.g., Cl–Cl or O–O) leads to a greater partial positive charge (δ+), increasing electrophilicity (e.g., $^{\delta+}C{=}O^{\delta-} \leftrightarrow {}^{+}C{-}O^{-}$ is a better electrophile than $^{\delta+}C{-}O^{\delta-}H$).
- Empty orbitals (e.g., metal ions) or a heterolytically cleavable bond to a leaving group (i.e., the leaving group retains the electron pair) increase electrophilicity (e.g., $C{-}^{+}OH_2$ is a better electrophile than C–OH because H_2O is a better leaving group than OH^-).

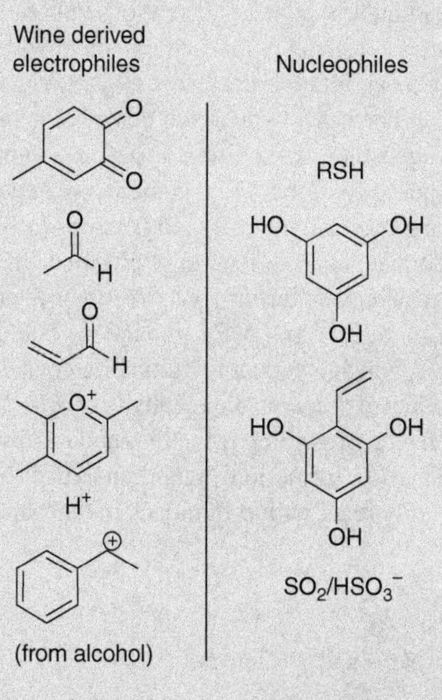

Common electrophiles and nucleophiles in wine

The majority of wine electrophiles arise from oxidation, but the proton is an important one in the acid environment of wine. The most important nucleophiles are thiols and the phloroglucinol A-ring of flavonoids. Conjugated versions of either thiols or phloroglucinols, for example, vinylphloroglucinols, are also effective nucleophiles. Although amines are important nucleophiles in many other food systems, the low pH of wine (and protonation of the amine group) eliminates its nucleophilicity.

References

1. McGorran, R.J. (2011) The significance of volatile sulfur compounds in food flavors, in *Volatile sulfur compounds in food* (eds Qian, M.C., Fan, X., Mahattanatawee, K.), American Chemical Society, Washington, DC, pp. 3–31.
2. Mestres, M., Busto, O., Guasch, J. (2000) Analysis of organic sulfur compounds in wine aroma. *Journal of Chromatography A*, **881** (1–2), 569–581.
3. Lund, C.M., Thompson, M.K., Benkwitz, F., *et al.* (2009) New Zealand Sauvignon Blanc distinct flavor characteristics: sensory, chemical, and consumer aspects. *American Journal of Enology and Viticulture*, **60** (1), 1–12.
4. Siebert, T.E., Solomon, M.R., Pollnitz, A.P., Jeffery, D.W. (2010) Selective determination of volatile sulfur compounds in wine by gas chromatography with sulfur chemiluminescence detection. *Journal of Agricultural and Food Chemistry*, **58** (17), 9454–9462.
5. Mateo-Vivaracho, L., Zapata, J., Cacho, J., Ferreira, V. (2010) Analysis, occurrence, and potential sensory significance of five polyfunctional mercaptans in white wines. *Journal of Agricultural and Food Chemistry*, **58** (18), 10184–10194.
6. Tominaga, T. and Dubourdieu, D. (2006) A novel method for quantification of 2-methyl-3-furanthiol and 2-furanmethanethiol in wines made from *Vitis vinifera* grape varieties. *Journal of Agricultural and Food Chemistry*, **54** (1), 29–33.
7. Bailly, S., Jerkovic, V., Marchand-Brynaert, J., Collin, S. (2006) Aroma extraction dilution analysis of Sauternes wines. Key role of polyfunctional thiols. *Journal of Agricultural and Food Chemistry*, **54** (19), 7227–7234.
8. Roland, A., Schneider, R., Razungles, A., Cavelier, F. (2011) Varietal thiols in wine: discovery, analysis and applications. *Chemical Reviews*, **111** (11), 7355–7376.
9. Tominaga, T., Guimbertau, G., Dubourdieu, D. (2003) Role of certain volatile thiols in the bouquet of aged Champagne wines. *Journal of Agricultural and Food Chemistry*, **51** (4), 1016–1020.
10. Dubourdieu, D. and Tominaga, T. (2009) Polyfunctional thiol compounds, in *Wine chemistry and biochemistry* (eds Moreno-Arribas, M.V. and Polo, M.C.), Springer, New York, pp. 275–293.
11. Coetzee, C. and du Toit, W.J. (2012) A comprehensive review on Sauvignon Blanc aroma with a focus on certain positive volatile thiols. *Food Research International*, **45** (1), 287–298.
12. Capone, D.L., Sefton, M.A., Jeffery, D.W. (2012) Analytical investigations of wine odorant 3-mercaptohexan-1-ol and its precursors, in *Flavor chemistry of wine and other alcoholic beverages*, American Chemical Society, pp. 15–35.
13. Peña-Gallego, A., Hernández-Orte, P., Cacho, J., Ferreira, V. (2012) *S*-cysteinylated and *S*-glutathionylated thiol precursors in grapes. A review. *Food Chemistry*, **131** (1), 1–13.
14. Tominaga, T., Niclass, Y., Frerot, E., Dubourdieu, D. (2006) Stereoisomeric distribution of 3-mercaptohexan-1-ol and 3-mercaptohexyl acetate in dry and sweet white wines made from *Vitis vinifera* (Var. Sauvignon Blanc and Semillon). *Journal of Agricultural and Food Chemistry*, **54** (19), 7251–7255.
15. King, E.S., Osidacz, P., Curtin, C., *et al.* (2011) Assessing desirable levels of sensory properties in Sauvignon Blanc wines – consumer preferences and contribution of key aroma compounds. *Australian Journal of Grape and Wine Research*, **17** (2), 169–180.
16. Benkwitz, F., Tominaga, T., Kilmartin, P.A., *et al.* (2012) Identifying the chemical composition related to the distinct aroma characteristics of New Zealand Sauvignon Blanc wines. *American Journal of Enology and Viticulture*, **63** (1), 62–72.
17. Tominaga, T., Baltenweck-Guyot, R., Peyrot des Gachons, C., Dubourdieu, D. (2000) Contribution of volatile thiols to the aromas of white wines made from several *Vitis vinifera* grape varieties. *American Journal of Enology and Viticulture*, **51** (2), 178–181.
18. Rigou, P., Triay, A., Razungles, A. (2014) Influence of volatile thiols in the development of blackcurrant aroma in red wine. *Food Chemistry*, **142**, 242–248.
19. Sarrazin, E., Shinkaruk, S., Tominaga, T., *et al.* (2007) Odorous impact of volatile thiols on the aroma of young botrytized sweet wines: identification and quantification of new sulfanyl alcohols. *Journal of Agricultural and Food Chemistry*, **55** (4), 1437–1444.

20. Nikolantonaki, M., Chichuc, I., Teissedre, P.-L., Darriet, P. (2010) Reactivity of volatile thiols with polyphenols in a wine-model medium: impact of oxygen, iron, and sulfur dioxide. *Analytica Chimica Acta*, **660** (1–2), 102–109.
21. Laurie, V.F., Zúñiga, M.C., Carrasco-Sánchez, V., *et al.* (2012) Reactivity of 3-sulfanyl-1-hexanol and catechol-containing phenolics *in vitro*. *Food Chemistry*, **131** (4), 1510–1516.
22. Kritzinger, E.C., Bauer, F.F., du Toit, W.J. (2013) Role of glutathione in winemaking: a review. *Journal of Agricultural and Food Chemistry*, **61** (2), 269–277.
23. Sarrazin, E., Shinkaruk, S., Pons, M., *et al.* (2010) Elucidation of the 1,3-sulfanylalcohol oxidation mechanism: an unusual identification of the disulfide of 3-sulfanylhexanol in Sauternes botrytized wines. *Journal of Agricultural and Food Chemistry*, **58** (19), 10606–10613.
24. Hofmann, T., Schieberle, P., Grosch, W. (1996) Model studies on the oxidative stability of odor-active thiols occurring in food flavors. *Journal of Agricultural and Food Chemistry*, **44** (1), 251–255.
25. Makhotkina, O. and Kilmartin, P.A. (2012) Hydrolysis and formation of volatile esters in New Zealand Sauvignon Blanc wine. *Food Chemistry*, **135** (2), 486–493.
26. Rauhut, D. (2009) Usage and formation of sulphur compounds, in *Biology of microorganisms on grapes, in must and in wine* (eds König, H., Unden, G., Fröhlich, J.), Springer-Verlag, Berlin and Heidelberg, pp. 181–207.
27. Ugliano, M. and Henschke, P.A. (2009) Yeasts and wine flavour, in *Wine chemistry and biochemistry* (eds Moreno-Arribas, M.V. and Polo, M.C.), Springer, New York, pp. 313–392.
28. Limmer, A. (2005) Suggestions for dealing with post-bottling sulfides. *The Australian and New Zealand Grapegrower and Winemaker*, **503**, 67–76.
29. Vermeulen, C., Gijs, L., Collin, S. (2005) Sensorial contribution and formation pathways of thiols in foods: a review. *Food Reviews International*, **21** (1), 69–137.
30. Leppänen, O.A., Denslow, J., Ronkainen, P.P. (1980) Determination of thiolacetates and some other volatile sulfur compounds in alcoholic beverages. *Journal of Agricultural and Food Chemistry*, **28** (2), 359–362.
31. Nikolantonaki, M. and Waterhouse, A.L. (2012) A method to quantify quinone reaction rates with wine relevant nucleophiles: a key to the understanding of oxidative loss of varietal thiols. *Journal of Agricultural and Food Chemistry*, **60** (34), 8484–8491.
32. Ugliano, M., Henschke, P.A., Waters, E.J. (2012) Fermentation and post-fermentation factors affecting odor-active sulfur compounds during wine bottle storage, in *Flavor chemistry of wine and other alcoholic beverages*, American Chemical Society, Washington, DC, pp. 189–200.
33. Segurel, M.A., Razungles, A.J., Riou, C., *et al.* (2005) Ability of possible DMS precursors to release DMS during wine aging and in the conditions of heat-alkaline treatment. *Journal of Agricultural and Food Chemistry*, **53** (7), 2637–2645.
34. Blanchard, L., Tominaga, T., Dubourdieu, D. (2001) Formation of furfurylthiol exhibiting a strong coffee aroma during oak barrel fermentation from furfural released by toasted staves. *Journal of Agricultural and Food Chemistry*, **49** (10), 4833–4835.
35. Pereira, V., Albuquerque, F.M., Ferreira, A.C., *et al.* (2011) Evolution of 5-hydroxymethylfurfural (HMF) and furfural (F) in fortified wines submitted to overheating conditions. *Food Research International*, **44** (1), 71–76.
36. Tominaga, T., Guimbertau, G., Dubourdieu, D. (2003) Contribution of benzenemethanethiol to smoky aroma of certain *Vitis vinifera* L. wines. *Journal of Agricultural and Food Chemistry*, **51** (5), 1373–1376.
37. Mottram, D.S. (2007) The Maillard reaction: source of flavour in thermally processed foods, in *Flavours and fragrances* (ed. Berger, R.G.), Springer, Berlin, pp. 269–283.
38. Cerny, C. and Guntz-Dubini, R. (2013) Formation of cysteine-*S*-conjugates in the Maillard reaction of cysteine and xylose. *Food Chemistry*, **141** (2), 1078–1086.

11

Introduction to Phenolics

11.1 Introduction

The grape berry is the source of the vast majority of the many *phenolic* compounds found in wine, although oak or other woods used in the production or aging of the wine may be a minor source. Because of their importance to wine stability and organoleptic properties, reviews of grape and wine phenolics are common in the literature [1–7]. Most phenolic compounds are non-volatile, but a few volatile and odorous phenols exist, such as 4-ethylphenol, and are addressed separately in Chapter 12. As phenolics are ubiquitous in the plant world, all plant foods contain at least small quantities. Aside from grapes and wine, other major dietary sources are coffee, tea, chocolate, fruits, and, in lesser quantities, vegetables. Distilled spirits will lack phenolics unless there has been an infusion process after distillation. For instance, whiskeys contain some phenolics from the barrel used in aging.

Like aliphatic alcohols and water, phenols partake in hydrogen bonding and are weak acids. However, due to the inductive electron withdrawing effect of the benzene ring, a phenolic proton is more acidic, with a pK_a of about 10 (cf. ethanol pK_a = 15.9), but at wine pH 3–4, the ionization of phenolics is too small to be a factor in reactions. Oxidation of phenols to quinones is facile for *ortho*-diphenols, and this important chemistry is detailed elsewhere (Chapter 24). The aromatic ring of phenols is susceptible to electrophilic aromatic substitution (Chapter 10 for overview of wine electrophiles and nucleophiles). For example, electrophiles such as chlorine (Figure 11.1) can add to the ring to yield halophenols. Subsequent microbial activity can methylate the phenolic groups to form haloanisoles, for example, 2,4,6-trichloroanisole, the well-known wine taint (Chapter 18). The electron-rich phloroglucinol moiety in the flavanols is quite reactive with electrophiles, as detailed in Chapter 14.

The naming system is rather complex, so a few definitions are in order.

- The terms *phenols* and *phenolics* encompass all compounds with hydroxyl groups attached to aromatic rings. The simplest phenols are those substances that have a single aromatic ring with one or more hydroxyl groups, and examples include guaiacol and caffeic acid.
- *Polyphenols* (or *polyphenolics*) are a class of compounds with multiple phenol rings within a single structure. This would include most of the phenolic substances in wine; epicatechin and resveratrol are two such compounds. Flavonoids are polyphenols with a very specific C_6–C_3–C_6 3-ring structure elaborated below (Section 11.3).

Understanding Wine Chemistry, First Edition. Andrew L. Waterhouse, Gavin L. Sacks, and David W. Jeffery.

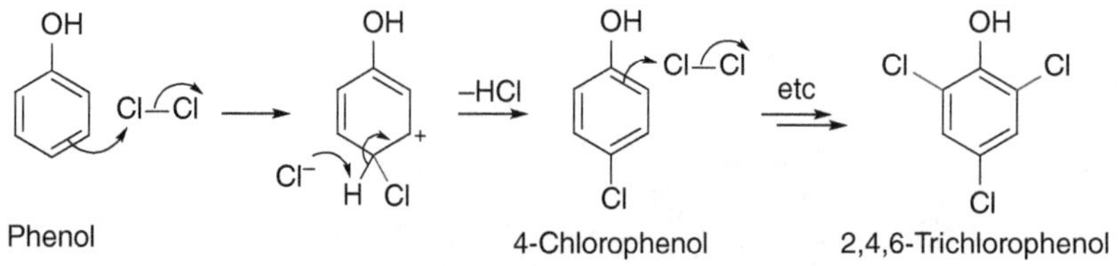

Figure 11.1 *Electrophilic aromatic substitution reaction of phenol with sources of halogens such as chlorine contained in bleach used for cleaning*

- *Tannin* is widely used (including in this text) to describe all polymeric polyphenols. However, from a strict chemical perspective, the term *tannin* does not define a chemical structure but instead declares a specific function for a substance – the ability to tan animal hide (through crosslinking) into leather. These are generally high molecular weight (molecular weight > 1200 daltons) polyphenolic compounds traditionally obtained from plants. These extracts typically contain a complex combination of flavan-3-ol oligomeric and polymeric polyphenols, correctly described as *condensed tannin*, found in grapes, or *hydrolyzable tannin*, found in oak. Condensed tannins are also called *proanthocyanidins*, since they will produce anthocyanidins following acid hydrolysis. The hydrolyzable tannins (e.g., tannic acid) are oligomeric forms of gallic acid and can be specified as gallotannins or ellagitannins depending on whether they are constituted of gallic or ellagic acid moieties. These oligomers are not found in *V. vinifera*, but are in oak wood and *V. rotundifolia* (muscadine grapes). The root names for many phenolic moieties encountered in grapes and wine phenolics are listed in Table 11.1.

The more detailed descriptions below and in following chapters are based on the compounds and levels found in grapes and wine from *V. vinfera*, the European wine grape, which dominates global wine production. The factors that alter phenolic content include [8–10]:

- Grape variety (genetics)
- Site, particularly climate, soil, aspect, sun exposure, elevation (environment)
- Vine, cluster, berry (within and between plants on the same site, there is high variability).

Grape phenolics are largely found in the skin and seed of the berry, while the juice and pulp have much lower concentrations. Red grapes have higher total phenolic concentrations than white grapes, due to the presence of the red anthocyanins in the skin (Figure 11.2). Typically, white wines are made by quickly pressing the juice away from the grape solids, so have lower concentrations of phenolics than reds, which are made by fermenting the juice in the presence of the grape skins and seeds (Chapter 21). Due to the sensory impact of phenolics, winemakers attempt to control the amount and relative extraction of different phenolic classes during winemaking. Control is exercised by manipulation of extraction, called maceration (Chapter 21), as well as additions of protein and other agents to bind and precipitate tannins, a process called fining (Chapter 26.2). Overall, wine phenolics are grouped into two categories, the flavonoids and non-flavonoids.

Many grape phenolics have their structures altered by fermentation reactions and subsequent chemical reactions during storage, for example, glycoside hydrolysis, acid-catalyzed rearrangements, oxidation reactions, and reactions with oxidation products. All phenolic compounds are susceptible to oxidation reactions, and in the presence of iron react with oxygen to produce quinones and hydrogen peroxide, both of which continue the oxidation process (Chapter 24). The reactivity of phenolics towards oxidation products contributes to their complex role as both pro-oxidant and antioxidant – their presence accelerates the rate of oxygen consumption and formation of some oxidation products, but can concurrently decrease the occurrence of some (often undesirable) oxidative changes (Chapter 24).

Table 11.1 The simple phenolic structural motifs found in wine phenolic compounds. Their nomenclature provides guidance to understanding more complex names

Structural moiety	Name	Notes
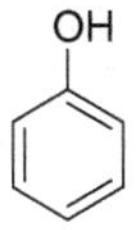	Phenol	Root name for phenolics; functional group found only in a few classes of phenolics in grapes
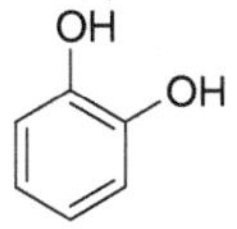	Catechol or pyrocatechol	Catechol functional group – common; susceptibility to oxidation makes this group very important (Chapter 24)
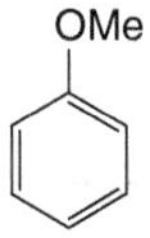	Anisole	Not generally found in wine-related phenolics
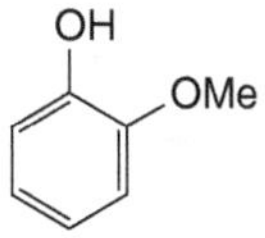	Guaiacol	Common in most classes of phenolics found in grapes and wine
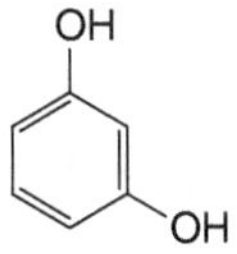	Resorcinol	The term *res* is commonly applied to compounds that include this functional group, such as resveratrol
	Hydroquinone	Not found in grapes and wine
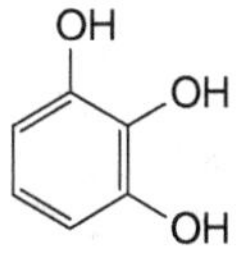	Pyrogallol	Common in grape and wine phenolics; easily oxidized like catechol
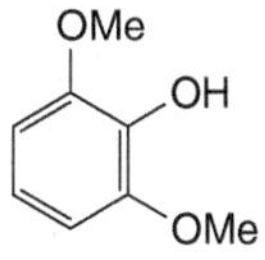	Syringol	Found on the B-ring of anthocyanins, in malvidin
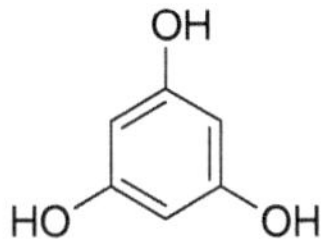	Phloroglucinol	Present in the A-ring of flavonoids; strong nucleophilic character due to π-donation by oxygens

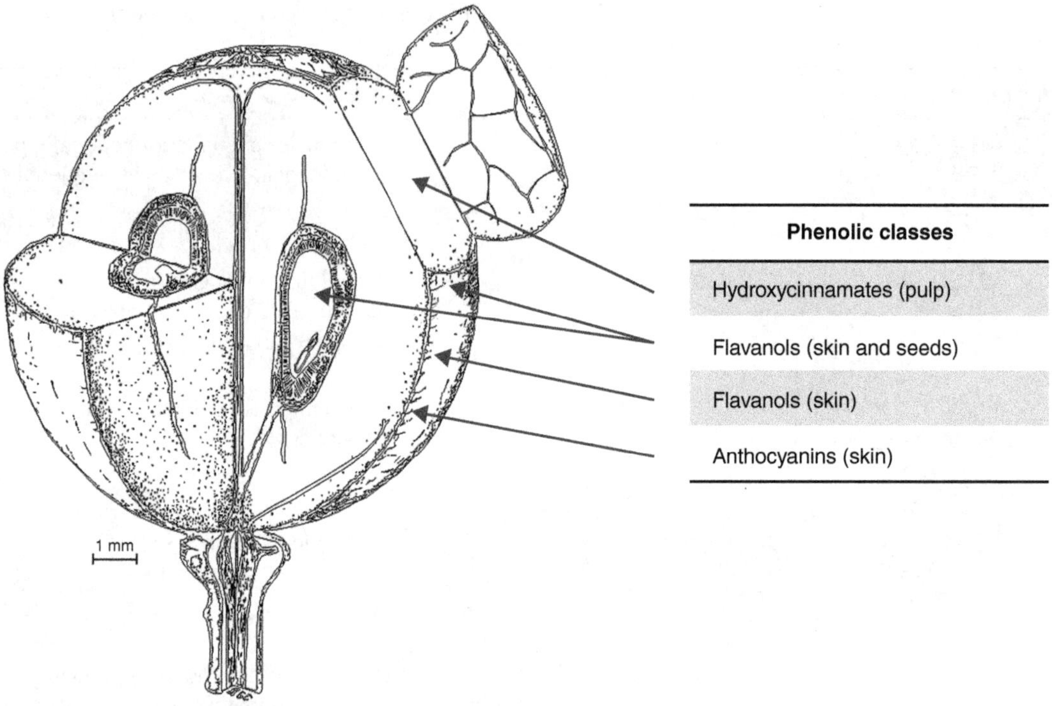

Figure 11.2 Tissue sources of grape phenolics. Source: Coombe 1987 [11]. Reproduced with permission of American Journal of Enology and Viticulture

The total phenolic content of wines varies widely, but is generally about 200 mg/L of gallic acid equivalents (GAEs) in white wines and 2000 mg/L in reds that are ready to drink. Reds destined for long aging will have as much as 3500 mg/L or more, but at the high end of the range, the wines are very astringent until aging reduces the perceived tannin (Chapter 33). While analyses can be done chromatographically, the analysis of total phenolics is typically conducted by the Folin–Ciocalteu method or by UV absorbance at 280 nm [12, 13].

- Measurement of absorbance at 280 nm (Abs 280) is very quick and is blind to sugar, and is thus useful during fermentations or for routine authenticity checks. Quantitative comparison of Abs 280 values across wines is challenging because there is considerable variation in molar absorbance (extinction coefficients) across phenolic compounds.
- The Folin method gives quantitative results that better reflect true phenolic concentrations and are more useful for cross-wine comparisons. However, sugar is an interference.

Both methods have problems with interferences for wines with low phenolic concentrations, which includes many white wines.

11.2 Non-flavonoids

Hydroxycinnamates are the major phenols in grape juice and the major class of phenolics in white wine; similar levels are found in red and white wine (Chapter 13). These components – particularly caffeic acid – are also the first to be enzymatically oxidized during crushing and subsequently initiate browning in must

(Chapters 13 and 24). There are three common hydroxycinnamates in grapes and wine, those based on coumaric acid (4-hydroxycinnamic acid, derived from phenol), caffeic acid (catechol substitution in place of phenol), and ferulic acid (guaiacol substitution).

Benzoic acids are a minor component in most newly fermented wines. The major benzoic acid – gallic acid – can appear during storage via extraction from oak or by hydrolysis of oak hydrolyzable tannins or grape proanthocyanidins during aging.

Stilbenes include the most famous individual wine phenolic compound, resveratrol. These are produced in response to fungal infection and other stresses and are found throughout the grapevine, as well as in the berry skin. Resveratrol is produced as a glucoside and the aglycone appears to oligomerize to the active antifungal viniferins.

11.3 Flavonoids

Flavonoids contain multiple aromatic rings possessing hydroxyl groups and are thus polyphenols. Flavonoids are characterized by a three-ring system with a central oxygen-containing ring (C ring) that has different oxidation states defining the flavonoid class. It is fused to an aromatic ring (A ring) along one bond and attached to another aromatic ring with a single bond (B ring), as seen in Figure 11.3. The flavonoids found in grapes and wine all have the same hydroxyl substitution groups on ring A, at positions 5 and 7. Differences in the oxidation state and substitution on ring C define the different classes of flavonoids.

- *Flavan-3-ols* are the most abundant class of flavonoids. The term "flavan" indicates a saturated C ring, and the "-3-ol" suffix indicates an –OH group at the C3 position. They include simple monomeric catechins, but most exist in the oligomeric and polymeric proanthocyanidin forms from the skins and seeds. These larger compounds constitute about half the phenolics in red wine but negligible concentrations in white wines (Chapter 14).
- *Flavonols* possess a keto group at position C4 and unsaturation between C2 and C3 (a "flavone") as well as an –OH group at C3. Flavonols are found in the berry skin, where they appear to function as sun screen, and are increased by high sunlight exposure of pre-veraison fruit (Chapter 15).
- *Anthocyanins* are red-colored polyphenols characterized by a fully aromatic, positively charged ring. While their direct effect on taste is minimal, better quality red wines often have higher levels of anthocyanins, and their reactions with the condensed tannins create stable red wine pigments (Chapter 16).

The substitution pattern on ring B defines the member of the class. Normal substitution patterns are a hydroxyl at the 4′ position with additional oxygen substitution at 3′ and/or 5′, and those oxygens can be hydroxyls or methoxyls. Thus the number of class members is relatively short; however, the "free" flavonoid structure can also be substituted further (usually with sugar conjugation on the oxygens), which gives rise to many additional compounds.

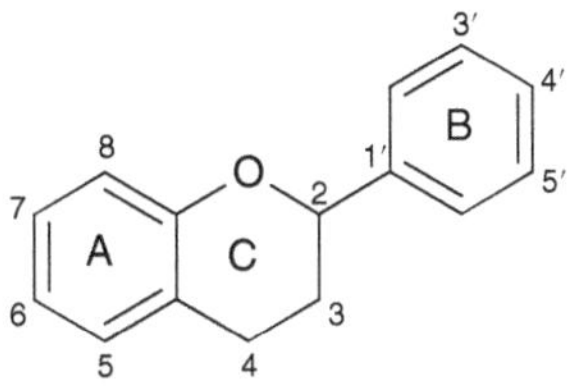

Figure 11.3 Core flavonoid ring system with a C_6–C_3–C_6 structure

The flavonoids are derived from extraction of grape skins and seeds during fermentation and other maceration steps, and comprise the majority of the phenols in red wine (Chapter 21). Ethanol is a good solvent for polyphenol extraction, and over a typical 4–10 day fermentation a good portion of polyphenols are extracted into red wine. As white wine is derived solely from juice, flavonoid levels are very low, less than 5% that of red wine. In general, about half the polyphenols from the skins and seeds are extracted into red wine during the maceration process, leaving the remainder behind in the pomace.

References

1. Singleton, V.L. and Esau, P. (1969) *Phenolic substances in grapes and wine, and their significance*, Academic, New York.
2. Singleton, V.L. (1982) Grape and wine phenolics; background and prospects, in *Proceedings of the Symposium of the University of Califiornia, Davis, Grape and Wine Centennial, 1980*, Department of Viticulture and Enology, University of California, Davis, CA, pp. 215–227.
3. Somers, T.C. and Vérette, E. (1988) Phenolic composition of natural wine types, in *Modern methods of plant analysis, wine analysis* (eds Liskens, H.G. and Jackson, J.F.), Springer-Verlag, Berlin, pp. 219.
4. Ribereau-Gayon, P. and Glories, Y. (1991) Phenolics in grapes and wines, in *Proceedings of the Sixth Australian Wine Industry Technical Conference*, pp. 247–256.
5. Fulcrand, H., Duenas, M., Salas, E., Cheynier, V. (2006) Phenolic reactions during winemaking and aging. *American Journal of Enology and Viticulture*, **57** (3), 289–297.
6. Kennedy, J.A. (2008) Grape and wine phenolics: observations and recent findings. *Ciencia e Investigacion Agraria*, **35** (2), 107–120.
7. Casassa, L.F. and Harbertson, J.F. (2014) Extraction, evolution, and sensory impact of phenolic compounds during red wine maceration. *Annual Review of Food Science and Technology*, **5** (1), 83–109.
8. Reynolds, A.G. (2010) Viticultural and vineyard management practices and their effects on grape and wine quality, in *Managing wine quality*, Vol. **1**, Viticulture and wine quality (ed. Reynolds, A.G.), Woodhead Publishing and CRC Press, Oxford and Boca Raton, FL.
9. Young, P.R. and Vivier, M.A. (2010) Genetics and genomic approaches to improve wine quality, in *Managing wine quality,* Vol. **1**, Viticulture and wine quality (ed. Reynolds, A.G.), Woodhead Publishing and CRC Press, Oxford and Boca Raton, FL.
10. Reynolds, A.G. and Heuvel, J.E.V. (2009) Influence of grapevine training systems on vine growth and fruit composition: a review. *American Journal of Enology and Viticulture*, **60** (3), 251–268.
11. Coombe, B.G. (1987) Distribution of solutes within the developing grape berry in relation to its morphology. *American Journal of Enology and Viticulture*, **38** (2), 120–127.
12. Zoecklein, B.W., Fugelsang, K.C., Gump, B.H., Nury, F.S. (1999) *Wine analysis and production*, Springer, New York.
13. Harbertson, J.F. and Spayd, S. (2006) Measuring phenolics in the winery. *American Journal of Enology and Viticulture*, **57** (3), 280–288.

12
Volatile Phenols

12.1 Introduction

Volatile phenols contribute characteristic odors to various foods and beverages, including wine. They are of particular importance to the flavor and appeal of smoked foods [1] and historically the application of wood smoke was a useful way of preventing food from spoiling.[1] In wine and alcoholic beverages, barrels or tanks made from wood (particularly oak) are frequently used for maturation and storage, and volatile phenols are among the most important organoleptic compounds extracted from toasted oak wood [2] (Chapter 25). Volatile phenols in wines may also arise from precursors originating in the grapes that are transformed by microbiological or chemical processes. A number of highly potent, halogenated volatile phenols, which are detrimental to wine aroma are derived from additional sources and are presented elsewhere (Chapter 18).

12.2 Structure and chemical properties

Volatile phenols are low molecular weight, aromatic[2] alcohols with some similarities in reactivity to other phenols (Chapter 11). The term encompasses phenol and its derivatives containing alkyl, methoxyl, vinyl, allyl, aldehyde (Table 12.1), and halide substituents (Chapter 18). Volatile phenols have hydrogen bonding capability due to the hydroxyl group, but in general are of intermediate hydrophobicity, with influences from various ring substituents (e.g., guaiacol has log $P = 1.3$ and water solubility = 15.3 g/L whereas 4-vinylguaiacol has log $P = 2.2$ and water solubility = 3.0 g/L). Compared to other phenolic compounds in wine (e.g., polyphenols and hydroxycinnamic acids (HCAs)), volatile phenols are much lower in concentration and simple *o*-diphenols (i.e., catechols) are rare, meaning this class of compounds is less likely to play a role in oxidation phenomenon (via quinone formation, Chapter 24). Apart from electrophilic aromatic substitution reactions (Chapter 11), volatile phenols are generally stable molecules, although reactive functional groups can undergo

[1] Volatile phenols present in smoke have antimicrobial and antioxidant properties that help to preserve food (primarily meat or fish), especially in conjunction with drying and curing.
[2] Aromatic in the chemical sense, that is, relating to aromaticity of a benzene ring, not in the context of odor (aroma).

Understanding Wine Chemistry, First Edition. Andrew L. Waterhouse, Gavin L. Sacks, and David W. Jeffery.

Table 12.1 *Indicative odor descriptors, detection thresholds, and odor activity values of selected volatile phenols encountered in wine [8, 11–18]*

Compound[a]	Structure	Odor descriptor[b]	Threshold (μg/L)[b]	OAV (max.)	Major source[c]
Phenol	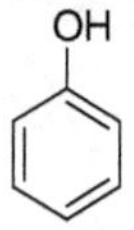	Chemical, ink	30	4	L
Guaiacol (2-methoxyphenol)	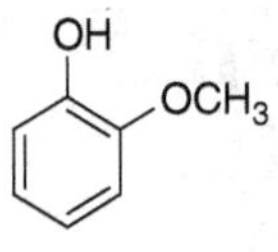	Smoke, sweet	23	3	L
4-Methylguaiacol (2-Methoxy-4-methylphenol)	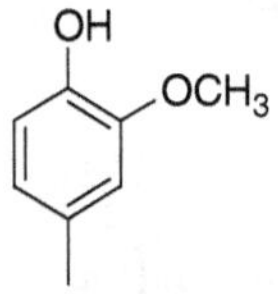	Smoke, ash	21	3	L
Syringol (2,6-dimethoxyphenol)	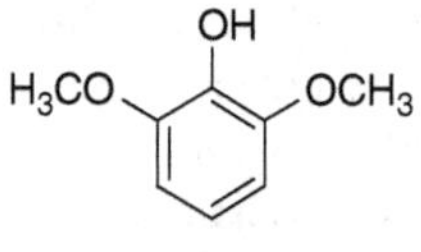	Smoke, medicinal	57	8	L
Eugenol (4-Allyl-2-methoxyphenol)	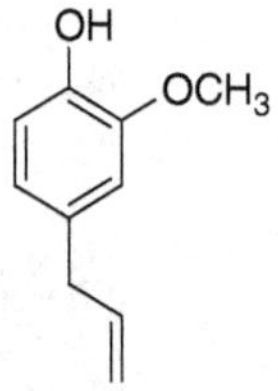	Spice, clove	6	15	L
Vanillin (4-hydroxy-3-methoxybenzaldehyde)	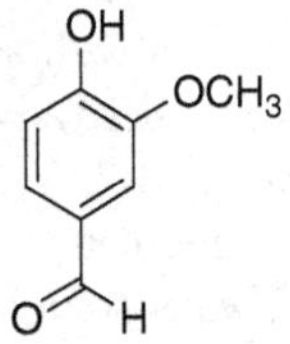	Vanilla	200	5	L
m-Cresol (3-methylphenol)	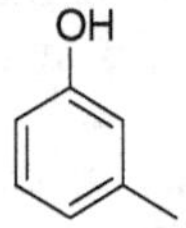	Leather	20	8	L
4-Ethylphenol (4-EP)	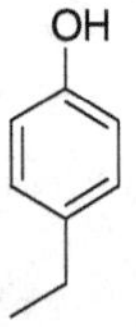	Leather, horse stable	440	6	B
4-Vinylphenol (4-VP)	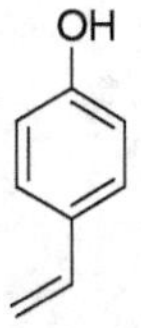	Medicinal, phenolic,	180	27	S

Table 12.1 (continued)

Compound[a]	Structure	Odor descriptor[b]	Threshold (μg/L)[b]	OAV (max.)	Major source[c]
4-Ethylguaiacol (4-EG)	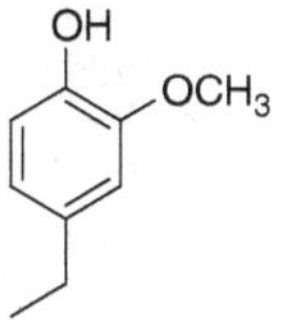	Spice, clove	33	13	B
4-Vinylguaiacol (4-VG)	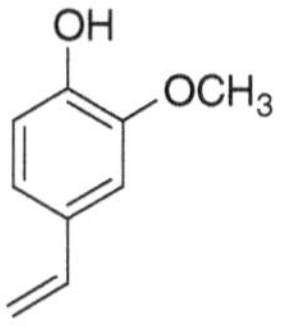	Smoke, phenolic	40	10	S

[a] Common names are widely used in the literature for these compounds; systematic names (or common abbreviations) can also be encountered and are given in parentheses.
[b] Descriptor and threshold data refer to different matrices, including water and red or white wine.
[c] L, lignin degradation (i.e., oak or smoke); B, *Brettanomyces* yeast activity; S, *Saccharomyces* yeast activity.

the usual enzymatic transformations (i.e., reduction of aldehyde or alkene), and they can be bound as phenolic *O*-glycosides (which may undergo hydrolysis, Chapter 23.1). Based on the extent of hydrophobicity of the different volatiles phenols, they may be adsorbed onto hydrophobic surfaces such as wood [3], yeast cell walls [4] or lees [5], or solid-phase polymers used for amelioration [6, 7].

12.3 Concentrations in wine and sensory effects

Volatile phenols can be found at μg/L–mg/L (i.e., suprathreshold) concentrations in wine and impart odors reminiscent of smoke, medicinal/cleaning products, vanilla, spice, and leather (Table 12.1). The presence of the phenolic hydroxyl group has a major impact on the odor of benzenoid substances; the hydroxylated phenols have lower thresholds compared to the methyl (toluene) analogs. The position of ring substituents relative to the hydroxyl group plays an important role in the aroma characteristics and detection thresholds of these compounds. For instance, monoalkyl groups *meta* to the hydroxyl (e.g. *m*-cresol) result in lower odor thresholds compared to their *ortho* and *para* analogues, and odor descriptors differ for the various isomers and as a function of alkyl chain length [8] (e.g., 4-methylguaiacol versus eugenol). In comparison, *ortho* halogenated compounds (Chapter 18) tend to have lower thresholds than their isomeric or methylated counterparts, with odor qualities again dependent on substitution pattern [9, 10].

12.4 Origins in wine and effects on volatile phenol profile

12.4.1 Oak storage

Several volatile phenols, including phenol, guaiacol, syringol, vanillin, cresols, and eugenol (Table 12.1), are extracted when storing wine in contact with toasted oak wood (e.g., oak barrels, chips, or staves). These compounds arise from thermal degradation of lignin and are generally positive contributors to wine aroma when

Table 12.2 Volatile phenols extracted into model wine (12% v/v/ethanol, 5 g/L tartaric acid, adjusted to pH 3.5) after two weeks of maceration of French oak wood shavings obtained from light, medium, and high toasting of barrels. Data from Reference [21]

Compound	Average µg/g (% RSD) of dry wood at toasting level		
	Light	Medium	High
Phenol	0.47 (102)	0.83 (17)	0.41 (101)
Guaiacol	0.45 (72)	1.26 (13)	0.40 (60)
4-Methylguaiacol	0.80 (28)	1.72 (23)	0.74 (20)
Syringol	0.94 (74)	3.66 (10)	0.83 (61)
Eugenol	1.87 (84)	1.27 (58)	1.65 (75)
Vanillin	27.17 (18)	49.81 (47)	25.45 (12)
4-Ethylphenol	0.30 (78)	0.29 (13)	0.26 (72)
4-Ethylguaiacol	0.04 (55)	0.14 (23)	0.04 (38)
4-Vinylguaiacol	0.13 (42)	0.17 (66)	0.12 (37)

present at appropriate concentrations. Analogously, heating other lignin-containing foodstuffs (e.g. roasting coffee or peanuts, kilning of malt for beer and whiskey) will generate volatile phenols that contribute to the characteristic odors of these products.[3] The extent of heating (i.e., amount of oak barrel toasting, Table 12.2) and time in contact with toasted oak influences the level of lignin degradation compounds in wine, as does wine alcohol content, barrel age, oak origin, and seasoning (Chapter 25) [19, 20]. Typically, dry red wines will be subjected to oak contact but few white wines receive such treatment, with Chardonnay, Semillon, and botrytised white wines being the main exceptions. Table 12.3 provides a guide to the concentrations that can be encountered for a range of volatile phenols in oaked wines.

12.4.2 Fermentation – release from grape-derived glycosides

Compounds formed by lignin degradation in toasted oak are also naturally present in wood smoke. Volatile phenols can therefore give rise to undesirable aroma and flavor characters (medicinal, smoky, ashy-aftertaste [24]) in wine when grapes become contaminated due to bushfire (wildfire) smoke in the vicinity of vineyards [25]. This *smoke taint* is of most concern in winegrape growing regions in proximity to areas where bushfires or forest fires are common, such as parts of Australia, North America, and South Africa. Originally, the focus was on the volatile compounds themselves, and guaiacol and 4-methylguaiacol were used as indicators of contamination by smoke, but these compounds can be found at higher concentrations in oaked wines [17] (e.g., compare Tables 12.3 and 12.4). Further research discovered that other volatile phenols from smoke were also involved (Table 12.4), and that volatile phenols could be glycosylated in grape leaves or berries. These non-volatile glycosides could be stored in grapes [26] and hydrolyzed by acid or enzymes during fermentation or storage, subsequently releasing the volatile compounds into wine [27] (Chapter 23.1). These glycosides can also be released in-mouth due to oral microflora (and therefore perceived retronasally) [17, 28], and the phenomenon of glycosylation of exogenous volatiles by grape berries and release during fermentation and storage (or enzymatically) has since been extended to studies involving the deliberate application of oak extracts to grapevines in the vineyard [29–31].

[3] These same compounds are naturally found in wood smoke and give smoked foods their unique flavor. For example, the peat fires used in drying malt for Scotch whisky production yield *o*-, *m*-, and *p*-cresol at low mg/L concentrations. These same compounds are also important to the "barnyard" smell of pig and cattle farms.

Table 12.3 Concentrations of selected volatile phenols in a number of wine varieties aged in contact with toasted oak wood [14, 18, 22, 23]

Compound	Concentration range (μg/L)			
	Chardonnay[a]	Tempranillo[b]	Cabernet Sauvignon[c]	Merlot[d]
Guaiacol	3–28	25–76	5–44	16–139
4-Methylguaiacol	1–8	16–66	<1–16	5–32
Syringol	–[e]	–	–	68–488
Eugenol	10–26	30–96	13–52	ND[f]–19[g]
Vanillin	198–388	347–854	9–369	77–236
4-Ethylphenol	ND–1	8–251	630–1850	–
4-Vinylphenol	7–76	1406–2521	1–5	–
4-Ethylguaiacol	ND–5	5–105	22–306	–
4-Vinylguaiacol	5–30	178–409	1–3	–

[a] Aged in medium toasted French and American oak barrels for 55 weeks.
[b] Aged in medium toasted Spanish, French, and American oak barrels for 52 weeks.
[c] Aged in medium toasted French and American oak barrels for 52 and 93 weeks.
[d] Aged in contact with light to heavy toasted French oak wood staves in stainless steel tanks for 52 weeks.
[e] –, not reported.
[f] ND, not detected.
[g] Eugenol + isoeugenol.

Table 12.4 Average concentration (and standard deviation, SD) of volatile phenols in unoaked wines elaborated from grapes affected by bushfire smoke. Data from Reference [17]

Compound	Average concentration (SD) (μg/L)			
	Control wines ($N=3$)[a]	Pinot Noir ($N=9$)	Cabernet Sauvignon ($N=2$)	Shiraz ($N=3$)
Guaiacol	5.0 (1.7)	18.2 (14.7)	23.5 (10.6)	32.7 (4.9)
4-Methylguaiacol	2.3 (3.2)	3.8 (2.7)	5.0 (0)	4.7 (4.0)
4-Vinylguaiacol	ND[b]	4.9 (4.6)	2.0 (2.8)	5.7 (3.1)
Syringol	9.7 (5.5)	16.9 (6.0)	19.5 (4.9)	19.3 (4.0)
4-Methylsyringol	2.0 (2.0)	5.1 (2.3)	7.5 (3.5)	6.0 (5.2)
4-Allylsyringol	8.3 (9.1)	10.7 (5.6)	4.0 (2.8)	8.7 (7.4)
Phenol	2.7 (3.1)	24.3 (15.9)	34.5 (7.8)	28.7 (13.2)
o-Cresol	2.7 (3.1)	10.1 (6.4)	7.5 (2.1)	5.0 (1.7)
m-Cresol	2.0 (1.7)	7.1 (2.9)	7.5 (0.7)	2.3 (0.6)
p-Cresol	0.7 (1.2)	5.0 (0.9)	3.5 (0.7)	2.0 (1.0)

[a] N, number of wines; the control wines were Pinot Noir ($N=2$) and Cabernet Sauvignon ($N=1$).
[b] ND, not detected.

12.4.3 Fermentation – metabolism of hydroxycinnamic acids

Volatile phenols can also arise from yeast activity acting on grape-derived components, presenting a significant challenge to winemakers. Metabolism of HCAs in wine (Chapter 13) gives rise to "Brett" off-flavor compounds such as 4-ethylphenol (4-EP) and 4-ethylguaiacol (4-EG, Table 12.1) from their 4-vinyl analogs [32], which contrasts with the small amounts formed during lignin pyrolysis. The route from precursor acids involves enzymatic decarboxylation (by *Saccharomyces* or *Brettanomyces*) of HCAs to give 4-vinylphenol (4-VP) and 4-vinylguaiacol (4-VG, Table 12.1) from *p*-coumaric and ferulic acids, respectively, and subsequent reduction of the vinyl group by *Brettanomyces* to form 4-EP/4-EG (Chapter 23.3). Although dependent on grape variety and wine style, the relative concentrations of Brett spoilage compounds tend to reflect the concentrations of the precursor acids (Chapter 23.3). Typically, this leads to an approximately 8:1 ratio of 4-EP to 4-EG [33]. These compounds demonstrate additive effects and are major contributors to "Brett" aromas, although predicting sensory effects can be complicated by the presence of other volatile phenols, or due to masking, for example, by isovaleric and isobutyric acids [34]. Table 12.3 shows representative concentrations of Brett-related volatile phenols in different wines, with the higher levels of ethyl analogs in particular being an indication of *Brettanomyces* activity during barrel aging. Wines are particularly susceptible to Brett spoilage during this time (and also while undergoing MLF) due to the low levels of SO_2, slow ingress of O_2 and potential for poor sanitation, especially of older barrels.

References

1. Maga, J.A. (1987) The flavor chemistry of wood smoke. *Food Reviews International*, **3** (1–2), 139–183.
2. Maga, J.A. (1989) The contribution of wood to the flavor of alcoholic beverages. *Food Reviews International*, **5** (1), 39–99.
3. Barrera-García, V.D., Gougeon, R.D., Voilley, A., Chassagne, D. (2006) Sorption behavior of volatile phenols at the oak wood/wine interface in a model system. *Journal of Agricultural and Food Chemistry*, **54** (11), 3982–3989.
4. Jiménez-Moreno, N. and Ancín-Azpilicueta, C. (2009) Sorption of volatile phenols by yeast cell walls. *International Journal of Wine Research*, **1**, 11–18.
5. Chassagne, D., Guilloux-Benatier, M., Alexandre, H., Voilley, A. (2005) Sorption of wine volatile phenols by yeast lees. *Food Chemistry*, **91** (1), 39–44.
6. Fudge, A.L., Ristic, R., Wollan, D., Wilkinson, K.L. (2011) Amelioration of smoke taint in wine by reverse osmosis and solid phase adsorption. *Australian Journal of Grape and Wine Research*, **17** (2), S41–S48.
7. Larcher, R., Puecher, C., Rohregger, S., *et al.* (2012) 4-Ethylphenol and 4-ethylguaiacol depletion in wine using esterified cellulose. *Food Chemistry*, **132** (4), 2126–2130.
8. Czerny, M., Brueckner, R., Kirchhoff, E., *et al.* (2011) The Influence of molecular structure on odor qualities and odor detection thresholds of volatile alkylated phenols. *Chemical Senses*, **36** (6), 539–553.
9. Strube, A., Buettner, A., Czerny, M. (2012) Influence of chemical structure on absolute odour thresholds and odour characteristics of *ortho*- and *para*-halogenated phenols and cresols. *Flavour and Fragrance Journal*, **27** (4), 304–312.
10. Capone D.L., Van Leeuwen K.A., Pardon K.H., et al. (2010) Identification and analysis of 2-chloro-6-methylphenol, 2,6-dichlorophenol and indole: causes of taints and off-flavours in wines. *Australian Journal of Grape and Wine Research*, **16** (1), 210–217.
11. Chatonnet, P., Dubourdieu, D., Boidron, J.-N., Lavigne, V. (1993) Synthesis of volatile phenols by *Saccharomyces cerevisiae* in wines. *Journal of the Science of Food and Agriculture*, **62** (2), 191–202.
12. López, R., Aznar, M., Cacho, J., Ferreira, V. (2002) Determination of minor and trace volatile compounds in wine by solid-phase extraction and gas chromatography with mass spectrometric detection. *Journal of Chromatography A*, **966** (1–2), 167–177.
13. Culleré L., Escudero A., Cacho J., Ferreira V. (2004) Gas chromatography–olfactometry and chemical quantitative study of the aroma of six premium quality Spanish aged red wines. *Journal of Agricultural and Food Chemistry*, **52** (6), 1653–1660.
14. Fernandez de Simon, B., Cadahia, E., Sanz, M., *et al.* (2008) Volatile compounds and sensorial characterization of wines from four Spanish denominations of origin, aged in Spanish Rebollo (*Quercus pyrenaica* Willd.) oak wood barrels. *Journal of Agricultural and Food Chemistry*, **56** (19), 9046–9055.
15. Prida, A. and Chatonnet, P. (2010) Impact of oak-derived compounds on the olfactory perception of barrel-aged wines. *American Journal of Enology and Viticulture*, **61** (3), 408–413.

16. Fernández de Simón, B., Cadahía, E., Muiño, I., *et al.* (2010) Volatile composition of toasted oak chips and staves and of red wine aged with them. *American Journal of Enology and Viticulture*, **61** (2), 157–165.
17. Parker, M., Osidacz, P., Baldock, G.A., *et al.* (2012) Contribution of several volatile phenols and their glycoconjugates to smoke-related sensory properties of red wine. *Journal of Agricultural and Food Chemistry*, **60** (10), 2629–2637.
18. Chira, K. and Teissedre, P.-L. (2013) Extraction of oak volatiles and ellagitannins compounds and sensory profile of wine aged with French winewoods subjected to different toasting methods: behaviour during storage. *Food Chemistry*, **140** (1–2), 168–177.
19. Garde-Cerdán, T. and Ancín-Azpilicueta, C. (2006) Review of quality factors on wine ageing in oak barrels. *Trends in Food Science and Technology*, **17** (8), 438–447.
20. Pérez-Coello, M.S. and Díaz-Maroto, M.C. (2009) Volatile compounds and wine aging, in *Wine chemistry and biochemistry* (eds Moreno-Arribas, M.V. and Polo, M.C.), Springer, New York, pp. 295–311.
21. Chatonnet, P., Cutzach, I., Pons, M., Dubourdieu, D. (1999) Monitoring toasting intensity of barrels by chromatographic analysis of volatile compounds from toasted oak wood. *Journal of Agricultural and Food Chemistry*, **47** (10), 4310–4318.
22. Spillman, P.J., Sefton, M.A., Gawel, R. (2004) The effect of oak wood source, location of seasoning and coopering on the composition of volatile compounds in oak-matured wines. *Australian Journal of Grape and Wine Research*, **10** (3), 216–226.
23. Garde Cerdán, T., Rodríguez Mozaz, S., Ancín Azpilicueta, C. (2002) Volatile composition of aged wine in used barrels of French oak and of American oak. *Food Research International*, **35** (7), 603–610.
24. Ristic, R., Boss, P., Wilkinson, K. (2015) Influence of fruit maturity at harvest on the intensity of smoke taint in wine. *Molecules*, **20** (5), 8913.
25. Jiranek V. (2011) Smoke taint compounds in wine: nature, origin, measurement and amelioration of affected wines. *Australian Journal of Grape and Wine Research*, **17** (2), S2–S4.
26. Hayasaka, Y., Baldock, G.A., Parker, M., *et al.* (2010) Glycosylation of smoke-derived volatile phenols in grapes as a consequence of grapevine exposure to bushfire smoke. *Journal of Agricultural and Food Chemistry*, **58** (20), 10989–10998.
27. Kennison, K.R., Gibberd, M.R., Pollnitz, A.P., Wilkinson, K.L. (2008) Smoke-derived taint in wine: the release of smoke-derived volatile phenols during fermentation of Merlot juice following grapevine exposure to smoke. *Journal of Agricultural and Food Chemistry*, **56** (16), 7379–7383.
28. Mayr, C.M., Parker, M., Baldock, G.A., *et al.* (2014) Determination of the importance of in-mouth release of volatile phenol glycoconjugates to the flavor of smoke-tainted wines. *Journal of Agricultural and Food Chemistry*, **62** (11), 2327–2336.
29. Martínez-Gil, A.M., Garde-Cerdán, T., Martínez, L., *et al.* (2011) Effect of oak extract application to Verdejo grapevines on grape and wine aroma. *Journal of Agricultural and Food Chemistry*, **59** (7), 3253–3263.
30. Martínez-Gil, A.M., Garde-Cerdán, T., Zalacain, A., *et al.* (2012) Applications of an oak extract on Petit Verdot grapevines. Influence on grape and wine volatile compounds. *Food Chemistry*, **132** (4), 1836–1845.
31. Martínez-Gil, A.M., Angenieux, M., Pardo-García, A.I., *et al.* (2013) Glycosidic aroma precursors of Syrah and Chardonnay grapes after an oak extract application to the grapevines. *Food Chemistry*, **138** (2–3), 956–965.
32. Oelofse, A., Pretorius, I.S., du Toit, M. (2008) Significance of *Brettanomyces* and *Dekkera* during winemaking: a synoptic review. *South African Journal for Enology and Viticulture*, **29** (2), 128–144.
33. Chatonnet, P., Dubourdieu, D., Boidron, J.N., Pons, M. (1992) The origin of ethylphenols in wines. *Journal of the Science of Food and Agriculture*, **60** (2), 165–178.
34. Romano, A., Perello, M.C., Lonvaud-Funel, A., *et al.* (2009) Sensory and analytical re-evaluation of "Brett character." *Food Chemistry*, **114** (1), 15–19.

13

Non-flavonoid Phenolics

13.1 Introduction

Non-flavonoid phenolics include several subclasses of importance to wine, in particular the hydroxycinnamates, stilbenes, and benzoic acids. The hydroxycinnamates and stilbenes are found in the grape, while the benzoic acids are found in the grape and in oak, so oak-treated wines will include additional phenolics from that source.

13.2 Hydroxycinnamates

The hydroxycinnamates are phenolic acids that include a conjugated double bond between the phenolic ring and the carboxylate group. Three acids are commonly found – coumaric, caffeic, and ferulic acids – but these simple hydroxycinnamic acids are not found in grape berries; they instead exist as their tartaric acid esters. Enologists have adopted trivial names for these esters, those being *p*-coutaric acid, caftaric acid, and fertaric acid, respectively (Table 13.1). These substances are found in the flesh of the fruit, and thus are found in all grape juices and consequently in all wines. Plants predominantly produce the *trans* form of hydroxycinnamic acids, but isomerization to the *cis* form is induced by light and thus varying amounts are found in wine [1, 2]. All plant foods contain hydroxycinnamates, though the tartrate esters are relatively unusual – their presence is evidence that a juice was sourced from grapes.

Considering their functionality, the grape-derived tartrate esters are susceptible to hydrolysis, which is accelerated via hydroxycinnamate ester hydrolase enzymes produced by lactic acid bacteria and other organisms and continues, albeit very slowly, in the aqueous acid environment of finished wine (Chapter 23.3). The action of hydrolysis releases the simple hydroxycinnamic acids, which can be readily detected in newly fermented wines [3]. Conversely, the free acids and tartrate derivatives (also bearing free carboxylic acids) will partially esterify with the ethanol in wine [4] (Chapter 7 and 25). In addition, caffeic acid and its tartrate ester in both juice and wine are susceptible to oxidation and follow-on reactions that lead to browning and many other outcomes (Chapter 24).

Understanding Wine Chemistry, First Edition. Andrew L. Waterhouse, Gavin L. Sacks, and David W. Jeffery.

Table 13.1 The hydroxycinnamic acids shown in their tartrate ester form as found in the grape. Enzymatic hydrolysis of the ester leads to the free acid form in wine

Name	Structure	Representative levels (juice/wine, mg/L)
Coutaric acid, coumaric acid, tartrate ester	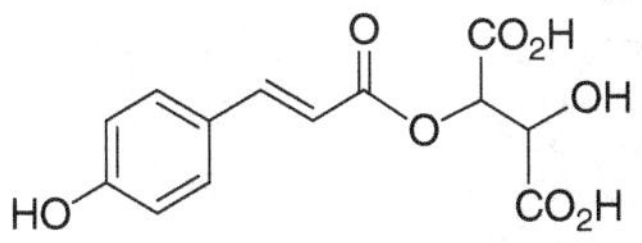	20/15
Caftaric acid, caffeic acid, tartrate ester	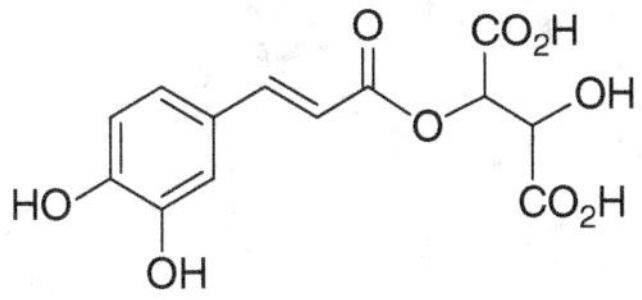	170/40
Fertaric acid, ferulic acid, tartrate ester	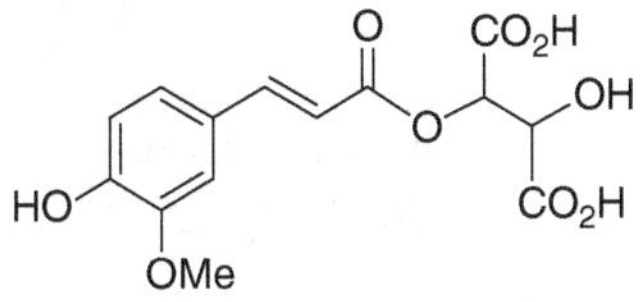	5/4

The hydroxycinnamates in various forms have been reported to have bitterness and astringency in water [5], but work by Noble and others [6] showed that the levels of these compounds were below the threshold in wine.

13.3 Hydroxybenzoic acids

Gallic acid is not found in grapes, but is generated in wine by the hydrolysis of the gallate esters found in condensed and hydrolyzable tannins (Figure 13.1) [7]. On long aging, gallic acid is persistent and can be observed in older wines. In typical red table wines it can be found at about 70 mg/L, but white wines contain less, at about 10 mg/L [8]. Small amounts of syringic, protocatechuic, and vanillic acids are also found. The presence of syringic acids has been used to identify the prior presence of malvidin (from grape wine) in archeological samples [9]. As carboxylic acids, these can undergo acid catalyzed esterification in the presence of ethanol to yield a fraction of the ethyl ester (Chapters 7 and 25). In a manner similar to the hydroxycinnamates, the acids with catechol or galloyl functionality (Chapter 11) are also susceptible to oxidation.

13.3.1 Hydrolyzable tannins

The hydrolyzable tannins are composed of gallic acid and ellagic acid esters of glucose (Figure 13.1) or related sugars. Hydrolyzable tannins are further categorized as gallotannins or ellagitannins based on their constituent phenolic acid, but most plant sources consist of a mixture of the two. The term "hydrolyzable" results from the fact that the ester linkage is more susceptible to hydrolysis under mild conditions than the interflavan linkages of condensed tannins,[1] and hydrolyzable tannins will form their constituent gallic and

[1] Condensed tannins (proanthocyanidins, Chapter 14) can also be hydrolyzed, but require harsher conditions (higher temperature, lower pH) or longer times to achieve the same degree of degradation

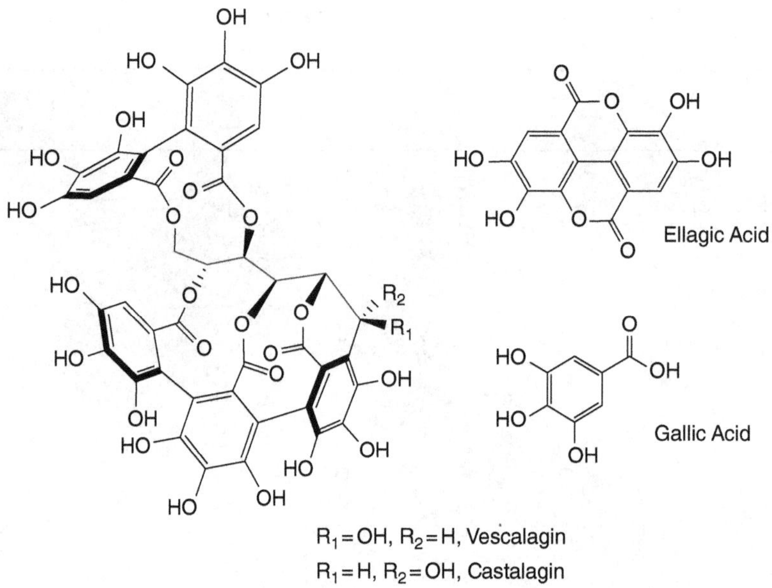

Figure 13.1 *Oak tannins, vescalagin and castalagin, and ellagic acid and gallic acid. The sugar moiety is highlighted*

ellagic acids on aging. These compounds, largely castalagin, vescalagin and the roburins, are extracted into wine from oak, affording levels near a few mg/L for white wines after 6 months in new barrels, while red wines will have levels in the range of 2–20 mg/L after aging two or more years [10] (Chapter 25). Hydrolyzable tannins are not found in *V. vinifera*, but can be found in other fruits, such as raspberries or muscadine (*V. rotundifolia*) grapes. The taste impact of oak-derived tannins in wine is likely to be minor [11]. When hydrolyzed in wine from ellagitannins, ellagic acid will precipitate if it is at high levels, as in muscadine wine.

13.4 Stilbenes

The principal stilbene in grapes, resveratrol, is produced by vines, as glucosides, in response to *Botrytis* infection and other fungal attacks [12]. The actual antifungal compounds appear to be the oligomers of resveratrol, called the viniferins. Several forms of resveratrol exist including the *cis* and *trans* isomers as well as their glucosides (Figure 13.2). Grapes appear to biosynthesize the *trans* form and light causes *cis*/*trans* isomerization [13]. Resveratrol derivatives are found largely in the skin of the grape, and consequently greater amounts are found in red wine, and although significant amounts can also be found in the vine shoots, this is largely irrelevant to winemaking. The total levels of all forms average about 7 mg/L for reds [14], 2 mg/L for rosés, and 0.5 mg/L for white wines [15]. Resveratrol has been implicated as a wine component that may reduce heart disease or cancer, with interest greatly stimulated by a 1997 report in *Science* [16], and it is a known marker in urine for wine consumption [17]. Numerous follow-on studies have shown other benefits as well, but generally the therapeutic effects (in animals) occur only at doses 10–100 times that found in a glass of wine, in part because absorption is poor. More important to winemaking, resveratrol, and its *O*-methylated derivative pterostilbene in particular (Figure 13.2), have potent antimicrobial activity against wild yeasts and *Acetobacter* [18].

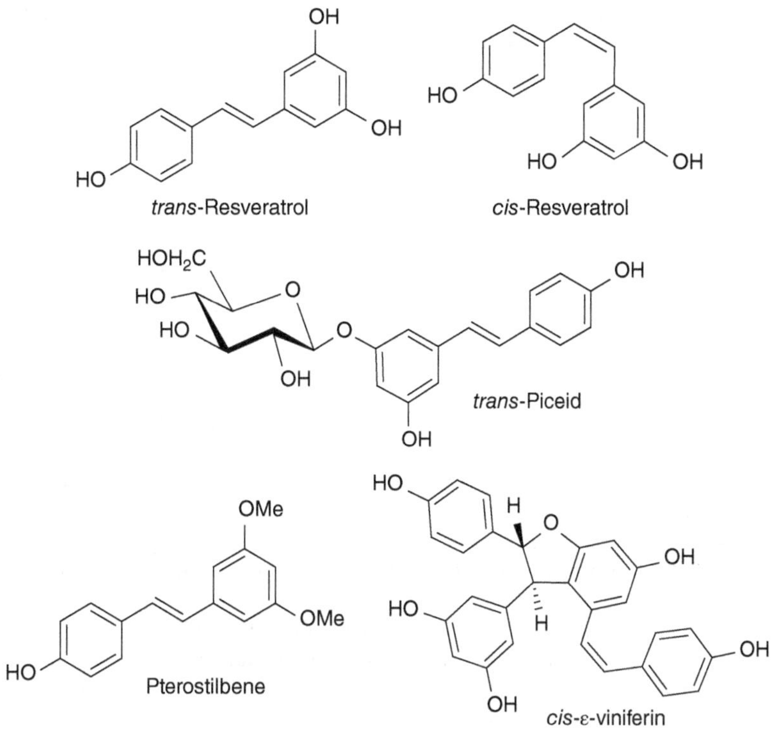

*Figure 13.2 Selected stilbenes in wine, including the more abundant resveratrol isomers, along with an example of a resveratrol glucoside (*trans*-piceid), a dimethoxylated analog (pterostilbene), and a dimeric viniferin*

References

1. Baranowski, J.D. and Nagel, C.W. (1981) Isolation and identification of the hydroxycinnamic acid-derivatives in white Riesling wine. *American Journal of Enology and Viticulture*, **32** (1), 5–13.
2. Ritchey, J.G. and Waterhouse, A.L. (1999) A standard red wine: monomeric phenolic analysis of commercial Cabernet Sauvignon wines. *American Journal of Enology and Viticulture*, **50** (1), 91–100.
3. Rentzsch, M., Schwarz, M., Winterhalter, P., Hermosin-Gutierrez, I. (2007) Formation of hydroxyphenyl-pyranoanthocyanins in Grenache wines: precursor levels and evolution during aging. *Journal of Agricultural and Food Chemistry*, **55** (12), 4883–4888.
4. Somers, T.C., Vérette, E., Pocock, K.F. (1987) Hydroxycinnamate esters of *V. vinifera*: changes during white vinification and effects of exogenous enzyme hydrolysis. *Journal of the Science of Food and Agriculture*, **40** (1), 67–78.
5. Hufnagel, J.C. and Hofmann, T. (2008) Orosensory-directed identification of astringent mouthfeel and bitter-tasting compounds in red wine. *Journal of Agricultural and Food Chemistry*, **56** (4), 1376–1386.
6. Verette, E., Noble, A.C., Somers, C.T. (1988) Hydroxycinnamates of *Vitis vinifera*: sensory assessment in relation to bitterness in white wine. *Journal of the Science of Food and Agriculture*, **45** (3), 267–272.
7. Chira, K., Pacella, N., Jourdes, M., Teissedre, P.L. (2011) Chemical and sensory evaluation of Bordeaux wines (Cabernet-Sauvignon and Merlot) and correlation with wine age. *Food Chemistry*, **126** (4), 1971–1977.
8. Teissedre, P.L., Frankel, E.N., Waterhouse, A.L., *et al.* (1996) Inhibition of *in vitro* human LDL oxidation by phenolic antioxidants from grapes and wine. *Journal of the Science of Food and Agriculture*, **70** (1), 55–61.
9. Guasch-Jane, M.R., Andres-Lacueva, C., Jauregui, O., Lamuela-Raventos, R.M. (2006) The origin of the ancient Egyptian drink Shedeh revealed using LC/MS/MS. *Journal of Archaeological Science*, **33** (1), 98–101.

10. Garcia-Estevez, I., Escribano-Bailon, M.T., Rivas-Gonzalo, J.C., Alcalde-Eon, C. (2012) Validation of a mass spectrometry method to quantify oak ellagitannins in wine samples. *Journal of Agricultural and Food Chemistry*, **60** (6), 1373–1379.
11. Glabasnia, A. and Hofmann, T. (2006) Sensory-directed identification of taste-active ellagitannins in American (*Quercus alba* L.) and European oak wood (*Quercus robur* L.) and quantitative analysis in bourbon whiskey and oak-matured red wines. *Journal of Agricultural and Food Chemistry*, **54** (9), 3380–3390.
12. Joshi, V.K. and Devi, M.P. (2009) Resveratrol: importance, role, contents in wine and factors influencing its production. *Proceedings of the National Academy of Sciences India Section B-Biological Sciences*, **79**, 212–226.
13. Trela, B.C. and Waterhouse, A.L. (1996) Resveratrol: isomeric molar absorptivities and stability. *Journal of Agricultural and Food Chemistry*, **44** (5), 1253–1257.
14. Lamuela-Raventos, R.M., Romero Perez, A.I., Waterhouse, A.L., de la Torre-Boronat, M.C. (1995) Direct HPLC analysis of *cis*- and *trans*-resveratrol and piceid isomers in Spanish Red *Vitis vinifera* wines. *Journal of Agricultural and Food Chemistry*, **42** (2), 281–283.
15. Romero-Perez, A.I., Lamuela-Raventos, R.M., Waterhouse, A.L., de la Torre-Boronat, M.C. (1996) Levels of *cis*- and *trans*-resveratrol and their glycosides in white and rose *Vitis vinifera* wines from Spain. *Journal of Agricultural and Food Chemistry*, **44** (8), 2124–2128.
16. Jang, M., Cai, L., Udeani, G.O., *et al.* (1997) Cancer chemopreventive activity of resveratrol, a natural product derived from grapes. *Science*, **275** (5297), 218–220.
17. Zamora-Ros, R., Urpi-Sarda, M., Lamuela-Raventos, R.M., *et al.* (2009) Resveratrol metabolites in urine as a biomarker of wine intake in free-living subjects: the PREDIMED Study. *Free Radical Biology and Medicine*, **46** (12), 1562–1566.
18. Pastorkova, E., Zakova, T., Landa, P., *et al.* (2013) Growth inhibitory effect of grape phenolics against wine spoilage yeasts and acetic acid bacteria. *International Journal of Food Microbiology*, **161** (3), 209–213.

14

Flavan-3-ols and Condensed Tannin

14.1 Introduction

Amongst flavonoids, the class present at the largest quantity in grapes is the flavan-3-ols, with significant amounts in the seeds as well as in the berry skins. Flavanols are widespread in foods, with the major sources in the Western diet being chocolate, tea, and apples. The use of the number "3" in the name refers to the position of the alcohol (Figure 14.1). Flavanols with an alcohol group at position 4 do exist at trace levels, though these compounds are highly reactive and rarely reported. The flavan-3-ols are notable for the fact that both positions 2 and 3 can have isomers with *cis* and *trans* forms relative to the attached B ring, a situation unique to this subclass. This gives rise to the existence of variants of the C ring, with the *cis* form noted by the prefix "epi-" when two natural forms exist. The natural forms have (*2R*) stereochemistry, but in the acid conditions of wine, C2 can equilibrate due to acid protonation on the C-ring oxygen. Thus, in wine, the (*2S*) form exists as well, leading to partial racemization of the catechins [1].

On the B ring, there are only two variants observed among flavanols the "normal" 3′,4′-dihydroxy substitution, and the "gallo" 3′,4,′5′- version. Consequently, the flavan-3-ol with *cis* substitution on the C ring and trihydroxy substituents on the B ring is epigallocatechin. There is also substitution at the 3 position, and the flavan-3-ols have gallic acid esters depicted in some examples in Figure 14.2. Glycosides of flavan-3-ols have been detected but not quantified, and appear to be present at low levels [2]. There are five different monomeric flavanols found in grapes (Figure 14.2). The distribution of the flavanols in grape berries is not the same in all varieties, and will also vary between seed and skin [3] (Table 14.1).

14.2 Monomeric catechins

The levels of catechin and epicatechin in Cabernet Sauvignon wine have been reported to be 37–80 mg/L, with the major proportion usually being catechin (Table 14.1, Figure 14.2) [4]. In contrast, epigallocatechin, gallocatechin, and epicatechin gallate have been reported in small amounts in wine [5]. While these are relatively minor components of wine, they are useful as markers for phenolic extraction, that is, from skins and seeds, since the analysis of these compounds is specific and can be precisely measured. This is in contrast

Understanding Wine Chemistry, First Edition. Andrew L. Waterhouse, Gavin L. Sacks, and David W. Jeffery.

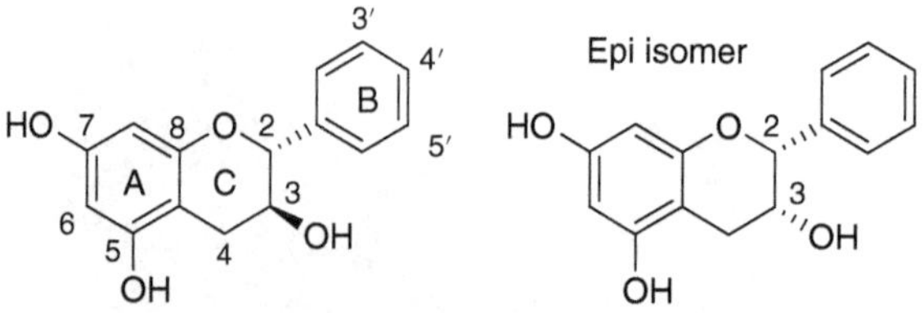

Figure 14.1 Parent ring system for flavan-3-ols

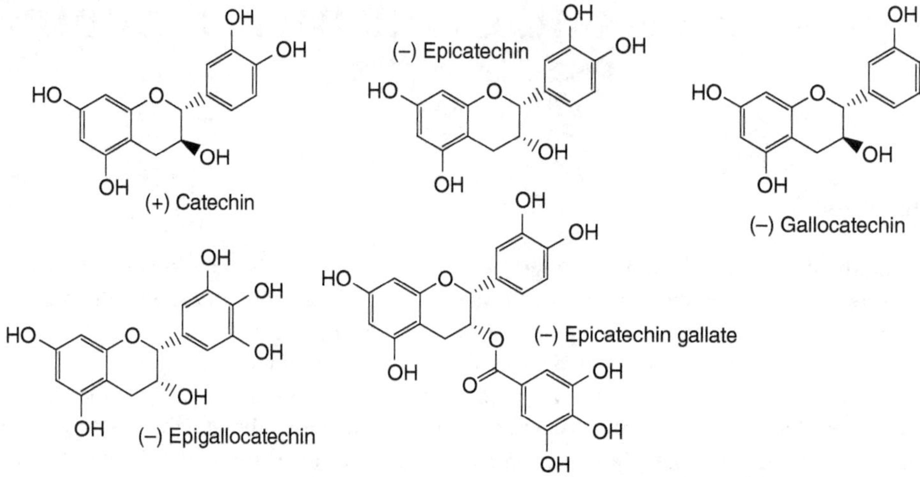

Figure 14.2 Grape-derived monomeric flavan-3-ols in wine

Table 14.1 Flavan-3-ol monomer wine composition

Winegrape variety	Catechin	Epicatechin	EGC	ECG	GC	Reference
Tempranillo	16	10	nr	nr	nr	[6]
Graciano	33	34	nr	nr	nr	
Cabernet Sauvignon	42	20	nr	nr	nr	
Merlot	27	19	nr	nr	nr	
Cabernet Sauvignon	19	58	20	2	nr	[7]
Tannat	43	65	60	0	14	[8]
Nero D'Avola	25	32	nr	nr	nr	[9]
Cabernet Sauvignon	38	16	nr	nr	nr	[10]
Syrah	43	51	nr	nr	nr	
Tempranillo	27	54	nr	nr	nr	

EGC = epigallocatechin, ECG = epicatechin gallate, GC = gallocatechin, nr = not reported.

with measurements of the oligomers and especially polymers of flavan-3-ols, see below. The analysis of these monomeric flavanols is typically via direct HPLC separation of wine samples [4].

While the catechins are minor components, they are also often used as models for the reactions expected of the abundant oligomers and polymers. Many of the reactions described for flavan-3-ols in general are based on reactions of the catechins. For instance, reactions with acetaldehyde were first demonstrated with catechin to form bridged dimers [11], and catechin was used to describe the oxidation of flavanols [12]. Catechin and epicatechin have been reported to isomerize via ring-C opening at high temperatures to yield the enantiomer of the epimer. For instance, (+) catechin isomerizes to (+) epicatechin [13]; however, no enantiomers were observed in new wine pomace [14].

14.3 Oligomeric proanthocyanidins and polymeric condensed tannins

A significant portion, 25–50%, of phenolic compounds in a typical red wine exist as oligomers (proanthocyanidins with distinguishable components of different degrees of polymerization) and polymers (condensed tannins) of flavan-3-ols [15]. These are likely formed through the biochemical condensation of flavan-3-ol units, although the enzymes responsible for forming these linkages, if they exist, are still not elucidated [16], although the genes related to proanthocyanidin production are coming into focus [17]. The condensation forms covalent bonds between flavan-3-ol subunits, the most common linkages being 4→8 (i.e., procyanidin B1) and 4→6 positions (i.e., procyanidin B5) (Figure 14.3). Epicatechin is the predominant unit in condensed tannins from grapes and wine and catechin is the next most abundant (often found at the terminal position, that is, those units with no bond at the 4 position).

Proanthocyanidin is the overarching name given to the class that encapsulates the procyanidins and the prodelphinidins. These names arise from the fact that when these substances are treated with strong mineral acid, they break down into two specific anthocyanidins with the related substitution on the B ring, that is, the 3′,4′-dihydroxy catechin and epicatechin yield cyanidin (procyanidins), while the 3′,4′,5′-trihydroxy subunits

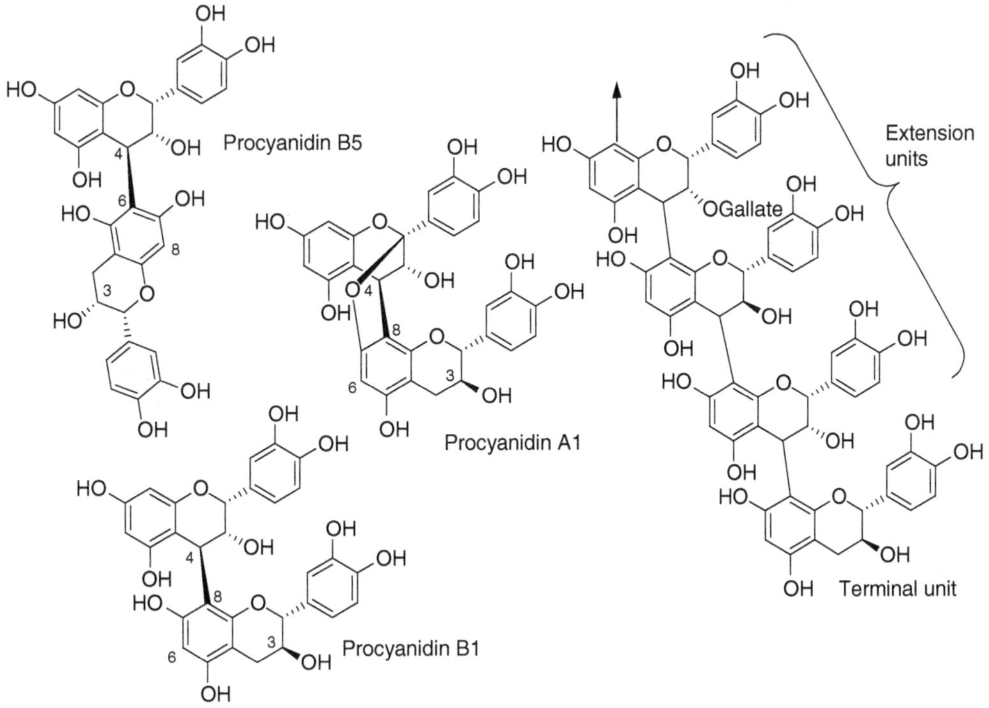

Figure 14.3 Condensed forms of flavan-3-ols

in gallocatechin yield delphinidin (prodelphinidins) (Chapter 16) (Figure 14.4) [18].[1] Only two products are found because this treatment hydrolyzes the gallate esters and eliminates stereochemical differences (normal versus epi) on the C ring (Figure 14.4). This type of method is commonly used to measure the amount of proanthocyanidin in supplements, to give "OPC" values that are calibrated based on cyanidin absorbance. A milder degradation, based on dilute acid-catalyzed cleavage of the 4→8 and 4→6 interflavan bonds, followed by nucleophilic trapping of the reactive carbocation at position 4, yields all subunits with their original stereochemical configuration and C3 ester substitution retained [19] (see Table 14.2 for sample data). Apart from subunit composition, this method can also provide an estimate of the average degree of polymerization (DP) by comparing the amount of released terminal units that are not modified versus the released extension units that are modified at C4. For instance, a tetramer will yield three modified extension units and one terminal unit. Due to the greater amount of information, this method is preferred and widely used. Indeed, proanthocyanidin cleavage in the presence of a nucleophile (e.g., phloroglucinol = phloroglucinolysis or benzyl mercaptan = thiolysis) can be used to compare the ratio of seed-derived gallate ester and skin-derived epigallocatechin in order to assess seed versus skin extraction in winemaking [20]. A limitation of the method is seen with samples where the proanthocyanidin chain has a large amount of A-type linkages or been modified by oxidation or anthocyanin reactions, and in these cases the treatment has low yields of the normal

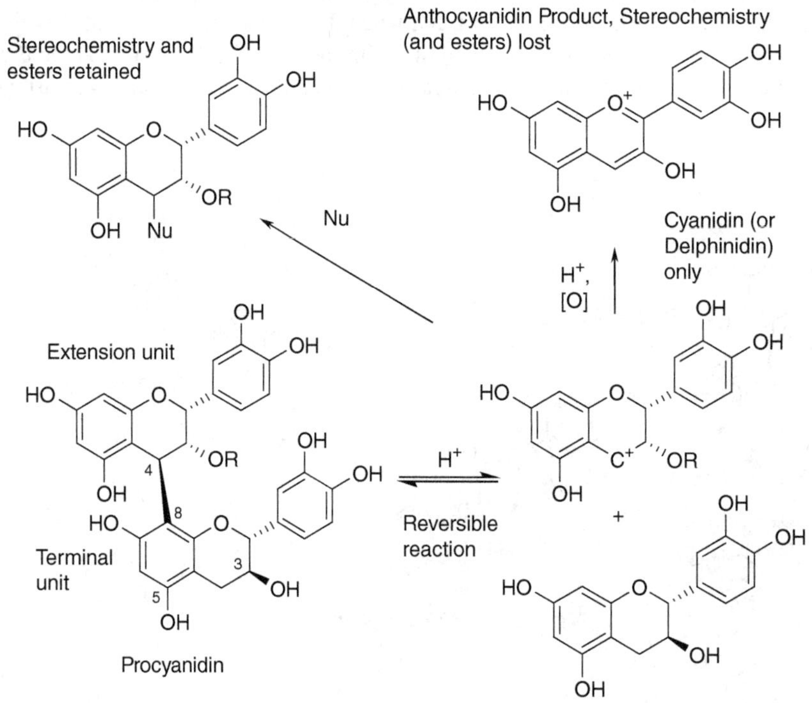

Figure 14.4 Reactions to analyze the proanthocyanidin structure. Strong acid product with oxidation shown on top right, nucleophile (thiolysis or phloroglucinolysis) product on top left. R = H or gallic acid

[1] A dramatic culinary example of this phenomenon is the production of Spanish membrillo, where colorless quince fruit is transformed into a red paste following cooking due to proanthocyanidin hydrolysis. However, oxidation is necessary for anthocyanidin formation, so a trace of iron is essential, and the use of purified water can defeat color formation.

Table 14.2 Proanthocyanidin composition of Cabernet Sauvignon wine, subunit composition of fractions by degree of polymerization [32]. Separation by polar LC followed by phloroglucinolysis of the fractions

DP	mg/L[a]	Extension units				Terminal	
		EGC	Cat	EC	ECG	Cat	EC
5	37.6	33.0	4.0	60.4	2.6	81.6	18.4
6	60.1	32.5	4.0	61.1	2.5	81.3	18.7
7	60.7	32.3	4.2	60.9	2.7	82.0	18.0
8	39.0	32.4	3.4	61.5	2.8	82.3	17.7
9	189.3	36.0	2.9	58.4	2.8	80.0	20.0
10	18.9	36.6	3.0	57.6	2.8	79.3	20.7
11	14.7	35.6	3.3	58.0	3.2	76.2	23.8
12	6.4	34.2	2.8	60.1	2.9	75.0	25.0
13	26.8	34.0	2.9	60.0	3.1	75.7	24.3
14	16.0	37.6	2.9	56.6	2.9	77.5	22.5
15	124.0	39.6	2.4	54.7	3.3	81.0	19.0
Total	593.5						

[a] By HPLC.

products (Chapter 24). The same chemical reaction pathway is important in the continuous rearrangement of proanthocyanidins in wine. See the later boxed section on electrophilic aromatic substitution.

Grape seeds can also contain A-type proanthocyanidins, where there is an additional bond between the oxygen at the C-7 position of one subunit and the C-2 position on the unit "above" [21,22] (Figure 14.3). A-type are also found as a result of oxidation under model conditions [23], so oxidative aging may contribute to additional A-type proanthocyanidins.

14.3.1 Role of proanthocyanidins in wine

The proanthocyadins are quality indicators in grape and wine samples and have several key roles:

- They may react with anthocyanins to form some of the stable wine pigments found in aged red wine (Chapters 16 and 25).
- As high-concentration phenolic compounds, they can both accelerate the rate of oxygen consumption and react with products of oxidation (Chapter 24). Oxidation yields *o*-quinone products from the catechol groups on the B ring, which then react with other phenols, primarily via the nucleophilic A ring to create new bonds.
- With the nucleophilicity of the A ring, these compounds will react with numerous oxidation products, such as quinones and aldehydes, essentially scavenging oxidation products and avoiding their accumulation.
- They are highly correlated with the perception of *astringency* in red wines (i.e., the perceived loss of lubrication in the mouth), as described in Section 14.4.

14.3.2 Proanthocyanidin concentrations and measurements in grapes and wine

Concentrations of proanthocyanidins in grapes are typically reported to be in the range of 0.5–1.5 g/L, with the caveat that the amount is method dependent – see below. Proanthocyanidins are incompletely extracted during fermentation (Chapter 21), and red wine proanthocyanidin concentrations are typically <50% of concentrations in grapes. As discussed later, red wine proanthocyanidin concentrations are often poorly correlated with their corresponding grape concentrations, due to differences in maceration approach and tannin extractability across grapes. White wine proanthocyanidin concentrations are substantially lower, being in the range of 10–50 mg/L, with higher concentrations usually observed in harder press fractions.

As a caveat, comparison of proanthocyanidin values across the literature is complicated because several methods for their measurement exist [24]. This problem arises not only because of the complexity of proanthocyanidins, but also because there is no standard reference material available or even a standardized definition. Although proanthocyanidins absorb strongly at 280 nm, direct measurement using a single UV wavelength is not appropriate because of the presence of other phenolics. Published methods usually specify an isolation or separation step [25], followed by one of several types of procedures:

- Classic tannin analyses rely on condensation reactions with aldehydes such as vanillin to yield a colored product [26]. These approaches will detect monomers as well, and are not often reported for wine analysis.
- Methods that precipitate and isolate proanthocyanidins using protein [27] or polysaccharides [28] prior to UV-vis measurement. These methods correlate well to perceived astringency, described in more detail below.
- Direct HPLC measurement using polar or gel permeation columns [29, 30], both which separate based on molecular size.
- Approaches that degrade the proanthocyanidin into modified monomers prior to HPLC analysis, for example, phloroglucinolysis as described above.
- Finally, secondary methods that use UV-vis spectral measurements combined with multivariate statistics, and are calibrated against the aforementioned methods (Chapter 32) [31].

The correlation between any two of these analytical methods among a group of wines varies – for example, normal phase HPLC and protein precipitation show a strong correlation ($r=0.93$), but much lower correlations ($r<0.7$) are observed with an aldehyde-based reagent [33]. Additionally, even when methods show strong correlations, they can still differ by a constant factor, for example, using methylcellulose in place of a protein as a precipitant to isolate proanthocyanidins results in 3-fold higher measured values [34]. Correlations among methods are even weaker in grapes [35], presumably because of greater variation in proanthocyanidin extraction in the presence of grape solids. As a result, we recommend caution in comparing absolute proanthocyanidin concentrations across the numerous studies regarding the effects of winemaking or grapegrowing conditions on these compounds, for example, Reference [36], especially when methods differ.

14.4 Sensory effects

The monomers are bitter and astringent, and as the DP increases to dimers and trimers, bitterness decreases and astringency increases [37, 38]. LC-Taste studies of red wine have identified the polymeric fraction, termed "procyanidins," as responsible for the majority of wine's astringency, but also noted different sensory responses in the smaller and larger molecular weight subfractions [39]. As mentioned above, some analytical methods for tannins are based on the interaction between protein [27] or modified carbohydrates [28] – these methods typically show excellent correlation ($r^2>0.8$) between perceived astringency and measured tannin [34]. Aging results in the hydrolysis of tannins and their reaction with other wine components (Chapter 25). The resulting modified tannins are more hydrophobic, yield low DP values by phloroglucinolysis, and are less astringent [40]. Other challenges in tannin sensory are discussed later (Chapter 33).

Chemical principles: electrophilic aromatic substitution

One of the most important reactions for aromatic rings is electrophilic aromatic substitution (EAS). In this reaction an electrophile, usually possessing a positive charge, reacts at an unsubstituted position on a benzene ring to displace a proton. The reverse can also occur, with a proton displacing an electrophile.

The tendency for this reaction to occur is based on the reactivity of the electrophile, but also on the nucleophilic properties of the aromatic ring. The latter is based on the other substituents on the ring, and whether they contribute to and activate or withdraw from and deactivate the electron density of the benzene ring. Electrophiles will react more quickly with molecules that have a higher electron density. Oxygen as a phenol or alkylated substituent is a strong activator, with anisole reacting 100 times faster than benzene.

On the A-ring of flavonoids, there are three oxygens, making this very reactive towards electrophiles. In addition, their alternating positions enhance the effect, which can be rationalized by the resonance structures of the transition state of the reaction. At the transition state, the electrophile has bonded to the ring and the positive charge is now distributed around the aromatic ring. Since the charge is concentrated at positions 2, 4, and 6 relative to the point of reaction, substituents that can stabilize the positive charge at those positions have the most impact. Oxygen can provide stabilization from its lone pair electrons into the aromatic pi orbitals. Thus, the alternating oxygens on the A ring create one of the most reactive aromatic rings to electrophilic aromatic substitution.

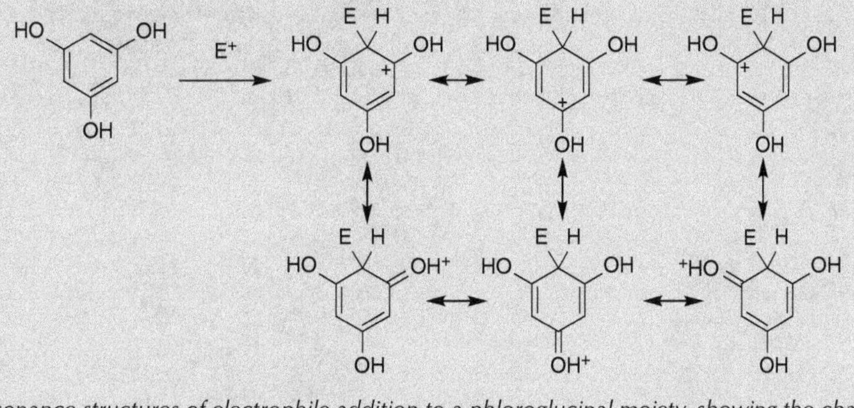

Resonance structures of electrophile addition to a phloroglucinal moiety, showing the charge delocalization that stabilizes the intermediate

Two important examples of EAS in wine chemistry involve the flavan-3-ols, the most reactive to electrophiles. When protonated by the acid in wine, acetaldehyde becomes a potent electrophile and reacts at the 6 and 8 positions of flavan-3-ols, creating a new covalent bond. The benzylic alcohol product will be protonated at wine pH, resulting in a benzylic carbon. This carbon is a very good electrophile, so a second EAS occurs, resulting in a bridge between two flavanol systems. Nearly any aldehyde can react in a similar manner, and this EAS reaction is the basis of the vanillin method for quantifying condensed tannin (although the vanillin does not create bridged structures). A second example is the cleavage and re-formation of proanthocyanidin chains [41]. The bond from the 4 position on one catechin unit to the 8 position on another can be cleaved by EAS by a proton from acid. The released benzylic cation can then react with the A ring of a different catechin unit elsewhere, resulting in rearrangement of the proanthocyanidin chains. When this reaction occurs with a large amount of a monomer, the monomer will dominate the terminal position in the equilibrated sample [42], and the mean degree of polymerization of the proanthocyanidins will decrease. If other nucleophiles are present, additional products can be formed [43]. One example would be anthocyanin (see Chapter 16).

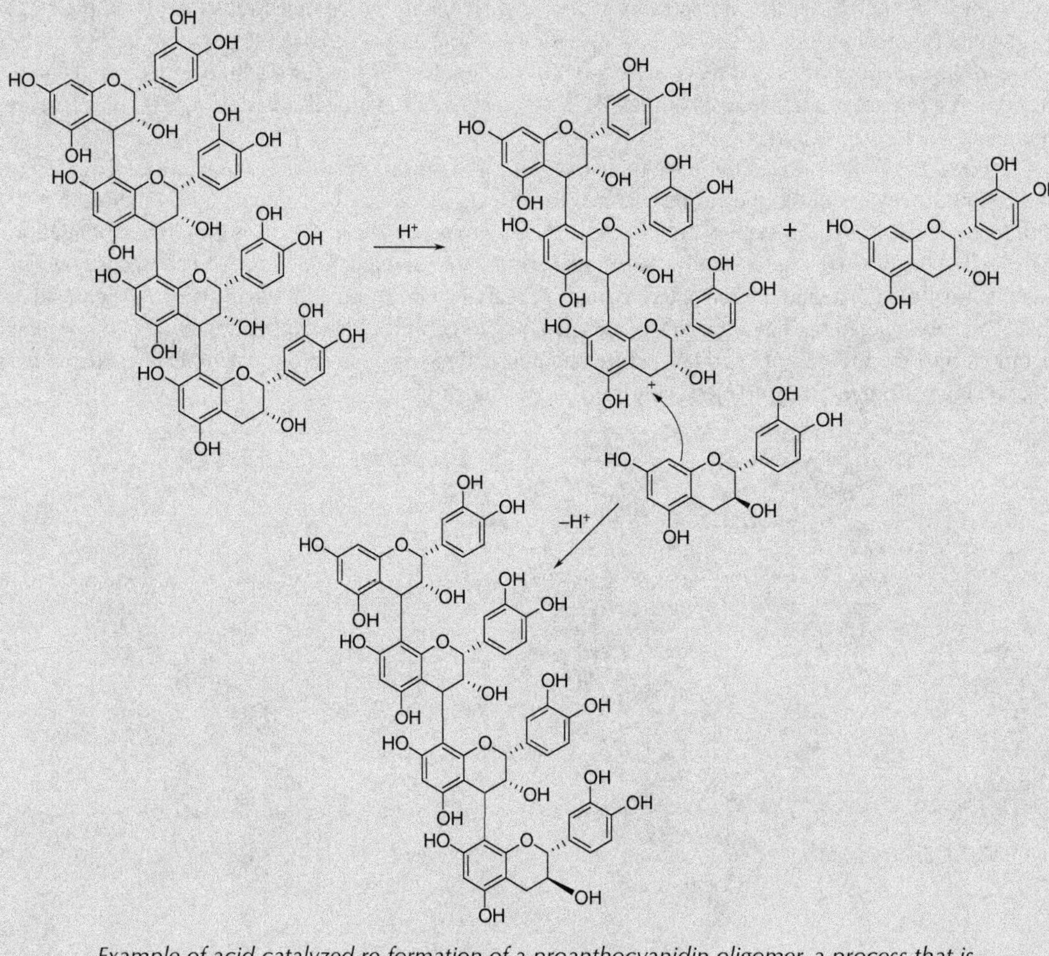

Example of acid-catalyzed re-formation of a proanthocyanidin oligomer, a process that is happening constantly in wine

References

1. Nay, B., Monti, J.P., Nuhrich, A., *et al.* (2000) Methods in synthesis of flavonoids. Part 2: High yield access to both enantiomers of catechin. *Tetrahedron Letters*, **41** (47), 9049–9051.
2. Delcambre, A. and Saucier, C. (2012) Identification of new flavan-3-ol monoglycosides by UHPLC-ESI-Q-TOF in grapes and wine. *Journal of Mass Spectrometry*, **47** (6), 727–736.
3. Mattivi, F., Vrhovsek, U., Masuero, D., Trainotti, D. (2009) Differences in the amount and structure of extractable skin and seed tannins amongst red grape varieties. *Australian Journal of Grape and Wine Research*, **15** (1), 27–35.
4. Ritchey, J.G. and Waterhouse, A.L. (1999) A standard red wine: monomeric phenolic analysis of commercial Cabernet Sauvignon wines. *American Journal of Enology and Viticulture*, **50** (1), 91–100.
5. Ricardo da Silva, J.M., Rosec, J.P., Bourzeix, M., Heredia, N. (1990) Separation and quantitative determination of grape and wine procyanidins by high-performance reversed phase liquid chromatography. *Journal of the Science of Food and Agriculture*, **53** (1), 85–92.
6. Monagas, M., Suarez, R., Gomez-Cordoves, C., Bartolome, B. (2005) Simultaneous determination of nonanthocyanin phenolic compounds in red wines by HPLC-DAD/ESI-MS. *American Journal of Enology and Viticulture*, **56** (2), 139–147.
7. Canals, R., Llaudy, M.D.C., Canals, J.M., Zamora, F. (2008) Influence of the elimination and addition of seeds on the colour, phenolic composition and astringency of red wine. *European Food Research and Technology*, **226** (5), 1183–1190.
8. Boido, E., Garcia-Marino, M., Dellacassa, E., *et al.* (2011) Characterisation and evolution of grape polyphenol profiles of *Vitis vinifera* L. cv. Tannat during ripening and vinification. *Australian Journal of Grape and Wine Research*, **17** (3), 383–393.
9. La Torre, G.L., Saitta, M., Vilasi, F., *et al.* (2006) Direct determination of phenolic compounds in Sicilian wines by liquid chromatography with PDA and MS detection. *Food Chemistry*, **94** (4), 640–650.
10. Gutierrez, I.H., Lorenzo, E.S.P., Espinosa, A.V. (2005) Phenolic composition and magnitude of copigmentation in young and shortly aged red wines made from the cultivars, Cabernet Sauvignon, Cencibel, and Syrah. *Food Chemistry*, **92** (2), 269–283.
11. Saucier, C., Little, D., Golries, Y. (1997) First evidence of acetaldehyde-flavanol condensation products in red wine. *American Journal of Enology and Viticulture*, **48** (3), 370–373.
12. Guyot, S., Cheynier, V., Souquet, J.-M., Moutounet, M. (1995) Influence of pH on the enzymatic oxidation of (+)-catechin in model systems. *Journal of Agricultural and Food Chemistry*, **43** (9), 2458–2462.
13. Wang, H.F. and Helliwell, K. (2000) Epimerisation of catechins in green tea infusions. *Food Chemistry*, **70** (3), 337–344.
14. Rockenbach, I.I., Jungfer, E., Ritter, C., *et al.* (2012) Characterization of flavan-3-ols in seeds of grape pomace by CE, HPLC-DAD-MS^n and LC-ESI-FTICR-MS. *Food Research International*, **48** (2), 848–855.
15. Singleton, V.L. (1992) Tannins and the qualities of wines, in *Plant Polyphenols* (eds Hemingway, R.W. and Laks, P.E.), Plenum Press, New York, pp. 859–880.
16. Dixon, R.A., Xie, D.Y., Sharma, S.B. (2005) Proanthocyanidins – a final frontier in flavonoid research? *New Phytologist*, **165** (1), 9–28.
17. Hancock, K.R., Collette, V., Fraser, K., *et al.* (2012) Expression of the R2R3-MYB transcription factor TaMYB14 from *Trifolium arvense* activates proanthocyanidin biosynthesis in the legumes *Trifolium repens* and *Medicago sativa*. *Plant Physiology*, **159** (3), 1204–1220.
18. Porter, L.J., Hrstich, L.N., Chan, B.G. (1986) The conversion of procyanidins and prodelphinidins to cyanidin and delphinidin. *Phytochemistry*, **25** (1), 223–230.
19. Kennedy, J.A. and Jones, G.P. (2001) Analysis of proanthocyanidin cleavage products following acid-catalysis in the presence of excess phloroglucinol. *Journal of Agricultural and Food Chemistry*, **49** (4), 1740–1746.
20. Des Gachons, C.P. and Kennedy, J.A. (2003) Direct method for determining seed and skin proanthocyanidin extraction into red wine. *Journal of Agricultural and Food Chemistry*, **51** (20), 5877–5881.
21. Passos, C.P., Cardoso, S.M., Domingues, M.R.M., *et al.* (2007) Evidence for galloylated type-A procyanidins in grape seeds. *Food Chemistry*, **105** (4), 1457–1467.

22. De Marchi, F., Seraglia, R., Molin, L., *et al.* (2014) Study of isobaric grape seed proanthocyanidins by MALDI-TOF MS. *Journal of Mass Spectrometry*, **49** (9), 826–830.
23. He, F., Pan, Q.H., Shi, Y., Duan, C.Q. (2008) Chemical synthesis of proanthocyanidins *in vitro* and their reactions in aging wines. *Molecules*, **13** (12), 3007–3032.
24. Herderich, M.J. and Smith, P.A. (2005) Analysis of grape and wine tannins: methods, applications and challenges. *Australian Journal of Grape and Wine Research*, **11** (2), 205–214.
25. Jeffery, D.W., Mercurio, M.D., Herderich, M.J., *et al.* (2008) Rapid isolation of red wine polymeric polyphenols by solid-phase extraction. *Journal of Agricultural and Food Chemistry*, **56** (8), 2571–2580.
26. Caceres-Mella, A., Pena-Neira, A., Narvaez-Bastias, J., *et al.* (2013) Comparison of analytical methods for measuring proanthocyanidins in wines and their relationship with perceived astringency. *International Journal of Food Science and Technology*, **48** (12), 2588–2594.
27. Harbertson, J.F., Picciotto, E.A., Adams, D.O. (2003) Measurement of polymeric pigments in grape berry extracts and wines using a protein precipitation assay combined with bisulfite bleaching. *American Journal of Enology and Viticulture*, **54** (4), 301–306.
28. Sarneckis, C.J., Dambergs, R.G., Jones, P., *et al.* (2006) Quantification of condensed tannins by precipitation with methyl cellulose: development and validation of an optimised tool for grape and wine analysis. *Australian Journal of Grape and Wine Research*, **12** (1), 39–49.
29. Waterhouse, A.L., Ignelzi, S., Shirley, J.R. (2000) A comparison of methods for quantifying oligomeric proanthocyanidins from grape seed extracts. *American Journal of Enology and Viticulture*, **51** (4), 383–389.
30. Kennedy, J.A. and Taylor, A.W. (2003) Analysis of proanthocyanidins by high-performance gel permeation chromatography. *Journal of Chromatography A*, **995** (1–2), 99–107.
31. Dambergs, R.G., Mercurio, M.D., Kassara, S., *et al.* (2012) Rapid measurement of methyl cellulose precipitable tannins using ultraviolet spectroscopy with chemometrics: application to red wine and inter-laboratory calibration transfer. *Applied Spectroscopy*, **66** (6), 656–664.
32. Hanlin, R.L., Kelm, M.A., Wilkinson, K.L., Downey, M.O. (2011) Detailed characterization of proanthocyanidins in skin, seeds, and wine of Shiraz and Cabernet Sauvignon wine grapes (*Vitis vinifera*). *Journal of Agricultural and Food Chemistry*, **59** (24), 13265–13276.
33. De Beer, D., Harbertson, J.F., Kilmartin, P.A., *et al.* (2004) Phenolics: a comparison of diverse analytical methods. *American Journal of Enology and Viticulture*, **55** (4), 389–400.
34. Mercurio, M.D. and Smith, P.A. (2008) Tannin quantification in red grapes and wine: comparison of polysaccharide- and protein-based tannin precipitation techniques and their ability to model wine astringency. *Journal of Agricultural and Food Chemistry*, **56** (14), 5528–5537.
35. Seddon, T.J. and Downey, M.O. (2008) Comparison of analytical methods for the determination of condensed tannins in grape skin. *Australian Journal of Grape and Wine Research*, **14** (1), 54–61.
36. Downey, M., Mazza, M., Seddon, T.J., *et al.* (2012) Variation in condensed tannin content, composition and polymer length distribution in the skin of 36 grape cultivars. *Current Bioactive Compounds*, **8** (3), 200–217.
37. Peleg, H., Gacon, K., Schlich, P., Noble, A.C. (1999) Bitterness and astringency of flavan-3-ol monomers, dimers and trimers. *Journal of the Science of Food and Agriculture*, **79** (8), 1123–1128.
38. Robichaud, J.L. and Noble, A.C. (1990) Astringency and bitterness of selected phenolics in wine. *Journal of the Science of Food and Agriculture*, **53** (3), 343–353.
39. Hufnagel, J.C. and Hofmann, T. (2008) Orosensory-directed identification of astringent mouthfeel and bitter-tasting compounds in red wine. *Journal of Agricultural and Food Chemistry*, **56** (4), 1376–1386.
40. McRae, J.M., Schulkin, A., Kassara, S., *et al.* (2013) Sensory properties of wine tannin fractions: implications for in-mouth sensory properties. *Journal of Agricultural and Food Chemistry*, **61** (3), 719–727.
41. Lea, A.G.H. (1978) The phenolics of ciders: oligomeric and polymeric procyanidins. *Journal of the Science of Food and Agriculture*, **29** (5), 471–477.
42. Vidal, S., Cartalade, D., Souquet, J.-M., *et al.* (2002) Changes in proanthocyanidin chain length in winelike model solutions. *Journal of Agricultural and Food Chemistry*, **50** (8), 2261–2266.
43. Foo, L.Y., McGraw, G.W., Hemingway, R.W. (1983) Condensed tannins – preferential substitution at the interflavanoid bond by sulfite ion. *Journal of the Chemical Society – Chemical Communications*, (12), 672–673.

15

Flavonols

15.1 Introduction

In nature, flavonols are perhaps the most diverse of the non-polymeric flavonoids. Although the number of flavonol aglycones is limited (only six are known in grapes and wines), they possess a tremendous diversity of glycosides, both in terms of the position and type of sugar substituent. The principal aglycones in grape include quercetin, myricetin, laricitrin, and kaempferol with generally lesser levels of isorhamnetin and syringetin (Figure 15.1), although there are varieties with substantial amounts of the latter compounds. Additionally, because of their photoprotective effect in unripe fruit, their total concentrations show one of the strongest and most consistent increases in response to fruit sunlight exposure of any grape compound.

15.2 Concentrations of flavonols

Flavonols occur in a very wide range of vegetable food sources. This class of compounds is always present in glycosidic form in plants including grape berries, where it is found in grape skin. The major glycosides in grapes include the 3-*O*-glucosides as well as the 3-*O*-glucuronides, and these appear in the wine [1].

Flavonol concentrations in wines are dependent on skin extraction, so white wines have much less than reds. The hydrolysis of the glycosides during wine storage (Chapter 25) complicates studies of typical glycoside and aglycone concentrations found in wine and the amount of aglycone increases with time; however, there are a number of survey reports with an adequate number of wines to provide a good sense of typical values (Table 15.1).

A metabolomic study of 91 grape varieties, conducted with hydrolysis of all the glycosides such that the flavonols were quantified solely as the aglycones, showed that the amount of myricetin averaged about the same as that of quercetin, at about 12 mg/kg, with lesser amounts of the other four aglycones (1–2 mg/kg). However, some varieties had much greater amounts of either myricetin or quercetin, with ratios ranging as high as 3–4 in either direction [3]. The same study observed that white varieties lack aglycones possessing B-rings with three oxygenated substituents, suggesting a loss of the flavonoid 3′,5′-hydroxylase enzyme activity [3].

Understanding Wine Chemistry, First Edition. Andrew L. Waterhouse, Gavin L. Sacks, and David W. Jeffery.

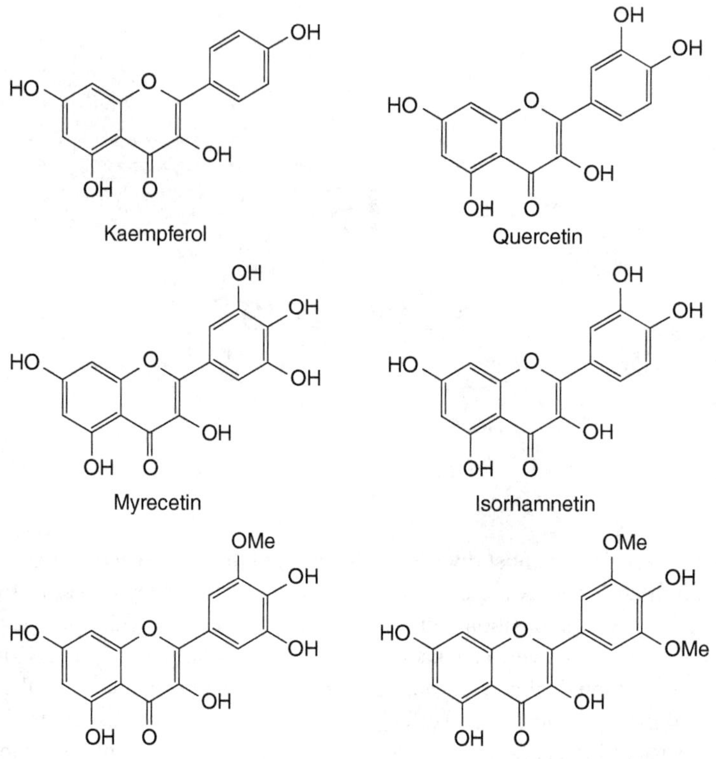

Figure 15.1 Flavonol aglycones found in wine

Table 15.1 Levels of flavonols in red and white wines. Data from References [1] and [2]

	Red wine		White wine	
	Range	Mean	Range	Mean
Quercetin-3-galactoside	ND–6	2	ND	ND
Quercetin-3-glucoside	ND–14	2	ND	ND
Quercetin-3-glucuronide	ND–113	17	ND	ND
Isorhamnetin-3-glucoside	ND–4	1	ND–0.14	ND
Syringetin-3-glucoside	1–27	7	ND	ND
Laricitrin-glucoside	1–23	6		
Myricetin-3-glucuronide	1–12	3.5		
Myricetin-3-glucoside	1–57	15		
Myricetin	2.70–28.78	11.59	ND	ND
Quercetin	3.49–37.36	16.18	ND–2.05	ND
Laricitrin	0.32–4.77	1.81	ND	ND
Kaempferol	ND–2.46	0.82	ND	ND
Isorhamnetin	ND–18.12	3.45	ND	ND
Total glycosides	1.42–55.69	16.16	ND–0.80	0.05
Total aglycones	7.42–76.51	33.85	ND–2.26	0.12

ND: not detected or quantified, blank: no data.

15.3 Effects of growing conditions and winemaking

Initial work on Pinot Noir indicated that berry sunlight exposure strongly enhanced the levels of the flavonols [4]. This observation was later confirmed in Merlot, where a 10-fold increase was observed between shaded and sun-exposed clusters, while temperature had no effect [5]. Sunlight exposure upregulates genes encoding for flavonol synthase [6]. Since flavonols absorb UV light strongly at 360 nm, and they are located mostly in the outermost layer of cells in the berry, it appears that the plant produces these compounds as a natural sunscreen. Because flavonols are correlated with grape sun exposure, and sun exposure has been shown to correlate with many other quality parameters, flavonol concentrations have been proposed as a general quality marker for red grapes [7].

As mentioned earlier, flavonols must be extracted from skins during maceration, and are thus at higher concentrations in red wines. While the glycosides have good solubility, the aglycone quercetin is particularly insoluble in aqueous solutions. As wine ages, the glycosides are hydrolyzed (Chapter 25) and the resulting quercetin is then liable to precipitate and form a haze or deposit in the bottle. Rutin, the 3-*O*-rutinoside of quercetin is occasionally reported in wine, but a detailed study has shown that while rutin is found in grapes, it hydrolyzes in wine within 1–2 days, while other glycosides are stable during that time [2]. The aglycones can be removed by polyvinylpolypyrrolidone (PVPP) treatment, but this treatment is not very effective for glycoside removal [8]. However, it is possible to measure levels of quercetin in wine that are much higher than the point at which saturation occurs in model wine solutions. Some wines, for example, Sangiovese, can throw a precipitate of quercetin on aging, presumably from the hydrolysis of glycosides and resulting supersaturation. The flavonols, particularly quercetin, are well documented to provide a strong co-pigmentation effect with anthocyanins (Chapter 16), which may explain their stability at supersaturated levels, although the relative importance of flavonols to co-pigmentation under wine conditions is debated [9].

The flavonols are bitter, but at the levels found in wine it is not clear if these substances make a large contribution to flavor. One multivariate statistical study of different wines did not observe a correlation between bitterness and higher levels of flavonols [10] although it is possible that other compounds could have overwhelmed their effect. A different study that added back phenolic fractions found some bitterness associated with the fractions high in flavonols [11]. Others have found that flavonols possess "velvety astringency," a concept discussed later (Chapter 33) [12].

While not often discussed in detail, grapes also contain small amounts of dihydroflavonols (flavanonols), first reported by Trousdale and Singleton [13], and by others more recently [2, 14].

References

1. Castillo-Munoz, N., Gomez-Alonso, S., Garcia-Romero, E., Hermosin-Gutierrez, I. (2007) Flavonol profiles of *Vitis vinifera* red grapes and their single-cultivar wines. *Journal of Agricultural and Food Chemistry*, **55** (3), 992–1002.
2. Jeffery, D.W., Parker, M., Smith, P.A. (2008) Flavonol composition of Australian red and white wines determined by high-performance liquid chromatography. *Australian Journal of Grape and Wine Research*, **14** (3), 153–161.
3. Mattivi, F., Guzzon, R., Vrhovsek, U., *et al.* (2006) Metabolite profiling of grape: flavonols and anthocyanins. *Journal of Agricultural and Food Chemistry*, **54** (20), 7692–7702.
4. Price, S.F., Breen, P.J., Valladao, M., Watson, B.T. (1995) Cluster sun exposure and quercetin in Pinot noir grapes and wine. *American Journal of Enology and Viticulture*, **46** (2), 187–194.
5. Spayd, S.E., Tarara, J.M., Mee, D.L., Ferguson, J.C. (2002) Separation of sunlight and temperature effects on the composition of *Vitis vinifera* cv. Merlot berries. *American Journal of Enology and Viticulture*, **53** (3), 171–182.
6. Downey, M.O., Harvey, J.S., Robinson, S.P. (2004) The effect of bunch shading on berry development and flavonoid accumulation in Shiraz grapes. *Australian Journal of Grape and Wine Research*, **10** (1), 55–73.
7. Ritchey, J.G. and Waterhouse, A.L. (1999) A standard red wine: monomeric phenolic analysis of commercial Cabernet Sauvignon wines. *American Journal of Enology and Viticulture*, **50** (1), 91–100.

8. Laborde, B., Moine-Ledoux, V., Richard, T., *et al.* (2006) PVPP-polyphenol complexes: a molecular approach. *Journal of Agricultural and Food Chemistry*, **54** (12), 4383–4389.
9. Lambert, S.G., Asenstorfer, R.E., Williamson, N.M., *et al.* (2011) Copigmentation between malvidin-3-glucoside and some wine constituents and its importance to colour expression in red wine. *Food Chemistry*, **125** (1), 106–115.
10. Saenz-Navajas, M.P., Ferreira, V., Dizy, M., Fernandez-Zurbano, P. (2010) Characterization of taste-active fractions in red wine combining HPLC fractionation, sensory analysis and ultra performance liquid chromatography coupled with mass spectrometry detection. *Analytica Chimica Acta*, **673** (2), 151–159.
11. Preys, S., Mazerolles, G., Courcoux, P., *et al.* (2006) Relationship between polyphenolic composition and some sensory properties in red wines using multiway analyses. *Analytica Chimica Acta*, **563** (1–2), 126–136.
12. Hufnagel, J.C. and Hofmann, T. (2008) Orosensory-directed identification of astringent mouthfeel and bitter-tasting compounds in red wine. *Journal of Agricultural and Food Chemistry*, **56** (4), 1376–1386.
13. Trousdale, E.K. and Singleton, V.L. (1983) Astilbin and engeletin in grapes and wine. *Phytochemistry*, **22** (2), 619–620.
14. Vitrac, X., Castagnino, C., Waffo-Teguo, P., *et al.* (2001) Polyphenols newly extracted in red wine from south-western France by centrifugal partition chromatography. *Journal of Agricultural and Food Chemistry*, **49** (12), 5934–5938.

16

Anthocyanins

16.1 Introduction

Anthocyanins are the compounds responsible for the color of red and black grapes and red wines; the red and blue colors found in many plants are also attributed to anthocyanins, especially in the flowers and fruits. The perceived red color is due to absorption of green visible light (520 nm), which results from the fully conjugated 10 π-electron (i.e., aromatic) A–C flavonoid ring system that is also cross-conjugated into the B ring.[1] If that conjugation is disrupted the color is lost, such as when anthocyanins react with bisulfite, water, or other nucleophiles. Anthocyanin structure and chemistry is further complicated by the ability of anthocyanins to bind both covalently and non-covalently to common wine constituents. The monomeric anthocyanins react with carbonyl compounds and tannins to produce "stabilized" wine pigments, also called polymeric pigments in some texts. Although monomeric anthocyanins are detectable in small amounts in some wines beyond 5 years, the primary sources of color in most red wines past a few years of age are stabilized pigments.

16.2 Structures and forms

The term for the simple flavonoid ring system is *anthocyanidin*, but anthocyanidins are not found in grapes or wine except in trace quantities because they are unstable. *Anthocyanin* refers to glycosylated anthocyanidins. These anthocyanins are also referred to as monomeric pigments to distinguish them from wine pigments formed from their reaction with condensed tannins and other compounds (see Section 16.5). In *V. vinifera*, the 3-*O*-glucoside is the predominant form of anthocyanins (Figure 16.1). In American species and hybrids (Chapter 31), however, the 3,5-di-*O*-glucoside is also found and its presence is the basis for identifying the use of non-*vinifera* grapes to make wines. The glucose conjugated to anthocyanidins may be further

[1] Organic pigment compounds often have high degrees of alternating single and double bonds ("conjugation"). The red and yellow colors of tomatoes, peppers, watermelons, and related members of the nightshade or curcubit family are due to carotenoids (see Chapter 10), composed of 40 carbons arranged in mostly conjugated double bonds. Carotenoids are very non-polar, and a "wine" produced from tomatoes would be nearly colorless.

Understanding Wine Chemistry, First Edition. Andrew L. Waterhouse, Gavin L. Sacks, and David W. Jeffery.

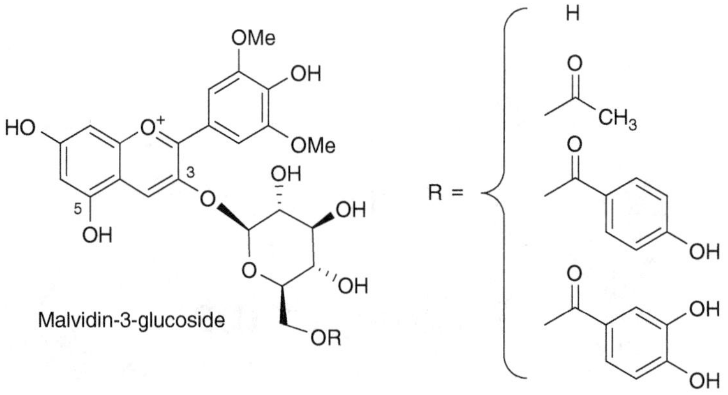

Figure 16.1 Malvidin-3-glucoside (M3G), the dominant anthocyanin in many grapes, is shown as a general example of the anthocyanin structure. The simple sugar (R = H, malvidin-3-glucoside) is most abundant, but varying amounts of acetyl and coumaroyl substituents are found, with traces of caffeoyl, depending on grape variety

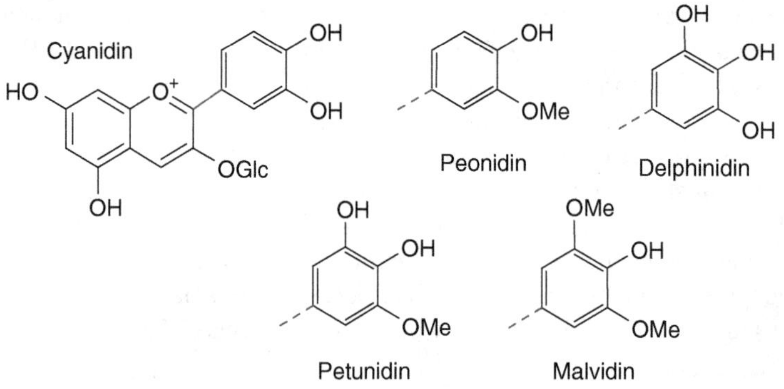

Figure 16.2 The five versions of the B-ring found in grape anthocyanins

substituted through esterification at the 6-position of the sugar moiety, either to an acetyl or a coumaroyl group; a small amount of caffeoyl substitution is also observed [1] (Figure 16.1).

There are five major aglycones in red grapes, which differ in their substitution patterns on the B ring (Figure 16.2); the form with only one hydroxyl at the 4′ position, apigenin, is absent in grapes. Considering the B-ring substitution and acylation noted above, it is possible to observe 10–15 anthocyanins in any particular red grape. Anthocyanin profiles will vary among cultivars, which has led to attempts to use these profiles to authenticate the grapes used to produce young red wines [2]. For instance, Pinot Noir is notable for its lack of acylated glucoside forms, but unfortunately the loss of these grape-derived monomeric anthocyanins during aging negates the application of this approach to differentiate aged wines (Chapter 28). Malvidin-3-glucoside (M3G) and its derivatives dominate the anthocyanin profile in most red grapes and red wines, so most investigations focus on the reactivity and interactions of this anthocyanin alone. However, information from those studies can generally apply to the other anthocyanins.

Anthocyanins present in wine and similar solutions exist as several forms in a pH dependent equilibrium, and the relative proportions strongly affect the color of the solution. The positively charged C-ring of the flavylium (2-phenylchromenylium or benzopyrylium) cation is electrophilic, and the C2 and C4 positions can react with nucleophiles present in wine. Common reactions are with water and bisulfite; in both cases, the red color of the

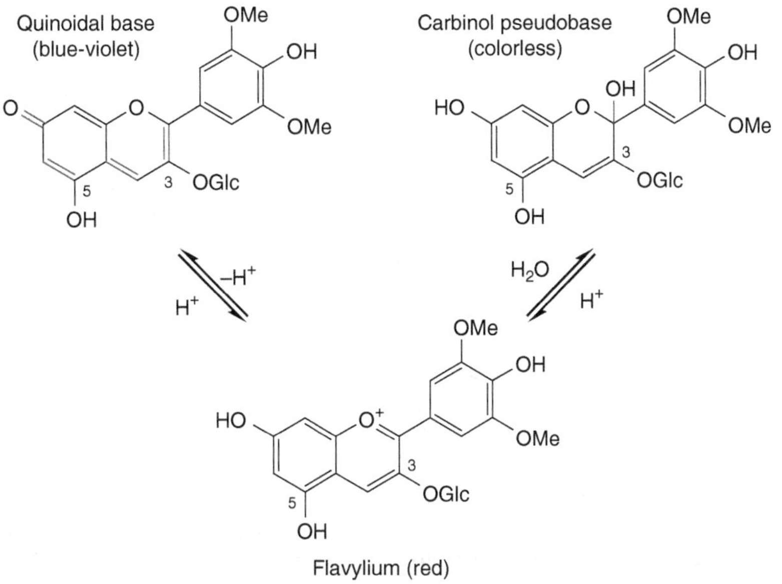

Figure 16.3 *Equilibrating forms of anthocyanins*

flavylium form is lost when the double-bond conjugation is disrupted. As pH of a solution rises the flavylium cation equilibrates like an acid, reacting with water, to become a colorless neutral species, the carbinol pseudobase, after losing a proton from the nucleophilic water molecule (Figure 16.3). The pK_a of the flavylium–pseudobase equilibrium is 2.7, so at a typical wine pH of 3.7, 90% of the anthocyanin pool is rendered colorless [3]. At low pH, all forms are converted to the flavylium cation, facilitating quantitation of the anthocyanins by light absorbance at 520 nm. In addition, there is a quinone form, which has a violet hue and a pK_a of 4.7, so it is present in small amounts at high wine pH values. One research team has proposed an alternate explanation of anthocyanin color based on the observation that at wine pH, wine pigments appear to have zero charge [4].

16.3 Non-covalent interactions: co-pigmentation

Anthocyanins can form non-covalent interactions with other phenolic compounds in solution to create an effect known as *co-pigmentation*. This results in stabilization of the pigmented, aromatic forms of anthocyanins as compared to the non-aromatic carbinol base. The outcome of co-pigmentation is that absorbance of anthocyanin-containing solutions, for example, red wine, are often greater than what is predicted by anthocyanin concentration and the solution pH alone [5]. Co-factor–anthocyanin interactions can be described by the following relationship:

$$\text{Anthocyanin} + \text{Co-factor} \xrightleftharpoons{K_d} [\text{Co-pigmented Complex}]$$

K_d is the binding constant of the anthocyanin and co-factor. Typically, the best co-factors (greatest enhancement of color) are planar aromatic structures, since non-planar structures will be disfavored by steric effects. For example, the planar quercetin has a binding constant of 2900 M^{-1}, while the non-planar catechin has a binding constant of 90 M^{-1}. Because the co-pigmentation reaction is bimolecular, co-pigmentation complexes

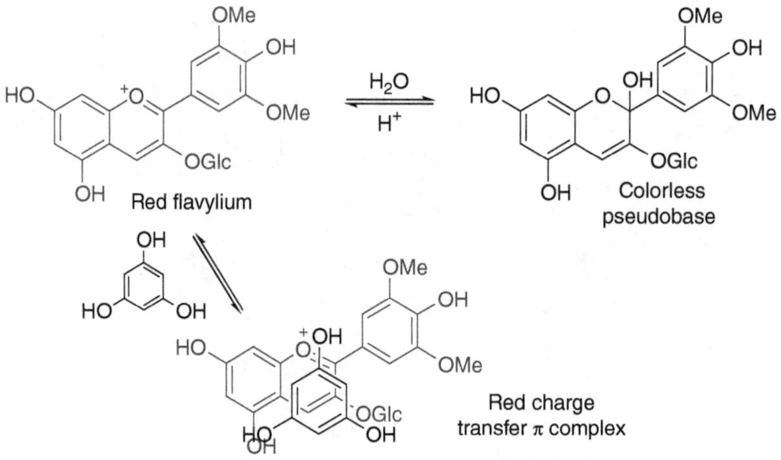

Figure 16.4 Formation of a model charge transfer complex with malvidin-3-glucoside and phloroglucinol, which has the effect of shifting anthocyanin equilibria towards the red colored form. The red flavylium form is shown with highlighting. Phloroglucinol is used as the electron-rich partner for simplicity, but it is not found in wine

will dissipate rapidly as wine is diluted (e.g., a 2-fold dilution will cause a 4-fold dilution in complexes). This results in a deviation of wine color from Beer's Law.

Multiple explanations have been proposed for the nature of chemical bonds in co-pigmentated complexes. These complexes increase the proportion of anthocyanins that exist in the flavylium form as compared to a solution of anthocyanins alone, in the same manner as decreasing the pH. One possible explanation is charge transfer complexation, which occurs when two aromatic-ring substances in solution have different electron densities (Figure 16.4) [6]. In wine, the flavylium form of an anthocyanin has a positive charge and is electron poor, and the other phenolic compounds in wine are generally electron rich because the phenol groups are strong electron donors. An alternative explanation is that the co-pigmentation interactions are primarily hydrophobic – that is, the faces of the planar aromatic molecules arrange in a π–π stacked manner [7]. For complexes found in wine, both charge transfer and hydrophobic interactions may contribute.

At wine pH, there is a large pool of the colorless pseudobase (ca. 90%), so the observed co-pigmentation effect can be very significant – doubling of absorbance is not unusual – and colorimetric analyses of anthocyanin content must be corrected for this effect. However, at low pH, only the flavylium form is present, so under these circumstances, the observed co-pigmentation contribution to absorbance and bathochromic shift is small. A final complicating issue is that anthocyanins can self-associate, resulting in strong co-pigmentation when anthocyanin concentrations are high [8].

16.4 Bisulfite bleaching

Anthocyanins in the flavylium form will react with the bisulfite nucleophile (Chapter 17) on the C-ring at the C4 position (Figure 16.5), unlike water, which will add at the C2 position to form a pseudobase (cf. Figure 16.3). However, the overall effect is the same as water addition in that pigmentation is lost due to disruption of conjugation. The resulting loss of red color is referred to as "bleaching" [9]. Formation of covalent bonds at the C4 position can block bisulfite addition and decrease this bleaching effect, as described below. Because wine samples often have sulfites present, it is essential to eliminate this bleaching effect when quantifying anthocyanins via spectral analysis [5].

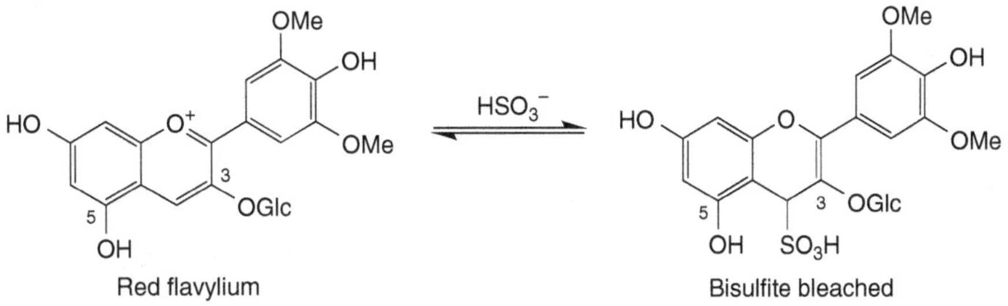

Figure 16.5 *Bleaching of malvidin-3-glucoside through addition of bisulfite to the flavylium cation at C4*

16.5 Wine pigments

During and after fermentation, anthocyanins react with various compounds in the red wine matrix to form modified pigments. Some of the reactions are acid-catalyzed, while others require an oxidation step. The identity of some specific products is known, but comprehensive characterization of all wine pigments is not yet possible, and the relative significance of any of the pathways described below is not established. Additional details are provided in Chapter 25.

Although the average molar absorptivity of the total anthocyanin pool has decreased as a result of conversion to wine pigments [10], the resulting pigments play an important role in wine color because they are "stabilized," a descriptor that covers a wide range of phenomena:

- Wine pigments are less prone to degradation during long-term storage.
- They absorb more strongly at wine pH and demonstrate less pH dependence in their absorbance behavior.
- They are less bleachable by bisulfite.

Modified wine pigments are also referred to in some texts as *polymeric pigments* (and occasionally other terms), although this term suffers from ambiguity. Initially, "polymeric pigment" was applied to any wine pigment that was not bleachable by bisulfite, as it was believed that the non-bleachable species were exclusively reaction products of polymeric condensed tannins and anthocyanins [11]. This operational definition is still used in some assays [12]. However, as described below, stable wine pigments like the vitisins (see below), can be formed without the need for reacting with the "polymeric" condensed tannins. Polymeric pigment is also used to describe high molecular weight pigments separated chromatographically, for example, the late eluting hump in an HPLC chromatogram. While this is a more grammatically correct definition, not all of these high molecular weight species demonstrate behavior associated with key wine pigments, for example, resistance to SO_2 bleaching [13] (see below).

The simplest forms of wine pigments are those formed by electrophilic aromatic substitution on the A ring of a flavonoid. In the first case, referred to as the T-A (i.e., tannin-anthocyanin) type, the electrophilic cation formed in the course of proanthocyanidin interflavan bond cleavage (Chapter 14) reacts with the A ring of an anthocyanin pseudobase to form a series of pigments from dimer to much larger. Figure 16.6 shows an example of a T-A type pigment formed from a proanthocyanidin trimer [14]. This occurs since ca. 90% of the anthocyanins in wine are in the neutral pseudobase form (Figure 16.4) and therefore available to act as a nucleophile, attacking the electrophilic C4 carbocation of the flavonoid.

In the flavylium form, anthocyanins display electrophilic character, as described above, and react directly at the C4 position with the nucleophilic A ring of a proanthocyanidin to create the A-T (i.e., anthocyanin-tannin) type pigment (Figure 16.7). This addition disrupts the aromaticity of the flavylium cation, resulting in a colorless flavene product. This could potentially oxidize to generate the aromatic C-ring and recover the

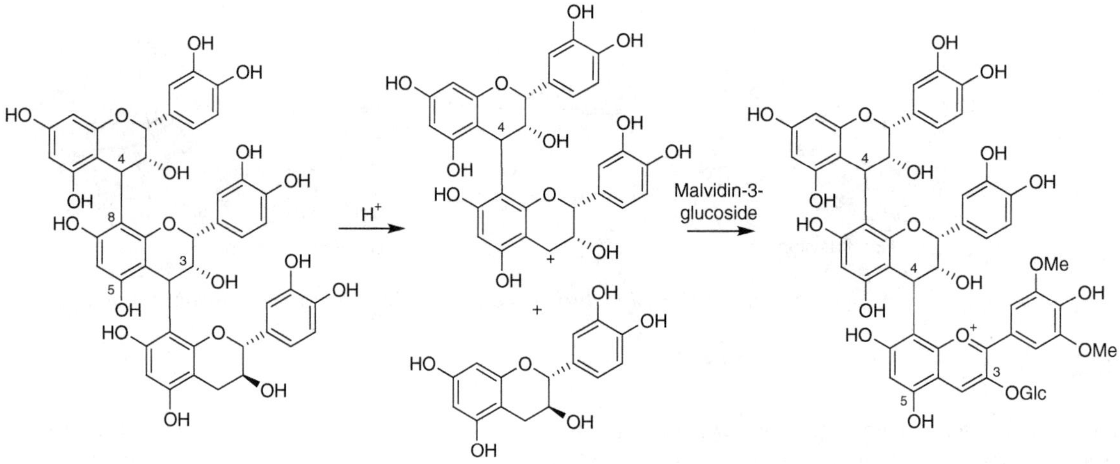

Figure 16.6 Formation of T-A type wine pigments via condensation of anthocyanins and proanthocyanidins

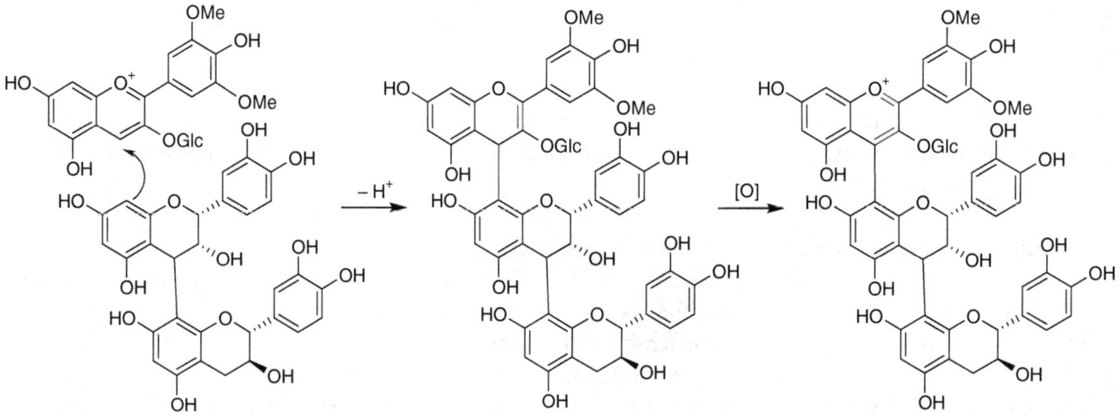

Figure 16.7 Formation of A-T type wine pigment resulting in the colorless flavene form of the anthocyanin in the center. Oxidation could potentially return the anthocyanin to its colored flavylium form

pigmentation, and then react to form a xanthylium cation (Chapter 25), although there is some suggestion that the reaction stops at the flavene, leading to a net loss of color [14].

Wine oxidation (Chapter 24) and fermentation produce reactive compounds that modify anthocyanins in a number of ways to produce new pigments. The simplest is acetaldehyde, which is produced by yeast sugar metabolism during fermentation or through ethanol oxidation during wine storage. Acetaldehyde is a good electrophile and can act to bridge two flavonoid A rings (Figure 16.8). These can be two flavanols, a flavanol and anthocyanin, or two anthocyanins [15].[2] These ethylene-bridged pigments show a bathochromic shift (violet color) and are less susceptible to nucleophilic addition (and color loss) by SO_2 and water. However, this reaction is reversible and could release the anthocyanin. Other aldehydes can initiate related reactions,

[2] In the US (27 CFR 24.246), addition of acetaldehyde to juice (but not wine) for the purposes of stabilizing color is legal, although not widely practiced.

such as glyceraldehyde formed by oxidation of glycerol (Chapter 3), which can react with anthocyanins and flavanols to form analogous pigments [16].

Electrophiles formed during fermentation or storage are also capable of forming *pyranoanthocyanins*, which contain an additional pyran ring. A representative structure of vitisin A, formed by reaction of pyruvic acid and malvidin-3-glucoside is shown in Figure 16.8 [17]. The pyranoanthocyanin pigments have a blocked

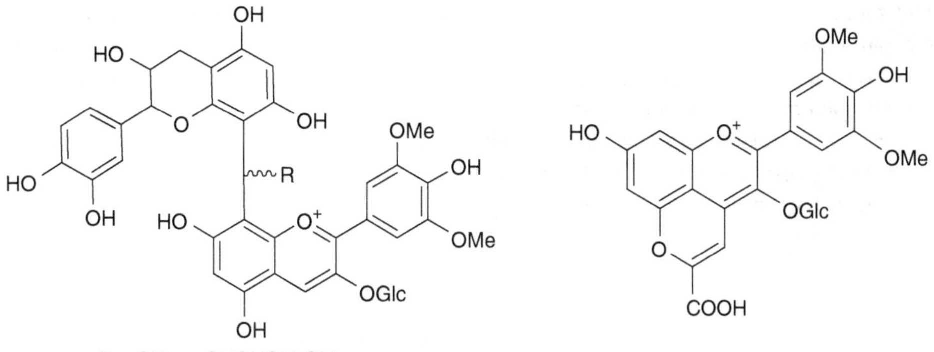

Figure 16.8 *Anthocyanin-derived wine pigments such as the ethyl-linked M3G-flavan-3-ol dimer incorporating acetaldehyde, and vitisin A, which results from M3G and pyruvic acid condensation*

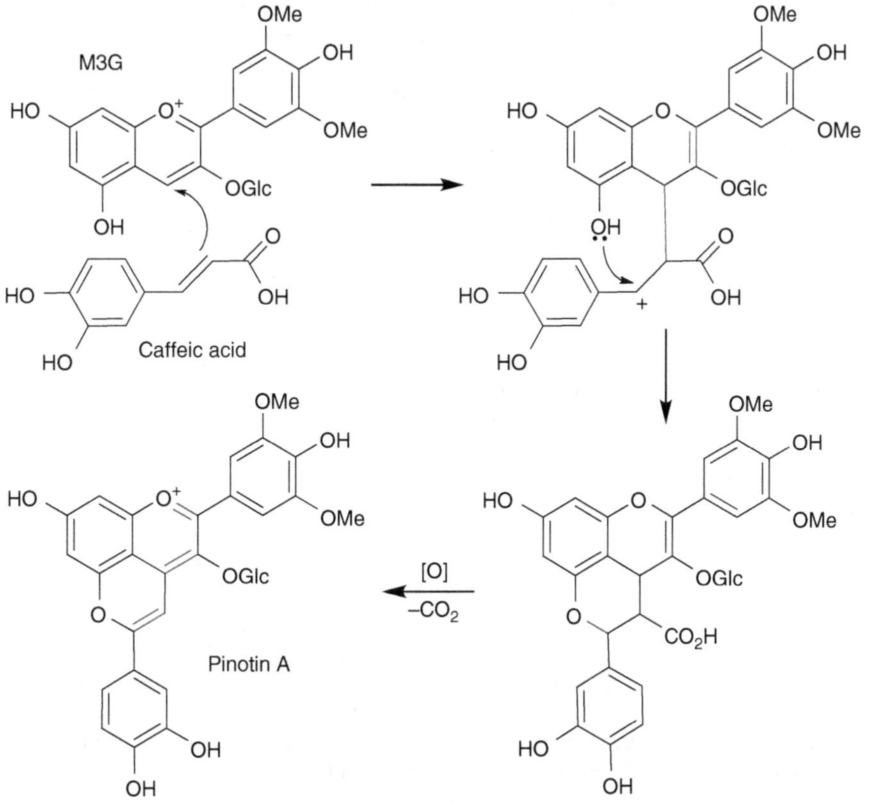

Figure 16.9 *Pinotin A, arising from reaction of M3G with caffeic acid*

C4 position – vitisin A appears to be totally resistant to bleaching up to 250 mg/L of SO_2 and vitisin B (from acetaldehyde addition instead of pyruvic acid) is half bleached at 200 mg/L of SO_2, while malvidin-3-glucoside is 80% bleached at 50 mg/L. Both vitisins show only about 10% variation in absorbance intensity between pH 2 and 4, while the anthocyanin shows a 40% change [18]. These pyranoanthocyanins also appear to be more stable during long-term storage.

Aside from oxidation products, hydroxycinnamates can react directly with anthocyanins to form pinotins [19] in a cyclization with the 5-OH group of the anthocyanin, forming another pyranoanthocyanin (Figure 16.9). The pinotins are distinguished by the phenolic ring attached to the new pyran ring of the derived pigment. These products are also fairly long lived and certainly contribute to the red pigmentation of aged wine.

The multiple members of the anthocyanin class, along with the varying substitution of the glucoside, makes anthocyanin chemistry complex in wine. However, further details of equilibration with water and bisulfite, plus co-pigmentation and the multiple reactions to form wine pigments, make it difficult to predict anthocyanin chemistry during aging (Chapter 25).

References

1. Mattivi, F., Guzzon, R., Vrhovsek, U., *et al.* (2006) Metabolite profiling of grape: flavonols and anthocyanins. *Journal of Agricultural and Food Chemistry*, **54** (20), 7692–7702.
2. Eder, R., Wendelin, S., Barna, J. (1994) Classification of red wine cultivars by means of anthocyanin analysis. 1st report: application of multivariate statistical methods for differentiation of grape samples. *Mitteilungen Klosterneuburg*, **44** (6), 201–212.
3. Brouillard, R. and Delaporte, B. (1977) Chemistry of anthocyanin pigments. 2. Kinetic and thermodynamic study of proton-transfer, hydration, and tautomeric reactions of malvidin 3-glucoside. *Journal of the American Chemical Society*, **99** (26), 8461–8468.
4. Asenstorfer, R.E., Iland, P.G., Tate, M.E., Jones, G.P. (2003) Charge equilibria and p*K*(a) of malvidin-3-glucoside by electrophoresis. *Analytical Biochemistry*, **318** (2), 291–299.
5. Boulton, R. (2001) The copigmentation of anthocyanins and its role in the color of red wine: a critical review. *American Journal of Enology and Viticulture*, **52** (2), 67–87.
6. Castaneda-Ovando, A., Pacheco-Hernandez, M.D., Paez-Hernandez, M.E., *et al.* (2009) Chemical studies of anthocyanins: a review. *Food Chemistry*, **113** (4), 859–871.
7. Kunsagi-Mate, S., May, B., Tschiersch, C., *et al.* (2011) Transformation of stacked pi-pi-stabilized malvidin-3-*O*-glucoside–catechin complexes towards polymeric structures followed by anisotropy decay study. *Food Research International*, **44** (1), 23–27.
8. Lambert, S.G., Asenstorfer, R.E., Williamson, N.M., *et al.* (2011) Copigmentation between malvidin-3-glucoside and some wine constituents and its importance to colour expression in red wine. *Food Chemistry*, **125** (1), 106–115.
9. Timberlake, C.F. and Bridle, P. (1967) Flavylium salts anthocyanidins and anthocyanins. 2. Reactions with sulphur dioxide. *Journal of the Science of Food and Agriculture*, **18** (10), 479–485.
10. Zimman, A. and Waterhouse, A.L. (2004) Incorporation of malvidin-3-glucoside into high molecular weight polyphenols during fermentation and wine aging. *American Journal of Enology and Viticulture*, **55** (2), 139–146.
11. Somers, T.C. (1971) The polymeric nature of wine pigments. *Phytochemistry*, **10** (9), 2175–2186.
12. Harbertson, J.F. and Spayd, S. (2006) Measuring phenolics in the winery. *American Journal of Enology and Viticulture*, **57** (3), 280–288.
13. Versari, A., Boulton, R.B., Parpinello, G.P. (2007) Analysis of SO_2-resistant polymeric pigments in red wines by high-performance liquid chromatography. *American Journal of Enology and Viticulture*, **58** (4), 523–525.
14. Hayasaka, Y. and Kennedy, J.A. (2003) Mass spectrometric evidence for the formation of pigmented polymers in red wine. *Australian Journal of Grape and Wine Research*, **9** (3), 210–220.
15. Escribano-Bailon, T., Alvarez-Garcia, M., Rivas-Gonzalo, J.C., *et al.* (2001) Color and stability of pigments derived from the acetaldehyde-mediated condensation between malvidin 3-*O*-glucoside and (+)-catechin. *Journal of Agricultural and Food Chemistry*, **49** (3), 1213–1217.

16. Laurie, V.F. and Waterhouse, A.L. (2006) Glyceraldehyde bridging between flavanols and malvidin-3-glucoside in model solutions. *Journal of Agricultural and Food Chemistry*, **54** (24), 9105–9111.
17. Fulcrand, H., Benabdeljalil, C., Rigaud, J., *et al.* (1998) A new class of wine pigments generated by reaction between pyruvic acid and grape anthocyanins. *Phytochemistry*, **47** (7), 1401–1407.
18. Bakker, J. and Timberlake, C.F. (1997) Isolation, identification, and characterization of new color-stable anthocyanins occurring in some red wines. *Journal of Agricultural and Food Chemistry*, **45** (1), 35–43.
19. Schwarz, M., Wabnitz, T.C., Winterhalter, P. (2003) Pathway leading to the formation of anthocyanin-vinylphenol adducts and related pigments in red wines. *Journal of Agricultural and Food Chemistry*, **51** (12), 3682–3687.

17

Sulfur Dioxide

17.1 Introduction and terminology

Sulfur dioxide (SO_2) has been used for centuries [1] by winemakers as a preservative due to its antimicrobial and antioxidant properties. In wine, the terms "sulfites," "SO_2," and "sulfur dioxide" are often used interchangeably. From a strict chemical perspective, SO_2 refers only to the neutral, volatile species. However, SO_2 behaves as a weak, diprotic acid in aqueous environments and the major species at wine pH is generally bisulfite (HSO_3^-). The major roles of SO_2 in wine are summarized below:

- *Nucleophile.* HSO_3^- can form covalent adducts with aldehydes and other electrophilic wine components.
- *Reducing agent/antioxidant.* SO_2 is one of the most strongly reducing compounds found in wine. SO_2 does not react directly with O_2, but instead reacts in the form of HSO_3^- with byproducts of oxidation. The specific chemistry of HSO_3^- in this role is discussed in Chapter 24.
- *Enzymatic deactivation.* SO_2, in the form of HSO_3^-, can inhibit the activity of many enzymes. For example, SO_2 inhibits polyphenoloxidase (PPO) activity, slowing the reaction responsible for oxidative browning of musts (Chapter 24).
- *Antimicrobial.* Molecular SO_2 inhibits growth of a wide range of microorganisms, including yeasts and bacteria. The mechanism is likely multitarget and may include reduction of co-factors and vitamins (NAD^+, FAD, thiamin), reduction of disulfide bridges in proteins, and reaction with nucleic acids [2].

SO_2 can exist as several species in wine, summarized in Table 17.1. The definitions of these different species are described in more detail later in the chapter.

Understanding Wine Chemistry, First Edition. Andrew L. Waterhouse, Gavin L. Sacks, and David W. Jeffery.

17.2 Acid–base chemistry of SO_2

Following hydration under aqueous conditions, SO_2 acts as a weak acid and will form conjugate bases (bisulfite, sulfite),[1] as shown in Figure 17.1.

From pK_{a1} it is possible to calculate the relative concentrations of the different species using the Henderson–Hasselbalch equation (Chapter 3). The pK_{a1} and pK_{a2} of SO_2 in practical wine texts (~1.8 and 7.0,

Table 17.1 *Properties of different species of SO_2 in wine*

SO_2 fraction	Importance	Typical target or constraint (as SO_2 equivalents)
Molecular	Antimicrobial	Microbial stability: typical targets are in range of 0.6–0.8 mg/L for wines [2] Sensory threshold for irritation: 2 mg/L [3]
Free bisulfite (HSO_3^-)	Antioxidant Regulatory (less common)	Preventing wine oxidation: typical target for free SO_2 (mostly bisulfite) is 20–40 mg/L Free SO_2 is regulated in a few wine regions, e.g., <70 mg/L in Canada
Sulfite (SO_3^{2-})	Of negligible importance, represents <1% of SO_2 species at wine pH	
Bound bisulfite adducts	Contributes to total SO_2; may have minor antimicrobial activity [4]. Weakly bound adducts can contribute to free SO_2 pool following loss of free bisulfite	
Total (free + bound SO_2)	Regulatory	Regulated in most countries: In US, < 350 mg/L for all wines, <10 mg/L for "no sulfites added." In Australia, <300 mg/L total SO_2 for low sugar wines (<35 g/L sugar), and <250 mg/L for sweet wines

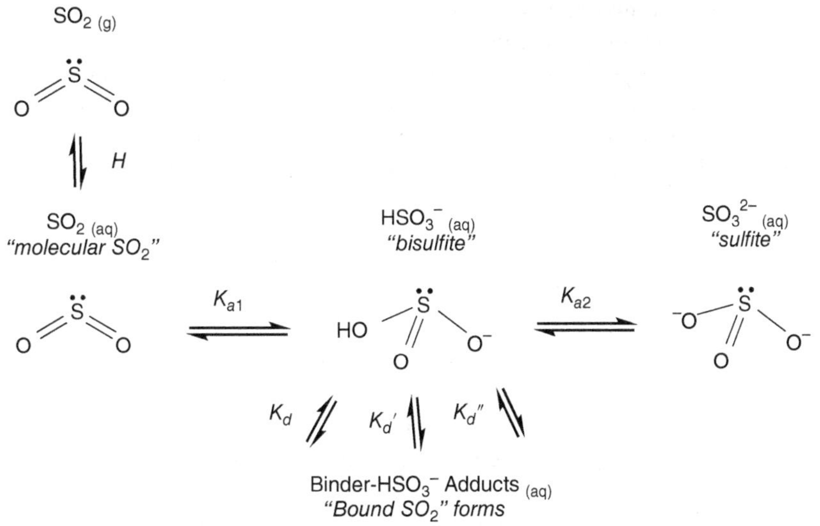

Figure 17.1 *Equilibria of sulfur dioxide species. At 20 °C in H_2O, $pK_{a1} = 1.8$ and $pK_{a2} = 7.0$ [5], and Henry's Law coefficient (H) is 0.38 atm/M [6]*

[1] More properly, bisulfite and sulfite are the conjugate bases of H_2SO_3 (sulfurous acid), formed by reaction of H_2O and SO_2. However, H_2SO_3 decomposes rapidly, and has not been characterized or isolated on Earth, although there is spectroscopic evidence for its existence on Io, one of Jupiter's moons.

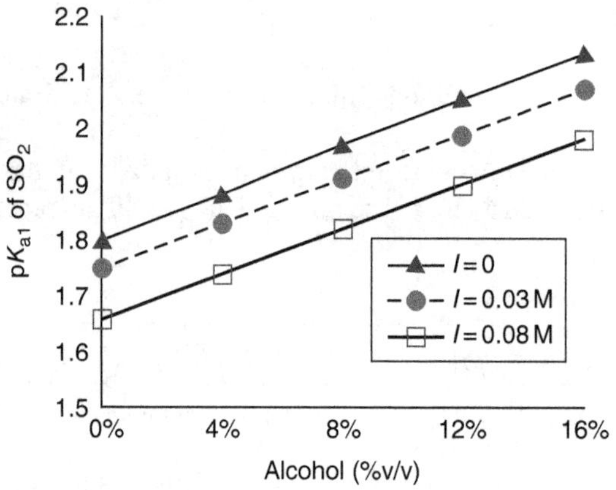

Figure 17.2 Effect of alcohol content on pK_{a1} of SO_2 for three different ionic strengths (I), using data shown in Reference [8]. I=0 corresponds to pure hydroalcoholic solutions, I=0.03M a typical value from Reference [8], and I=0.08M a typical value from Reference [9]

respectively) are often based on their values at 20 °C in H_2O [7]. However, as mentioned in Chapter 3, pK_a values for weak acids (including SO_2) increase with increasing ethanol and temperature and decrease with increasing ionic strength. Empirically derived correction factors and tables exist for pK_a values of SO_2 in wine-like media over a range of temperatures [8]. At 20 °C, for a typical wine alcohol (10–14% v/v) and ionic strength (0.03–0.08 M), pK_{a1} will be in the range of 1.85–2.00, or 0.05–0.2 units higher than typically used values (Figure 17.2). While this may seem like a trivial difference, an error of 0.1 pH units will lead to a 30% error in molecular SO_2, as calculated from Equation (17.1).

At normal wine pH (3–4), the predominant free SO_2 species (>90%) will be bisulfite, with minor contributions from the neutral "molecular SO_2" form. Concentrations of the sulfite ion (SO_3^{2-}) are negligible at wine pH and are generally ignored. In winery settings, it is most common to measure so-called "free SO_2" – that is, the sum of molecular SO_2 and bisulfite. The relationship between molecular and free SO_2 can be derived from the Henderson–Hasselbalch expression:

$$\left[\text{Molecular SO}_2\right] = \frac{\left[\text{Free SO}_2\right]}{1+10^{(\text{pH}-\text{p}K_{a1})}} \qquad (17.1)$$

Equation (17.1) can be used to determine the amount of free SO_2 necessary to achieve different molecular SO_2 concentrations as a function of pH (Figure 17.3). A higher percentage of SO_2 will exist in the molecular form at a lower pH than at a higher pH. For example, using, $pK_a = 1.9$ (12% alcohol, 20 °C, $I = 0.08$ M), the % free SO_2 existing as molecular SO_2 decreases when going from pH 3 (7.7%) to pH 3.3 (3.8%) to pH 3.6 (1.9%).

As summarized in Table 17.1, free SO_2 and molecular SO_2 have different roles in wine. Typical targets are 20–40 mg/L free SO_2 to avoid oxidation, at least 0.6 mg/L molecular SO_2 to prevent microbial spoilage of dry wines, and at least 0.8 mg/L molecular SO_2 to prevent spoilage of sweet wines. The sensory threshold of molecular SO_2 is 2 mg/L, so winemakers usually aim to have molecular SO_2 below this level. These constraints are particularly challenging to achieve simultaneously in wines with low pH values. As shown in Figure 17.3, a pH 2.9 wine would need to have a free SO_2 level < 20 mg/L (risking oxidation) to achieve a molecular SO_2 < 2 mg/L (below the sensory threshold).

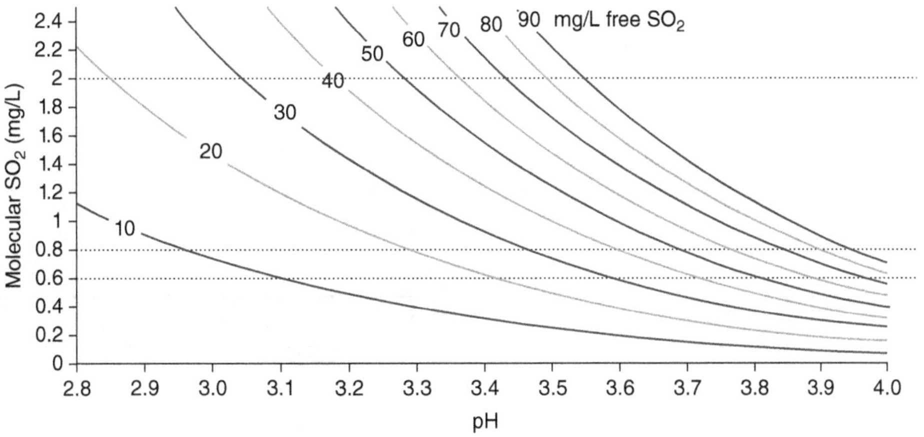

Figure 17.3 Iso-concentration curves for molecular SO_2 as a function of pH for different free SO_2 concentrations, determined in pure water (pK_a =1.8) from the Henderson–Hasselbalch equation. Dotted lines indicate the typical target range for molecular SO_2 (0.6–0.8 mg/L) and the sensory threshold (2 mg/L)

17.3 Sulfonate adducts, "bound SO_2," and antioxidant effects

HSO_3^- is a "soft" nucleophile that will readily form covalent adducts with soft electrophiles to yield sulfonates. The formation of the adduct results in a decrease in the activity of both the SO_2 fraction and the electrophile. The equilibrium between electrophiles and their bisulfite adducts is often represented as a dissociation rather than a formation (Figure 17.4). K_d values in aqueous systems have been determined for the major SO_2 binding species in wine, as well as for lower concentration odorants capable of binding (Table 17.2).

K_d values can vary considerably in the literature – sometimes by as much as a factor of 10 – potentially due to approaches used to quantify free SO_2. Furthermore, these values will vary with temperature, ethanol content, and other factors. However, some general conclusions can be drawn. With the exception of anthocyanins, whose reactions with SO_2 were discussed in Chapter 16, the most important SO_2 binders in wine (those with the smallest K_d values) are carbonyl compounds, and particularly saturated aldehydes like acetaldehyde. The presence of electron-donating groups or conjugation adjacent to the carbonyl will increase K_d. Bisulfite addition to carbonyls may occur at one of two sites, shown in Figure 17.5:

- Direct nucleophilic addition to the carbonyl group (1,2-addition) – observed for saturated carbonyls like acetaldehyde.
- Michael addition, observed for unsaturated conjugated carbonyls (1,4-addition), for example, (*E*)-2-alkenals and β-damascenone.

SO_2 binders can be classified as "weak" or "strong" based on the dissociation constant. The distinction is somewhat arbitrary, but generally SO_2 binders with $K_d < 1 \times 10^{-5}$ are referred to as strong binders, which equates to greater than 95% in the bound form for typical free SO_2 concentrations in bottled wine (20–40 mg/L, or about 0.5 mM). In white wines, the only strong SO_2 binder is acetaldehyde, while other species (pyruvic acid, ketoglutarate) will be weak binders. In red wines, some anthocyanins likely act as strong binders, but this evaluation is complicated by the large number of compounds and pH dependence of anthocyanin

$$\text{Bisulfite adduct} \overset{K_d}{\rightleftharpoons} HSO_3^- \text{ (bisulfite)} + \text{E (electrophile)}$$

$$K_d = \frac{[E][HSO_3^-]}{[\text{Adduct}]}$$

Figure 17.4 Equilibrium of bisulfite with electrophilic binders

Table 17.2 Dissociation constants (K_d) for important SO_2 binding electrophiles (E) from References [10] to [12]

Major SO_2 binders		Other odor-active SO_2 binders	
E	K_d (M^{-1})	*E*	K_d (M^{-1})
Glucose	2.2×10^{-1}		
Fructose	1.5		
Acetoin	8.0×10^{-2}	Diacetyl	1.4×10^{-4}
Galacturonic acid	1.6×10^{-2}	β-Damascenone	Unknown
α-Ketoglutarate	4.9×10^{-4}	β-Ionone	2.1×10^{-4}
Pyruvate	1.4×10^{-4}	Hexanal	3.5×10^{-6}
		(*E*)-2-Pentenal	8.3×10^{-3}
Acetaldehyde	1.5×10^{-6}	(*E*)-2-Nonenal	Unknown
Anthocyanin[a]	1×10^{-5}		

[a] For flavylium form of cyanidin-3-glucoside [5].

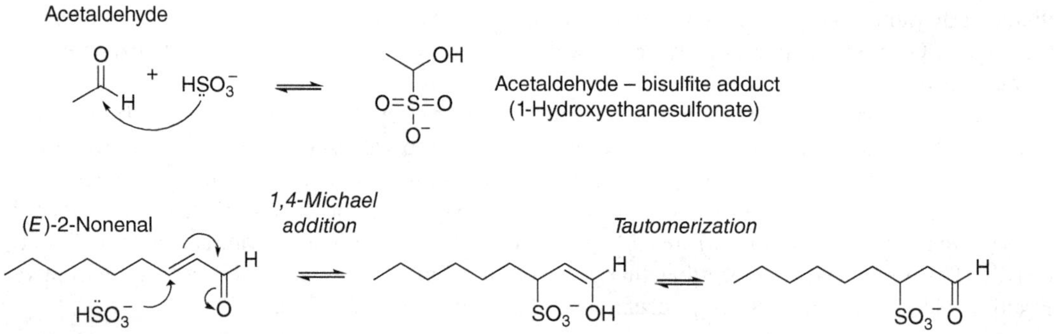

Figure 17.5 Examples of bisulfite adduct formation to the carbonyl group of a saturated aldehyde (acetaldehyde) and via a 1,4-Michael addition to an unsaturated aldehyde ((E)-2-nonenal)

species (Chapter 16). Using K_d it is possible to calculate the relative distributions of bound and free forms for a given binder in a wine and to predict the amount of binding that will occur following addition or removal of a binder or SO_2 to solution (Table 17.3). The major SO_2 binder in most wines is acetaldehyde, due to its strong binding properties and relatively high concentrations [13]. One exception may be in some sweet wines (or musts), where glucose concentrations can exceed 50 g/L. At these concentrations, about 50% of the total SO_2 will be bound to glucose.

Table 17.3 Concentrations of major SO_2 binders in white wine, from data on 127 white wines from Reference [13]

SO_2 binder	Mean concentration (mg/L)	% of binder in bound form, calculated[a]	% of bound SO_2 accounted for, calculated[a]
Acetaldehyde	40 ± 3	99.6	71.5 ± 4.3
Pyruvic acid	25 ± 2	77	16.7 ± 1.6
α-Ketoglutaric acid	31 ± 3	49	7.6 ± 0.9
Galacturonic acid	267 ± 13	2.8	2.7 ± 0.2
Glucose	4750 ± 648	0.22	1.5 ± 0.3
			Total = 100%
			Mean = 81.2 mg/L of bound SO_2

[a] Calculations on SO_2 binding are based on a wine containing the mean values of the binders listed in this table, 30 mg/L free SO_2. K_d values are from Table 17.2.

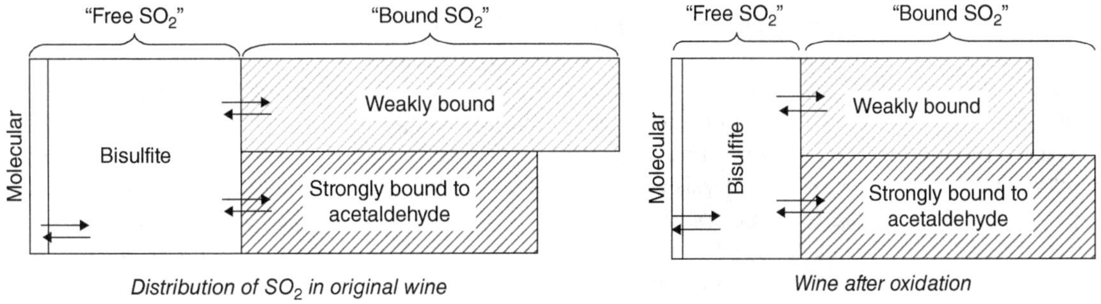

Figure 17.6 Schematic of different SO_2 pools in a hypothetical white wine before and after oxidation

As described in subsequent chapters (Chapter 24), free bisulfite can be lost following wine oxidation, either by reaction with H_2O_2 or by forming adducts with acetaldehyde or other electrophiles. SO_2 adduct formation is reversible and thus formation and dissociation reactions are happening continuously. Figure 17.6 illustrates how SO_2 species in a wine will redistribute following an oxidative event (e.g., addition of H_2O_2 to a wine) which will decrease both the bisulfite and molecular SO_2 pools. Eventually, the weakly bound SO_2 adducts will hydrolyze, resulting in a partial replenishment of the free SO_2 forms. Thus, weakly bound SO_2 can be thought of as a *reservoir* of free SO_2, which can provide some antioxidant protection beyond the "true" free SO_2 concentration, albeit at very low concentrations of free SO_2.

Because there are SO_2 binding compounds in wine of varying strength, addition of SO_2 to a wine will result in a non-quantitative and somewhat unpredictable increase in free SO_2 (see Figure 17.7). Initial additions of SO_2 to a newly fermented wine are typically on the order of 50 mg/L, and it is not uncommon for more than half of this addition to become bound, especially if high concentrations of free acetaldehyde are present. Further addition of SO_2 will result in increasing proportions of free versus bound SO_2 as pools of SO_2 binders become saturated.

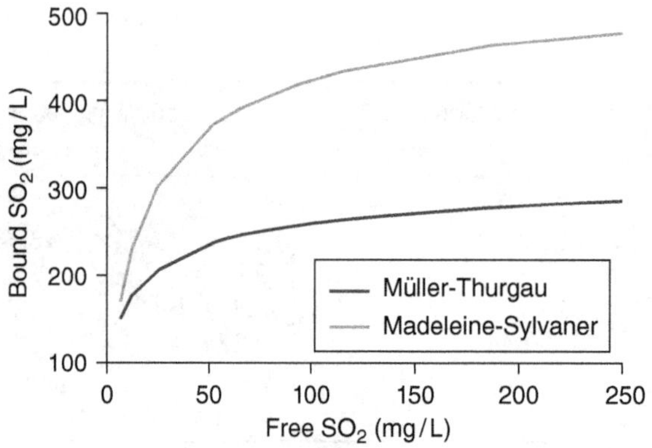

Figure 17.7 *Changes in bound SO_2 concentration following additions of free SO_2 for two representative wines (Muller-Thurgau and Madeleine-Sylvaner). Data from Reference [11]*

17.4 Typical sources and concentrations of SO_2 in wine

Even in wines without SO_2 additions, low amounts of total SO_2 (typically 10–20 mg/L) are present at the end of alcoholic fermentation as a result of SO_2 formation during amino acid biosynthesis [14] (Chapter 22.3). Concentrations of free SO_2 at the end of fermentation are often below detection limits, since most SO_2 added or released as a result of yeast biosynthesis will be bound to acetaldehyde and other fermentation metabolites. Typically, for every 10 mg/L addition of SO_2, a 5 mg/L increase in total SO_2 will be observed at the end of fermentation due to binding to acetaldehyde [15]. Pre-fermentation SO_2 additions will thus also result in higher concentrations of aldehydes post-fermentation.

However, in most wines, the majority of SO_2 is added exogenously either before or after fermentation, often in the form of potassium metabisulfite (KMBS) or as SO_2 gas (Chapter 27). Average total concentrations of SO_2 in commercial wines are about 80 mg/L and 60 mg/L for white and red wines, respectively [16]. Despite some anecdotes to the contrary, white wines on average have higher total concentrations of SO_2 than red wines. White wines are on average bottled with higher SO_2 either because they are more likely to contain residual sugar or because oxidation has more deleterious effects on the aroma and color of white wines than on red wines. Furthermore, acetaldehyde, the most important binder, is on average at higher concentrations in white wines than in red wines (40 versus 25 mg/L) [15].

17.5 Measurement of molecular, free, and total SO_2

In winery settings, the free and molecular SO_2 concentrations are likely to be determined routinely (on a weekly or monthly basis) as a quality control measure. There are three general approaches to measure free and molecular SO_2 summarized below. More detailed explanations of these techniques can be found in wine analysis textbooks, for example, References [17] and [18].

A. Perform titrimetric or colorimetric analysis in which oxidizing reagents are added directly to the wine. The best known of these methods is iodometric titration with the endpoint determined by a starch indicator ("Ripper method") or potentiometer.

B. Acidify a sample to convert to molecular SO_2, and then isolate and quantify molecular SO_2. The best known of these methods is aeration-oxidation (A-O), in which SO_2 in an acidified sample is distilled into a receiver flask containing hydrogen peroxide to generate sulfuric acid, which can then be titrated.
C. Separate and quantify a free SO_2 form (molecular or bisulfite) without pH adjustment or dilution. These methods are not widely practiced in the wine industry, and include capillary electrophoresis of HSO_3^- or quantification of headspace SO_2 using GC-MS or colorimetric methods [6].

Methods that involve dilution, pH changes, and/or rely on the destruction of free SO_2 (approaches A or B) likely overestimate free SO_2, since weakly bound SO_2 adducts will dissociate and be measured as free. The dissociation of bound acetaldehyde reportedly has a half-life about an order of magnitude longer than weak adducts, particularly anthocyanin-bisulfite adducts (1.5 h versus ~10 min) [5, 19]. As a result, the major methods used in the wine industry appear to overestimate free SO_2 in real wines – by an average of 15% in white wines and by a factor of 2 in red wines [6]. The reason for the severe measurement error in red wines is likely due to dissolution of anthocyanin-bisulfite adducts. For typical SO_2 concentration and pH values (3.0–3.8) approximately 70–85% of monomeric anthocyanins are reported to be bound [20] and, as mentioned above, these adducts can dissociate relatively quickly during analysis.

The ramifications of these incorrect measurements are still unknown. A-O (Approach B) and related methods are possibly better suited than non-disrupting methods for predicting the antioxidant capacity of SO_2 in a wine because weakly bound forms will partially replenish free SO_2 pools. However, the anthocyanin-bisulfite fraction measured by A-O and other classic methods has negligible antimicrobial activity, and thus these methods appear to be poorly suited for predicting red wine microbial stability [21]. In spite of these problems, A-O and related methods (e.g., flow-injection analysis) are reproducible across labs [22].[2] Analysis of aldehydes and ketones can have complementary problems when SO_2 is present, as described previously (Chapter 9).

Total SO_2 can be determined by disrupting bound SO_2 forms prior to or during analysis with any of the aforementioned methods. This can be accomplished by treatment of the sample with base to favor SO_3^{2-} and hydrolyze adducts prior to sample acidification (used with iodometric titrations) or heating the sample to disrupt adducts (used in A-O).

17.6 Sensory effects

The aqueous form of molecular SO_2 is in equilibrium with its volatilized form, with a Henry's Law coefficient of $H = 0.38$ atm M^{-1} or $K_{g,l} = 0.016$ at 21 °C in H_2O [6]. Volatilized SO_2 can cause an "irritating, burning" sensation in the nose,[3] and the sensory threshold for molecular SO_2 in wine is reportedly 2 mg/L [3]. HSO_3^- and bound SO_2 forms are reported to have minimal direct sensory effects. However, there is an important indirect effect of SO_2 binding – many odor active compounds are carbonyl compounds, and their bound sulfonate forms will be non-volatile. This binding may be desirable, as is often the case for acetaldehyde and other aldehydes with oxidative aromas, but it may also mean loss of aroma activity for more pleasant smelling compounds, for example, the fruity smelling β-damascenone (Chapter 8) [23].

[2] Which brings to mind the adage "often in error, never in doubt."

[3] The receptor(s) responsible for this chemesthetic sensation is not yet determined, but it may be similar to CO_2, that is, triggering of TRPA1 thermal receptors by diffusion and intracellular acidification.

References

1. McGovern, P.E. (2003) *Ancient wine: the search for the origins of viniculture*, Princeton University Press, Princeton, NJ.
2. Fugelsang, K.C. and Edwards, C.G. (2007) *Wine microbiology practical applications and procedures*, Springer, New York.
3. Ribereau-Gayon, P., Glories, Y., Maujean, A., Dubourdieu, D. (2006) *Handbook of enology*, Vol. 2, *The chemistry of wine stabilization and treatments*, 2nd edn, John Wiley & Sons, Chichester, UK.
4. Wells, A. and Osborne, J.P. (2012) Impact of acetaldehyde- and pyruvic acid-bound sulphur dioxide on wine lactic acid bacteria. *Letters in Applied Microbiology*, **54** (3), 187–194.
5. Brouillard, R. and El Hage Chahine, J.M. (1980) Chemistry of anthocyanin pigments. 6. Kinetic and thermodynamic study of hydrogen sulfite addition to cyanin – formation of a highly stable Meisenheimer-type adduct derived from a 2-phenylbenzopyrylium salt. *Journal of the American Chemical Society*, **102** (16), 5375–5378.
6. Coelho, J.M., Howe, P.A., Sacks, G.L. (2015) A headspace gas detection tube method for measurement of SO_2 in wine without disruption of sulfur dioxide equilibria. *American Journal of Enology and Viticulture*, **66** (3), 257–265.
7. Zoecklein, B.W., Fugelsang, K.C., Gump, B.H., Nury, F.S. (1999) *Wine analysis and production*, Kluwer Academic/Plenum Publishers, New York.
8. Usseglio-Tomasset, L. and Bosia, P. (1984) La prima costante di dissociazione dell'acido solforoso [nei vini]. *Vini d'Italia*, **26** (5), 7–14.
9. Abguéguen, O. and Boulton, R.B. (1993) The crystallization kinetics of calcium tartrate from model solutions and wines. *American Journal of Enology and Viticulture*, **44** (1), 65–75.
10. Blouin, J. (1966) Contribution to study of binding of sulphur dioxide in musts and wines. *Annales de Technologie Agricole*, **15** (3), 223–287.
11. Burroughs, L.F. and Sparks, A.H. (1973) Sulphite-binding power of wines and ciders. I. Equilibrium constants for the dissociation of carbonyl bisulphite compounds. *Journal of the Science of Food and Agriculture*, **24** (2), 187–198.
12. de Azevedo, L.C., Reis, M.M., Motta, L.F., *et al.* (2007) Evaluation of the formation and stability of hydroxyalkylsulfonic acids in wines. *Journal of Agricultural and Food Chemistry*, **55** (21), 8670–8680.
13. Jackowetz, J.N. and Mira de Orduña, R. (2013) Survey of SO_2 binding carbonyls in 237 red and white table wines. *Food Control*, **32** (2), 687–692.
14. Fleet, G.H. (1993) *Wine microbiology and biotechnology*, Harwood Academic Publishers, Chur, Philadelphia, PA.
15. Jackowetz, J.N., Dierschke, S., de Orduna, R.M. (2011) Multifactorial analysis of acetaldehyde kinetics during alcoholic fermentation by *Saccharomyces cerevisiae*. *Food Research International*, **44** (1), 310–316.
16. Peterson, G.F., Kirrane, M., Hill, N., Agapito, A. (2000) A comprehensive survey of the total sulfur dioxide concentrations of American wines. *American Journal of Enology and Viticulture*, **51** (2), 189–191.
17. Zoecklein, B.W., Fugelsang, K.C., Gump, B.H., Nury, F.S. (1995) *Wine analysis and production*, Chapman & Hall, New York.
18. Iland, P. (2004) *Chemical analysis of grapes and wine: techniques and concepts*. Patrick Iland Wine Promotions Pty Ltd, Campbelltown, SA, Australia.
19. Boulton, R.B., Singleton, V.L., Bisson, L.F., Kunkee, R.E. (1999) *Principles and practices of winemaking*, Kluwer Academic/Plenum Publishers, New York.
20. Usseglio-Tomasset, L., Ciolfi, G., di Stefano, R. (1982) The influence of the presence of anthocyanins on the antiseptic activity of sulfur dioxide towards yeasts. *Vini d'Italia*, **24** (137), 86–94.
21. Howe, P.A. (2015) Re-thinking free and molecular sulfur dioxide measurements in wine. PhD Thesis, Department of Food Science, Cornell University, Ithaca, NY.
22. Sullivan, J.J., Hollingworth, T.A., Wekell, M.M., *et al.* (1990) Determination of free (pH 2.2) sulfite in wines by flow injection analysis: collaborative study. *Journal of the Association of Official Analytical Chemists*, **73** (2), 223–226.
23. Daniel, M.A., Elsey, G.M., Capone, D.L., *et al.* (2004) Fate of damascenone in wine: the role of SO_2. *Journal of Agricultural and Food Chemistry*, **52** (26), 8127–8131.

18

Taints, Off-flavors, and Mycotoxins

18.1 Introduction

Undesirable flavors and aromas are a major reason for consumer rejection of wines and other foodstuffs. Many, if not all, flavor compounds can contribute either positively or negatively, with a given compound being indicative of inferior product quality or becoming disagreeable when the concentration exceeds norms for a particular product.[1] Typically, these flavors are classified as either:

- Taints, when the responsible flavor compound is from external contamination, that is, the environment. Taints frequently arise from packaging contamination, but also from equipment, additives, and processing aids, air (or other gases), or microbial activity external to the product.
- Off-flavors, when the flavor is from biochemical or non-enzymatic changes to compounds in the original product.[2] Off-flavors are typically encountered due to inappropriate handling and storage (e.g., light, oxidation, temperature, microbial load, processing operations) and are perhaps more easily avoided than the often accidental contamination leading to taints.

This chapter will primarily focus on taint compounds arising from external (exogenous) contamination. Taint compounds are generally more potent than off-flavors, but either defect often results in a decrease in desirable product attributes rather than the creation of a distinct character from the offending compound(s). Ridgway *et al.* [1] review common food "tainting" compounds and include their origins, descriptors, and thresholds, whereas Boutou and Chatonnet [2] provide data specific to wine. Off-flavors from fermentative sulfur compounds, oxidation, and related compounds ordinarily formed in wine production are described in more detail elsewhere in this book (and in the literature, e.g., see Reference [3]), but are compiled in a cross-referenced table at the end of this chapter (see Table 18.2), for convenience.

[1] An analogy can be drawn between a "good" versus "bad" flavor compound and a "drug" – the dose makes the poison.

[2] A distinction is not necessarily made between taints and off-flavors as the cause of an undesired aroma or flavor is not always clear.

Understanding Wine Chemistry, First Edition. Andrew L. Waterhouse, Gavin L. Sacks, and David W. Jeffery.

The majority of compounds responsible for wine taints have been identified in other foodstuffs. In spite of extensive research, however, it is not always easy to determine the source/s of taints, and compounds presented in this chapter have been categorized based on their suspected or known origins. Compounds causing aroma and flavor defects can possess a wide range of functional groups, but often contain common structural motifs known for low sensory detection thresholds (e.g., sulfur, nitrogen, or halogen atoms). Infrequent taints arising from pesticides, hydraulic oils and other hydrocarbons, paints, resins and plasticizers, metals, and brine used for tank cooling, or deliberate adulterations to manipulate sensory attributes, are not discussed here (for more information see, for example, References [3] to [6]). While this chapter primarily considers compounds that affect organoleptic qualities, some contaminants with negative health effects, for example, mycotoxins from diseased grapes, will also be discussed.

18.2 Common wine taints

18.2.1 Taints from cork or wood

"Cork taint" is the commonest and probably most well-known taint encountered in wine and consequently the taint of interest when evaluating a wine in a restaurant. The major cause is the potent and earthy/musty/moldy-smelling 2,4,6-trichloroanisole (TCA, Table 18.1). In addition to having an undesirable aroma, TCA can inhibit olfactory signal transduction, thereby interfering with the perception of other odorants [14]. The requirements for formation of TCA are:

- 2,4,6-Trichlorophenol (TCP), whose origin is described in more detail below.
- Microorganisms capable of *O*-methylating TCP to produce TCA.

Although cork closures are an obvious (and eponymous) source of TCA, all wood-containing materials are potential sources, including oak barrels (Figure 18.1) or processing aids/equipment, cardboard packaging materials, or even the winery itself (e.g., wooden structures). Related haloanisoles, such as di, tetra, and pentachloroanisoles or 2,4,6-tribromoanisole (TBA), arising from their respective halophenols in the same manner as TCA, can also contribute to cork taint. Several non-haloanisole compounds have been identified in corks, such as geosmin, 1-octen-3-one and related 1-octen-3-ol, 2-methylisoborneol (MIB), and 2-methoxy-3,5-dimethylpyrazine (MDMP); these taints are typically produced by molds and have musty or earthy aromas analogous to TCA [2,15] (Table 18.1). 3-Isopropyl-2-methoxypyrazine (IPMP) and guaiacol can also contribute to cork-derived taints, but are most often derived from other sources and are discussed elsewhere (Chapters 5 and 12). Because cork taint compounds are microbially derived, it is common for more than one of the aforementioned compounds to be present when cork taint is detected, although TCA has received by far the most attention (see Reference [15] for a comprehensive review).

The transfer of TCA from corks and other contaminated materials into wine has been well studied, and TCA is expected to be a good analog for other haloanisoles. In general, TCA-containing material must be in direct contact with wine, or at least in close proximity to allow aerial transfer (TCA vapor can also taint corks in this way [16]), and there is no evidence that TCA can otherwise form in bottled wine. Similarly, TBA is known to contaminate wine and other materials such as corks, barrels, and winemaking equipment via the atmosphere [7].

Several factors will affect the amount of TCA transferred into wine from a cork. TCA concentration varies within cork and other materials (i.e., on the surface or internal), and migration down the full length of a cork in a sealed bottle is slow enough to be an improbable route of transmission. The rate of transfer will also be

Table 18.1 Indicative odor descriptors, detection thresholds, and odor activity values of the main compounds that impart a taint to wine [2,7–13][a]

Compound[b]	Structure	Odor descriptor	Threshold (ng/L)	OAV (max.)
2,4,6-Trichloroanisole (TCA)	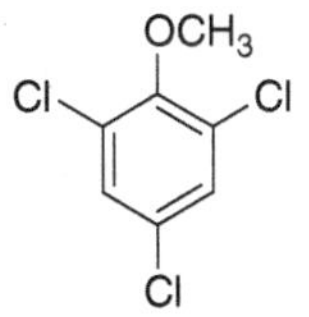	Mold, earth	3	12
2,4,6-Tribromoanisole (TBA)	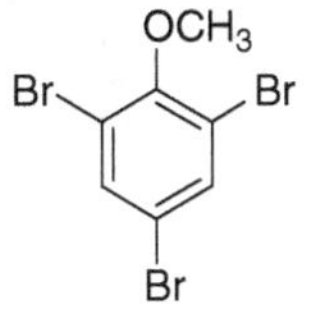	Mold, earth	8	5
2,6-Dichlorophenol (2,6-DCP)	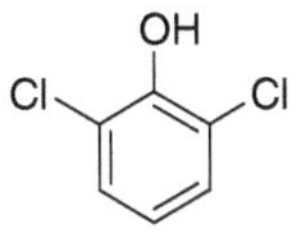	Plastic, medicinal	32	6
6-Chloro-*o*-cresol (6-CC)	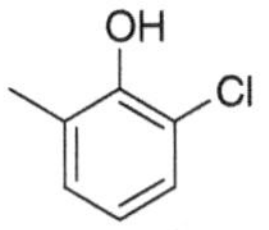	Disinfectant	70	7
(–)-Geosmin	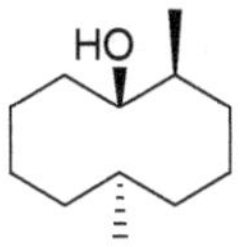	Earth	50	6
1-Octen-3-one	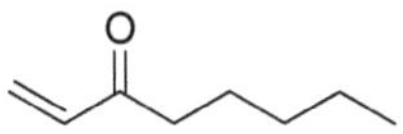	Mushroom	70	6
1-Octen-3-ol	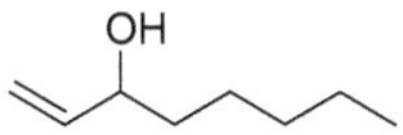	Mushroom	40 000	5
(–)-2-Methylisoborneol (MIB)	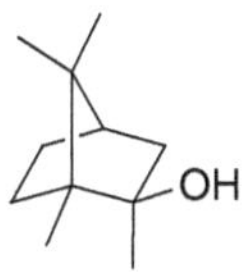	Earth	55	3[c]
2-Methoxy-3,5-dimethylpyrazine (MDMP)	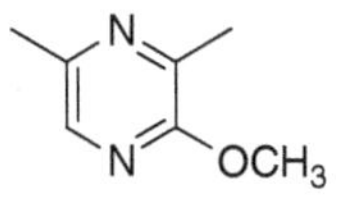	Fungal, earth	2	2 (7[d])

[a] Descriptor and threshold data refer to different matrices, including water, model wine, and red or white wine.

[b] Common names are widely used in the literature since the systematic names are more complex; for example, geosmin is (4*S*,4a*S*,8a*R*)-4,8a-dimethyl-1,2,3,4,5,6,7,8-octahydronaphthalen-4a-ol.

[c] For red grape must. Although often found in rotten grapes or musts, MIB is not detectable after fermentation, either due to metabolism by yeast, binding to lees, or other mechanisms.

[d] For a wine stored in a stainless steel tank in contact with slightly toasted oak chips. Temperatures encountered with traditional barrel-making readily eliminate MDMP if present in the wood.

Figure 18.1 Potential origins of taints include (a) cork wood with the outer ring of bark used to make (b) cork closures (shown in a piece of cork bark), and (c) an oak wood barrel (the discoloration around the bung is mold growth). Source: (b) Reproduced with permission of Duc-Truc Pham. (c) Reproduced with permission of Paul Grbin

affected by the ease of TCA migration within and polarity of the contaminated matrix[3] [15]. Because haloanisoles are highly non-polar (e.g., TCA has log $P = 3.7$ and water solubility = 0.01 g/L), uncontaminated cork and other similarly hydrophobic materials such as plastics are effective at removing TCA from contaminated wine. In contrast, more polar cork taint compounds bearing pyrazine, alcohol, and ketone functionalities would be more readily transferred into wine due to their greater solubility and lower affinity for cork [15]. TCA and geosmin are able to persist in bottled wine and it seems that other cork taint compounds would be quite stable as well.

18.2.2 Taints from other winery sources

Halophenols, the parent compounds of TCA and other haloanisoles, are widely used as wood preservatives due to their biocidal properties. As a result, 2,4,6-trichlorophenol (TCP) and its brominated equivalent, 2,4,6-tribromophenol (TBP, also used as a flame retardant in numerous products), are ubiquitous environmental pollutants. Alternatively, halophenols arise from the combination of phenol and halogen sources via electrophilic aromatic substitution (Chapter 11), leading to a range of substituted chloro- and bromophenols. For instance, chlorophenols can be formed when electrophilic chlorine (e.g., chlorine bleach, chlorine gas) comes into contact with materials containing phenols (e.g., wood, cardboard, plastics, cleaning products), with subsequent contamination of winemaking aids and additives or the wines directly. As an example, bleaching (i.e., hypochlorite washing) of corks has been reported to be a common source of TCP (and eventually TCA) [8], although this process is no longer used by major cork manufacturers. Beyond serving as haloanisole precursors, halophenols are implicated as taint compounds in their own right. They tend to have plastic/chemical/medicinal aromas but higher thresholds than haloanisoles, although potent halophenols with low

[3] A common quality control for TCA (and other taint compounds) in corks is a "soak test" – a sample of corks are soaked in aqueous ethanol or neutral wine and the extract is analysed sensorially or instrumentally. However, there are limitations to this process. Often the number of corks screened from a batch is relatively small (e.g., 5 corks) so does not account for cork variability. Additionally, taint compounds may not be uniformly distributed in the cork – a short soak test will not account for migration from within the cork matrix over many months of storage, leading to an underestimation of the potential to taint a wine. Conversely, more polar compounds such as MDMP can be more easily extracted during soaking trials, meaning cork screening may overestimate their potential importance. Thus, soaking trial conditions do not fully emulate what occurs in bottled wine, but rather give an indication of the danger presented by TCA (or similarly hydrophobic compounds) following short-term storage.

ng/L thresholds do exist, such as 2,6-dichlorophenol (2,6-DCP) and 6-chloro-*o*-cresol (6-CC) [17] (Table 18.1).[4] Similar to TCA, the most likely route of wine contamination is direct contact with tainted materials, although airborne transmission could be possible for wines handled in a contaminated atmosphere.[5]

18.2.3 Taints from the vineyard

In addition to appearing in tainted corks, the mold metabolites geosmin, 1-octen-3-one, 1-octen-3-ol and MIB can also be found at high levels in infected grapes (e.g., *Botrytis cinerea*, Figure 18.2) [9]. Concentrations of many of these compounds will decrease during fermentation by >90% [10]:

- Carbonyl containing mold-derived taint compounds, such as the ketone, 1-octen-3-one, will be reduced almost completely to corresponding and less potent alcohols, for example, 1-octen-3-ol (Chapter 22.1).
- The tertiary alcohol, MIB, decreases to undetectable levels, possibly as a result of carbocation formation and subsequent dehydration/degradation.
- Losses of the other mold metabolites can be around 50%, likely due at least in part to volatilization during fermentation or adsorption on to yeast lees.

Overall, geosmin and 1-octen-3-ol seem to be the most stable of these malodorous compounds, and the most likely to cause a taint in wine made from diseased grapes. Several other earthy-smelling compounds arising from rotting grapes (e.g., fenchone, fenchol, 2-octen-1-ol, heptan-1-ol) can potentially contribute to taint problems if present at high enough concentrations after fermentation ([2, 9, 18]). Apart from eliminating rotten grape bunches prior to fermentation, wine amelioration techniques are limited to fining with activated charcoal or other typical fining agents and these compounds appear to be challenging to remove selectively – although thermal treatments could potentially be used for geosmin remediation [19] (Chapter 26.2).

Contamination of grapes in the vineyard by insects can lead to taints, most notably multicolored Asian ladybeetle (MALB, Figure 18.2) taint caused by extraction of 3-isopropyl-2-methoxypyrazine (IPMP). *Insect-derived taints* contributing methoxypyrazines were covered earlier (Chapter 5); other insects such as millipedes and earwigs and their feces [20] can conceivably taint grapes and wines, although there is a lack of information on the extent of such contamination and the compounds responsible.

Smoke taint occurs in the vineyard as a result of contamination of grapes by bushfire (wildfire) smoke (Figure 18.2), thereby conferring undesirable smoke, ash, and medicinal characters to wine. This outcome is a result of volatile phenols present in smoke being transferred to the grape berry and subsequently glycosylated. This is covered in more detail in Chapter 12.

Eucalyptus aromas can arise in red wine from exposure of grapes in the vineyard to 1,8-cineole emanating from nearby *Eucalyptus* trees (Chapter 8). Although this illustrates the potential for exogenous volatiles in the vineyard to contaminate grapes and wine (anecdotal information exists about grapes being tainted by nearby food processing plants, for example), it is unclear if 1,8-cineole contamination results in increased consumer rejection. A limited number of consumer studies using red wines spiked with different amounts of 1,8-cineole indicate that the concentrations encountered in eucalyptus-affected wines are not objectionable to most consumers (see Reference [21] and references therein for more information).

---

[4] Often a number of chlorophenols may be identified in suspect wines, but their high thresholds relative to the concentrations found make it difficult to ascribe a primary role (and they are not included in Table 18.1). Based on their potency, compounds such as 2,6-DCP and 6-CC may be the principal cause of chlorophenol taints, but they are not necessarily detected in every analysis of tainted wines.

[5] Potable water supplies and distribution systems can be a source of halophenol (and haloanisole) contamination. This route of transmission seems less likely in a winery setting amongst all the other potential sources, in comparison to a brewery, for example, because the amount of water incorporated during winemaking is comparatively low.

Figure 18.2 Sources of taints from the vineyard, shown in images of (a) Botrytis on a grape berry, (b) powdery mildew on a grape cluster, (c) MALB on a grapevine leaf, (d) MALB floating in a red grape must, and (e) bushfire smoke drifting in the vicinity of a vineyard. Photographs courtesy of Tijana Petrovic (a), Bruce Bordelon (b and d), Erik Glemser (c), and Tony Mills (e)

Mycotoxins are another class of compounds arising in the vineyard due to fungal diseases [9]; their primary concern is not due to any adverse effect on flavor but because of concerns for human health. Ochratoxin A (OTA), produced by *Aspergillus* spp., is one such metabolite implicated as a human carcinogen and associated with liver and kidney toxicity [22]. In the European Union (EU), wine consumption is estimated to be the second greatest dietary source of OTA after cereals. Conditions that promote *Aspergillus* growth (warmer, more humid) typically have higher OTA concentrations. OTA from wine can be limited through physical (e.g., removal of moldy grape bunches, lighter pressing, or filtration – the most effective treatment), chemical (e.g., fining agents such as bentonite, chitosan, and charcoal, or oak products – currently the most studied remedy), or microbiological (e.g., yeast or bacteria, through adsorption rather than metabolism) processes [19]. However, a range of studies have shown that average concentrations are well below the designated EU limit (2 μg/L) and the need for OTA remediation is expected to be rare.

18.3 Off-flavors in wine

Off-flavor compounds arise from chemical or microbiological transformations of wine components, and thus can often be identified even in sound wines, but at sufficiently low concentrations so as not to be sensorially detectable or decrease consumer acceptance. One example is indole, which is only found at suprathreshold concentrations as a result of suboptimal (i.e., stuck or sluggish) fermentation (Chapter 5). More common wine off-flavors include acetic acid and ethyl acetate associated with volatile acidity, sulfur compounds such as H_2S related to reductive characters, nitrogenous compounds imparting vegetal or mousy flavors, carbonyl compounds from oxidation, and ethylphenols indicative of spoilage by *Brettanomyces* (*Dekkera*) yeast. These compounds are described in other chapters, as summarized in Table 18.2. As a caveat, these compounds are ordinary constituents of wine, and they may not be considered off-flavors in certain wine styles or by particular groups of consumers or wine experts (e.g., see References [23] and [24]). For example, the ethylphenols responsible for "Brett" aromas can be at concentrations well in excess of threshold in certain very expensive French red wines, where they may considered part of the wine's style rather than a fault. Similarly, oxidized aromas are entirely appropriate in Sherry wines, as are high levels of volatile acidity in botrytized wines such as Sauternes.

A final defect worthy of discussion is the off-odor reminiscent of crushed geranium leaves ("geranium taint"), attributable to 2-ethoxyhexa-3,5-diene [25]. This unsaturated ethyl ether, having a threshold of 100 ng/L [26], arises from reduction and subsequent rearrangement (cf. isoprenoid rearrangements, Chapter 8; also see nucleophiles/electrophiles, Chapter 10) of the preservative sorbic acid (i.e., (2*E*,4*E*)-2,4-hexadienoic acid,

Table 18.2 Cross-references to chapters that describe compounds causing some of the more common off-flavors in wine

Off-flavor	Odor descriptor	Compounds	Chapter
Volatile acidity	Vinegar	Acetic acid	3
	Nail polish remover	Ethyl acetate	6
Mousy	Mousy, cardboard, cracker	Cyclic imines	5
Atypical aging Foxy	Acacia, mothball	*ortho*-Aminoacetophenone	5
Oxidized	Rotting apple, green, earthy	Various aldehydes, sotolon	9
Reduced	Rotten egg, putrid, onion, vegetal	Various sulfur compounds	10
Brett	Leather, horse stable, spice	4-Ethylphenol 4-Ethylguaiacol	12

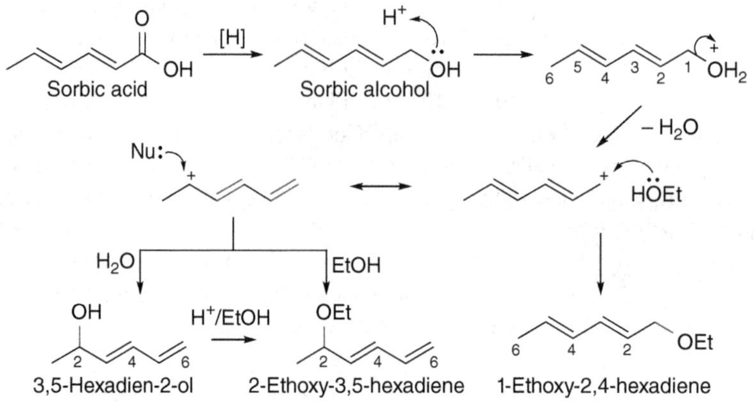

Figure 18.3 Formation of geranium off-odor through microbial reduction (indicated by [H]) of sorbic acid to the corresponding alcohol and acid-catalyzed rearrangement to the malodorous ether, 2-ethoxy-3,5-hexadiene. This may occur directly by addition of ethanol as a nucleophile (Nu:) to the secondary carbocation intermediate arising from protonated sorbic alcohol, or indirectly by addition of ethanol to C2 of protonated 3,5-hexadien-2-ol formed when Nu: is water. Addition of ethanol to the primary carbocation intermediate leads to another ether, 1-ethoxy-2,4-hexadiene (i.e., the ethyl ether of sorbic alcohol) – this ether and the two diene alcohols may partly contribute to the off-odor [25]

Figure 18.3) by lactic acid bacteria (Chapter 22.5). Sorbic acid or its salts may be used by winemakers to inhibit yeast growth in sweet wine, but adequate levels of SO_2 and sterile conditions at bottling are necessary to prevent bacterial growth and possible development of a geranium off-odor. It follows that sorbic acid should not be added to a wine that will subsequently undergo malolactic fermentation or be subjected to any other bacterial activity (even inadvertently).

References

1. Ridgway, K., Lalljie, S.P.D., Smith, R.M. (2010) Analysis of food taints and off-flavours: a review. *Food Additives and Contaminants: Part A: Chemistry, Analysis, Control, Exposure & Risk Assessment*, **27** (2), 146–168.
2. Boutou, S. and Chatonnet, P. (2007) Rapid headspace solid-phase microextraction/gas chromatographic/mass spectrometric assay for the quantitative determination of some of the main odorants causing off-flavours in wine. *Journal of Chromatography A*, **1141** (1), 19.
3. Hudelson, J. (2011) *Wine faults: causes, effects, cures*, The Wine Appreciation Guild, San Francisco, CA.
4. Jackson, R.S. (2008) Postfermentation treatments and related topics, in *Wine science: principles and applications*, Academic Press, San Diego, CA, pp. 418–519.
5. Cowey, G., Coulter, A., Holdstock, M. (2009) Don't get contaminated this vintage! *AWRI Technical Review*, **178**, 22–27.
6. Ribéreau-Gayon, P., Glories, Y., Maujean, A., Dubourdieu, D. (2006) Chemical nature, origins and consequences of the main organoleptic defects, in *Handbook of enology: the chemistry of wine stabilization and treatments*, 2nd edn (eds Ribéreau-Gayon, P., Glories, Y., Maujean, A., Dubourdieu, D.), John Wiley & Sons, Ltd, Chichester, UK, pp. 233–284.
7. Chatonnet, P., Bonnet, S., Boutou, S., Labadie, M.D. (2004) Identification and responsibility of 2,4,6-tribromoanisole in musty, corked odors in wine. *Journal of Agricultural and Food Chemistry*, **52** (5), 1255–1262.
8. Pollnitz, A.P., Pardon, K.H., Liacopoulos, D., *et al.* (1996) The analysis of 2,4,6-trichloroanisole and other chloroanisoles in tainted wines and corks. *Australian Journal of Grape and Wine Research*, **2** (3), 184–190.
9. Steel, C.C., Blackman, J.W., Schmidtke, L.M. (2013) Grapevine bunch rots: impacts on wine composition, quality, and potential procedures for the removal of wine faults. *Journal of Agricultural and Food Chemistry*, **61** (22), 5189–5206.

10. La Guerche, S., Dauphin, B., Pons, M., *et al.* (2006) Characterization of some mushroom and earthy off-odors microbially induced by the development of rot on grapes. *Journal of Agricultural and Food Chemistry*, **54** (24), 9193–9200.
11. Darriet, P., Pons, M., Lamy, S., Dubourdieu, D. (2000) Identification and quantification of geosmin, an earthy odorant contaminating wines. *Journal of Agricultural and Food Chemistry*, **48** (10), 4835–4838.
12. Pons, M., Dauphin, B., La Guerche, S., *et al.* (2011) Identification of impact odorants contributing to fresh mushroom off-flavor in wines: incidence of their reactivity with nitrogen compounds on the decrease of the olfactory defect. *Journal of Agricultural and Food Chemistry*, **59** (7), 3264–3272.
13. Bowen, A.J. and Reynolds, A.G. (2012) Odor potency of aroma compounds in Riesling and Vidal blanc table wines and icewines by gas chromatography–olfactometry–mass spectrometry. *Journal of Agricultural and Food Chemistry*, **60** (11), 2874–2883.
14. Takeuchi, H., Kato, H., Kurahashi, T. (2013) 2,4,6-Trichloroanisole is a potent suppressor of olfactory signal transduction. *Proceedings of the National Academy of Sciences*, **110** (40), 16235–16240.
15. Sefton, M.A. and Simpson, R.F. (2005) Compounds causing cork taint and the factors affecting their transfer from natural cork closures to wine – a review. *Australian Journal of Grape and Wine Research*, **11** (2), 226–240.
16. Barker, D.A., Capone, D.L., Pollnitz, A.P., *et al.* (2001) Absorption of 2,4,6-trichloroanisole by wine corks via the vapour phase in an enclosed environment. *Australian Journal of Grape and Wine Research*, **7** (1), 40–46.
17. Capone, D.L., Van Leeuwen, K.A., Pardon, K.H., *et al.* (2010) Identification and analysis of 2-chloro-6-methylphenol, 2,6-dichlorophenol and indole: causes of taints and off-flavours in wines. *Australian Journal of Grape and Wine Research*, **16** (1), 210–217.
18. Sadoughi, N., Schmidtke, L.M., Antalick, G., *et al.* (2015) Gas chromatography–mass spectrometry method optimized using response surface modeling for the quantitation of fungal off-flavors in grapes and wine. *Journal of Agricultural and Food Chemistry*, **63** (11), 2877–2885.
19. La Guerche, S. (2004) Recherches sur les déviations organoleptiques des moûts et des vins associées au développement de pourritures sur les raisins: étude particulière de la géosmine. PhD Thesis, Universitéde Bordeaux II, France.
20. Kehrli, P., Karp, J., Burdet, J.P., *et al.* (2012) Impact of processed earwigs and their faeces on the aroma and taste of "Chasselas" and "Pinot Noir" wines. *Vitis*, **51** (2), 87–93.
21. Capone, D.L., Jeffery, D.W., Sefton, M.A. (2012) Vineyard and fermentation studies to elucidate the origin of 1,8-cineole in Australian red wine. *Journal of Agricultural and Food Chemistry*, **60** (9), 2281–2287.
22. Quintela, S., Villarán, M.C., López de Armentia, I., Elejalde, E. (2013) Ochratoxin A removal in wine: a review. *Food Control*, **30** (2), 439–445.
23. Goode, J. (2005) Brettanomyces, in *The science of wine: from vine to glass*, University of California Press, Los Angeles, CA, pp. 136–143.
24. Jackson, R.S. (2009) Olfactory sensations, in *Wine tasting: a professional handbook*, 2nd edn, Academic Press, San Diego, CA, pp. 55–128.
25. Crowell, E.A. and Guymon, J.F. (1975) Wine constituents arising from sorbic acid addition, and identification of 2-ethoxyhexa-3,5-diene as source of geranium-like off-odor. *American Journal of Enology and Viticulture*, **26** (2), 97–102.
26. Wurdig, G. (1977) Technologie du traitement a l'acide sorbique. *Bulletin de l'OIV*, **50**, 547–558.

Part B

Chemistry of Wine Production Processes

19

Outline of Wine Production

19.1 Introduction

Wine production (winemaking) encompasses the techniques and technologies used in the transformation of grapes into wines of a target style, of which there are many. This primarily involves *vinification* (alcoholic fermentation) – the conversion of grape sugars into alcohol by yeast – but is accompanied by a variety of other important changes due to extraction and microbial metabolism of a multitude of other grape components. A secondary, malolactic fermentation (MLF), which can occur simultaneously with primary fermentation or sequentially, is also promoted for some wine styles. Furthermore, production techniques differ depending on whether white or red grapes are used and the type of wine being produced (Table 19.1).

Certain *specialty wine* styles with unique production steps exist. Notably, sparkling wine involves a secondary alcoholic fermentation stage, or the addition of CO_2 under pressure, to introduce the carbonation (fizz) associated with this category of wine. Other common specialty wines are the fortified wines, which have grape-derived ethanol added either during the primary fermentation phase (which arrests fermentation) or once dry, thereby enhancing the final alcohol content.

Finally, winemaking does not end after fermentation – a range of operations are used to clarify, stabilize, mature, age, and package the wine. This chapter offers an overview of wine production and links to later chapters, which detail the processes and chemical changes that occur at various stages during winemaking.

19.2 Basic workflow

The basic steps for winemaking are outlined in Figures 19.1 and 19.2 for white and red wines. The fundamental differences between the typical production of white and red wines can be briefly summarized as follows:

- White grapes are crushed/destemmed and pressed to juice and do not typically spend much time in contact with grape solids (i.e., there is no maceration step).
- Red grapes are crushed/destemmed and the must (juice and grape solids) undergoes maceration and fermentation in the presence of skins, seeds and juice (Chapter 21).

Understanding Wine Chemistry, First Edition. Andrew L. Waterhouse, Gavin L. Sacks, and David W. Jeffery.

Table 19.1 Classification of major wine styles found throughout the world. Adapted from Reference [1]

Description		Color	Residual sugar (g/L)	Alcohol (% v/v)	Examples
White wines					
Still	Dry	Pale straw to gold	<9	8 to 14.5	Riesling, Chardonnay, Semillon, Sauvignon Blanc, Colombard, Grüner Veltliner, Trebbiano, Chenin Blanc
	Sweet	Light yellow to gold	9 to 30 (semi-sweet)		Riesling, Gerwurztraminer, Semillon
			30 to 200 (sweet)		Riesling, Ice wines, Sauternes, Tokay
Sparkling	Dry to semi-sweet	Pale straw to amber (pink to light red for rosé wines)	0 to 50		Champagne, Chardonnay/Pinot Noir/Pinot Meunier[a], Riesling, Sauvignon Blanc, Cava, Prosecco
Fortified	Dry	Pale straw to amber	0 to 30	15 to 20.5	Fino and Amontillado Sherries
	Sweet		100 to 300		Oloroso and Pedro Ximénez[b] Sherries, Topaque
Red wines					
Still	Dry	Dark red to red/brown	<7.5	8 to 14.5	Grenache, Merlot, Cabernet Sauvignon, Tempranillo, Zinfandel, Sangiovese, Malbec, Pinot Noir, Shiraz
Sparkling	Semi-sweet		7.5 to 30		Pinot Noir, Shiraz, Cabernet Sauvignon
Fortified	Sweet	Red/gold to deep brown	100 to 300	18 to 22	Ruby, Tawny and Vintage Ports, Brown Muscat

[a] Of these classic Champagne grape varieties, the first two tend to be used more frequently in sparkling wines.
[b] Pedro Ximénez (PX) can be very sweet and may contain up to 450 g/L of residual sugar.

- As a result of the above, extraction of polyphenols is mostly avoided with white wines and encouraged in red wine production.
- Cooler temperatures are used for white fermentations to control aroma characteristics and warmer temperatures enhance extraction of solids in reds.
- White fermentations mostly exclude oxygen, whereas some aeration is encouraged during red winemaking through the maceration techniques employed.
- The majority of red wines undergo malolactic fermentation (MLF) and incorporation of oak in contrast to white wines, where only certain styles experience these treatments.
- Red wines undergo a period of maturation in tanks or oak barrels to promote stabilization of color and modification of mouthfeel.
- White wines can be released much earlier than red wines due to typically greater need for maturation with reds.

Additional variations to the basic process occur for mainstream production of rosé, sparkling, and fortified wines, as detailed below and elsewhere (for example, see References [2], [3], and [4]). More specific details of techniques employed for red winemaking are outlined in Chapter 21.

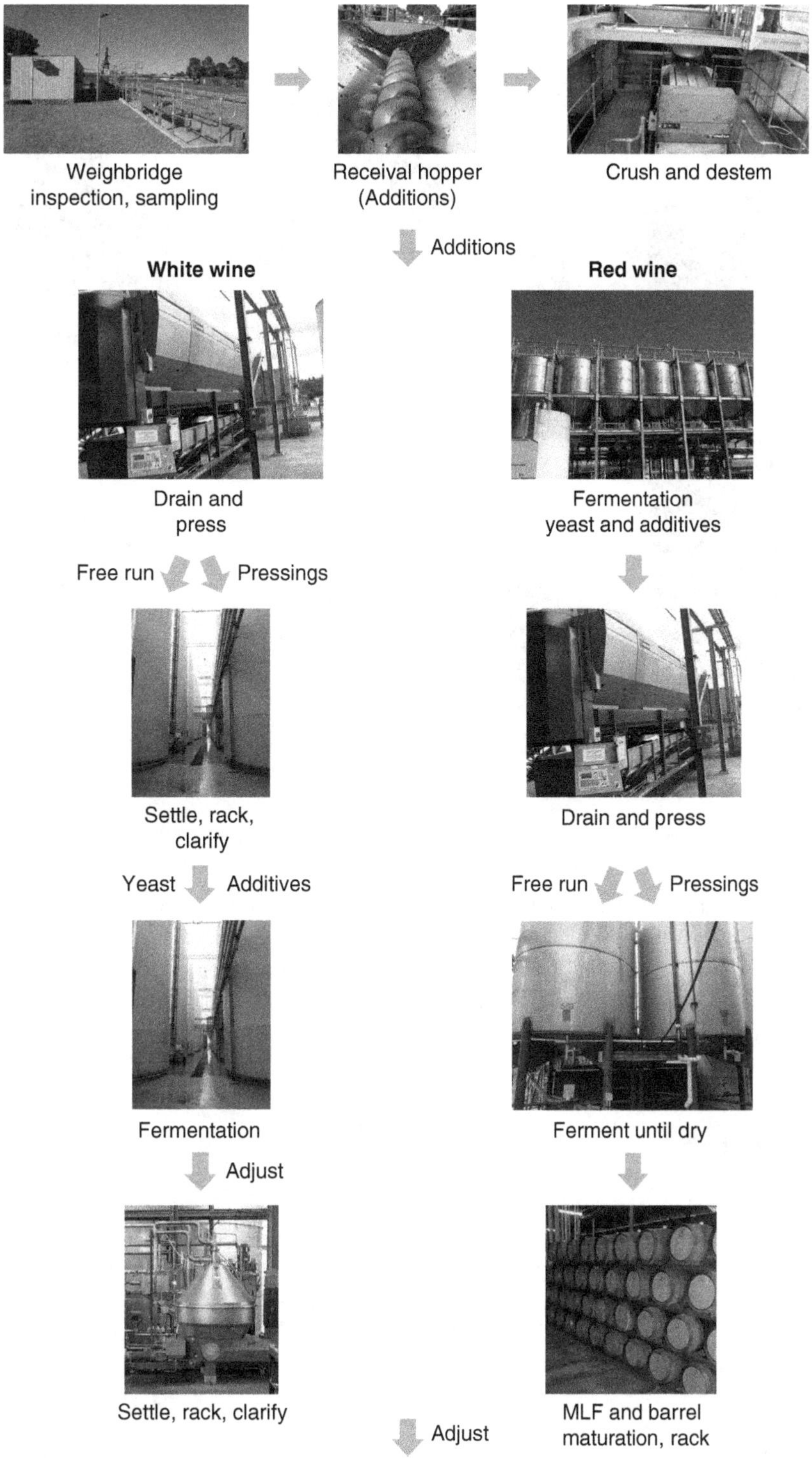

Figure 19.1 *Indicative flowchart for commercial winery fermentation operations. Additions may include tartaric acid, tannin, sugar, SO_2, and enzymes. Additives mostly include nutrients in the form of diammoniun phosphate, and adjustments mainly involve the use of tartaric acid, carbonate salts, and SO_2. Additional clarification steps beyond those shown are likely, and incorporation of oak (and MLF in some white wines) can occur in a number of ways and at different stages of the process. The remainder of the operations is outlined in Figure 19.2*

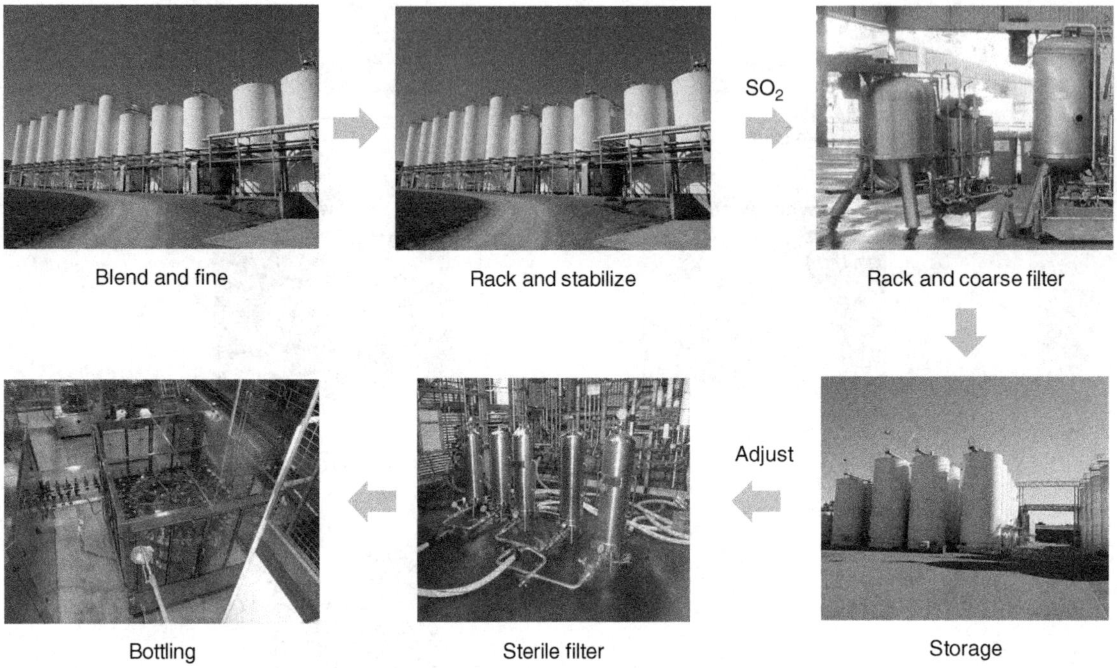

Figure 19.2 *Indicative flowchart for commercial winery operations after fermentation is complete. Adjustments primarily involve the use of tartaric acid, carbonate salts, and* SO_2

In most wineries, a range of parameters are monitored during winemaking to ensure a level of quality control over the production process. The main measurements relate to temperature, pH, titratable acidity, residual sugars, fermentation kinetics, assimilable nitrogen, alcohol, free and bound SO_2, malic acid, and volatile acidity. Comprehensive information on these tests and many others is provided in a variety of separate sources (for example, see References [5] to [9]).

19.3 Processes involved

Once grapes are harvested and transported to the winery, various processes are set in motion to transform those grapes into wine. The processes involve an array of equipment as depicted in Figures 19.1 and 19.2; the general equipment is further outlined in Table 19.2.[1] Because a range of tanks and fermentation vessels are used in the production of different wines throughout the world, these are described in more detail in Section 19.4.

[1] A refrigeration source is also common, among a raft of other equipment not specified in the table, such as pumps, hoses, sight glasses, valves, racking plates, equipment for cleaning/sanitizing, mixers/agitators, plungers, and other winery infrastructure (e.g., static must transfer lines, manifolds, conveyor belts, sorting tables, variable volume tanks, boilers, reticulated gases and water, etc.).

Table 19.2 Winemaking processes, description of objectives being achieved, and the equipment required for each processing stage

Process	Objectives	Typical Equipment
Grape weighing, inspection and sampling	Weigh and assess grapes (e.g. examine for disease or matter other than grapes), determine basic compositional parameters (e.g. pH, TA, TSS, color) and stream fruit for a specific wine style/quality grade	Weighbridge, automated grape sampler/analyzer (e.g. Maselli Misure) or basic lab equipment (e.g., refractometer, pH meter, autotitrator, spectrophotometer, thermometer)
Grapes tipped into hopper	Transfer grapes from picking bins so processing can begin	Crane or forklift, receival hopper with auger to move grapes to next stage
Crushing and destemming grapes[a]	Remove grapes from stems and crush to provide must (i.e., grape juice and solids)	Crusher/destemmer, dosing pump to include additives, heat exchanger to alter temperature of must
Pressing to produce free run and pressed fractions[b]	Extract and separate liquid (juice or wine) from grape solids (marc)	Screens or drainers, press (different kinds available)
Clarifying white juice	Promote precipitation and eliminate the majority of suspended solids (e.g., proteins, polysaccharides, grape debris, microflora, fatty acids) from the juice	Tanks (for cold settling) and clarification equipment (e.g., centrifuge, filter units)
Fermentation[b]	Conversion of grape sugars to alcohol (or malic to lactic acid for MLF)	Fermentation vessels
Maturation of red wine	Stabilize color and promote changes to mouthfeel	Tanks or barrels, possibly a microoxygenation system
Settling and racking	Remove yeast lees and other precipitated matter (e.g. phosphates, tartrates, colloids)	Tanks and clarification equipment (e.g., centrifuge, filter units)[c]
Blending and fining	Produce a desired wine style from different components and moderate sensory properties (e.g. removal of phenolics)	Tanks and clarification equipment
Stabilization	Achieve heat and cold stability of wine	Tanks or specialized equipment for tartrate stabilization (e.g., to perform electrodialysis, ion exchange, continuous processes)
Coarse filtration	Decrease turbidity of wine	Filter units
Storage	Maintain wine in good condition prior to bottling	Tanks
Sterile filteration and bottling	Produce a clear and microbiologically stable product in its package	Filter units, bottling line

[a] Grapes may be "whole bunch" pressed for some wine styles, eliminating this step.

[b] The order of these processes depends on whether red or white wine is being produced. The order presented here (i.e., pressing before fermentation) is for white wine production; fermentation would precede pressing for red wine production.

[c] This stage may include further clarification operations beyond settling in tanks, especially for white wine.

In contrast to red wines, most white wines and rosé wines are fermented in the absence of grape solids (skins, seeds). Prior to fermentation, the juice is pressed and separated from the rest of the grape material. Immediately following pressing, the juice is turbid due to the presence of pulp, pieces of skin, and other suspended insoluble materials. Production of high-quality white wine requires the juice to be clarified prior to fermentation – clarified juices produce wines with higher concentrations of "fruity" esters and lower concentrations of higher alcohols and low molecular weight sulfur compounds, among other effects (Chapters 22.2 to 22.4). However, excessive clarification will result in sluggish or stuck fermentations,[2] and winemakers will typically target a small proportion of suspended solids (e.g., 0.5% by volume) to ensure a good balance of fermentation characteristics. Clarification can be achieved through several means:

- Most commonly, clarification is achieved through cold settling, where juice is held cold (e.g., 5 °C) in a tank for a period of time. This process is simple and relies on the greater density of most insoluble solids (polysaccharides, proteins, polyphenols) as compared to the juice. Once a sufficient degree of clarification occurs, the juice can be racked to a separate tank. This process can be enhanced through the use of pectic enzymes, which break down cell wall polysaccharides and increase their settling rate (Chapter 21).
- Centrifugation achieves similar results to cold settling but in far less time, although it requires more equipment.
- Flotation involves addition of gelatin (Chapter 26.2) to turbid juice to flocculate with the polysaccharide and polyphenolic constituents. The gelatin-containing precipitate is then floated to the surface after addition of gas and slowly allowing the pressure to release in a tank – a result similar to the formation of a proteinaceous "scum" on the surface of a soup pot when making stock. The clear juice can then be racked from the bottom of the tank.
- Filtration may also be employed, typically using high capacity DE-based depth filters (Chapter 26.3).
- Following cold settling, juice may also be recovered from tank bottoms by applying similar clarification treatments, but this juice is often of lower quality.

19.4 Tanks and fermenters

Tank sizes and the types of fermenters used for wine production vary depending on the size of the winery (i.e., tonnages crushed) and styles of wine being produced (e.g., red, white, or sparkling wine). Tanks in modern wineries are typically made of stainless steel and are used to store juice or wine for purposes such as cold settling, stabilizing, maturing, blending, and while awaiting bottling. Tanks vary in their capacity and range from several hundred liters (including variable volume designs) to hundreds of kiloliters; some very large wineries have tanks that can hold a million liters or more. Smaller tanks may be housed indoors but bigger wineries will have large outdoor "tank farms" (e.g., see images in Figure 19.2) consisting of insulated and refrigerated tanks to maintain temperatures. Different size tanks are used not only to separate different batches of wine but also so they can be kept full to minimize *ullage* (i.e., the headspace above the liquid) and limit contact with air, which decreases the incidence of oxidation or spoilage issues (Chapters 22.5 and 24).

---

[2] The reason for the slower fermentation kinetics of overclarified juices is not resolved, but may be because suspended solids provide a nucleation site for CO_2 (which can be toxic at high concentrations) or because they help distribute yeast in the fermenting must and prevent localized nutrient deficiencies. Regardless, the effect does not appear to be because of nutrient removal, since the fermentation rate in overclarified musts can be restored through addition of inert solids like diatomaceous earth (DE).

Inert gas coverage (commonly CO_2 or N_2, or their combinations[3]) is used to exclude air when transferring wine and during storage, and solid CO_2 (dry ice) powder may be employed to displace air in the headspace after opening the lid of a tank.

Oak barrels and vats, being the traditional forms of a "tank", are still commonly used for storage during maturation, primarily to incorporate oak flavor components (Chapters 12 and 25), but also to allow a slow ingress of oxygen that facilitates the maturation and stabilization process for red wines (Chapters 24 and 25). In contrast to seeking a flavor contribution from oak, certain spirits and most fortified wines are aged in old oak vessels for many years to allow for concentration of components and slow chemical reactions, and the promotion of oxidative changes in some cases. In recent years, food-grade polyethylene has been used in the construction of smaller tanks for storage or maturation; the plastic allows for a slow ingress of air in order to emulate the effect obtained with oak barrels. Desirable oak components can be incorporated to an extent with the use of oak alternatives such as chips, staves, and powders (Chapter 27), and considering the cost of purchasing and maintaining oak barrels, this approach may be more suitable for some wineries.

Fermentation vessels may be the same as the tanks used for storage, especially with white table wine production, or may be more specific, as is the case for tanks used for maceration during red winemaking (Figure 19.3). In this case, there is a need to promote extraction of color and tannins from the *cap* (the grape skins that rise to the top during fermentation) through different vessel designs. This can be achieved in open (e.g., Figure 19.3a and d) or rotary (e.g., Figure 19.3b) fermenters, where the cap can be submerged and macerated through periodic mechanical action. Alternatively, closed static fermenters (e.g., Figure 19.3c) are used to obtain a similar effect, as these allow for wetting of the cap via mechanisms that pump fermenting juice over the top (i.e., *pump over* as depicted for an open fermenter in Figure 19.3e) at certain intervals to facilitate extraction. In most cases the operations can be automated and winemakers may use a combination of cap management techniques depending on the fermenter in use. With red wine production there is also a need to remove the grape solids (*marc* or *pomace*) once the maceration stage is complete and the tank is drained, and design features assist with this (e.g., large door at base, sweeping arm, open top, ability to tilt or lift fermenter). Additionally, it is necessary to control fermentation temperatures by applying heating or cooling, either to the building, directly to the tank, or via a heat exchanger.

Fermenters as described above are often made from stainless steel and can be quite large (e.g., capable of holding 50 t/50 kL or more), but oak is still very commonly used in the construction of (smaller) vessels for fermentation, and other materials such as concrete or stone (e.g., a square open fermenter usually sealed with wax or lined in some way, or other fermenters such as egg- or oblong-shaped), plastic (e.g., cube or egg-shaped polyethylene as mentioned above), or terracotta (e.g., amphora often lined with beeswax) also feature in some wineries, but are necessarily much smaller than stainless steel tanks. Furthermore, certain types of sparkling wine production call for the use of pressure vessels after alcoholic fermentation to incorporate or maintain a high level of CO_2 in the product, whereas in other cases the bottle acts as the vessel for the fermentation stage, which imparts the carbonation (see Section 19.6 below).

[3] Note that these gases have different solubilities in wine as a function of temperature. CO_2 can dissolve in wine much more readily than N_2 (by a factor of almost 90 on a g/L basis) and some dissolved CO_2 is often desirable for palatability. Still white wines will usually require more dissolved CO_2 than red wines, so CO_2 will often be employed with white wines and N_2 with reds, but a mixture of these gases can be used to obtain the required sensory outcome for a particular wine.

Figure 19.3 Examples of red wine fermentation showing (a) 5 tonne open fermenters, which can be lifted by a crane and taken to a press, (b) 60 tonne rotary fermenter, (c) 50 tonne closed static sweeping arm Potter (SWAP) fermenters, (d) 6 tonne concrete open fermenters with cap submerged using "heading down" boards, and (e) a "pump over" to wet the cap in an open fermenter

19.5 Beyond fermentation

During the winemaking process, and particularly post-fermentation, there are a number of operations that may occur, as depicted in Figures 19.1 and 19.2, and Table 19.2. The major activities are described more thoroughly in other chapters as follows:

- Maceration: Chapter 21
- Aging: Chapter 25
- Cold stabilization: Chapter 26.1
- Fining: Chapter 26.2
- Filtration: Chapter 26.3

19.6 Specialty wines

Certain specialty wines undergo processing steps that lead to distinctive changes in final chemical composition and sensory properties. The two most common specialty wines are sparkling and fortified.

19.6.1 Sparkling wine

As the most widely consumed specialty wine, sparkling wine is produced by addition of yeast and sugar to a dry *base wine* (most often a blend, known as the cuvée) prior to carrying out a second fermentation in a closed container so that both the ethanol and carbon dioxide are captured and retained (e.g., in a sealed bottle or tank).[4]

The traditional method for sparkling wine production (*méthode champenoise*)[5] involves the following steps [2]:

- Fruit is harvested early, 16–18 °Brix, and fermented to dryness to make the base wine.
- Base wine is blended to produce a cuvée, and added to bottles along with sugar and yeast, the *tirage* (liqueur de tirage) – every 4 g/L of sugar yields 1 bar of pressure, and 32 g/L is generally added to generate 8 bar following secondary fermentation.
- The second fermentation occurs in the bottles under crown seal, and there is a period of 1–4 or more years of aging on lees in the same bottle. During this period, yeast will undergo autolysis (Chapter 25).
- The yeast is dislodged from the bottle and collected in the neck of the bottle by *riddling* (remuage), in which the bottle is rotated in small increments at regular intervals.[6]
- The wine is chilled and the yeast is removed by freezing and *disgorging*. In addition to the yeast, a portion of CO_2 (accounting for about 2 bar) will also be lost.

[4] For those in a hurry, it is also possible to carbonate a wine by bottling under carbon dioxide pressure (similar to soft drink production). In most countries, such wine must be labeled "carbonated wine" to distinguish it from sparkling wine, where the carbonation arises from a second fermentation; for example, see 27CFR 24.10 (2014) for US regulations.

[5] The term derives from the practices of the Champagne region of France. Sparkling wine from this *Appellation d'Origine Contrôlée* can legally use the term Champagne on the bottle, while sparkling wine produced elsewhere cannot. In the US, it is still legal to use the term with an additional geographic qualification, such as "California Champagne," though few producers take advantage of the allowance.

[6] Riddling was invented at the winery of the veuve (widow) Clicquot, and the classic process involves placing bottles in holes on a board, and rotating them 1/8th of a turn each day while slowly moving from a horizontal to vertical position. Perhaps apocryphally, the first riddling board is supposed to have been created by cutting holes in the widow's dining room table. The process was used unchanged until the end of the twentieth century, when machines were developed to replicate the bottle agitation needed to clear the yeast off the bottle.

- A small amount of sugar is added, the *dosage* (liqueur d'expedition, may also contain additives such as citric/ascorbic acid, brandy, and SO_2), and the bottle is resealed at a pressure of about 6 bar.

Alternative methods to fermentation in the bottle exist. In the Charmat process, the secondary fermentation is conducted in a large pressurized tank prior to bottling. In the transfer process, secondary fermentation occurs in the bottle, but the individual bottles are emptied under pressure into a tank prior to rebottling.

In addition to "prickling" sensations and sour taste imparted by CO_2, sparkling wines that undergo long periods of lees aging will have modified wine aroma and flavor due to yeast autolysis, such as an increase in "toasty" aromas. These effects are discussed in more detail in Chapter 25.

19.6.2 Fortified wines

Wines may also be fortified with distilled spirits during production [3] (see Chapter 26.4 for more information on distillation). The addition of spirits can halt fermentation and increases the final alcohol content of the wine (typically, 15–22% v/v at the time of fortification), thereby serving as a preservative. As such, fortified wines are at low risk of microbial spoilage (e.g., due to acetic acid bacteria, Chapter 22.5), even in the presence of oxygen. These wines were historically of great importance because of their ability to be shipped and stored for long periods of time. Fortified wines can be further classified by the timing of spirit addition, which can be:

- Prior to or soon after the beginning of fermentation (Pineau de Charentes, Moscatel de Valencia, Angelica). Early fortification leads to sweet wines with moderately high alcohol, often referred to as *mistelles*. There is negligible contribution of ethanol from fermentation, low extraction of grape constituents, and little or no production of fermentation-related aroma compounds. These wines are often aged in old oak casks, as mentioned below.
- Prior to the end of fermentation (Port, Madeira). Known as *mutage*, this process leads to sweet wines with high alcohol. Since these wines are usually stored in partially filled old oak casks, they typically possess high concentrations of compounds associated with oxidation reactions (Chapter 24). Some styles (e.g., Madeira) will also have high concentrations of sugar degradation products due to high storage temperatures.
- At the end of fermentation (Sherry, Vin Jaune). This process is typically used to produce dry fortified wines. In production of certain styles (such as Fino Sherry) the alcohol content is increased to only ~15% v/v to permit growth of a surface (*flor*) yeast. This results in high concentrations of biotically formed acetaldehyde via ethanol oxidation, which may in turn react with other components to yield further characteristic compounds like sotolon (Chapter 9). Other styles with higher alcohol concentrations will be fully exposed to air and generate oxidation products abiotically (Chapter 24).

References

1. Iland, P., Gago, P., Caillard, A., Dry, P. (2009) *A taste of the world of wine*, Patrick Iland Wine Promotions, Adelaide, SA, Australia.
2. Howe, P. (2003) Sparkling wines, in *Fermented beverage production*, 2nd edn (eds Lea, A.G.H. and Piggott, J.R.), Kluwer Academic/Plenum Publishers, New York, pp. 139–155.
3. Reader, H.P. and Dominguez, M. (2003) Fortified wines: Sherry, Port and Madeira, in *Fermented beverage production*, 2nd edn (eds Lea, A.G.H. and Piggott, J.R.), Kluwer Academic/Plenum Publishers, New York, pp. 157–194.

4. Rankine, B.C. (2004) Winemaking procedures, in *Making good wine*, Pan Macmillan Australia, Sydney, NSW, Australia, pp. 44–78.
5. Iland, P., Bruer, N., Ewart, A., *et al.* (2004) *Monitoring the winemaking process from grapes to wine: techniques and concepts*, Patrick Iland Wine Promotions, Adelaide, SA, Australia.
6. Iland, P., Bruer, N., Edwards, G., *et al.* (2004) *Chemical analysis of grapes and wine: techniques and concepts*, Patrick Iland Wine Promotions, Adelaide, SA, Australia.
7. International Organisation of Vine and Wine (2012) *Compendium of international methods of wine and must analysis*, Vol. **1**, OIV, Paris, France.
8. International Organisation of Vine and Wine (2012) *Compendium of international methods of wine and must analysis*, Vol. **2**, OIV, Paris, France.
9. Zoecklein, B.W., Fugelsang, K.C., Gump, B.H., Nury, F.S. (1995) *Wine analysis and production*, 1st edn, Chapman and Hall, New York.

20

Grape Must Composition Overview

20.1 Sampling

A typical winegrape berry sugar content is in the range of 18–25% w/w, but the standard deviation among berries in a vineyard, and even within a cluster, is often ± 2% w/w (Figure 20.1), likely due to a range of flowering dates and subsequent fruit-set dates for individual berries [1]. This variation in sugar can even be seen within a single berry [2]. The result is that not all berries are at the same degree of ripeness at any one time. Similar variations have been reported for other compositional aspects like color [1], and some aroma compounds (e.g., rotundone, Chapter 8) are reported to vary by 10-fold within a vineyard [3]. Predicting the concentrations of sugars and other parameters in a population of berries, that is, a vineyard block, based on a small or non-representative subsample is thus challenging for researchers and winemakers alike. Berry sample sizes (either as individual berries or whole clusters) for any particular component must be evaluated to determine if they will lead to adequate precision, and sampling must be appropriately representative to achieve acceptable accuracy [4, 5].

20.2 Sugars

After water, the most abundant substances in grapes at ripeness are sugars, in the form of fructose and glucose. These are found in near equimolar amounts since they arise from hydrolysis of the disaccharide, sucrose, produced through leaf photosynthesis. Small amounts of pentoses and other sugars are also detected (Chapter 2). Sugar concentrations are very low prior to veraison, but accumulate rapidly afterwards, and may reach 25% w/w or higher by harvest time. Sugar concentration is arguably the most widely used parameter for assessing grape composition and ripeness, in part because the amount of sugar present determines the eventual alcohol concentration. Within a region, the sugar attained during ripening can be predicted by the ratio of vegetative growth (a proxy for leaf area and photosynthetic activity) to the quantity of grapes to be ripened [6], and warmer temperatures and longer growing seasons will result in more photosynthesis. Thus, a limiting factor in achieving adequate sugar in winegrapes are temperatures during the growing season (degree-days is one such measure) [7]; hence, sugar additions are more common in cool regions.

Understanding Wine Chemistry, First Edition. Andrew L. Waterhouse, Gavin L. Sacks, and David W. Jeffery.

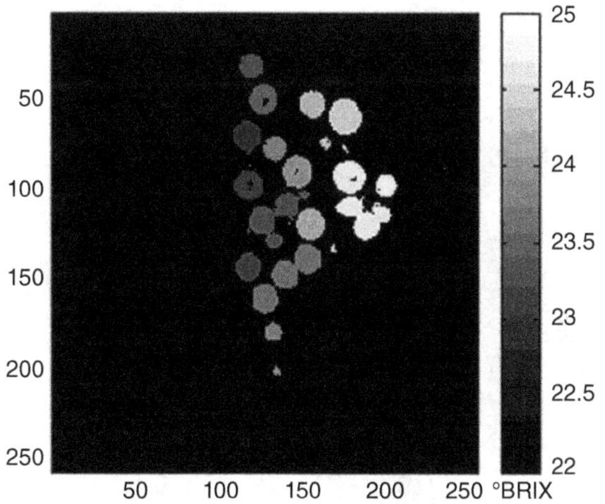

Figure 20.1 Magnetic resonance image of berry sugar levels (°Brix) in a grape cluster. Source: Andaur 2004 [2]. Reproduced with permission of American Chemical Society

Related to sugars are the polysaccharides, carbohydrate-containing polymers originating from berry cell wall material (Chapter 2). Although pectins and related structural polysaccharides may approach 1% in *V. labruscana* grapes, they are generally at values closer to 0.1–0.2% in *vinifera* juices. Enzymatic treatments can hydrolyze some polysaccharides and will increase their release into must and the resulting wine [8] (Chapters 21 and 23.1). Common pectinases will yield as much as 1 g/L of galacturonic acid from pectin hydrolysis (Chapter 2), which is unfermentable and can persist into finished wine. Aside from pectins, grape skin contains hemicellulose and cellulose, with some of the former being extracted into wine. The combination of polysaccharides is sometimes referred to as fiber [9] and is a major constituent (a few percent) of the pomace on a fresh weight basis (Figure 20.2).

20.3 Acids

Grapes contain substantial quantities of organic acids (Chapter 3), 10 grams or more per kg, and these are largely retained in the wine.[1] The presence of organic acids is essential to the taste of wine and juice, and the resulting low pH results in exclusion of many spoilage and pathogenic microbes. Tartaric acid is generally the major acid and a key marker for grape juice, while malic acid will also contribute substantially to acidity. Minor acids in grapes include citric and ascorbic, although the former will also be produced through yeast metabolism, and traces of other acids have also been documented (Chapter 3) [10]. Both malic acid and citric acid can be metabolized by lactic acid bacteria if they are present during winemaking, such as when inoculated for malolactic fermentation (Chapter 22.5). Malic and tartaric acids are accumulated pre-veraison. The acid concentration decreases during ripening and sugar accumulation, primarily due to respiration and loss of malic acid, but also because of berry expansion (and dilution of both acids). The rate of malic acid degradation increases at higher temperatures – thus grapes grown in cooler climates or harvested at earlier dates generally yield higher acidity levels in the must and wine [11]. These factors explain why acid additions are

[1] As discussed in Chapter 3, this 10 g/kg value for "total" acids is not the same as the titratable acidity (TA). The TA is usually lower, 6–9 g/kg, because the grape has exchanged titratable protons for K^+ and other minerals.

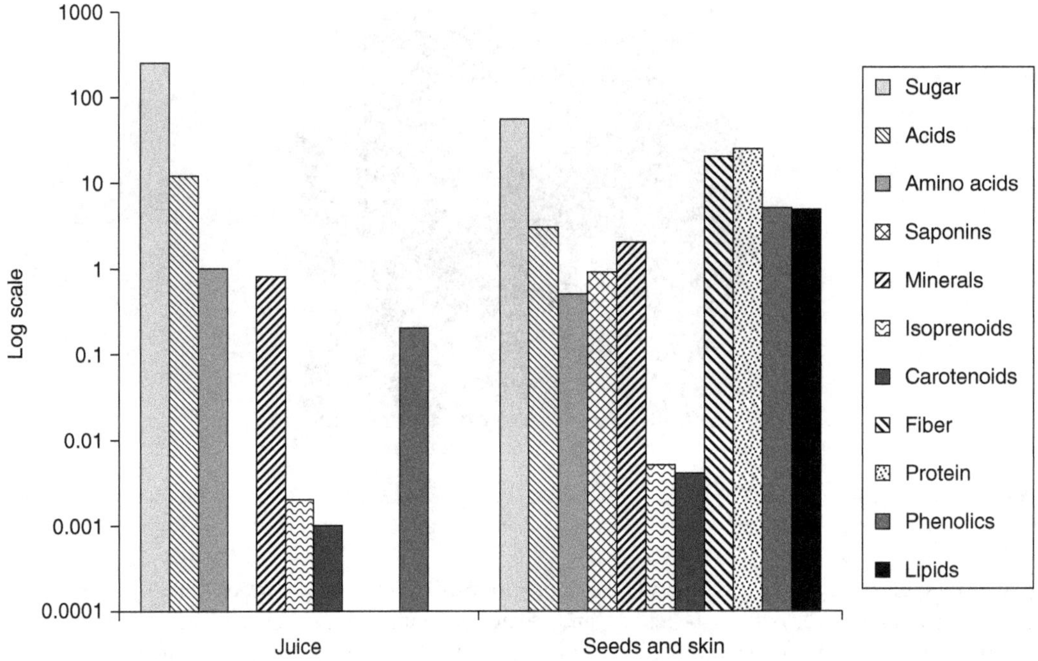

Figure 20.2 Components of grape juice (g/L) versus seeds and skin (g/kg)

common in warm regions, whereas levels of acidity are often lowered by various treatments when using fruit from cooler environments (Chapter 3).

20.4 Phenolics

Phenolic compounds are found largely in the skins and seeds of grape berries (Chapter 11), though additional material can be extracted from the rachis if whole cluster fermentation is practiced. While considerable variation exists among varieties, a representative study reported red berries to have a total phenolic content of 5.6 g/kg, with one-third (1.9 g/kg) in the skins, two-thirds (3.5 g/kg) in the seeds and less than 5% in the pulp and juice. White berries have about 3.8 g/kg, with less than 1 g/kg in skins due to the absence of the anthocyanins, but also about 2.8 g/kg in the seeds, with a similarly small amount in the pulp and juice [12]. Conventional red winemaking only extracts about half of the phenolic substances, with the maceration protocol having a significant impact on their extraction into wine (Chapter 21).

The phenolics are comprised of four major classes and numerous minor ones.

- The skin of grapes contains flavonols, whose formation is induced by sunlight as a UV protective material [13] with typical concentrations in the range of 1–80 mg/kg (Chapter 15) [14].
- Red grape skins are pigmented by the presence of anthocyanins, with 5–15 or more constituent compounds, varying by B-ring substitution, glucosylation, and further acylation of the glucose moiety (Chapter 16). Anthocyanin concentrations in red varieties vary widely, from 200 to 6000 mg/kg [14], and darker grapes or those with red flesh (*teinturier*) having higher levels.

- Flavan-3-ol monomers and their polymers (proanthocyanidins, also called condensed tannin) are a major phenolic constituent. These are found in both skin and seed, and comprise approximately half the phenolics in grape berries (Chapter 14). During ripening, extractable seed tannin (and flavan-3-ol monomers) declines as the result of oxidative reactions in the seed coat – this manifests as a change in the seed color from green to brown [15]. There are many ways to measure tannin (Chapters 14 and 33), but a chromatographic method comparing 37 varieties showed an average of 1.3 g tannin per kg of fruit [16].
- Finally, the hydroxycinnamates are the major phenolic compounds in the pulp and juice of the berry and are the dominant phenolics in white grape juice and wine; one survey of 28 varieties (both red and white) reported an average of 178 mg/kg, with a range from 85 to 400 mg/kg [17].

20.5 Nitrogen species

Amino acids and ammonium salts are the major nitrogenous compounds present in grapes, although their concentrations can vary considerably (300–5000 mg/L in juice or 40–700 mg/L as N) (Chapter 5). These compounds are necessary for yeast nitrogen metabolism and synthesis of proteins and other key macromolecules. However, only about half of this pool are α-amino acids – which can be metabolized during fermentation (yeast assimilable nitrogen (YAN)) – while the remainder is proline and cannot be utilized (Chapter 22.3). The amino acid profile varies depending on grape variety and environmental factors [18], and musts may be supplemented with nitrogen, typically as diammonium phosphate, if available YAN levels are too low for a successful fermentation (Chapter 22.3).

Proteomics analyses of grape berries shows many different enzymes and other proteins [19], but these are usually present only at low concentrations (<50 mg/kg). Several berry proteins have enzymatic activity, including oxidases that will affect juice browning upon grape berry damage or crushing at the winery [20], as well as chitinases, esterases, glucosidases, pectinases, and glucanases [21], some of which are discussed elsewhere (Chapters 21 and 23.1). Thaumatin-like proteins and chitinases are produced in response to fungal infections (i.e., pathogenesis-related), and when induced these become the major proteins present [22], with amounts approaching 300 mg/kg fresh weight (FW) of berry (Chapter 5). These proteins can cause white wine haze (Chapter 26.2) and may also bind to grape tannins and decrease their extractability (Chapter 21). Grapes also contain oligopeptides, and one of importance is glutathione, a key antioxidant due to its thiol functional group (Chapter 5). One report describes levels in grapes as varying from approximately 15 to 100 mg/L with an average of 44 mg/L [23]. Finally, grapes contain small amounts of biogenic amines, including isopentylamine, ethylamine, agmatine, diaminopropane, spermidine, and spermine, at approximately 3–5 mg/kg [24].

20.6 Lipids and waxes

Lipids are a key constituent of grape seeds, but are found in low amounts in skins and pulp. Totals vary from 1.5 to 2.5 g/kg FW in must, with the major lipids in Cabernet Sauvignon being glycolipids and phospholipids containing palmitic, stearic, linoleic, and linolenic acids [25]. However, over 20 different fatty acids have been detected in the neutral and polar lipid fractions, and a *V. vinifera* cultivar (Cabernet Sauvignon) had much higher levels of unsaturated fatty acids than hybrids, where saturated fatty acids dominate. While they are poorly extracted, the unsaturated fatty acids – for example, oleic, linolenic, and linoleic acids – are of particular importance because (i) they can serve as substrates for formation of the C_6 alcohols and aldehydes (Chapter 23.3) and (ii) they are critical for the yeast cell membrane and thus yeast growth (Chapter 22.2).

The exterior of the grape skin is covered with a waxy cuticle, which is a protective water vapor barrier several μm thick composed of triterpenoids (sometimes called saponins or phytosterols). The mixture is complex and includes mostly oleanolic acid and related compounds such as ursolic acid, α-amyrin, and others [26]. In addition, the wax also contains long-chain alcohols, esters, aldehydes, hydrocarbons, and other substances [27]. The amount of wax on grape berries is quite high, 1–2 g/kg in fresh fruit, but owing to their hydrophobicity these compounds are weakly soluble in wine and only low amounts are extracted, even in red wine.

20.7 Minerals and vitamins

Minerals found in wine serve primarily as counter ions to deprotonated acids (Chapter 3).

- The dominant juice cation is potassium (1–2 g/kg in grapes), with much lower amounts of calcium (100 mg/kg), magnesium (70 mg/kg), sodium (20 mg/kg), and iron (3 mg/kg) [28].
- Grape juices also contain several anions – phosphate (200 mg/L), sulfate (260 mg/L as K_2SO_4) and chloride (232 mg/L as NaCl) [29] and small amounts of nitrate. Minerals in juice can vary over 10-fold with NaCl levels as high as 1800 mg/L and K_2SO_4 as high as 1200 mg/L resulting from saline soil [29].
- Grapes contain many vitamins, most of which are utilized by yeast but are then returned to the wine at similar levels [30]. The amounts of vitamins as reported in the USDA nutritional analysis of fresh grapes include ascorbic acid (32 mg/kg), niacin (1.8 mg/kg), vitamin B_6 (0.86 mg/kg), riboflavin (0.7 mg/kg), thiamin (0.69 mg/kg), folate (20 μg/kg), and vitamin A (30 μg/kg) [28]. Data on winegrape juice shows that levels are similar, with biotin reported at 1–3 μg/L and pantothenic acid at 0.5 mg/L [31].

20.8 Isoprenoids

Carotenoids are found in most plants, but only at low levels in grape berries. They are of interest because of their role as precursors to C_{13}–norisoprenoid aroma compounds (Chapters 8 and 23.1). Grapes contain 5 major carotenoids, and at ripeness have total carotenoid levels in the range of 0.4 to 2.5 mg/kg [32].

Isoprenoids are a large class of substances that include monoterpenoids, sesquiterpenoids, and C_{13}–norisoprenoids, and many are important aroma contributors in wine (Chapter 8). C_{13}–Norisoprenoid accumulation starts shortly after veraison, or shortly after carotenoid degradation completes, and peaks within a few weeks [33]. Accumulation of other isoprenoids starts a few weeks after veraison and can continue well past commercial ripeness. While many isoprenoids – particularly C_{13}–norisoprenoids – exist as non-volatile glycosides, a fraction of monoterpenoids exists in free form, accounting for the distinctive floral aroma of Muscat-type grapes (Chapter 8) [34]. The amounts of these substances vary widely depending on grape variety for *vinifera* vines. For instance, grapes of the Muscat family have high levels, in the range of 1–6 mg/kg of bound and free monoterpenoids at ripeness [35], with free linalool and geraniol well in excess of sensory thresholds (Chapter 8). Other varieties, such as Riesling, have perithreshold concentrations (0.05–0.2 mg/kg total), and most *vinifera* varieties have levels that are too low to affect sensory perception.

20.9 Insoluble materials

Insoluble grape tissues, including the skin and seeds, are present in grapes but these components are not extracted into wine. On average, seeds comprise approximately 4% of grape weight and skins about 11% [16]. A total analysis of grape skins showed that on a dry weight basis, 22.6% of the grape skin was insoluble in strong sulfuric acid [36]. The authors studied the residue by solid phase NMR and suggested that it was composed of cellulose and waxy material, and did not contain lignin. Other studies of pomace have yielded similar results, showing that in skins from white pomace, soluble sugars compose a large fraction of the mass, approximately 50%, while the insoluble polysaccharides are about 25% of the dry matter. On the other hand, in red pomace, the soluble sugars are only a few percent of the total while insoluble polysaccharides are 50% of the dry matter. Tannins and other polyphenols comprise most of the remaining characterized material [37]. About half of the mass of seeds results from polysaccharides, referred to as neutral detergent fiber (NDF), but due to the highly lignified state of this material, it is not very useful as a source of calories (in feeding animals, for example). The seeds also contain about 12% crude protein, 12% fat, and 5% ash, largely composed of minerals including potassium, calcium, phosphate, sulfate, and magnesium [38].

References

1. Pagay, V. and Cheng, L. (2010) Variability in berry maturation of Concord and Cabernet franc in a cool climate. *American Journal of Enology and Viticulture*, **61** (1), 61–67.
2. Andaur, J.E., Guesalaga, A.R., Agosin, E.E., *et al.* (2004) Magnetic resonance imaging for nondestructive analysis of wine grapes. *Journal of Agricultural and Food Chemistry*, **52** (2), 165–170.
3. Zhang, P., Barlow, S., Krstic, M., *et al.* (2015) Within-vineyard, within-vine, and within-bunch variability of the rotundone concentration in berries of *Vitis vinifera L.* cv. Shiraz. *Journal of Agricultural and Food Chemistry*, **63** (17), 4276–4283.
4. Wolpert, J.A. and Howell, G.S. (1984) Sampling Vidal Blanc grapes. II. Sampling for precise estimates of soluble solids and titratable acidity of juice. *American Journal of Enology and Viticulture*, **35** (4), 242–246.
5. Meyers, J.M. and Vanden Heuvel, J.E. (2014) Use of normalized difference vegetation index images to optimize vineyard sampling protocols. *American Journal of Enology and Viticulture*, **65** (2), 250–253.
6. Bravdo, B., Hepner, Y., Loinger, C., *et al.* (1984) Effect of crop level on growth, yield and wine quality of a high yielding carignane vineyard. *American Journal of Enology and Viticulture*, **35** (4), 247–252.
7. McIntyre, G.N., Kliewer, W.M., Lider, L.A. (1987) Some limitations of the degree day system as used in viticulture in California. *American Journal of Enology and Viticulture*, **38** (2), 128–132.
8. Ducasse, M.A., Canal-Llauberes, R.M., de Lumley, M., *et al.* (2010) Effect of macerating enzyme treatment on the polyphenol and polysaccharide composition of red wines. *Food Chemistry*, **118** (2), 369–376.
9. Diaz-Rubio, M.E. and Saura-Calixto, F. (2006) Dietary fiber in wine. *American Journal of Enology and Viticulture*, **57** (1), 69–72.
10. Kliewer, W.M. (1966) Sugars and organic acids of *Vitis vinifera*. *Plant Physiology*, **41** (6), 923.
11. Lakso, A.N. and Kliewer, W.M. (1975) Influence of temperature on malic acid metabolism in grape berries. 1. Enzyme responses. *Plant Physiology*, **56** (3), 370–372.
12. Singleton, V.L. and Esau, P. (1969) *Phenolic substances in grapes and wine, and their significance*, Academic, New York.
13. Price, S.F., Breen, P.J., Valladao, M., Watson, B.T. (1995) Cluster sun exposure and quercetin in Pinot noir grapes and wine. *American Journal of Enology and Viticulture*, **46**, 187–194.
14. Mattivi, F., Guzzon, R., Vrhovsek, U., *et al.* (2006) Metabolite profiling of grape: flavonols and anthocyanins. *Journal of Agricultural and Food Chemistry*, **54** (20), 7692–7702.
15. Kennedy, J.A., Matthews, M.A., Waterhouse, A.L. (2000) Changes in grape seed polyphenols during fruit ripening. *Phytochemistry*, **55** (1), 77–85.
16. Travaglia, F., Bordiga, M., Locatelli, M., *et al.* (2011) Polymeric proanthocyanidins in skins and seeds of 37 *Vitis vinifera* L. cultivars: a methodological comparative study. *Journal of Food Science*, **76** (5), C742–C749.
17. Cheynier, V., Souquet, J.M., Moutounet, M. (1989) Glutathione content and glutathione to hydroxycinnamic acid ratio in *Vitis vinifera* grapes and musts. *American Journal of Enology and Viticulture*, **40**, 320–324.
18. Huang, Z. and Ough, C.S. (1991) Amino-acid profiles of commercial grape juices and wines. *American Journal of Enology and Viticulture*, **42** (3), 261–267.

19. Negri, A.S., Prinsi, B., Scienza, A., *et al.* (2008) Analysis of grape berry cell wall proteome: a comparative evaluation of extraction methods. *Journal of Plant Physiology*, **165** (13), 1379–1389.
20. Macheix, J.J., Sapis, J.C., Fleuriet, A. (1991) Phenolic-compounds and polyphenoloxidase in relation to browning in grapes and wines. *Critical Reviews in Food Science and Nutrition*, **30** (4), 441–486.
21. Vincenzi, S., Tolin, S., Cocolin, L., *et al.* (2012) Proteins and enzymatic activities in Erbaluce grape berries with different response to the withering process. *Analytica Chimica Acta*, **732**, 130–136.
22. Salzman, R.A., Tikhonova, I., Bordelon, B.P., *et al.* (1998) Coordinate accumulation of antifungal proteins and hexoses constitutes a developmentally controlled defense response during fruit ripening in grape. *Plant Physiology*, **117** (2), 465–472.
23. Fracassetti, D., Lawrence, N., Tredoux, A.G.J., *et al.* (2011) Quantification of glutathione, catechin and caffeic acid in grape juice and wine by a novel ultra-performance liquid chromatography method. *Food Chemistry*, **128** (4), 1136–1142.
24. Smit, I., Pfliehinger, M., Binner, A., *et al.* (2014) Nitrogen fertilisation increases biogenic amines and amino acid concentrations in *Vitis vinifera* var. Riesling musts and wines. *Journal of the Science of Food and Agriculture*, **94** (10), 2064–2072.
25. Gallander, J.F. and Peng, A.C. (1980) Lipid and fatty-acid compositions of different grape types. *American Journal of Enology and Viticulture*, **31** (1), 24–27.
26. Pensec, F., Paczkowski, C., Grabarczyk, M., *et al.* (2014) Changes in the triterpenoid content of cuticular waxes during fruit ripening of eight grape (*Vitis vinifera*) cultivars grown in the Upper Rhine Valley. *Journal of Agricultural and Food Chemistry*, **62** (32), 7998–8007.
27. Grncarevic, M. and Radler, F. (1971) Review of surface lipids of grapes and their importance in drying process. *American Journal of Enology and Viticulture*, **22** (2), 80–86.
28. US Department of Agriculture (2014) A.R.S. USDA National Nutrient Database for Standard Reference, Release 27, http://www.ars.usda.gov/ba/bhnrc/ndl.
29. Leske, P.A., Sas, A.N., Coulter, A.D., *et al.* (1997) The composition of Australian grape juice: chloride, sodium and sulfate ions. *Australian Journal of Grape and Wine Research*, **3** (1), 26–30.
30. Ough, C.S. and Amerine, M.A. (1988) *Methods for analysis of musts and wines*, 2nd edn, Wiley-Interscience, New York.
31. Hagen, K.M., Keller, M., Edwards, C.G. (2008) Survey of biotin, pantothenic acid, and assimilable nitrogen in winegrapes from the Pacific Northwest. *American Journal of Enology and Viticulture*, **59** (4), 432–436.
32. Oliveira, C., Ferreira, A.C.S., Pinto, M.M., *et al.* (2003) Carotenoid compounds in grapes and their relationship to plant water status. *Journal of Agricultural and Food Chemistry*, **51** (20), 5967–5971.
33. Ryona, I. and Sacks, G.L. (2013) Behavior of glycosylated monoterpenes, C_{13}-norisoprenoids, and benzenoids in *Vitis vinifera* cv. Riesling during ripening and following hedging, in *Carotenoid cleavage products*, American Chemical Society, pp. 109–124.
34. Hjelmeland, A.K. and Ebeler, S.E. (2015) Glycosidically bound volatile aroma compounds in grapes and wine: a review. *American Journal of Enology and Viticulture*, **66** (1), 1–11.
35. Mateo, J.J. and Jimenez, M. (2000) Monoterpenes in grape juice and wines. *Journal of Chromatography A*, **881** (1–2), 557–567.
36. Mendes, J.A.S., Prozil, S.O., Evtuguin, D.V., Lopes, L.P.C. (2013) Towards comprehensive utilization of winemaking residues: characterization of grape skins from red grape pomaces of variety Touriga Nacional. *Industrial Crops and Products*, **43** (0), 25–32.
37. Deng, Q., Penner, M.H., Zhao, Y. (2011) Chemical composition of dietary fiber and polyphenols of five different varieties of wine grape pomace skins. *Food Research International*, **44** (9), 2712–2720.
38. Spanghero, M., Salem, A.Z.M., Robinson, P.H. (2009) Chemical composition, including secondary metabolites, and rumen fermentability of seeds and pulp of Californian (USA) and Italian grape pomaces. *Animal Feed Science and Technology*, **152** (3–4), 243–255.

21

Maceration and Extraction of Grape Components

21.1 Introduction

Winemaking results in the incomplete and variable extraction of components from the skin, pulp and seed – thus, even when fermentation and its resultant production of ethanol and other metabolites are ignored, wine composition differs from initial grape composition. Winemakers refer to the extraction process as *maceration*, a description that includes the physical manipulation of the must as well as the chemical extraction that results from it. Maceration may also be referred to as "skin contact," especially when discussing short-term, pre-fermentation maceration often used for aromatic varieties (Chapter 23.1). For our purposes, we will focus on the extraction products that are unmodified by microbial metabolism, although discussion of some transformed products is unavoidable. In addition, while this section deals with the impact of the extraction process on wine composition, in many cases initial grape composition will have a greater effect than altering maceration practices, for example, for explaining differences in varietal aroma compounds between cultivars.

There are a number of variables that have an important impact on the extraction process, with time and temperature having the largest influences [1]. Beyond these basic parameters, extraction of components from the grape solids can be enhanced by a range of chemical and physical processes, such as enzymatic treatments, mixing, pressing, heating, freezing, high-pressure pulses, and application of microwave or even pulsed electric field (PEF) treatments. These processes are rarely selective and will result in increased extraction of a range of compounds. The following pages focus on the major constituents of grapes that largely remain unchanged by a simple extraction step, although the maceration/extraction process can also increase the yield of flavor precursors that are transformed later in winemaking (Chapter 23).

The principal components studied in maceration investigations are the phenolics, especially condensed tannins and anthocyanins. As seen in Figure 21.1, an extended maceration process following fermentation will increase condensed tannin, but decrease the anthocyanin levels, and by doing so the total phenolics increase or stay level. Polysaccharides – an often overlooked but important factor – also increase with increasing maceration time. The extraction of other substances will also be dependent on maceration conditions, and many of these cases are mentioned below where data are available.

Understanding Wine Chemistry, First Edition. Andrew L. Waterhouse, Gavin L. Sacks, and David W. Jeffery.

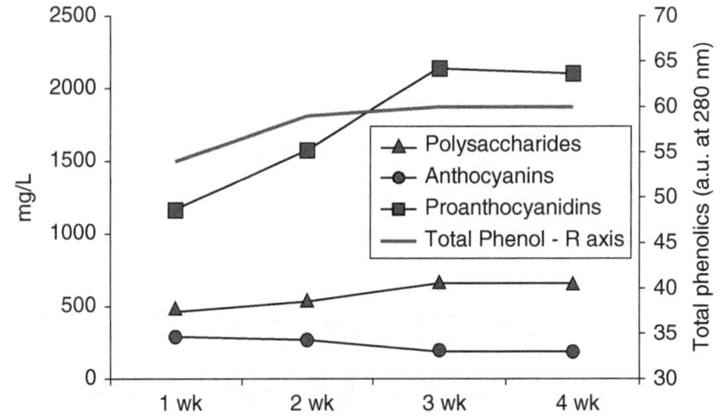

Figure 21.1 *Extraction of key components during extended maceration. Data from Gil et al., 2012 [2]*

21.1.1 General features of extraction during maceration

Some general principles should be recognized in interpreting extraction data from winemaking. It is possible to derive a fractional extraction coefficient (F) for compounds during maceration, where $F = m_{wine}/m_{grape}$, or the ratio of the amount of a compound in the wine to the initial amount in the grape. If m_{wine} is measured at multiple time points during maceration it is possible to derive an extraction rate function, and to model these by appropriate equations (e.g., Fick's Law of diffusion). Fitting the exponential model to the data can reveal an extraction constant and equilibrium tannin concentration (Figure 21.2) [3].

As described later, the extraction coefficient generally increases with increasing maceration time, temperature, and other parameters. However, assuming identical maceration conditions, the extraction rate and extraction coefficient for a given compound will depend largely on three factors:

- Location. Typically, compounds in the pulp are extracted immediately after crushing, while compounds from the skins will reach a maximum after several days, and compounds from the seeds may take weeks.
- Polarity. Hydroalcoholic solutions like wine are relatively polar solvents and readily dissolve polar substances such as sugars, acids, phenolics, alcohols, and so on. Non-polar substances can dissolve in wine, but will not be extracted as quickly, will have low limits of solubility, and the amount dissolved is generally much less than that limit. For some compounds, extraction will be minimal until sufficient ethanol is produced to solubilize the compound.
- Molecular size.

The apparent extraction coefficient will also be modified by these factors:

- Re-adsorption on to skin and other tissues.
- Reactive losses, for example, hydrolysis of bound forms, that is, glycosides, or reactions with other wine components such as acetaldehyde.

21.1.2 Extraction of phenolics

The grape components for which maceration is of greatest interest are the phenolic compounds (Chapter 11), and a number of reviews have addressed this topic [4]. In the berry, most of the phenolics are in the grape solids, skins, and seeds, and an extractive step is essential to their incorporation into wine. Of those, condensed

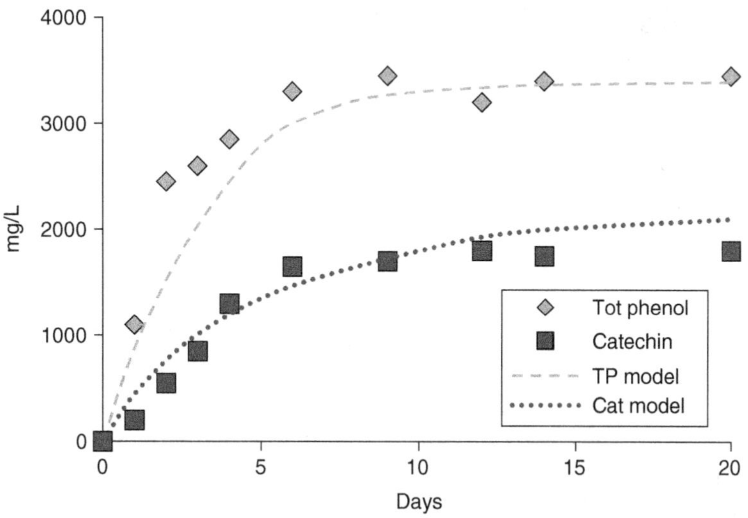

Figure 21.2 *Measured extraction of total phenolics and catechin versus modeling. Data from Andrich et al., 2005 [3]*

tannins and pigments receive the most attention because they are recognized as essential factors in red wine quality (Chapter 11), and are typically the targets when altering maceration regimes (Figure 21.1). Extraction of flavonols from skin is generally ignored, but some studies show the effects of particular maceration techniques. On the other hand, ~50% of hydroxycinnamates are found in juice (Chapter 13) and thus are rarely measured in maceration studies.

Condensed tannins are part of the flavan-3-ol family (Chapter 14), all of which are extracted together, so increases in monomeric catechins and oligomeric proanthocyanidins are coincident with any process that enhances tannin extraction [5]. While seed tannin extraction can extend over weeks, the extraction of tannin from the skin is very rapid, typically reaching a maximum within a few days [3]. Quantifying the extraction of phenolics (as well as other grape components) as a result of maceration is complicated by reactions that occur subsequent to extraction; for instance, the increase in anthocyanins during winemaking is less than the amount extracted from the solids due to the wide range of reactions in which these compounds can participate (Chapters 16 and 25). However, it is possible to compare the amount of a particular substance in wine to the amount in the grapes and derive a fractional extraction coefficient. Singleton first reported the grape phenolic content per mass of tissue at about 40 mg/kg (fresh berry weight) in pulp, 1800 mg/kg in red skin (and half as much for white), and 3500 mg/kg in seeds [6]. In typical red winemaking, 30–50% of these substances (2000–3500 mg/L of total phenols) end up in wine, assuming the mass of wine is 80% that of the grapes used [7].

While anthocyanins in grapes tend to be well correlated with anthocyanins in wine, assuming similar maceration conditions, grape tannin is often poorly correlated with wine tannin, especially across cultivars (Figure 21.3) [8]. Potential explanations for the variation in tannin extraction coefficients across grape varieties include:

- Grape tannin measurements do not always distinguish between seed and skin tannin, and the former is more slowly extracted than the latter.
- The "amount" of phenolic material present in the grape is generally determined by an aqueous acetone extraction (e.g., 70% v/v), but recent studies have shown some phenolics may be released via acid treatment [9], and the acidic environment of wine could promote their extraction.
- Grape cell wall material can re-bind tannins after it is initially extracted into the fermenting must [10].

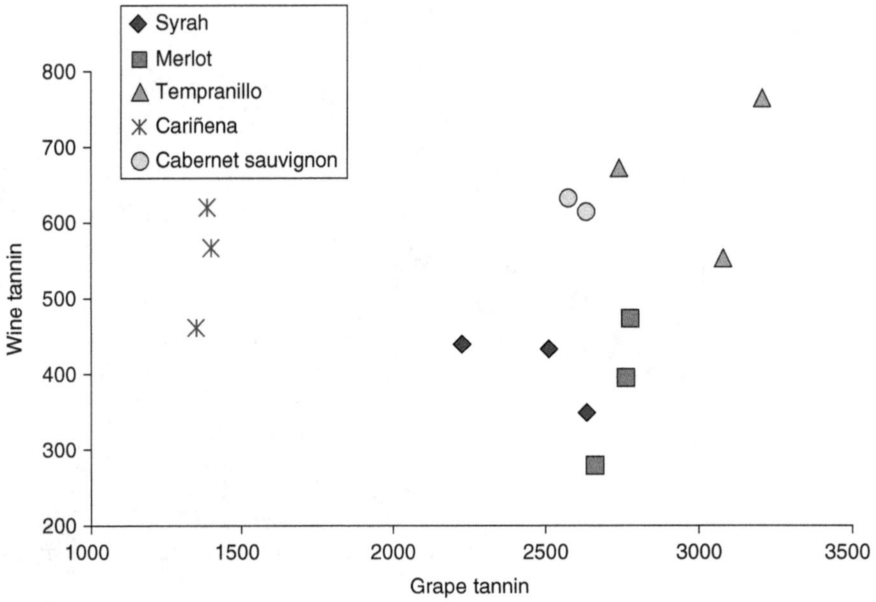

Figure 21.3 Data (mg/L and mg/kg) from Tables S2, S4 in Fragoso et al., 2011 [8]. Source: Casassa 2013 [8]. Reproduced with permission of John Wiley and Sons

- Anthocyanins can complex with tannins (Chapter 16), increasing the latter's extractability [11].
- The major soluble proteins in grapes can bind tannin and limit their extraction [12].

In contrast to current winemaking practices, in the distant past *white* grapes were frequently vinified with substantial maceration of the skin throughout fermentation. The resulting "orange wines" would have better longevity imparted by a higher antioxidant phenolic content (important considering the poor oxygen exclusion of the containers used at the time). However, enhanced phenolics in white wines cuts both ways, enhancing oxidative browning potential, and scavenging volatile thiols via the formation of quinones[1] (Chapter 24) [13]. Orange wines possess high astringency (more akin to red wines than white), but are rare in the current wine marketplace, and thus are unfamiliar to many white wine drinkers today.[2]

21.1.3 Extraction of polysaccharides

In comparison to phenolics, the effects of maceration on polysaccharides are often overlooked. Grape polysaccharides are derived from the cell walls of the berries, and appear to have minor effects on the physical sensation of wine in the mouth (i.e., mouthfeel) [14] as well as the volatility of some aroma compounds (Chapter 2). Polysaccharides, particularly pectin, hemicellulose, and cellulose, constitute the vast majority of grape cell wall structural compounds (Chapter 20), but the majority of these components are poorly recovered, with only a small portion of pectic substances (~0.1% w/w) extracted during fermentation. These are

[1] Named varieties of Muscat-type grapes, which rely on monoterpenoids for their characteristic aroma, have been known for millennia. By comparison, Sauvignon Blanc, which is largely dependent on varietal thiols for its characteristic aroma, has only been a named variety since the sixteenth century. The highly oxidative wine production and storage conditions in the age before cork stoppers, glass bottles, and sulfites would have made retention of thiols challenging, and provided a nondescript wine.

[2] This traditional style is found in Georgia and nearby regions of central Asia, as well as some Eastern European countries. There are also a few Western winemakers producing orange wines in response to market interest.

primarily rhamnogalacturonans (RGs) and arabinogalactan proteins (AGPs), whose effect and concentration in wine are described in Chapter 2. These two classes of polysaccharides share the properties of being both sufficiently soluble as well as being resistant to acid and enzymatic degradation during fermentation, and thus can persist into the finished wines. Similar to seed phenolics, extraction of both classes increases throughout maceration (Figure 21.1) [2]. Since most grape polysaccharides are retained in the pomace, the surface of the skins can act as an adsorbent and retain polyphenolics that would otherwise be solubilized into wine. Hydrolyzing the polysaccharides of the grape skin also increases the yield of juice from pressing, and alters juice clarity.

21.1.4 Extraction of odorants and their precursors

Most primary odorants in grapes (e.g. monoterpenoids, methyl anthranilate, rotundone, IBMP) are found predominantly in the skins and thus will increase with increasing maceration time, although maximal concentrations will usually be reached quickly (with 2–3 days) because of these compounds' low molecular weight and fast diffusivity. Maceration will also increase concentrations of volatile compound precursors, for example, glycosides, hydroxycinnamic acids, *S*-conjugates (Chapter 23). Similar to primary odorants, concentrations of their precursors will usually reach a maximum soon after fermentation commences.

21.1.5 Extraction of cations

Maceration will result in the exchange of metal cations for protons. In many literature reports, this results in an increase in both metal cations and pH (see Chapter 3). However, high potassium juices may show an initial decrease in potassium during skin contact, followed by stable levels. Testing at different pH levels suggested an ion-exchange effect with pectins in the skins and higher ethanol suppressed potassium extraction [15].

21.2 Pre-fermentative treatments

Many of the pre-fermentation maceration techniques involve physical or chemical changes to the structure of the grape tissue that are designed to enhance release of desired phytochemicals or increase juice yield. The treatments often respond differently depending on grape source. For example, Gonzáles-Neves *et al.* investigated the effects of enzymes and "cold soak" on multiple varieties (Merlot, Tannat, and Syrah) [16] – the large standard deviations indicate the variable response across the cultivars (Figure 21.4).

21.2.1 Use of maceration enzymes

The most common treatment to enhance juice yield is the use of enzymes to chemically degrade cell wall polysaccharides, with reported increases of 10–30% [17, 18]. These processing enzymes include polygalacturonase, pectin-esterase and pectin-lyase activity, but as these are industrial preparations, they include other activities (Chapter 23.1), which can have a major impact on aroma and/or phenolics (sometimes intentionally) [19] (Chapter 27). A key side-effect of pectinase treatments is the release of methanol from the methyl ester in pectin slightly enhancing methanol levels (Chapter 6), which can be concentrated on distillation (Chapter 26.4).

While the total picture is partially obscured by the presence of yeast polysaccharides (Chapter 2), pectinase results in a significant decrease in the high molecular weight fractions of the grape-sourced arabinogalactan proteins (AGPs) and arabinans, and these are replaced by lower molecular weight substances, including rhamnogalacturonan II (RGII) and other smaller molecules [20]. The composition of the polysaccharides is also affected by enzyme treatment – in other words, the constituent subunit composition changes. The use of an enzymatic treatment generally decreases the amount of arabinose and galactose in the remaining polysaccharides [21].

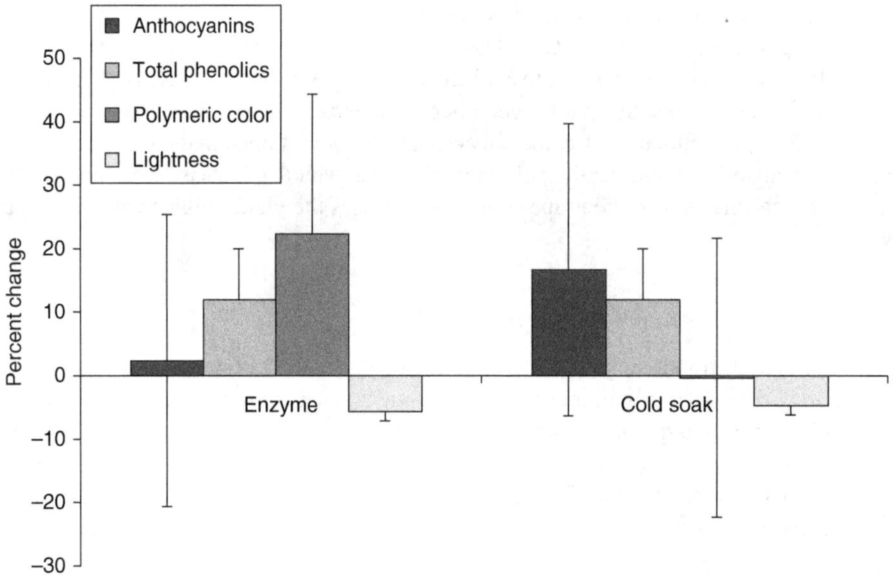

Figure 21.4 Comparison of average and standard deviation of wine phenolic and color parameters affected by cold soak and enzyme addition on three varieties. Data from Gonzáles-Neves et al., 2013 [16].

Enzymatic treatments of red grapes increase wine polyphenol content. Ducasse *et al.* showed total polyphenol concentration increased modestly, by about 10%, while color increased between 10 and 20%, largely from an increase in polymeric wine pigments. Tannin rose consistently by approximately 20%, explaining the common perception of increased astringency from these treatments [20]. There are at least two possible explanations (not mutually exclusive) for these observations: one is that polysaccharides, particularly those high in arabinose and galactose, have been shown to moderate astringency in wine [22], so degrading these by enzymes should increase astringency and give a higher response in protein-binding assays for tannin. Alternatively, degrading the polysaccharides may expose tannin-containing tissue to more facile extraction, thereby increasing tannin extraction.

21.2.2 Application of heat

Thermovinification involves heating either white or red must and then pressing before fermentation, and is a common treatment in some areas. One of its major applications is for decreasing the microbial load or inhibiting browning enzymes on mold-contaminated musts, and inactivating enzymes can also result in lower production of C_6 alcohols (Chapter 6) [23], although others have reported that thermal inactivation can preserve C_6 aldehydes by inactivating ADH enzymes that would reduce the aldehydes to alcohols (Chapter 22.1). Thermovinification also has a profound effect on phenolic extraction as compared to traditional maceration, leading to reductions in tannin when pressing is performed pre-fermentation – this approach also increases anthocyanin content, presumably by limiting reactions with tannins that would result in color loss [24]. On the other hand, significant increases are observed when pressing is delayed until after fermentation. A side-effect of thermal treatments as compared to conventional macerations is a fruitier aroma due to an increase in both acetate and ethyl esters (up to an order of magnitude) [23], although it is not clear if this is due to lower solids content or due to lower oxygen exposure during fermentation (Chapters 22.2 and 22.3).

Flash release (flash détente) is a related, modern maceration technique that involves heating the must and releasing it into a vacuum, causing grape tissue cells to rupture. This technique increased arabinose- and galactose-rich polysaccharides in the finished wine, while a similar thermal-only treatment had little effect [21]. This technique was very effective in increasing the amounts of polyphenolics in the juice (>5 times more) and in the subsequent wine by 1.2–2 times for some seed components [25]. As expected, there was no increase in juice-derived hydroxycinnamates or the anthocyanins as these are readily extracted by normal maceration, but the harder to extract seed-derived tannins and catechins did increase. Flash release is also reported to increase juice yield as compared to conventional maceration [18].

21.2.3 Extraction by other physical means

Several other physical treatments of grape musts have been explored for their ability to generally or selectively enhance extraction. One recent innovation in fruit and vegetable processing is the use of pulsed electric field (PEF) treatments to damage cell walls; in the case of grapes, this increases both yield [26] and extraction of tannins and anthocyanins [27]. Some reports suggest that tannin extraction can be higher but still yield wines with less astringency [28, 29], perhaps due to greater polysaccharide extraction, while others report effects on aromas [30]. Other related technologies that have been studied include ultrasound [31], microwave [32], and freezing (*cryomaceration*) [33], with varying and modest effects in wine and model systems. However, cryomaceration of the whole or crushed fruit has been shown to increase the extraction of volatile aroma substances [34].

21.2.4 Simple pre-fermentation skin contact

One basic approach to increasing extraction is to simply allow for solids/juice contact before fermentation occurs. In the case of white wine, this skin-contact step usually involves allowing the crushed grapes to stay in contact with the skins for several hours before pressing and fermenting. A major reason to carry out skin contact with white grapes is to increase extraction of aroma compounds and their precursors, but a few investigations have looked at phenolics from white grapes, showing that a cold soak increased flavan-3-ols and flavonols [35], as well as total polyphenolics and polysaccharides. However, the resulting differences had only a small impact on taste or mouthfeel [36]. There appears to be a pronounced effect of skin contact on grape-derived volatiles. A study of Airén showed that this approach increased monoterpenoids (Chapter 8) in the wine in addition to skin flavonols and minor increases of seed phenolics [37].

Several studies have investigated the effects of various pre-fermentation treatments of varietal thiols, particularly in Sauvignon Blanc. As described earlier, it is now well-established that machine harvesting followed by some hours of skin maceration time yields far more of the varietal thiol 3-mercaptohexanol (3-MH) in finished wine [38] (Chapter 10). This is likely due to extraction of precursors from grape skin and not from direct extraction of free volatiles, as grapes contain negligible 3-MH [39]. It is possible that enzymatic oxidation [40] encountered during mechanical harvesting and/or pre-fermentative skin contact converts α-linolenic acid in the berries to (*E*)-2-hexenal (Chapter 23.3) as the introduction of oxygen during the maceration step increases 3-MH yield [41]. The subsequent steps from 2-hexenal and leading to 3-MH are covered in Chapter 23.2. Regardless of the mechanism, both skin contact and pressing have been shown to increase the ultimate amount of 3-MH precursor in different Sauvignon Blanc juices [42], and several odorous varietal thiols were enhanced four-fold after 48 hours of skin contact using Muscadine grapes [43].

For red grapes, pre-fermentative skin contact (i.e., cold soak) is sometimes done prior to fermentation on skins. Cold soak may last several days, and is purportedly done to alter phenolic composition or sensory properties of finished wines. The practice is controversial among winemakers, with advocates and detractors [44]. A review of numerous literature reports concluded that at the low temperatures employed (10–15 °C), a cold soak had little to no measureable effect on the finished wine [4], Subsequent reports showed varying effects, with Alvarez *et al.* observing modest effects depending on variety [33], Apolinar-Valiente *et al.* making similar observations (+10% in only one case of four), and also noting little impact on polysaccharides [45], while Koyama saw lower seed extracton [46]. In contrast, other investigators showed that a cold soak increased the amount of seed and skin proanthocyanidins extracted in the finished wine [47, 48].

Finally, freezing, dry ice treatment, or cold maceration can have modest effects on fermentation-derived esters and higher alcohols (particularly C_6 alcohols like 1-hexanol) in several *V. vinifera* wines [49]. These factors are discussed in more detail in Chapters 22.2, 22.3, and 23.3.

Overall, pre-fermentation maceration includes many different techniques to alter extraction. However, the presence of constitutive and added enzymes, and the potential for non-enzymatic chemical reactions or unintended microbial activity, ensure that the process is not a simple chemical extraction but a complex interactive system.

21.3 Maceration treatments during fermentation

In red winemaking, the maceration process generally targets the extraction of anthocyanins and tannins. During fermentation anthocyanin extraction typically peaks at about 3–5 days and then declines, due to re-adsorption on to skin tissue or reactions to form wine pigments (Figure 21.5). Tannin extraction is also limited by re-adsorption on to grape skin tissues, with the larger proanthocyanidins tightly bound by flesh cell wall material [10].

21.3.1 Time and temperature

As with pre-fermentation maceration, increasing time and temperature will generally increase the extraction coefficient and rate. However, a complicating issue is that the formation of ethanol will also favor greater extraction. Since fermentation rate (and thus ethanol production) can be accelerated by higher temperatures and other factors, it is often difficult to decouple the variables of fermentation conditions on extraction rates. Generally, extraction of both total and individual polyphenolic species increases with increasing temperature (Table 21.1), although both increases and decreases are reported for anthocyanins in response to increasing fermentation temperature. The variable behavior of anthocyanins likely arises because (i) anthocyanin extraction is rapid in comparison to tannins and thus increasing temperature may have negligible effect on extraction per se and (ii) anthocyanins can be lost to side reactions that are promoted at higher temperatures. For example, new wine pigments can be formed due to increased production of carbonyl metabolites by the yeast, resulting in the reaction of these aldehydes with the anthocyanins and tannins (Chapter 24).

As shown in Figure 21.5, anthocyanin concentration increases and then decreases during the fermentation. By introducing ^{14}C labeled malvidin-3-glucoside at day three, Zimman and Waterhouse [50] showed that even at maximum extraction, half the anthocyanins were not in solution, but apparently bound to the grape solids. After pressing, anthocyanin levels continued to decrease but the constant level of radioactivity in solution indicated that this was due to formation of soluble anthocyanin derivatives (Chapters 16 and 25) [50].

21.3.2 Mixing solids and liquid

Techniques for mixing the grape solids into the fermenting must are diverse and very creative, and are designed to enhance or otherwise alter the extraction of tannins, color, and other phenolics (Chapter 19). As with fermentation temperature, little attention has been directed to understanding how these techniques might alter the extraction of non-phenolic compounds from grape solids. New techniques are typically

Table 21.1 *Fermentation maceration effects on red wine composition*

Treatment	Control[a]	Change relative to control[c]					Analysis time[e]	Notes[f]	Reference
		Total phenol	Tannin	Anthocyanin	Wine pigment[d]	Color			
Temperature, 27 °C	12 °C	–	–	–	–	+100%	130 d	38 L, PN	[51]
Temperature, 25 °C	15 °C	–23%	–	–	–	–1%	~1 mo	3 L, 6 da, CS	[52]
Temperature, 32 °C	24 °C	–	+11%	NS[b]	NS	NS	18 mo	22,800 L, 7 d, CS	[1]
Temperature, 30 °C	20 °C	–	–	NS	+50%	–50%	225 d	40 kg, 7, 5 d, PN	[53]
Temperature, 30 °C	15 °C	+32%	–	+21%	–	+20%	2.5 y	35 L, press at dryness, Pn	[54]
Delestage[g]	Conventional	–5%	+9%	NS	–9%	+26%	4 mo	250 L, 12 d, Mencia	[55]
Ganimede[h]	Conventional	–13%	–23%	NS	–9%	+16%	4 mo	250 L, 12 d, Mencia	[55]
Press 3 versus 6 d	3 d	–5%	–	–	–	+6%	~1 mo	3 L, 25 °C, CS	[52]
Press 7 versus 10 d	7 d	–	+13%	–	–	+19%	10 mo	38 L, 25 °C, CS	[56]
Press 4 versus 10 d	4 d	–	–	+13%	+88%	+54%	12 mo	120 L, Monastrell	[57]
Pumpover	Punchdown		–10%	–11%	–	–		180 kg, 25 °C, Pinotage	[58]
Rotary	Pumpover	–	–	–4%	–	–8%	24 mo	5 °C, Vinhão	[59]

[a] Control treatment.
[b] NS = No significant difference.
[c] Various methods.
[d] Non-bleachable pigments.
[e] Elapsed time after fermentation.
[f] Scale, time or temperature of fermentation, and grape variety (CS = Cabernet Sauvignon, PN = Pinot Noir).
[g] Delestage is draining the fermentation tank and pouring the liquid back over the cap as one large addition.
[h] Ganimede is a complex semi-continuous extraction device generally used for Port wine production.

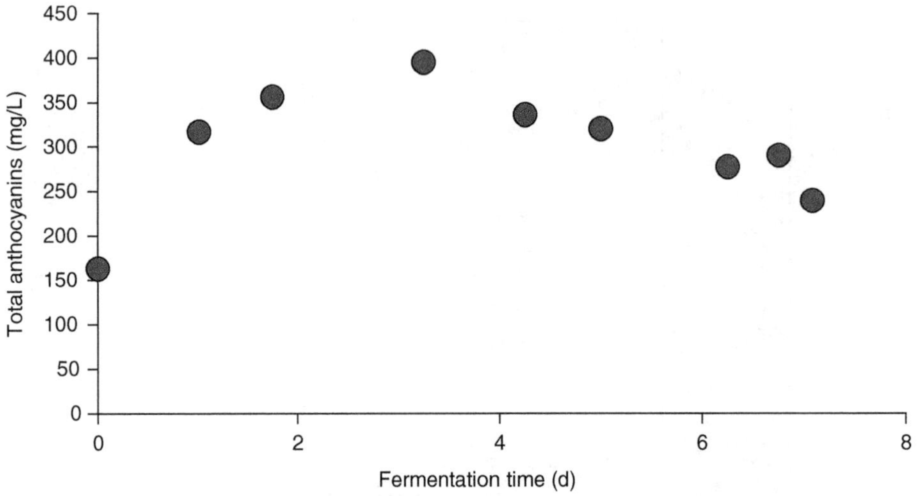

Figure 21.5 *Typical extraction curves for anthocyanin during fermentation. Data from Zimman and Waterhouse, 2004 [50]*

compared to the standard punchdown (*pigeage*), a physical mixing of the cap solids with the must, or pumpover, pumping the liquid must over the cap to facilitate an extraction process. Most variants have modest effects on overall phenolic extraction [1], and the observed effects are highly variable, presumably due to varying composition of the berries or difficulties in controlling these mechanical processes. However, extreme techniques, like the continuous maceration (i.e., Ganimede system) used in Port production, where extraction is limited to a few days early in the fermentation, can enhance the color and tannin extraction rate and coefficients [60]. It is most difficult to generalize the effects of specific treatments – appropriate use of these various fermentation protocols is obtained empirically by experimenting with specific fruit sources. One typical observation showed that different maceration techniques resulted in large differences immediately after fermentation or extended maceration, but during bottle aging the differences diminish [59], so observations just after the treatment are of questionable value. Table 21.1 shows the variation in response to a number of typical treatments and the time between fermentation and analysis.

Because the seeds and skins have different polyphenol compositions, altering the proportions of these tissues during maceration can have measureable effects. In a Cabernet Sauvignon fermentation, repeatedly racking the wine and returning it to the same tank (*delestage*) decreased the seed content by ~80%. The treatment also decreased the proanthocyanidin content and epicatechin gallate to epigallocatechin ratio, made the wine darker while diminishing the anthocyanin content, and moderated bitterness and astringency [61]. A similar study on Merlot showed little change in color, but the proanthocyanidins contained more epigallocatechin subunits [62]. Apart from phenolic extraction, removing the solids during fermentation of a red grape must has dramatic effects on some volatiles, and both increases and decreases have been observed [63].

21.3.3 Carbonic maceration

Aside from physical maceration methods, the practice of carbonic maceration (CM) is a well-known technique that alters extraction of red grape must. In CM, a small fraction of the grapes are crushed into a fermentation tank with the balance of the fruit added as whole clusters. The tank is sealed under a carbon dioxide atmosphere, which induces the fruit to ferment its sugars anaerobically, in addition to alcoholic fermentation brought about by yeast. After about one week, the fruit is crushed and pressed into another tank to complete fermentation by yeast. Wines made with this technique have a lower anthocyanin content and thus are lighter, and while the hue is similar, the CM wines have higher chroma values [64]. These wines also have very different aroma profiles due to high ester levels that contribute a strong fruity aroma [65], similar to what is observed with thermovinified wines.

21.4 Post-fermentation maceration

21.4.1 Extended maceration (EM)

The practice of continuing maceration after alcoholic fermentation is complete is common with particular wine styles where high levels of tannin are desired; seed tannin continues to dissolve into the wine, as the alcohol from fermentation is at the maximum concentration at this point (Figure 21.6). After alcoholic fermentation has ended, however, the wine is no longer protected from oxygen by carbon dioxide evolution. As such, this maceration step is typically conducted in a closed system with little air exposure because any solids floating on the wine can facilitate *Acetobacter* growth and acetic acid formation under aerobic conditions (Chapter 22.5).

A remarkably detailed study compared pressing at 10 days versus 30 days total maceration time. Of the tannin available in seeds and skins, extended maceration (EM) increased seed extraction from 11 to 18%, but skin tannin remained constant, and the extracted levels in the wine did not change up to 540 days later. EM resulted in much poorer recovery of anthocyanins, and while it accelerated polymeric pigment formation during the first year of aging, after 540 days the amount was similar to the control, so the fraction of color from polymeric pigment was higher in the EM wine at that time [66]. Similar results were observed by Yokotsuka *et al.* [67]. Yet another study also found that EM leads to increases in color and proanthocyanidins, but a lower yield from thiolysis, suggesting that more irregular linkages arise during the extended maceration and that EM was related to increases in astringency [68].

These results suggest that EM is primarily a physiochemical extraction process, with little effect from yeast or grape enzymatic activity. Components in the grape solids slowly equilibrate with the wine, although reactions with dissolved phenolics and aldehydes/ketones generated during fermentation will continue.

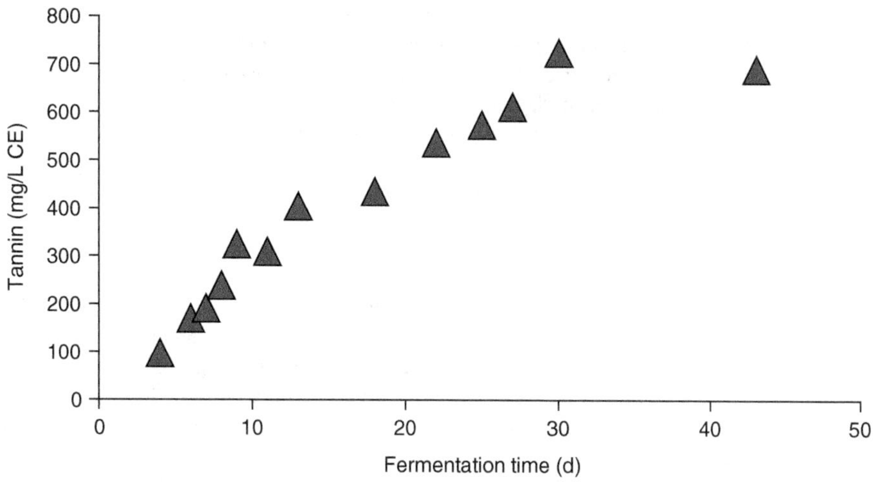

Figure 21.6 Tannin extraction during extended maceration. Data from Casassa et al., 2013 [66]

As for other components, EM time substantially increased the concentration of polysaccharides (Figure 21.1) in wine in nearly all cases, across a range of grape maturities [2]. Finally, extended skin contact was also suggested as a means to increase rotundone levels for varieties that contain this "peppery" compound, since rotundone accumulates predominantly in skins, is very non-polar, and so is poorly extracted (~10%) [69].

References

1. Zimman, A., Joslin, W.S., Lyon, M.L., *et al.* (2002) Maceration variables affecting phenolic composition in commercial-scale Cabernet Sauvignon winemaking trials. *American Journal of Enology and Viticulture*, **53** (2), 93–98.
2. Gil, M., Kontoudakis, N., Gonzalez, E., *et al.* (2012) Influence of grape maturity and maceration length on color, polyphenolic composition, and polysaccharide content of Cabernet Sauvignon and Tempranillo wines. *Journal of Agricultural and Food Chemistry*, **60** (32), 7988–8001.
3. Andrich, G., Zinnai, A., Venturi, E., Fiorentini, R. (2005) A tentative mathematical model to describe the evolution of phenolic compounds during the maceration of Sangiovese and Merlot grapes. *Italian Journal of Food Science*, **17** (1), 45–58.
4. Sacchi, K.L., Bisson, L.F., Adams, D.O. (2005) A review of the effect of winemaking techniques on phenolic extraction in red wines. *American Journal of Enology and Viticulture*, **56** (3), 197–206.
5. Sun, B.S., Pinto, T., Leandro, M.C., *et al.* (1999) Transfer of catechins and proanthocyanidins from solid parts of the grape cluster into wine. *American Journal of Enology and Viticulture*, **50** (2), 179–184.
6. Singleton, V.L. and Esau, P. (1969) *Phenolic substances in grapes and wine, and their significance*, Academic, New York.
7. Singleton, V.L. (1982) Grape and wine phenolics; background and prospects, in *Proceedings of the Symposium of the University of California, Davis, Grape and Wine Centennial, 1980*, Department of Viticulture and Enology, University of California, Davis, CA, pp. 215–227.
8. Fragoso, S., Guasch, J., Aceña, L., *et al.* (2011) Prediction of red wine colour and phenolic parameters from the analysis of its grape extract. *International Journal of Food Science and Technology*, **46** (12), 2569–2575.
9. Rustioni, L., Fiori, S., Failla, O. (2014) Evaluation of tannins interactions in grape (*Vitis vinifera* L.) skins. *Food Chemistry*, **159**, 323–327.
10. Bindon, K.A., Smith, P.A., Holt, H., Kennedy, J.A. (2010) Interaction between grape-derived proanthocyanidins and cell wall material. 2. Implications for vinification. *Journal of Agricultural and Food Chemistry*, **58** (19), 10736–10746.
11. Singleton, V.L. and Trousdale, E.K. (1992) Anthocyanin–tannin interactions explaining the differenced in polymeric phenols between white and red wines. *American Journal of Enology and Viticulture*, **43** (1), 63–70.
12. Springer, L.E., Stahlecker, A.C., Sherwood, R.W., Sacks, G.L. (2015) Limits on red wine tannin extraction and addition. Part II: Role of pathogenesis related proteins in Terroir, in American Society for Enology and Viticulture 66th National Conference, Portland, Oregon, p. 61.
13. Merida, J., Moyano, L., Millan, C., Medina, M. (1991) Extraction of phenolic compounds in controlled macerations of Pedro Ximenez grapes. *Vitis*, **30**, 117–127.
14. Vidal, S., Francis, L., Williams, P., *et al.* (2004) The mouth-feel properties of polysaccharides and anthocyanins in a wine like medium. *Food Chemistry*, **85** (4), 519–525.
15. Harbertson, J.F. and Harwood, E.D. (2009) Partitioning of potassium during commercial-scale red wine fermentations and model wine extractions. *American Journal of Enology and Viticulture*, **60** (1), 43–49.
16. Gonzalez-Neves, G., Gil, G., Favre, G., *et al.* (2013) Influence of winemaking procedure and grape variety on the colour and composition of young red wines. *South African Journal of Enology and Viticulture*, **34** (1), 138–146.
17. Haight, K.G. and Gump, B.H. (1994) The use of macerating enzymes in grape juice processing. *American Journal of Enology and Viticulture*, **45** (1), 113–116.
18. Paranjpe, S.S., Ferruzzi, M., Morgan, M.T. (2012) Effect of a flash vacuum expansion process on grape juice yield and quality. *LWT-Food Science and Technology*, **48** (2), 147–155.

19. Bisson, L.F. and Butzke, C.E. (1996) Technical enzymes for wine production. *Agro Food Industry Hi-Tech*, **7** (3), 11–14.
20. Ducasse, M.A., Canal-Llauberes, R.M., de Lumley, M., *et al.* (2010) Effect of macerating enzyme treatment on the polyphenol and polysaccharide composition of red wines. *Food Chemistry*, **118** (2), 369–376.
21. Doco, T., Williams, P., Cheynier, V. (2007) Effect of flash release and pectinolytic enzyme treatments on wine polysaccharide composition. *Journal of Agricultural and Food Chemistry*, **55** (16), 6643–6649.
22. Quijada-Morin, N., Williams, P., Rivas-Gonzalo, J.C., *et al.* (2014) Polyphenolic, polysaccharide and oligosaccharide composition of Tempranillo red wines and their relationship with the perceived astringency. *Food Chemistry*, **154**, 44–51.
23. Fischer, U., Strasser, M., Gutzler, K. (2000) Impact of fermentation technology on the phenolic and volatile composition of German red wines. *International Journal of Food Science and Technology*, **35** (1), 81–94.
24. Gao, L., Girard, B., Mazza, G., Reynolds, A.G. (1997) Changes in anthocyanins and color characteristics of Pinot noir wines during different vinification processes. *Journal of Agricultural and Food Chemistry*, **45** (6), 2003–2008.
25. Morel-Salmi, C., Souquet, J.M., Bes, M., Cheynier, V. (2006) Effect of flash release treatment on phenolic extraction and wine composition. *Journal of Agricultural and Food Chemistry*, **54** (12), 4270–4276.
26. Praporscic, I., Lebovka, N., Vorobiev, E., Mietton-Peuchot, M. (2007) Pulsed electric field enhanced expression and juice quality of white grapes. *Separation and Purification Technology*, **52** (3), 520–526.
27. Lopez, N., Puertolas, E., Hernandez-Orte, P., *et al.* (2009) Effect of a pulsed electric field treatment on the anthocyanins composition and other quality parameters of Cabernet Sauvignon freshly fermented model wines obtained after different maceration times. *LWT-Food Science and Technology*, **42** (7), 1225–1231.
28. Delsart, C., Ghidossi, R., Poupot, C., *et al.* (2012) Enhanced extraction of phenolic compounds from Merlot grapes by pulsed electric field treatment. *American Journal of Enology and Viticulture*, **63** (2), 205–211.
29. Luengo, E., Franco, E., Ballesteros, F., *et al.* (2014) Winery trial on application of pulsed electric fields for improving vinification of Garnacha grapes. *Food and Bioprocess Technology*, **7** (5), 1457–1464.
30. Garde-Cerdan, T., Gonzalez-Arenzana, L., Lopez, N., *et al.* (2013) Effect of different pulsed electric field treatments on the volatile composition of Graciano, Tempranillo and Grenache grape varieties. *Innovative Food Science and Emerging Technologies*, **20**, 91–99.
31. Tiwari, B.K., Patras, A., Brunton, N., *et al.* (2010) Effect of ultrasound processing on anthocyanins and color of red grape juice. *Ultrasonics Sonochemistry*, **17** (3), 598–604.
32. Ghassempour, A., Heydari, R., Talebpour, Z., *et al.* (2008) Study of new extraction methods for separation of anthocyanins from red grape skins: analysis by HPLC and LC-MS/MS. *Journal of Liquid Chromatography and Related Technologies*, **31** (17), 2686–2703.
33. Alvarez, I., Aleixandre, J.L., Garcia, M.J., Lizama, V. (2006) Impact of prefermentative maceration on the phenolic and volatile compounds in Monastrell red wines. *Analytica Chimica Acta*, **563** (1–2), 109–115.
34. Bavcar, D., Cesnik, H.B., Cus, F., *et al.* (2011) Impact of alternative skin contact procedures on the aroma composition of white wine. *South African Journal of Enology and Viticulture*, **32** (2), 190–203.
35. Caceres-Mella, A., Pena-Neira, A., Parraguez, J., *et al.* (2013) Effect of inert gas and prefermentative treatment with polyvinylpolypyrrolidone on the phenolic composition of Chilean Sauvignon blanc wines. *Journal of the Science of Food and Agriculture*, **93** (8), 1928–1934.
36. Gawel, R., Day, M., Van Sluyter, S.C., et al. (2014) White wine taste and mouthfeel as affected by juice extraction and processing. *Journal of Agricultural and Food Chemistry*, **62** (41), 10008–10014.
37. Cejudo-Bastante, M.J., Castro-Vazquez, L., Hermosin-Gutierrez, I., Perez-Coello, M.S. (2011) Combined effects of prefermentative skin maceration and oxygen addition of must on color-related phenolics, volatile composition, and sensory characteristics of Airen white wine. *Journal of Agricultural and Food Chemistry*, **59** (22), 12171–12182.
38. Allen, T., Herbst-Johnstone, M., Girault, M., *et al.* (2011) Influence of grape-harvesting steps on varietal thiol aromas in Sauvignon blanc wines. *Journal of Agricultural and Food Chemistry*, **59** (19), 10641–10650.
39. Capone, D.L., Sefton, M.A., Jeffery, D.W. (2012) Analytical investigations of wine odorant 3-mercaptohexan-1-ol and its precursors, in *Flavor chemistry of wine and other alcoholic beverages*, American Chemical Society, Washington, DC, pp. 15–35.
40. Podolyan, A., White, J., Jordan, B., Winefield, C. (2010) Identification of the lipoxygenase gene family from *Vitis vinifera* and biochemical characterisation of two 13-lipoxygenases expressed in grape berries of Sauvignon Blanc. *Functional Plant Biology*, **37** (8), 767–784.

41. Larcher, R., Nicolini, G., Tonidandel, L., *et al.* (2013) Influence of oxygen availability during skin-contact maceration on the formation of precursors of 3-mercaptohexan-1-ol in Muller-Thurgau and Sauvignon Blanc grapes. *Australian Journal of Grape and Wine Research*, **19** (3), 342–348.
42. Maggu, M., Winz, R., Kilmartin, P.A., *et al.* (2007) Effect of skin contact and pressure on the composition of Sauvignon blanc must. *Journal of Agricultural and Food Chemistry*, **55** (25), 10281–10288.
43. Gurbuz, O., Rouseff, J., Talcott, S.T., Rouseff, R. (2013) Identification of Muscadine wine sulfur volatiles: pectinase versus skin-contact maceration. *Journal of Agricultural and Food Chemistry*, **61** (3), 532–539.
44. Cutler, L. (2009) Industry Roundtable: Fermentation Techniques. Wine Communications Group [cited December 20, 2013]; Available from: http://www.winebusiness.com/wbm/?go=getArticle&dataId=62843.
45. Apolinar-Valiente, R., Williams, P., Romero-Cascales, I., *et al.* (2013) Polysaccharide composition of Monastrell red wines from four different Spanish Terroirs: effect of wine-making techniques. *Journal of Agricultural and Food Chemistry*, **61** (10), 2538–2547.
46. Koyama, K., Goto-Yamamoto, N., Hashizume, K. (2007) Influence of maceration temperature in red wine vinification on extraction of phenolics from berry skins and seeds of grape (*Vitis vinifera*). *Bioscience Biotechnology and Biochemistry*, **71** (4), 958–965.
47. Des Gachons, C.P. and Kennedy, J.A. (2003) Direct method for determining seed and skin proanthocyanidin extraction into red wine. *Journal of Agricultural and Food Chemistry*, **51** (20), 5877–5881.
48. Favre, G., Peña-Neira, Á., Baldi, C., *et al.* (2014) Low molecular-weight phenols in Tannat wines made by alternative winemaking procedures. *Food Chemistry*, **158** (0), 504–512.
49. Moreno-Perez, A., Vila-Lopez, R., Fernandez-Fernandez, J.I., *et al.* (2013) Influence of cold pre-fermentation treatments on the major volatile compounds of three wine varieties. *Food Chemistry*, **139** (1–4), 770–776.
50. Zimman, A. and Waterhouse, A.L. (2004) Incorporation of malvidin-3-glucoside into high molecular weight polyphenols during fermentation and wine aging. *American Journal of Enology and Viticulture*, **55** (2), 139–146.
51. Ough, C.S. and Amerine, M.A. (1961) Studies on controlled fermentation. V. Effects on color, composition, and quality of red wines. *American Journal of Enology and Viticulture*, **12** (1), 9–19.
52. Sener, H. and Yildirim, H.K. (2013) Influence of different maceration time and temperatures on total phenols, colour and sensory properties of Cabernet Sauvignon wines. *Food Science and Technology International*, **19** (6), 523–533.
53. Girard, B., Yuksel, D., Cliff, M.A., *et al.* (2001) Vinification effects on the sensory, colour and GC profiles of Pinot noir wines from British Columbia. *Food Research International*, **34** (6), 483–499.
54. Girard, B., Kopp, T.G., Reynolds, A.G., Cliff, M. (1997) Influence of vinification treatments on aroma constituents and sensory descriptors of Pinot noir wines. *American Journal of Enology and Viticulture*, **48** (2), 198–206.
55. Vazquez, E.S., Segade, S.R., Fernandez, I.O. (2010) Effect of the winemaking technique on phenolic composition and chromatic characteristics in young red wines. *European Food Research and Technology*, **231** (5), 789–802.
56. Ough, C.S. and Amerine, M.A. (1961) Studies with controlled fermentation. VI. Effects of temperature and handling on rates, composition and quality of wines. *American Journal of Enology and Viticulture*, **12** (3), 117–128.
57. Gomez-Plaza, E., Gil-Munoz, R., Lopez-Roca, J.M., *et al.* (2001) Phenolic compounds and color stability of red wines: effect of skin maceration time. *American Journal of Enology and Viticulture*, **52** (3), 266–270.
58. Marais, J. (2003) Effect of different wine-making techniques on the composition and quality of Pinotage wine. II. Juice/skin mixing practices. *South African Journal of Enology and Viticulture*, **24** (2), 76–79.
59. Castillo-Sanchez, J.J., Mejuto, J.C., Garrido, J., Garcia-Falcon, S. (2006) Influence of wine-making protocol and fining agents on the evolution of the anthocyanin content, colour and general organoleptic quality of Vinhao wines. *Food Chemistry*, **97** (1), 130–136.
60. Garde-Cerdan, T., Jarauta, I., Salinas, M.R., Ancin-Azpilicueta, C. (2008) Comparative study of the volatile composition in wines obtained from traditional vinification and from the Ganimede method. *Journal of the Science of Food and Agriculture*, **88** (10), 1777–1785.
61. Canals, R., Llaudy, M.D.C., Canals, J.M., Zamora, F. (2008) Influence of the elimination and addition of seeds on the colour, phenolic composition and astringency of red wine. *European Food Research and Technology*, **226** (5), 1183–1190.
62. Lee, J.M., Kennedy, J.A., Devlin, C., *et al.* (2008) Effect of early seed removal during fermentation on proanthocyanidin extraction in red wine: a commercial production example. *Food Chemistry*, **107** (3), 1270–1273.

63. Callejon, R.M., Margulies, B., Hirson, G.D., Ebeler, S.E. (2012) Dynamic changes in volatile compounds during fermentation of Cabernet Sauvignon grapes with and without skins. *American Journal of Enology and Viticulture*, **63** (3), 301–312.
64. Gomez-Miguez, M. and Heredia, F.J. (2004) Effect of the maceration technique on the relationships between anthocyanin composition and objective color of Syrah wines. *Journal of Agricultural and Food Chemistry*, **52** (16), 5117–5123.
65. Etaio, I., Elortondo, F.J.P., Albisu, M., *et al.* (2008) Effect of winemaking process and addition of white grapes on the sensory and physicochemical characteristics of young red wines. *Australian Journal of Grape and Wine Research*, **14** (3), 211–222.
66. Casassa, L.F., Beaver, C.W., Mireles, M.S., Harbertson, J.F. (2013) Effect of extended maceration and ethanol concentration on the extraction and evolution of phenolics, colour components and sensory attributes of Merlot wines. *Australian Journal of Grape and Wine Research*, **19** (1), 25–39.
67. Yokotsuka, K., Sato, M., Ueno, N., Singleton, V.L. (2000) Colour and sensory characteristics of Merlot red wines caused by prolonged pomace contact. *Journal of Wine Research*, **11** (1), 7–18.
68. Cadot, Y., Caillé, S., Samson, A., *et al.* (2012) Sensory representation of typicality of Cabernet franc wines related to phenolic composition: impact of ripening stage and maceration time. *Analytica Chimica Acta*, **732**, 91–99.
69. Caputi, L., Carlin, S., Ghiglieno, I., *et al.* (2011) Relationship of changes in rotundone content during grape ripening and winemaking to manipulation of the "peppery" character of wine. *Journal of Agricultural and Food Chemistry*, **59** (10), 5565–5571.

22

The Biochemistry of Wine Fermentations

The chapters encompassed under this heading address both the major (bio)chemical reactions that occur during alcoholic fermentation and their role in forming secondary odorants and other important compounds in wine (see the Introduction). The compounds discussed here are those that would be formed by fermentation of a media containing only basic nutrients, for example, sugars, amino acids, and essential vitamins and minerals – in other words, the compounds that are common to all alcoholic beverages. The chapters progress through the following:

- Glycolysis and Krebs cycle intermediates in the production of ethanol and organic acids
- Formation of co-enzyme A thioesters and metabolism leading to fatty acids and their esters
- Amino acid metabolism and production of higher alcohols and their acetates
- Sulfur assimilation, and particularly its role in amino acid biosynthesis and in generating noxious hydrogen sulfide
- Lactic acid bacteria metabolism of key substrates (e.g., malic acid, citric acid, sugars) and the effects of spoilage bacteria.

Understanding Wine Chemistry, First Edition. Andrew L. Waterhouse, Gavin L. Sacks, and David W. Jeffery.

22.1
Glycolysis

22.1.1 Introduction

Glycolysis refers to the metabolism of six-carbon hexoses (e.g., fructose, glucose) to pyruvate. The topic is covered in detail in biochemistry texts (e.g., Reference [1]) and only matters critical to wine chemistry are discussed in this book. Yeast will utilize glycolysis for two reasons:

- To generate energy, in the form of adenosine triphosphate (ATP)
- To produce compounds necessary for growth and other physiological functions, for example, lipids, polysaccharides, proteins, nucleic acids.

Glycolysis generates an excess of reducing compounds, particularly the reduced formed of nicotinamide adenine dinucleotide (NADH) and related metabolites. Under aerobic conditions, many eukaryotes will undergo *respiration*, in which NADH is converted back to its oxidized form (NAD^+) by an external electron acceptor, typically O_2. This metabolic process can generate a large amount of energy, approximately 36 ATP per hexose sugar. A related redox couple, nicotinamide adenine dinucleotide phosphate (NADPH) and its oxidized form ($NADP^+$) can be produced through hexose sugar oxidation (pentose phosphate pathway). In addition to energy production, these compounds have important roles as reductants, with NADH primarily involved in catabolism and NADPH in anabolism.

In *fermentation*, no external electron acceptor is involved. In humans and other mammals, pyruvate is reduced to regenerate NAD^+ during fermentation, yielding lactic acid. In yeast, pyruvate is decarboxylated to acetaldehyde, which is subsequently reduced to ethanol. The net equation for this reaction is:

$$\text{Hexose} + 2\,\text{ADP} \rightarrow 2\,\text{Ethanol} + 2\,\text{CO}_2 + 2\,\text{ATP} \qquad (22.1.1)$$

Considerably less ATP is generated during fermentation than during respiration, and most eukaryotes will rely on fermentation only under anaerobic conditions. However, *Saccharomyces* (from Greek–Latin for "sugar fungi") will commence fermentation even under aerobic conditions if sufficient glucose is present in the media.[1] The major outcome of glycolysis is, of course, production of ethanol from hexose sugars, but a portion

[1] This "aerobic fermentation" behavior, also referred to as the Crabtree effect, may have evolved in yeast as a way to provide a competitive advantage over other microorganisms due to the antiseptic properties of ethanol. Based on genetic analyses, it is reported that the trait likely evolved at the end of the Cretaceous Age around the same time as fleshy fruits [2].

Understanding Wine Chemistry, First Edition. Andrew L. Waterhouse, Gavin L. Sacks, and David W. Jeffery.

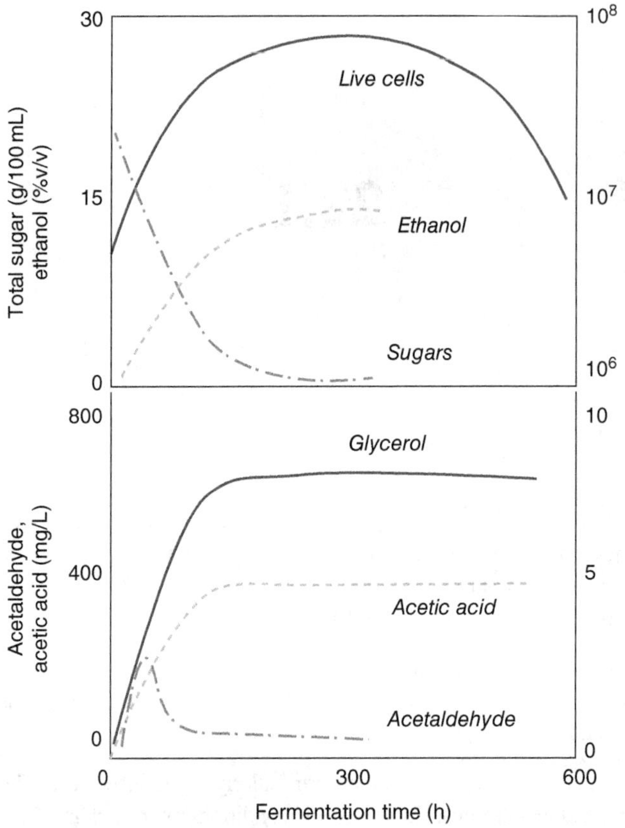

Figure 22.1.1 Change in concentration of sugars (hexose + glucose), live yeast cells, and key glycolysis metabolites during fermentation of a Chardonnay juice. Live cells increase during the growth phase, plateau during the stationary phase, and decrease during the decline phase. Note that an initial delay (lag phase) can also occur during which yeast adapt to must conditions [3]

of glycolysis products are diverted to biomass formation, yielding glycerol and acetic acid (Figure 22.1.1). Beyond this, several other minor metabolites important to flavor can be formed, and changes to the fermenting grape must will result from the reducing environment and entrainment of volatiles in CO_2 gas.

22.1.2 Glycolysis and alcoholic fermentation

S. cerevisae can grow on a limited number of carbohydrates including glucose, fructose, mannose, galactose, maltose, sucrose, and melibiose [4]. Glucose, and to a lesser extent fructose, are yeasts' preferred substrates for growth; conveniently, these are also the two major sugars in grape must. The presence of glucose in media results in major changes in yeast gene expression and metabolism, classified as either "glucose induction" or "glucose repression" [5]. Glucose induction results in increased expression of enzymes necessary for glycolysis and fermentation, while glucose repression results in decreases in enzymes necessary for respiration.

The first step of glucose and fructose utilization is the active transport of these sugars into the yeast cell by membrane-bound transporters, of which at least 17 exist in yeast [6]. Glucose transport is more efficient than

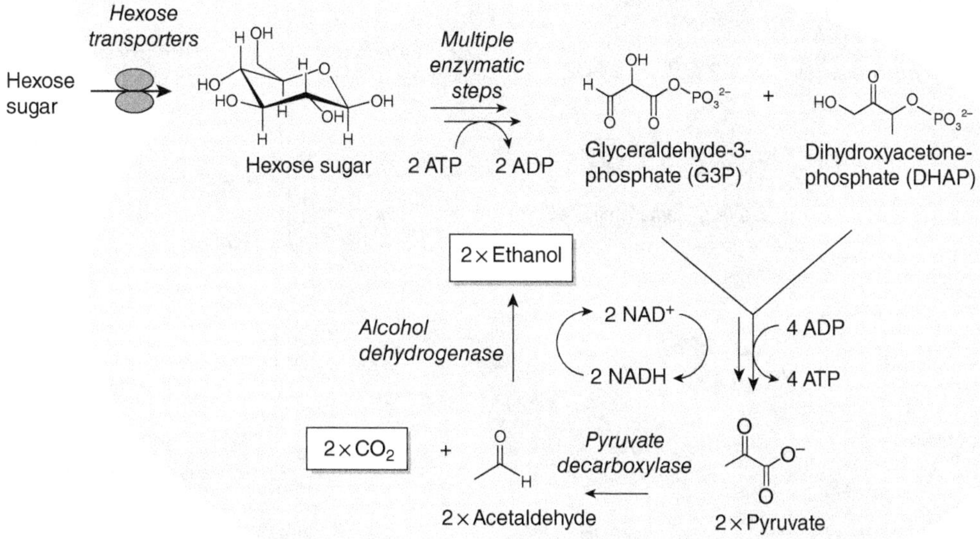

Figure 22.1.2 Overview of glycolysis (EMP pathway) and subsequent ethanol formation to restore the NADH:NAD⁺ redox balance. Utilization of hexose sugars requires active transport into the cell via hexose transporters

fructose transport in wine yeasts, and thus *S. cerevisae* is referred to as "glucophilic." Consequently, although musts start out with roughly equal concentrations of glucose and fructose, the concentration of fructose at the end of fermentation will be higher than glucose (Chapter 2).

Following transport into the cell, glucose or other hexoses will undergo glycolysis (Figure 22.1.2). The initial steps involve substrate phosphorylation to produce two C_3 compounds: dihydroxyacetone-phosphate (DHAP) and glyceraldehyde-3-phosphate (G3P). These steps consume energy in the form of two molecules of ATP. DHAP can then be converted enzymatically to G3P, and each G3P converted to pyruvate through another multistep process. These latter steps of glycolysis yield four ATP for a net energy formation of two ATP. These steps also generate two NADH and consume two NAD^+ for every equivalent of hexose. In biochemical texts, the steps from glucose to pyruvate are referred to as the Embden–Meyerhof–Parnas (EMP) pathway [1]. Under aerobic respiration conditions, the majority of pyruvate will enter the mitochondria, where it will be decarboxylated by pyruvate dehydrogenase (PDH) to form acetyl-CoA. Acetyl-CoA possesses an unstable, high-energy thioester bond (Figure 22.1.3), and thus has useful roles in the biosynthesis of several molecules, as described later in this chapter. During respiration, the majority of acetyl-CoA enters the Krebs cycle. More thorough descriptions of the EMP pathway and the Krebs cycle can be found in biochemistry textbooks (e.g., Reference [1]).

Under fermentation conditions, oxygen and other electron acceptors are not utilized. Glycolysis will therefore result in a *highly reducing environment* in which there is an excess of NADH with respect to NAD^+. To re-establish the redox balance, yeasts will enzymatically degrade pyruvate via pyruvate decarboxylase (PDC) to form acetaldehyde and CO_2 gas. Acetaldehyde is then reduced to ethanol by an alcohol dehydrogenase

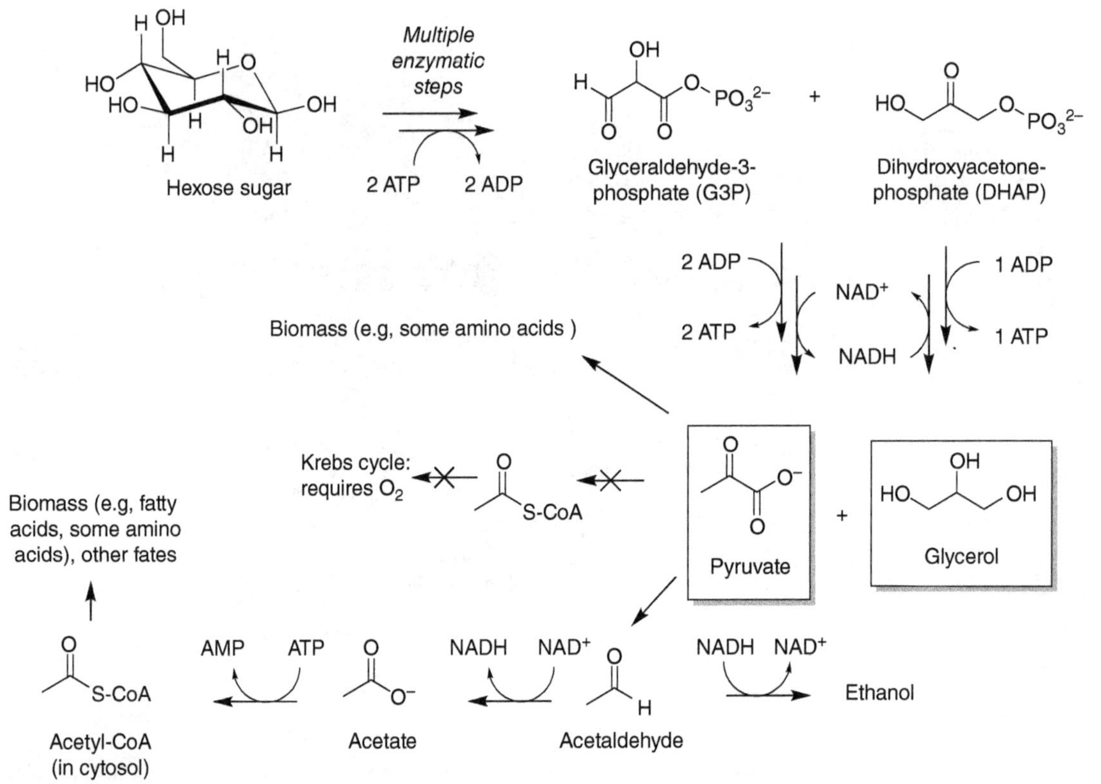

Figure 22.1.3 Overview of the glyceropyruvic pathway, showing the formation of glycerol and acetyl-CoA for utilization in biomass synthesis

(ADH), reforming NAD+ from NADH (Figure 22.1.1). CO_2 and ethanol will then diffuse from the yeast cell into the fermentation environment.

22.1.3 Glyceropyruvic fermentation

Based on the stoichiometry shown in the Gay–Lussac equation (i.e., 1 mole of hexose converts to 2 moles of ethanol and 2 moles of CO_2, Equation (22.1.1)), 1 L of grape must (1.07 kg) containing 200 g/L of hexose sugars should yield 970 mL of wine with an ethanol concentration of about 105 g/L (13.2% v/v) if fermented to dryness. Experimentally, the ethanol produced by commercial yeast strains for this example will vary from 11.8 to 12.1% v/v [7], or about 90% of the theoretical maximum.[2] Although ethanol and CO_2 represent the major products of fermentation, yeast cells must divert some carbon to form biomass compounds such as lipids, proteins, and DNA to grow, and will form other metabolites like glycerol and acetic acid as part of this process (Table 22.1.1).

[2] Predicting ethanol yield prior to fermentation in wineries is further complicated because typical sugar measurements on grapes are measured as "soluble solids," in units of Brix (g of sugar per 100 g of juice). These values are based on physical measurements like density and will usually over-report sugar concentrations by 5–10% due to the presence of other solutes like acids and minerals. Small amounts of ethanol (equivalent to 0.1–0.2% v/v) may also be lost as vapor due to CO_2 entrainment. A final complication is that sugar concentrations may be underestimated in shriveled grapes due to higher concentrations of sugar in the skins.

Table 22.1.1 Molar yield of key fermentation products from glucose (mol C of product/mol C of glucose) during exponential growth phase of an alcoholic fermentation at 28 °C and two different glucose concentrations (240 and 280 g/L) [8]

	240 g/L glucose	280 g/L glucose	Theoretical maximum (Gay–Lussac)
Ethanol	0.557	0.541	0.667
CO_2	0.292	0.287	0.333
Glycerol	0.054	0.064	
Acetic acid	0.006	0.009	
Biomass	0.086	0.095	
Carbohydrates	*0.028*	*0.025*	
Protein	*0.042*	*0.039*	
Other (e.g., succinic acid)	0.006	0.003	

Table 22.1.2 Balance of ATP energy and NADH:NAD+ ratio during alcoholic and glyceropyruvic fermentation

Sugar metabolite	Net ATP formation	NADH/NAD+ balance
Pyruvate	+2 ATP	–2 NAD+, +2 NADH
Ethanol, from pyruvate	0	+2 NAD+, –2 NADH
Total per mole of sugar during alcoholic fermentation	+2 ATP	Even
Glycerol	0 ATP	+1 NAD+, –1 NADH
Pyruvate (for biomass)	+1 ATP	–1 NAD+, +1 NADH
Total per mole of sugar during glyceropyruvic fermentation	+1 ATP	Even

Total biomass production will account for about 1% of sugar substrate that does not form CO_2 [4]. Many of the building blocks of these compounds, including fatty acids and most amino acids, will be formed from pyruvate. This diversion of carbon substrate away from acetaldehyde and ethanol is necessary for yeast growth, but creates an excess of NADH as compared to NAD+ [9]. The yeast's solution is to divert a portion of DHAP away from pyruvate formation, where it is reduced to glycerol in a two-step process (Figure 22.1.3). The coupled glycerol/pyruvate formation is referred to as *glyceropyruvic* fermentation. Glyceropyruvic fermentation generates only 1 ATP per hexose, but allows for the diversion of pyruvate to fates other than ethanol while still maintaining NADH:NAD+ balance (Table 22.1.2). For example, pyruvate can form Acetyl-CoA and fatty acids (Chapter 22.2), with concurrent production of ethanol to retain NADH:NAD+ balance (Figure 22.1.3).

Glycerol is also produced by yeast as a protectant against high osmotic stress. For example, increasing the sugar concentration of a must by roughly 1.5-fold from 224 g/L to 344 g/L resulted in a modest increase in ethanol (12.2 to 14.8%) but increased glycerol by over 50% (from 7.4 to 11.7 g/L) [3]. The increased production of glycerol would yield an excess of NAD+ to NADH if all pyruvate was converted to acetaldehyde and then to ethanol. Redox balance could be restored by formation of acetyl-CoA and its subsequent utilization of the Krebs cycle, but these pathways are not active under fermentation conditions, as described later in this chapter. Instead, a portion of acetaldehyde can be oxidized to acetic acid by *aldehyde dehydrogenase*, resulting in regeneration of NAD(P)H (Figure 22.1.3) [10]. Illustrating this point, the previous high gravity fermentation (344 g/L of sugar) produced 1.0 g/L of acetic acid, as compared to 0.39 g/L in the 224 g/L sugar fermentation. Because of this, strategies to develop genetically modified yeast for production of reduced alcohol wines by diverting substrate to glycerol have often suffered from excess acetic acid production [9].

Finally, glycerol can also be increased by the presence of high concentrations of bisulfite in the must, which will bind to acetaldehyde or other carbonyls and decrease their utilization as electron acceptors (Chapter 17).[3] As described in an earlier chapter, the practice of adding SO_2 to must to prevent browning will invariably lead to higher bound SO_2 in finished wines due to formation of bisulfite–carbonyl adducts (Chapter 9).

22.1.4 Succinic acid and other Krebs cycle intermediates

As mentioned above, glycolysis creates a strongly reducing environment with an excess of reduced vs. oxidized co-factors (e.g., an excess of NADH as compared to NAD^+). The reduction of acetaldehyde (to ethanol) and DHAP (to glycerol) represent the major routes for yeast to regenerate oxidized forms of NADH and other redox co-factors. Other minor routes to regenerate $NAD(P)^+$ include reduction of other aldehydes to alcohols, for example, higher alcohols formed during amino acid metabolism (Chapter 22.3), and reduction of acetoin to yield 2,3-butanediol (Chapter 22.5).

Another metabolite likely to be involved in recycling of redox co-factors is succinic acid. Succinic acid is formed by fermentation (Chapter 3), and at typical levels of 0.5–1.0 g/L, is often the most abundant fermentation metabolite after glycerol. Under aerobic respiration conditions, pyruvate derived from glycolysis is metabolized to acetyl-CoA, which is then incorporated into the Krebs cycle (Figure 22.1.3). The Krebs cycle uses the free energy of acetyl-CoA to generate a large excess of NADH and other reducing compounds (e.g., $FADH_2$). Eventually, these reducing compounds formed by glycolysis and the Krebs cycle can be oxidized in the electron transport chain (ETC) to generate a large amount of ATP [1]. Under fermentation conditions, several enzymes in the Krebs cycle have minimal activity, most notably *citrate lyase* and *succinic dehydrogenase* (Figure 22.1.3). As a result, the Krebs cycle acts like two independent branches under fermentation conditions, with most succinic acid being formed via either fumarate ("reductive branch") or succinyl-CoA ("oxidative branch").

In wine fermentations, the majority of succinic acid (75%) appears to be formed from the reductive branch of the Krebs cycle via pyruvate [11]. In brief, two moles of pyruvate are metabolized to form oxaloacetate, which can be subsequently converted to succinate via malate and fumarate. Oxaloacetate may also be formed from the amino acid aspartate via enzymatic transamination (Chapter 22.3), and high concentrations of aspartate in media will lead to an increase in succinic acid [12]. Additionally, a portion of grape-derived malic acid can enter the cell and be metabolized via the reductive branch to succinate [12]. The remainder of succinic acid appears to be formed from the oxidative branch of the Krebs cycle, and particularly from glutamate (Figure 22.1.4) [12]. Additionally, succinic acid can be formed via the "GABA shunt," which can convert γ-aminobutyric acid (GABA) to succinic acid at nearly a 1:1 molar ratio [13]. However, because GABA concentrations are usually at <100 mg/L (Chapter 5) this will represent a minor pathway to succinic acid in most fermentations.

Despite its high concentrations in wine and other fermented beverages, the reasons for succinate accumulation are still debated. Starting from glucose, production of succinic acid via the reductive branch results in a net production of four equivalents of NADH (reduced form), which appears to be contrary to the yeasts' need to regenerate oxidized co-factors. However, this pathway also generates two equivalents of FAD (oxidized form), which are necessary for the synthesis of fatty acids (Chapter 22.2). Yeast lacking *fumarate reductase* enzymes are unable to grow under fermentation conditions [14], further supporting the hypothesis that succinic acid formation is necessary for FAD regeneration and fatty acid production [11].

---

[3] In industrial microbiology settings, adding bisulfite to fermentations to induce glycerol formation has been used since 1915, where the glycerol was utilized in production of nitroglycerin explosives.

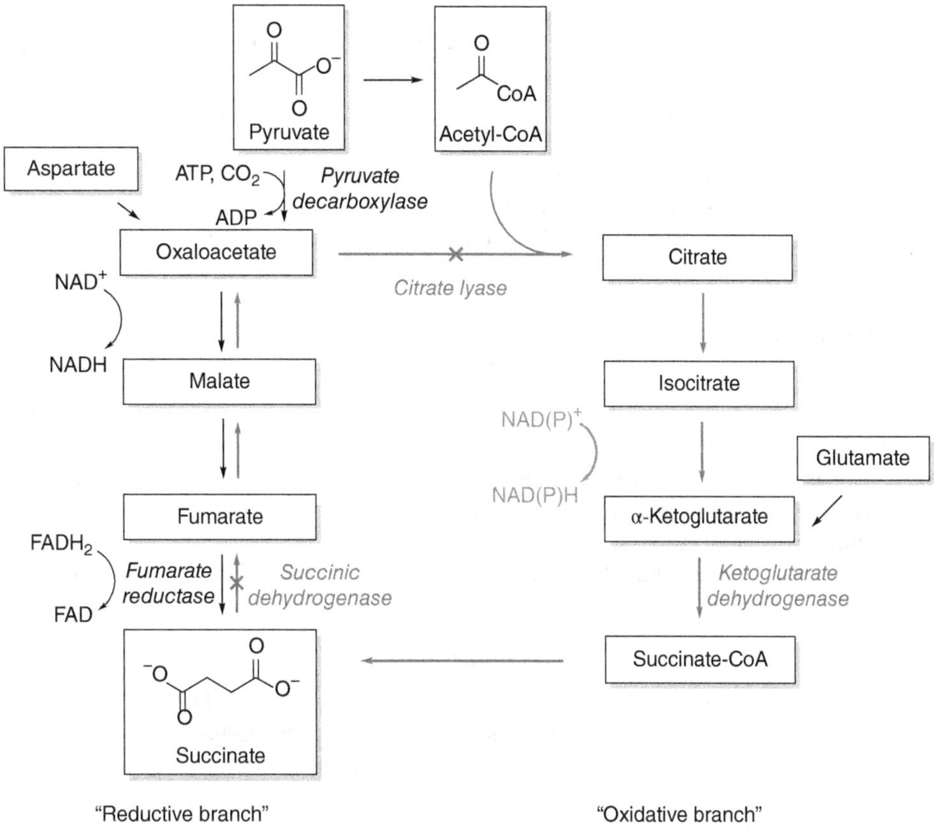

Figure 22.1.4 The Krebs cycle, aka TCA cycle or citric acid cycle. Red arrows show normal clockwise direction of the Krebs cycle under aerobic conditions, in which acetyl-CoA is incorporated into citrate as part of a cycle. Under anaerobic conditions, certain enzymes in the oxidative branch are inhibited (succinic dehydrogenase, citrate lyase), as indicated by "x" marks, and the cycle splits into two separate branches, with succinic acid accumulation primarily occurring through the reductive branch (black arrows). $NAD(P)^+/NAD(P)H$ indicates that either $NAD^+/NADH$ or $NADP^+/NADPH$ may participate

Yeast strain selection is of particular importance to succinic acid accumulation, with over an order of magnitude variation (0.2–1.7 g/L) reported across strains when fermenting the same grape must [12]. Even higher concentrations can be observed with genetically modified yeast [15]. Factors that stimulate yeast growth, such as aeration and unsaturated fatty acids, can also result in increases in succinate, and a modest correlation is observed between biomass production and succinic acid across yeast strains [16]. Also, as mentioned above, utilization of aspartate, GABA, and glutamate as nitrogen sources can also result in increased formation of succinic acid.

Other organic acid intermediates in the Krebs cycle, such as malic and citric acids, will also be formed during fermentation, but because their respective concentrations in grape juice are relatively high it is more common for these acids to decrease slightly during fermentation.[4] Conversion of malic acid to succinic acid

[4] In media with low endogenous acid concentrations, such as in sake production, succinic and other organic acids will represent the major contributors to titratable acidity.

is relatively inefficient in yeast (<5%), but may contribute slightly to elevated succinic levels in high malic acid musts [12]. A portion of malic acid may also be degraded by yeast through "maloethanolic fermentation," in which malic acid is converted back to pyruvate via oxaloacetate. The enzyme responsible for this transformation is called *malic enzyme*, and the process is comparable to glycolysis in that it generates both reduced co-factors (NAD(P)H) and pyruvate [17]. This process is inefficient in commercial *S. cereviseae* strains due to low malate transport and low *malic enzyme* activity, with a maximum malic degradation of 18% reported in one survey [18], but can approach 100% in other yeasts like *Sch. pombe*. Finally, both citric and malic acids may be metabolized by lactic acid bacteria, as described later (Chapter 22.5).

22.1.5 Consequences of glycolysis on wine chemistry

22.1.5.1 Loss of sugars and formation of fermentation metabolites

As described above, fermentation of sugars will produce a set of metabolites common to all alcoholic beverages (wine, cider, sake, etc.), most notably ethanol as a major product, but also glycerol and several organic acids as a byproduct of regenerating redox co-factors. Additional aroma compounds will be formed via acetyl-CoA, including volatile fatty acids and corresponding ethyl esters (from fatty acid metabolism, Chapter 22.2), and several higher alcohols and acetate esters (from amino acid metabolism, Chapter 22.3).

22.1.5.2 Enzymatic reduction of unsaturated compounds

The strongly reducing environment of fermentation[5] is brought about by the $NADH:NAD^+$ imbalance and the expulsion of oxygen by evolved CO_2. To restore redox balance, yeast will reduce many aldehydes (and, to a lesser extent, ketones) to their corresponding alcohols, among other redox reactions. Quantitatively, acetaldehyde is the major substrate for reduction, but seven alcohol dehydrogenase (Adh1–7*)* isozymes with a range of selectivities towards carbonyl substrates have been identified in yeast. For example, Adh1p is capable of reducing not only acetaldehyde, but also other straight-chain aldehydes containing up to ten carbons. Adh6p is capable of reducing both linear and branched-chain primary aldehydes, as well as substituted benzaldehydes and cinnamaldehydes [19]. Consequently, the concentrations of aldehydes or ketones initially present in the must or formed during early stages of fermentation will decrease by several orders of magnitude by the end of fermentation. These changes can have a major impact on wine aroma because carbonyl compounds typically have sensory thresholds 100–10 000 times lower than their corresponding alcohols (e.g., Chapter 9). Examples include:

- "Grassy" and "green" C_6 aldehydes (e.g., (*E*)-2-hexenal, hexanal) originating from enzymatic grape lipid oxidation will decrease from >500 mg/L to near undetectable concentrations (Chapter 6) [20].
- The fungal metabolite, 1-octen-3-one, imparting a "mushroom" aroma and present at concentrations up to 1 µg/L on rotten grapes, will be nearly undetectable following fermentation (Chapter 18) [21].
- Barrel fermented wines will have less intense oaky aromas than barrel-aged wines due to reduction of lignin-derived aldehydes (e.g., vanillin, furfural) to less potent alcohols (Chapters 9, 12, and 27) [22].

Aside from the usual biochemical and sensory changes related to the transformation of carbonyl compounds into alcohols in a fermenting must, the reductive nature of a fermentation is used occasionally by

[5] Note that this is the third time in this chapter that the reducing environment of fermentation has been mentioned – this is an important concept.

winemakers to "refresh" oxidized wines. To this end, when a small portion of oxidized wine is added to fresh must prior to its fermentation, carbonyl compounds present in the oxidized wine will be reduced, and the resultant wine rendered more saleable.

Some alkenes can also be reduced during fermentation, further altering the aroma profile of wine. For example, the monoterpenoid geraniol can be partially reduced to its monounsaturated and less potent analog, citronellol (Chapter 8) by the NADPH-dependent enzyme Oye2p [23].

22.1.5.3 Entrainment of volatile in CO_2

Based on the ideal gas law[6] and the stoichiometry identified in Equation (22.1.1), 1 L of must containing 200 g of hexoses/kg will produce approximately 55 L of CO_2 gas (at 25 °C) when fermented to dryness. As this gas is lost to the environment, it will strip ("entrain") volatiles from the fermenting must. The amount of a volatile lost to entrainment during fermentation can be modeled from two functions:

- The volatility of the compound, described by the Henry's Law coefficient (*H*). Generally, smaller and more non-polar compounds have higher vapor pressures. For a given compound, volatility will increase with higher temperatures and decrease with higher ethanol concentrations (Chapter 1).
- The timing of a compound's formation. For example, a compound that is present in the must initially is expected to be lost to a greater extent than a compound formed near the end of fermentation, assuming they have similar volatilities.

Differences in entrainment losses are well explained by differences in Henry's Law coefficients. For example, over 99% of the highly volatile H_2S ($H = 0.1\,\mathrm{M\,bar^{-1}}$) formed during fermentation can be lost due to entrainment. By comparison, 46% of ethyl hexanoate ($H = 3\,\mathrm{M\,bar^{-1}}$) and 0.6% of isobutanol ($H = 100\,\mathrm{M\,bar^{-1}}$) will be lost during a 20 °C fermentation [24]. Increasing the fermentation temperature to 30 °C will increase losses of ethyl hexanoate and isobutanol to 71% and 1.3%, respectively. Along with changes to yeast fatty acid biosynthesis (Chapter 22.2) these results explain the rationale behind using cooler fermentation temperatures in the production of some wines, particularly fruity white wines.

22.1.5.4 Heat formation

Fermentation is an exothermic process[7] and stoichiometric conversion of hexoses to ethanol and CO_2 will have an enthalpy change of $\Delta H = -67\,\mathrm{kJ/mol}$ hexose. In plainer terms, every 1% w/w (°Brix) of sugars fermented will raise the fermentation temperature by approximately 1.3 °C, assuming no loss of heat from the system. In practice, some heat will dissipate by convection due to the release of CO_2, but fermentation tanks are often fitted with cooling jackets to avoid excessive fermentation temperatures. Higher temperatures will have several effects on eventual wine chemistry, including:

- Increasing Henry's Law coefficients and volatilization losses, noted above
- Increasing extraction of compounds like polyphenols due to increases in extraction rates and changes to partition coefficients (Chapter 21)
- Altering formation of key metabolites like fatty acids (Chapter 22.2) and higher alcohols (Chapter 22.3).

---

[6] As a reminder $PV = nRT$; solved for V (in litres) given that $P = 1\,\mathrm{atm}$, $n \approx 2.22$ moles, $R = 0.08206\,\mathrm{L\,atm\,mol^{-1}\,K^{-1}}$ and $T = 298.15\,\mathrm{K}$ (i.e., 25 °C).

[7] The combination of heat and CO_2 generation explains the Latin root of fermentation (*fervere*, "to boil"), a mystical phenomenon until Louis Pasteur's identification of yeast as the responsible party in the 1860s.

References

1. Lehninger, A.L., Nelson, D.L., Cox, M.M. (2013) *Lehninger principles of biochemistry*, W.H. Freeman, New York.
2. Thomson, J.M., Gaucher, E.A., Burgan, M.F., *et al.* (2005) Resurrecting ancestral alcohol dehydrogenases from yeast. *Nature Genetics*, **37** (6), 630–635.
3. Frohman, C.A. and Mira de Orduña, R. (2013) Cellular viability and kinetics of osmotic stress associated metabolites of *Saccharomyces cerevisiae* during traditional batch and fed-batch alcoholic fermentations at constant sugar concentrations. *Food Research International*, **53** (1), 551–555.
4. Lea, A., Piggott, G.H., Raymond, J. (2003) *Fermented beverage production*, Kluwer Academic/Plenum, New York.
5. Horak, J. (2013) Regulations of sugar transporters: insights from yeast. *Current Genetics*, **59** (1–2), 1–31.
6. Barnett, J.A. and Entian, K.D. (2005) A history of research on yeasts. 9: Regulation of sugar metabolism. *Yeast*, **22** (11), 835–894.
7. Palacios, A., Raginel, F., Ortiz-Julien, A. (2007) Can the selection of *Saccharomyces cerevisiae* yeast lead to variations in the final alcohol degree of wines. *Australian and New Zealand Grapegrower and Winemaker*, **527**, 71–75.
8. Quirós, M., Martínez-Moreno, R., Albiol, J., *et al.* (2013) Metabolic flux analysis during the exponential growth phase of *Saccharomyces cerevisiae* in wine fermentations. *PloS One*, **8** (8), e71909.
9. Michnick, S., Roustan, J.-L., Remize, F., *et al.* (1997) Modulation of glycerol and ethanol yields during alcoholic fermentation in *Saccharomyces cerevisiae* strains overexpressed or disrupted for GPD1 encoding glycerol 3-phosphate dehydrogenase. *Yeast*, **13** (9), 783–793.
10. Eglinton, J.M., Heinrich, A.J., Pollnitz, A.P., *et al.* (2002) Decreasing acetic acid accumulation by a glycerol overproducing strain of *Saccharomyces cerevisiae* by deleting the ALD6 aldehyde dehydrogenase gene. *Yeast*, **19** (4), 295–301.
11. Camarasa, C., Grivet, J.-P., Dequin, S. (2003) Investigation by ^{13}C-NMR and tricarboxylic acid (TCA) deletion mutant analysis of pathways for succinate formation in *Saccharomyces cerevisiae* during anaerobic fermentation. *Microbiology*, **149** (9), 2669–2678.
12. Heerde, E. and Radler, F. (1978) Metabolism of the anaerobic formation of succinic acid by *Saccharomyces cerevisiae*. *Archives of Microbiology*, **117** (3), 269–276.
13. Bach, B., Sauvage, F.-X., Dequin, S., Camarasa, C. (2009) Role of γ-aminobutyric acid as a source of nitrogen and succinate in wine. *American Journal of Enology and Viticulture*, **60** (4), 508–516.
14. Arikawa, Y., Kuroyanagi, T., Shimosaka, M., *et al.* (1999) Effect of gene disruptions of the TCA cycle on production of succinic acid in *Saccharomyces cerevisiae*. *Journal of Bioscience and Bioengineering*, **87** (1), 28–36.
15. Raab, A.M., Gebhardt, G., Bolotina, N., *et al.* (2010) Metabolic engineering of *Saccharomyces cerevisiae* for the biotechnological production of succinic acid. *Metabolic Engineering*, **12** (6), 518–525.
16. Barbosa, C., Lage, P., Vilela, A., *et al.* (2014) Phenotypic and metabolic traits of commercial *Saccharomyces cerevisiae* yeasts. *AMB Express*, **4** (1), 1–14.
17. Volschenk, H., van Vuuren, H.J.J., Viljoen-Bloom, M. (2003) Malo-ethanolic fermentation in *Saccharomyces* and *Schizosaccharomyces*. *Current Genetics*, **43** (6), 379–391.
18. Redzepovic, S., Orlic, S., Majdak, A., *et al.* (2003) Differential malic acid degradation by selected strains of *Saccharomyces* during alcoholic fermentation. *International Journal of Food Microbiology*, **83** (1), 49–61.
19. De Smidt, O., Du Preez, J.C., Albertyn, J. (2008) The alcohol dehydrogenases of *Saccharomyces cerevisiae*: a comprehensive review. *FEMS Yeast Research*, **8** (7), 967–978.
20. Joslin, W.S. and Ough, C.S. (1978) Cause and fate of certain C_6 compounds formed enzymatically in macerated grape leaves during harvest and wine fermentation. *American Journal of Enology and Viticulture*, **29** (1), 11–17.
21. La Guerche, S., Dauphin, B., Pons, M., *et al.* (2006) Characterization of some mushroom and earthy off-odors microbially induced by the development of rot on grapes. *Journal of Agricultural and Food Chemistry*, **54** (24), 9193–9200.
22. Chatonnet, P., Dubourdieu, D., Boidron, J.N. (1992) Incidence of fermentation and aging conditions of dry white wines in barrels on their composition in substances yielded by oak wood. *Sciences Des Aliments*, **12** (4), 665–685.
23. Steyer, D., Erny, C., Claudel, P., *et al.* (2013) Genetic analysis of geraniol metabolism during fermentation. *Food Microbiology*, **33** (2), 228–234.
24. Mouret, J.R., Morakul, S., Nicolle, P., *et al.* (2012) Gas–liquid transfer of aroma compounds during winemaking fermentations. *LWT - Food Science and Technology*, **49** (2), 238–244.

22.2

Fatty Acid Metabolism

22.2.1 Introduction

Lipids are a diverse group of biomolecules with the common property of being soluble in non-polar solvents like chloroform: they represent about 7–15% of yeast biomass by dry weight [1]. While lipids constitute a broad class of compounds, fatty acids are of particular quantitative and sensory importance. In wine yeasts, the majority of fatty acids (>99%) are esterified with glycerol to form mono-, di-, and tri-acylglycerides, or are esterified with glycerophosphate groups to form glycerophospholipids (also called phospholipids, PL), with only a small amount existing as free fatty acids (FFAs) [2] (Figure 22.2.1). Although the mid-chain fatty acids (MCFAs) (Chapter 3) and their esters (Chapter 7) have the most profound effects on wine organoleptic properties, the majority of fatty acids in yeast PL are long-chain, particularly the saturated fatty acids (SFAs) palmitate (16 carbons, zero double bonds, designated 16:0) and stearate (18:0), and the unsaturated fatty acid (UFA) oleate (18:1) (Figure 22.2.1). The cell membrane of yeasts and other eukaryotes is composed largely (70%) of PL, which are organized into the well-known membrane bilayer structure in which the polar "heads" face the aqueous intracellular or extracellular matrices and the hydrophobic tails face each other in the bilayer interior (Figure 22.2.2) [3, 4]. The resulting cell membrane – approximately 5–9 nm thick – is largely impervious to passive diffusion of polar components.

22.2.2 Long-chain fatty acid metabolism

A detailed account of long-chain fatty acid biosynthesis can be found in biochemistry textbooks [5, 6]. The description in this chapter will focus only on key details that eventually impact wine flavor chemistry; the steps of fatty acid biosynthesis critical to an understanding of this chemistry are summarized in Figure 22.2.3, as adapted from elsewhere [7].

Understanding Wine Chemistry, First Edition. Andrew L. Waterhouse, Gavin L. Sacks, and David W. Jeffery.

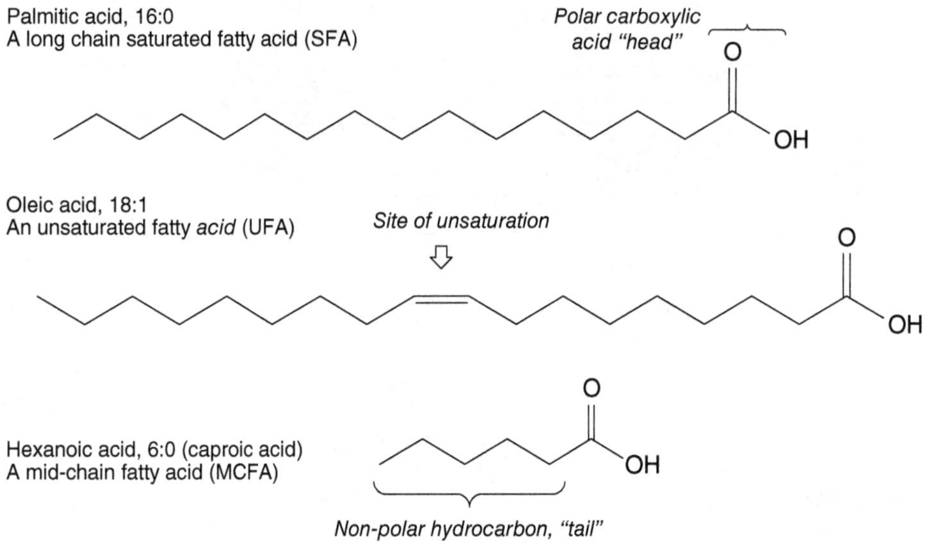

Figure 22.2.1 *Structures of representative free fatty acids (FFA) produced by yeast*

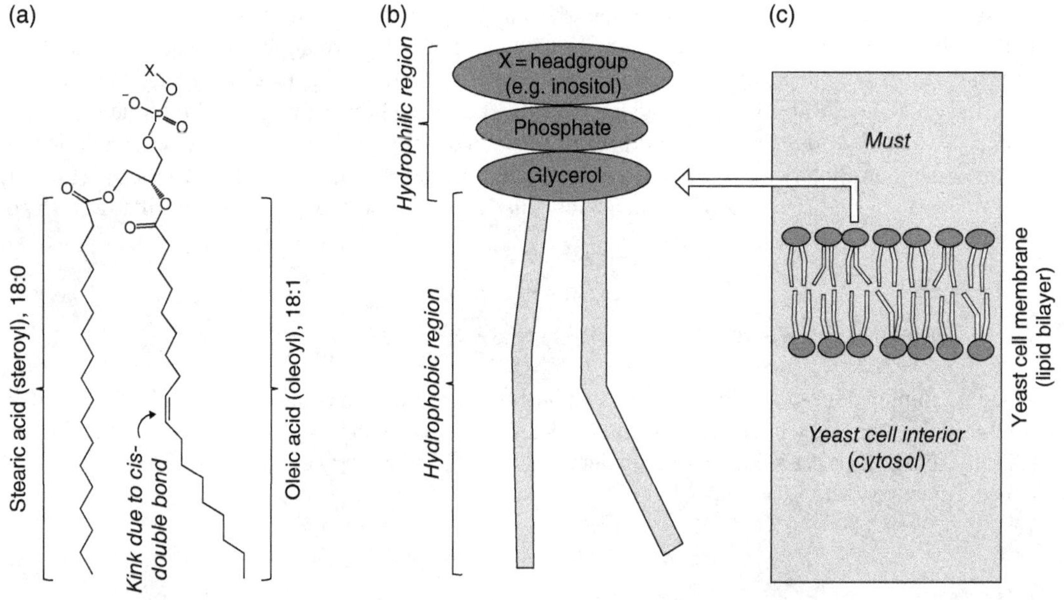

Figure 22.2.2 *(a) Structure and (b) cartoon of a representative phospholipid, showing the effect of saturated (stearic) and unsaturated (oleic) fatty acids on conformation, and the location of the hydrophobic and hydrophilic component, and (c) cartoon of the lipid membrane bilayer showing the separation of must and yeast cytosolic components. The presence of unsaturated fatty acids results in less tight packing of the bilayer. Adapted from Reference [4]*

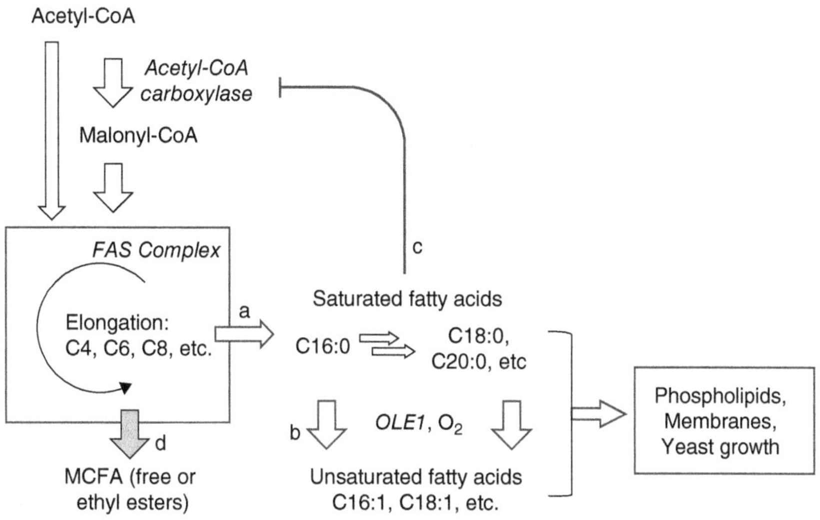

Figure 22.2.3 *Overview of fatty acid biosynthesis in yeast. (a) Palmitate (16:0) is synthesized in the fatty acid synthase (FAS) complex from acetyl-CoA and malonyl-CoA substrates in sequential 2-carbon elongations steps. (b) Following release from FAS, 16:0 can be elongated or desaturated, and used as substrates for formation of phospholipids (PL) during the growth phase. (c) Under anerobic conditions, desaturation is not possible and PL synthesis stops, resulting in accumulation of saturated fatty acids (SFA), which inhibit the initial acetyl-CoA formation steps. (d) This arrests activity in the FAS, resulting in release of intermediate mid-chain fatty acids (MCFA)*

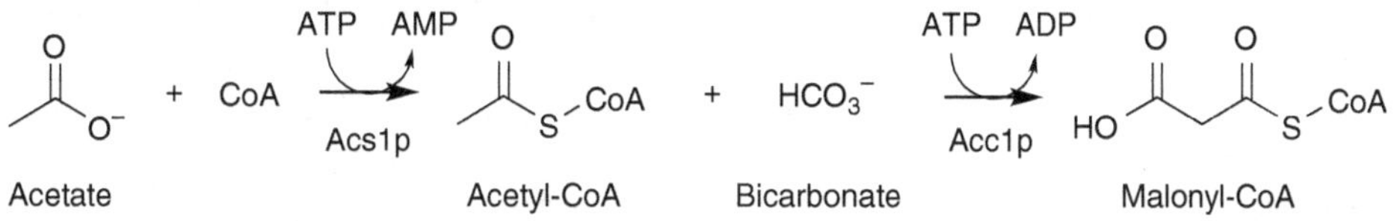

Figure 22.2.4 *ATP-dependent formation pathway for acetyl-CoA and malonyl-CoA, the building blocks of fatty acids. Acetyl-CoA formation is catalyzed by acetyl-CoA synthase (Acs1p) starting from acetic acid and coenzyme A (CoA). Malonyl-CoA is subsequently formed from acetyl-CoA and bicarbonate via acetyl-CoA carboxylase (Acc1p)*

22.2.2.1 Initial steps – formation of coenzyme A thioesters

Initial and intermediate steps of fatty acid biosynthesis utilize coenzyme A (CoA), a co-factor with a reactive sulfhydryl (–SH) group. The sulfhydryl group can be enzymatically esterified with carboxylic acids to yield thioester compounds, including acetyl-CoA (Figure 22.2.4). Acetyl-CoA was previously mentioned in the discussion of glycolysis (Chapter 22.1). While acetyl-CoA can be formed directly from pyruvate, this transformation occurs within the mitochondria and the resulting acetyl-CoA is not available for fatty acid synthesis in the cytosol under fermentation conditions [8]. Instead, the major source of acetyl-CoA during alcoholic fermentation is acetic acid, formed by oxidation of acetaldehyde. Acetyl-CoA can subsequently be transformed to malonyl-CoA (Figure 22.2.3).

22.2.2.2 Next steps – synthesis of palmitate by fatty acid synthase

Following synthesis of acetyl- and malonyl-CoA, palmitate (16:0) can be formed in the multienzyme *fatty acid synthase complex* (FAS, Figure 22.2.3). The pathway begins with the enzymatic elongation of acetyl-CoA to malonyl-CoA to form a 4-carbon intermediate (Figure 22.2.4). Analogous elongation steps by additional malonyl-CoA equivalents result in formation of intermediates that are 6-carbon, 8-carbon, etc., before eventual release of palmitate. Because palmitate and other fatty acids are synthesized in sequential steps involving two-carbon "chunks", the majority of fatty acids in yeast and other organisms are straight-chained and even-numbered. The reaction requires reducing equivalents (NADPH) derived primarily from the pentose phosphate pathway, which is also responsible for formation of nucleic acids [6]. The net reaction for formation of palmitate is

$$\begin{aligned}&\text{Acetyl}-\text{CoA}+7\,\text{Malonyl}-\text{CoA}+14\,\text{NADPH}+14\,\text{H}^{+}\rightarrow\text{Palmitate}(16:0)\\&+7\,\text{CO}_2+8\,\text{CoA}+14\,\text{NADP}^{+}+6\,\text{H}_2\text{O}\end{aligned} \tag{22.2.1}$$

22.2.2.3 Final steps – elongation and desaturation

Following palmitate synthesis, long-chain FFA with >16-carbons can be formed by elongation outside of the FAS complex and unsaturated fatty acids (particularly 18:1, oleate) can be formed by oxygen-dependent desaturase enzymes such as Ole1p [9]. Concentrations of unsaturated fatty acids may be close to 70% in commercial dry yeasts produced in (semi) aerobic conditions, but will be at much lower concentrations under anaerobic winemaking conditions [10]. Polyunsaturated fatty acids like linoleate (18:2) or linolenate (18:3) are not produced in significant concentrations by *S. cerevisiae* (<2% of fatty acids), even when oxygen is not limiting, although significant amounts may be formed by non-*Saccharomyces* yeasts [1].

Branched-chain and odd-numbered fatty acids can be formed by an analogous pathway to straight-chain fatty acids by substitution of an appropriate acyl-CoA group in place of acetyl-CoA [11]. For example, isovaleryl-CoA can serve as a starting point for longer branched-chain fatty acids. This compound can be formed by CoA acylation of isovaleric acid formed via Ehrlich degradation and subsequent oxidation of leucine, in a manner analogous to the formation of higher alcohols (Chapter 22.3). Fatty acids with an odd number of carbons can be formed starting from propionyl-CoA, likely formed via α-ketobutyric acid (2-oxobutanoic acid) as an intermediate in the Ehrlich degradation of threonine [12].

During the growth stage of fermentation, FFA will be converted to their corresponding fatty acid-CoA thioesters and incorporated into the PL and membrane bilayers of growing yeast cells (Figure 22.2.2). Throughout fermentation, yeast will adapt the FFA composition of their membranes to suit their changing environment [13]. In particular, the optimal functioning of membrane proteins and other cell membrane components is believed to require an appropriately fluid bilayer, with the term "membrane fluidity" referring to the degree of disorder [14]. Membrane fluidity increases roughly with the decreasing melting points of constituent fatty acids (i.e., the higher melting SFA have lower membrane fluidity), and will also increase with greater concentrations of sterols.[1] For a given bilayer composition, lower temperatures will result in a decrease in membrane fluidity. Poikilothermic organisms like yeast (i.e., internal temperature varies with that of the surrounding medium) will attempt to counteract this fluidity loss by modifying their membrane composition to regain the original fluidity properties. A common response of yeasts is to increase the ratio of lower-melting unsaturated fatty acids (UFA) to higher-melting saturated fatty acids (SFA),[2] and under cooler

[1] Sterol is an abbreviation of "steroid alcohol" and represents the other major class of lipids found in cell membranes other than phospholipids. Many organisms will produce sterols to modify lipid membrane function, but the specific structures will vary among kingdoms – yeast and other fungi produce ergosterols, plants produce phytosterols, and mammals (including humans) produce cholesterol. As described later in this chapter, yeast can incorporate plant phytosterols into their membranes, but the effects of cholesterol additions on fermentation efficiency have not been evaluated.

[2] This difference in melting behavior can also be observed between olive oil (mostly oleic acid – 18 carbons, 1 double bond, m.p. 13 °C) as compared to butter (mostly stearic acid – 18 carbons, no double bonds, m.p. 70 °C).

fermentation conditions (13 °C versus 25 °C) the UFA/SFA ratio in *S. cerevisiae* roughly doubles [10].[3] The presence of ethanol results in a similar increase of UFA/SFA in yeast [16], and thus the UFA content will often increase during fermentation [13].

Because of the benefits of UFA to membrane fluidity, the UFA/SFA ratio of commercially purchased *S. cerevisiae* produced under aerobic conditions may be 2:1 or greater [1]. However, UFA formation requires oxygen, as described above, and under typical fermentation conditions UFA/SFA will be closer to 1:1. The limited availability of UFA may eventually constrain PL biosynthesis and yeast growth (Figure 22.2.3), especially if growth is not limited by other nutritional deficiencies (e.g., insufficient nitrogen). As a result, FFA synthesis and yeast growth will often cease after the first 3–6 days of fermentation, especially under highly anaerobic conditions.

22.2.3 Mid-chain fatty acids (MCFAs) and ethyl esters

The major products of fatty acid metabolism – long-chain fatty acids and phospholipids – are only sparingly soluble in aqueous environments, possess low volatility, and are of negligible impact on finished wines. Conversely, mid-chain fatty acids (MCFAs, 4–12 carbons), which are minor byproducts of fatty acid metabolism, along with their corresponding esters, can have a significant flavor impact on wine due to their greater solubility and volatility.

The accumulation of MCFA during fermentation is hypothesized to be correlated with depletion of UFA and sterols, and the arrest of fatty acid biosynthesis [7], as described above. This will result in accumulation of long-chain saturated acyl-CoA compounds, which inhibits the initial stages of fatty acid synthesis. Under these conditions, MCFAs are released from the FAS complex in free form and/or as ethyl esters, which are subsequently excreted from the yeast cell [17].[4]

MCFAs are toxic to yeast and other microorganisms, and at sufficient concentrations can result in stuck or sluggish fermentations [20, 21]. MCFAs can permeate the yeast cell membrane at wine pH and cause intracellular acidification, and may also be incorporated into and adversely affect the properties of the membrane. Addition of UFA or sterols to stuck and sluggish fermentations can often increase the fermentation rate [22], likely by restarting phospholipid biosynthesis and removing MCFA from the fermentation media. In commercial wine production, these effects can be achieved through aeration [23] or by addition of yeast hulls (or yeast "ghosts," i.e. yeast cell wall material).

22.2.3.1 Ethyl esters of MCFA

The ethyl esters of MCFA are key contributors to "fruity" aromas in wines. As described earlier (Chapter 7), the expected molar ratio of an MCFA and its corresponding ethyl ester in a table wine-like matrix is approximately 6:1, and this equilibrium will slowly be approached during storage due to acid-catalyzed esterification and hydrolysis reactions. However, during fermentation, MCFA-ethyl esters can also be formed enzymatically by condensation of MCFA-CoA with ethanol [24] (Figure 22.2.5). Esterification reactions

[3] The inverse correlation of the UFA/SFA ratio and decreasing temperature extends to higher animals. Farkas *et al.* [15] observed that the ratio of stearic to oleic acid in the brain tissue of birds, fish, and mammals was well-correlated with organism body temperature over a range of 5–41 °C.

[4] The reason for MCFA release under these conditions is still debated, but may be a strategy for yeast to regenerate CoA. An alternative hypothesis for the release of MCFA is that they could substitute for UFA under conditions that require an increase in membrane fluidity [18]. However, this seems unlikely, as the majority of MCFA are released extracellularly [17], and no increase in MCFA content of yeast cell membranes is observed at cooler temperatures [19].

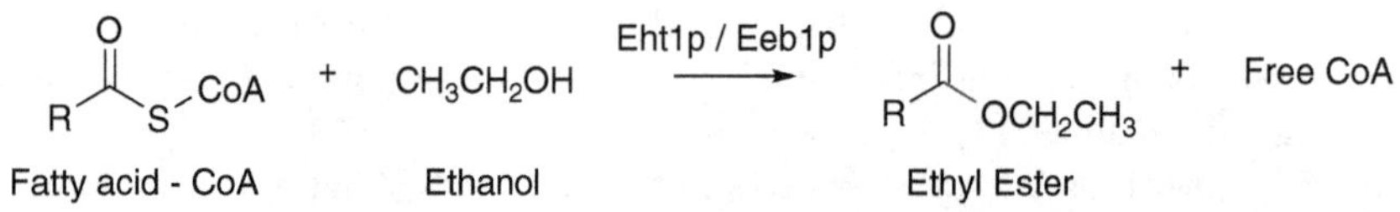

Figure 22.2.5 *Enzymatic formation of an ethyl ester from fatty acid–CoA and ethanol*

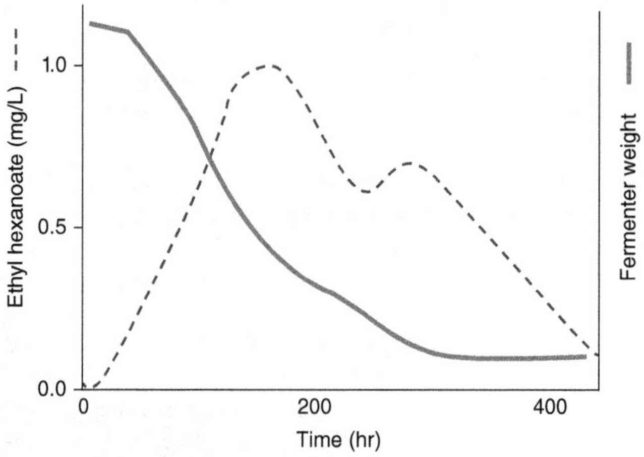

Figure 22.2.6 *Concentration of ethyl hexanoate in a fermentation as a function of fermentation time. Fermentation progress and sugar consumption are indicated by the decrease in fermenter weight. Data from Reference [26]*

involving MCFA-CoAs are energetically favorable as compared to those involving free MCFA, and as a result the ratios of ethyl esters to free acids at the end of fermentation are frequently in excess of equilibrium predictions (Chapter 7).

Two acyl-CoA:ethanol *O*-acyltransferase enzymes, Eht1p and Eeb1p, encoded by the genes *EHT1* and *EEB1*, respectively, have been characterized in yeast, and strains lacking these genes produce 10-fold lower concentrations of ethyl hexanoate and related ethyl esters [24]. However, gene overexpression results in only minor increases in ethyl esters (<50%, [25]) or no enhancement at all [24], possibly because Eht1p and Ehb1p also possess ester hydrolysis side activity.

22.2.3.2 Timing of MCFA and MCFA ethyl ester production

Under typical fermentation conditions, MCFA and MCFA ethyl ester production commences with yeast growth, and will peak at the end of the growth phase. Both MCFA and MCFA ethyl esters will decrease during the stationary phase once the majority of sugars are consumed, possibly because cell death results in release of UFA which can restart fatty acid biosynthesis in the FAS complex [7]. Exemplary data, demonstrating the peak and decline for ethyl hexanoate during fermentation, is shown in Figure 22.2.6. The second peak corresponds with the expected start of the decline phase (cell death) and may be indicative of a release in intracellular ethyl hexanoate following yeast cell autolysis.

22.2.4 Increasing MCFA and their ethyl esters in winemaking

Because MCFA ethyl ester concentrations are not highly sensitive to enzymatic activity, their concentrations in wine are largely dependent on MCFA concentrations [7]. This behavior is in contrast to acetate esters, whose formation is highly dependent on enzymatic activity (Atf1p and others) and less dependent on substrate concentration (Chapter 22.3). Winemakers interested in increasing MCFA production during fermentation can achieve this goal by increasing UFA demand and decreasing UFA supply as follows:

- Anaerobic conditions prevent enzymatic formation of UFA and will arrest fatty acid synthesis, resulting in greater MCFA release [27].
- Cooler temperatures increase the demand for UFA, resulting in earlier cessation of long-chain fatty acid synthesis and greater MCFA release (Figure 22.2.7).
- Ceasing fermentation before dryness can prevent further metabolism of MCFA ethyl esters (Figure 22.2.6), and presumably their corresponding MCFA, too.
- Decreasing sources of UFA or sterols, e.g. clarifying to remove grape solids, since these compounds will restart long-chain fatty acid synthesis and decrease MCFA release [28].
- Yeast strains can differ by roughly an order of magnitude in MCFA production under similar fermentation conditions, possibly because of differences in UFA requirements (Figure 22.2.7).

These fermentation conditions – lower oxygen, cooler temperatures, clarification to remove grape solids, appropriate yeast strain selection – are the standard tools used in white winemaking for the production of fruitier wines.[5]

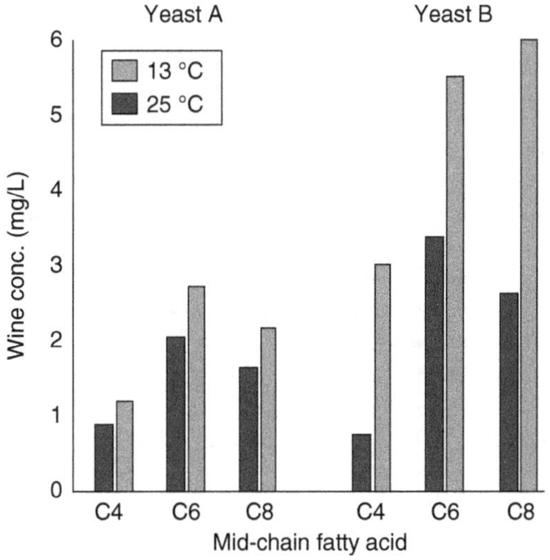

Figure 22.2.7 Effects of fermentation temperature (13 °C versus 25 °C) and yeast strain (A versus B) on MCFA concentrations in wine. C4 = butyric acid, C6 = hexanoic acid, C8 = octanoic acid. Data from Reference [10]

[5] One challenge for winemakers is that cool, low-solids, anaerobic conditions are also more stressful to yeast, and therefore present a greater risk of stuck and sluggish fermentations.

The effects of these fermentation parameters on MCFA ethyl esters immediately after fermentation are often complicated by physiochemical factors. For example, it is reported that increasing fermentation temperatures from 14 °C to 26 °C increases ethyl decanoate by a factor greater than 2, although ethyl hexanoate is unaffected, purportedly because diffusion of the more non-polar ethyl decanoate out of the cell is limited at cooler temperatures [29]. Cooler temperatures will also decrease losses of volatiles during fermentation due to CO_2 entrainment. Due to their low volatility, only negligible amounts of MCFA are expected to be lost through CO_2 entrainment. However, significant losses due to volatilization – over 50% – are possible for ethyl esters of MCFA during fermentation (Chapter 22.1). Cooler fermentation temperatures can have both physiochemical effects (less volatilization) and physiological effects (enhanced yeast MCFA biosynthesis), and the relative importance of each effect is not clear. During storage, the relative ratios of MCFA and MCFA ethyl esters will move towards equilibrium, and factors that affect MCFA production during fermentation are expected to be of greater long-term importance.

References

1. Halasz, A. and Lasztity, R. (1990) *Use of yeast biomass in food production*, Taylor & Francis.
2. Ramsay, A.M. and Douglas, L.J. (1979) Effects of phosphate limitation of growth on the cell-wall and lipid composition of *Saccharomyces cerevisiae*. *Journal of General Microbiology*, **110** (1), 185–191.
3. de Kroon, A.I.P.M., Rijken, P.J., De Smet, C.H. (2013) Checks and balances in membrane phospholipid class and acyl chain homeostasis, the yeast perspective. *Progress in Lipid Research*, **52** (4), 374–394.
4. Alberts, B., Johnson, A., Lewis, J. (2002) *Molecular biology of the cell*, 4th edn, Garland Science, New York.
5. Stipanuk, M.H. (2013) *Biochemical, physiological, and molecular aspects of human nutrition*, Elsevier/Saunders, St. Louis, MO.
6. Lehninger, A.L., Nelson, D.L., Cox, M.M. (2013) *Lehninger principles of biochemistry*, W.H. Freeman, New York.
7. Saerens, S.M.G., Delvaux, F.R., Verstrepen, K.J., Thevelein, J.M. (2010) Production and biological function of volatile esters in *Saccharomyces cerevisiae*. *Microbial Biotechnology*, **3** (2), 165–177.
8. Chen, Y., Siewers, V., Nielsen, J. (2012) Profiling of cytosolic and peroxisomal acetyl-CoA metabolism in *Saccharomyces cerevisiae*. *PloS One*, **7** (8), e42475.
9. Sajbidor, J. (1997) Effect of some environmental factors on the content and composition of microbial membrane lipids. *Critical Reviews in Biotechnology*, **17** (2), 87–103.
10. Torija, M.J., Betran, G., Novo, M., *et al.* (2003) Effects of fermentation temperature and *Saccharomyces* species on the cell fatty acid composition and presence of volatile compounds in wine. *International Journal of Food Microbiology*, **85** (1–2), 127–136.
11. Horning, M.G., Martin, D.B., Karmen, A., Vagelos, P.R. (1960) Synthesis of branched-chain and odd-numbered fatty acids from malonyl-CoA. *Biochemical and Biophysical Research Communications*, **3** (1), 101–106.
12. Luttik, M.A.H., Kötter, P., Salomons, F.A., *et al.* (2000) The *Saccharomyces cerevisiae* ICL2 gene encodes a mitochondrial 2-methylisocitrate lyase involved in propionyl-coenzyme A metabolism. *Journal of Bacteriology*, **182** (24), 7007–7013.
13. Henderson, C.M. and Block, D.E. (2014) Examining the role of membrane lipid composition in determining the ethanol tolerance of *Saccharomyces cerevisiae*. *Applied and Environmental Microbiology*, **80** (10), 2966–2972.
14. Los, D.A. and Murata, N. (2004) Membrane fluidity and its roles in the perception of environmental signals. *Biochimica et Biophysica Acta (BBA) – Biomembranes*, **1666** (1–2), 142–157.
15. Farkas, T., Kitajka, K., Fodor, E. *et al.* (2000) Docosahexaenoic acid-containing phospholipid molecular species in brains of vertebrates. *Proceedings of the National Academy of Sciences*, **97** (12), 6362–6366.
16. Weber, F.J. and de Bont, J.A.M. (1996) Adaptation mechanisms of microorganisms to the toxic effects of organic solvents on membranes. *Biochimica et Biophysica Acta (BBA) – Reviews on Biomembranes*, **1286** (3), 225–245.
17. Bardi, L., Cocito, C., Marzona, M. (1999) *Saccharomyces cerevisiae* cell fatty acid composition and release during fermentation without aeration and in absence of exogenous lipids. *International Journal of Food Microbiology*, **47** (1–2), 133–140.

18. Chapman, D. and Hoffmann, W. (1980) Enzyme function and membrane lipids. *Biochemical Society Transactions*, **8** (1), 32–34.
19. Torija, M.J., Beltran, G., Novo, M. *et al.* (2003) Effects of fermentation temperature and *Saccharomyces* species on the cell fatty acid composition and presence of volatile compounds in wine. *International Journal of Food Microbiology*, **85** (1), 127–136.
20. Fugelsang, K.C. and Edwards, C.G. (2007) *Wine microbiology practical applications and procedures*, Springer, New York.
21. Viegas, C.A., Rosa, M.F., Sa-Correia, I., Novais, J.M. (1989) Inhibition of yeast growth by octanoic and decanoic acids produced during ethanolic fermentation. *Applied and Environmental Microbiology*, **55** (1), 21–28.
22. Landolfo, S., Zara, G., Zara, S., *et al.* (2010) Oleic acid and ergosterol supplementation mitigates oxidative stress in wine strains of *Saccharomyces cerevisiae*. *International Journal of Food Microbiology*, **141** (3), 229–235.
23. Sablayrolles, J.-M., Dubois, C., Manginot, C., *et al.* (1996) Effectiveness of combined ammoniacal nitrogen and oxygen additions for completion of sluggish and stuck wine fermentations. *Journal of Fermentation and Bioengineering*, **82** (4), 377–381.
24. Saerens, S.M., Verstrepen, K.J., Van Laere, S.D., *et al.* (2006) The *Saccharomyces cerevisiae* EHT1 and EEB1 genes encode novel enzymes with medium-chain fatty acid ethyl ester synthesis and hydrolysis capacity. *Journal of Biological Chemistry*, **281** (7), 4446–4456.
25. Lilly, M., Bauer, F.F., Lambrechts, M.G., *et al.* (2006) The effect of increased yeast alcohol acetyltransferase and esterase activity on the flavour profiles of wine and distillates. *Yeast*, **23** (9), 641–659.
26. Vianna, E. and Ebeler, S.E. (2001) Monitoring ester formation in grape juice fermentations using solid phase microextraction coupled with gas chromatography–mass spectrometry. *Journal of Agricultural and Food Chemistry*, **49** (2), 589–595.
27. Dufour, J.P., Malcorps, P.H., Silcock, P. (2008) Control of ester synthesis during brewery fermentation, in *Brewing yeast fermentation performance*, Blackwell Science, pp. 213–233.
28. Edwards, C.G., Beelman, R.B., Bartley, C.E., McConnell, A.L. (1990) Production of decanoic acid and other volatile compounds and the growth of yeast and malolactic bacteria during vinification. *American Journal of Enology and Viticulture*, **41** (1), 48–56.
29. Saerens, S.M.G., Delvaux, F., Verstrepen, K.J., *et al.* (2008) Parameters affecting ethyl ester production by *Saccharomyces cerevisiae* during fermentation. *Applied and Environmental Microbiology*, **74** (2), 454–461.

22.3
Amino Acid Metabolism

22.3.1 Introduction

The biosynthesis of amino acids and their polymers (i.e., proteins) essential for yeast growth represents a major nitrogen sink during fermentation. Proteins are necessary for catalysis of key metabolic pathways (enzymes), and glycoproteins act as a structural component of yeast cell walls (mannoproteins). As compared to other nutrients such as carbohydrates, sulfur, and phosphorus, grape must usually contains low concentrations of nitrogen in a form that is assimilable (i.e., able to be incorporated) by yeast. Thus, available nitrogen is often a limiting factor for both yeast growth and fermentation rate, and its deficiency can result in suboptimal (sluggish or stuck) fermentations and off-aroma formation. Winemakers can address this inadequacy to an extent by supplementing musts with nitrogen (usually an inorganic form), but overcorrection may lead to microbial instability, production of toxic compounds, or undesirable changes to wine flavor.

22.3.2 Nitrogen uptake and catabolite repression

As described earlier (Chapter 5) and listed below, grape must contains several classes of nitrogenous compounds, but not all are equally useful to yeast as a nitrogen source under anaerobic conditions [1].

- Ammonium (NH_4^+) and most primary amino acids (e.g., leucine, glutamate, valine) can be utilized by yeast under fermentation conditions. The sum of these sources (i.e., inorganic nitrogen and *free amino nitrogen*, FAN) is referred to as *yeast assimilable nitrogen* (YAN).
- Secondary amino acids (proline and hydroxyproline) are not well utilized under anaerobic conditions because metabolism of these compounds requires oxygen. As an example, <10% proline was consumed as compared to >90% of the primary amino acids in a Chardonnay fermentation [2].
- Proteins and oligopeptides (with the exception of glutathione) are typically not well utilized as nitrogen sources, since *S. cerevisiae* has low protease activity at wine pH. However, certain yeast strains (particularly spoilage yeast like *Candida*) possess aspartic acid-rich proteases with higher activity under acidic conditions [3].

Understanding Wine Chemistry, First Edition. Andrew L. Waterhouse, Gavin L. Sacks, and David W. Jeffery.

YAN compounds do not enter yeast cells by passive diffusion, but are instead actively taken up through membrane proteins called *permeases*. These include *general amino acid permease* (Gaplp), which has a broad affinity for a range of amino acids, and several selective permeases with high affinity for particular nitrogen sources; for example, Mep1p, Mep2p, and Mep3p are selective for NH_4^+ [4,5]. Although yeast can grow on a large range or inorganic and organic nitrogen sources, not all forms of YAN in must are equally preferred, and different sources will be utilized at different rates [5]. Classically, assignment of preferred versus non-preferred nitrogen sources was based on relative growth rates or rates of amino acid depletion, but a more modern approach is to classify a nitrogen source as preferred if it decreases the expression of genes necessary for uptake or metabolism of other nitrogen sources, for example, *GAP1* [6]. These changes are collectively referred to as nitrogen regulation or *nitrogen catabolite repression* (NCR) [5].

The preferred sources that cause the strongest NCR effects are asparagine and glutamine, followed by NH_4^+ and glutamate [6]. The advantage of these nitrogen sources to yeast appears to be related to the fact that they have a central role in amino acid metabolism, and thus require fewer intermediary steps to be utilized for *de novo* synthesis of amino acids (Figure 22.3.1). More thorough treatments of yeast amino acid synthesis can be found elsewhere [1,5,7], and are summarized here. Briefly, glutamate (85%) and glutamine (15%) serve as the major nitrogen donors for amino acid biosynthesis. These two compounds can be regenerated from α-ketoglutarate and glutamate, respectively (Figure 22.3.1). Alternatively, other α-amino acids can act as nitrogen donors to regenerate glutamate from α-ketoglutarate, as described in the next paragraph. In combination, these reactions function as a continuous cycle to convert generic nitrogen sources into a desired amino acid product by incorporation of nitrogen into carbon skeletons, as described next.

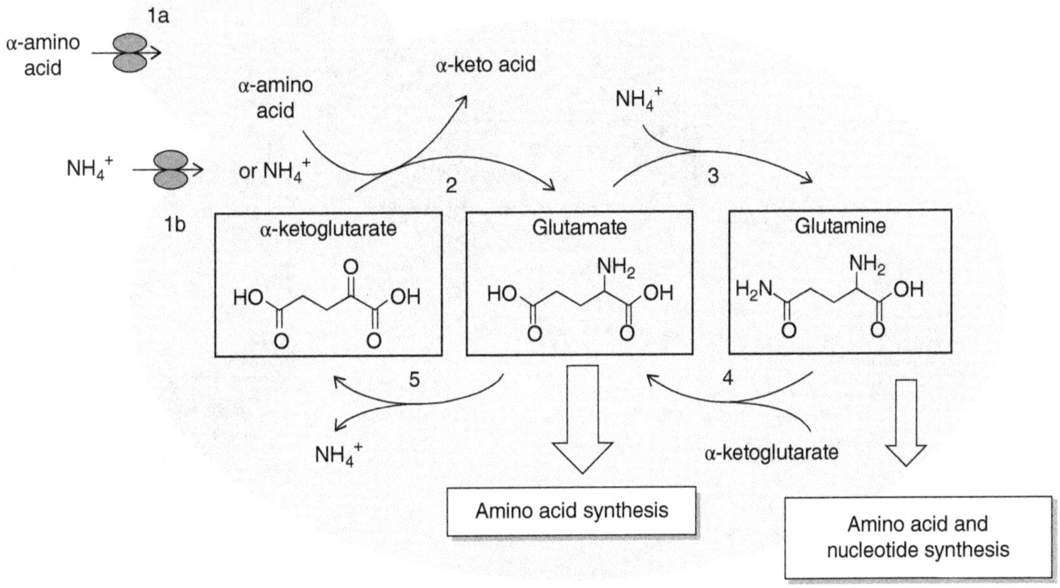

Figure 22.3.1 Schematic overview of nitrogen uptake and metabolism in yeast. 1a, Amino acid permease (e.g., Gap1p); 1b, ammonium permease (Mep1p/2p/3p); 2, aminotransferase (Bat1p/2p, Gdh1p); 3, glutamine synthetase (Gln1p); 4, glutamate synthetase, GOGAT (Glt1p); 5, glutamate dehydrogenase (Gdh2p)

22.3.3 Amino acid anabolism, catabolism, and carbon skeletons

Unlike humans who require certain "essential" amino acids in their diets, yeast can synthesize all amino acids necessary for their function and growth if they are provided with carbon and nitrogen sources and necessary co-factors. The anabolic pathway for amino acid biosynthesis begins with glucose and leads to a penultimate step of α-keto acid formation (also called a "carbon skeleton") [1,5,7]. The α-keto acid can subsequently accept an amine group from either glutamate or glutamine to form an amino acid (Figure 22.3.2).

The final transamination reaction in Figure 22.3.2 is reversible, and the equilibrium will shift to favor the *catabolism* of amino acids to corresponding α-keto acids under nitrogen-limited conditions [8]. The catabolic pathway liberates nitrogen (e.g., by reforming glutamate), which can then be utilized in the synthesis of different amino acids. The catabolic pathway thus allows yeast to utilize nearly any amino acid as a nitrogen source.

22.3.4 Higher alcohol formation

Both the anabolic and catabolic pathways will generate α-keto acids, but these compounds are not accumulated in appreciable amounts even under circumstances that limit amino acid formation (e.g., low YAN). Several α-keto acids including those corresponding to leucine, isoleucine, valine, phenylalanine, and methionine, are metabolized to their corresponding higher alcohols (aka fusel alcohols, Chapter 6) by the pathway

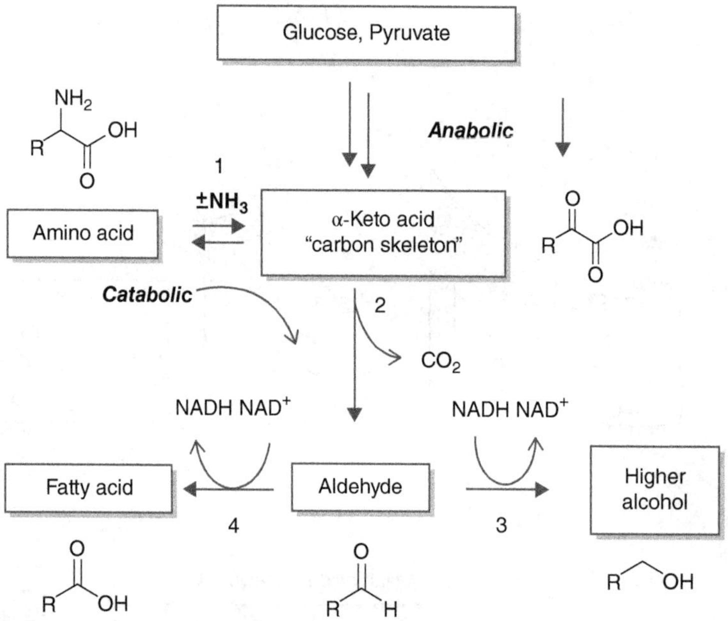

Figure 22.3.2 Schematic of anabolic and catabolic amino acid pathways leading to higher alcohol formation via α-keto acid carbon skeletons. A small percentage of carbon skeletons (~1%) will also be diverted to form corresponding fatty acids. 1, Aminotransferase (e.g., Bat1/2p); 2, α-keto acid decarboxylase (e.g., Pdc1p); 3, alcohol dehydrogenase (e.g., Adh1p); 4, aldehyde dehydrogenase (e.g., Ald1p)

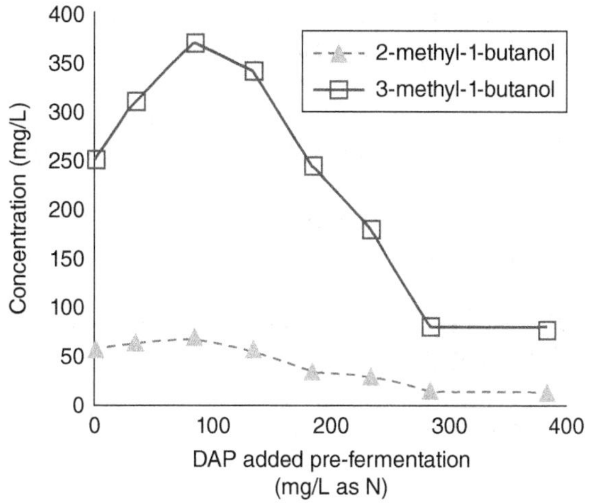

Figure 22.3.3 Effects of increasing YAN on higher alcohol formation. YAN was varied by adding diammonium phosphate (DAP) to a chemically defined medium containing 117 mg/L of YAN prior to fermentation using yeast strain AWRI 796. Other higher alcohols (2-phenylethanol, 2-methyl-1-propanol) show similar behavior. Data from Vilanova et al. [16]

shown in Figure 22.3.2.[1] Briefly, in an energetically favorable step, α-keto acids are decarboxylated to their corresponding aldehydes, which can then be enzymatically reduced to form a higher alcohol or enzymatically oxidized to form a carboxylic (fatty) acid [8]. Under anaerobic (reductive) fermentation conditions, the higher alcohol pathway is strongly favored. For example, in a model fermentation using phenylalanine as a YAN source, >99% of phenylpyruvate (α-keto acid) is diverted to 2-phenylethanol (higher alcohol) with the remainder forming phenylacetic acid (carboxylic acid) [10]. A review of typical concentrations and sensory properties of higher alcohols is provided earlier in the text (Chapter 6). Several factors are known to affect higher alcohol production as outlined below, and reviewed in more detail elsewhere [1].

- Low YAN increases higher alcohol production by favoring α-keto acids over amino acids at equilibrium (an example is shown in Figure 22.3.3). Fermentation conditions that stimulate yeast growth but limit nitrogen availability will also increase higher alcohol formation, presumably by increasing demand for amino acid production.
- Yeast strains will differ both in absolute and relative production of higher alcohols. For example, one group observed 2-fold differences in production of 3-methyl-1-butanol (isoamyl alcohol), 2-methyl-1-propanol (isobutanol), and methionol from the same must (Airén) using three different yeast strains [11].
- High concentrations of a specific precursor amino acid will result in increased formation of the corresponding higher alcohol (Chapter 6) through the catabolic pathway, i.e., using valine as a sole nitrogen source in media will result in formation of 2-methyl-1-propanol as the dominant higher alcohol [12], and methionol concentrations in wine are reported to correlate with methionine in must [13]. However, the concentration of higher alcohols in wine is often weakly correlated with their corresponding amino acid in must [14], and studies with ^{13}C-labeled tracers indicate that >75% of higher alcohols in wine are formed via the anabolic pathway [15].

[1] The catabolic reactions that lead from amino acid to higher alcohol are often referred to as the "Ehrlich Pathway" in honor of the researcher who first proposed the biochemical pathway, Felix Ehrlich [9].

The reason that the α-keto acid carbon skeletons are degraded to higher alcohols rather than recycled to other metabolites is unclear, but several hypotheses have been proposed to explain their benefits to yeast [8]:

- The equilibrium constant associated with the transamination reaction is roughly unity, and removing excess α-keto acids favors complete deamination of amino acids and thus is more efficient.
- Yeast growth results in excess NADH:NAD^+. To restore the redox balance, yeast will divert glycolysis products to glycerol at the expense of ATP production (Chapter 22.1). By using aldehydes formed through the catabolic pathway as electron acceptors, yeast can restore the redox balance without a loss of ATP production.
- Some higher alcohols (e.g., 2-phenylethanol, tryptophanol) may be needed by yeast for a hypothetical role in quorum sensing and signaling [8].

Finally, in addition to the Ehrlich pathway, two other catabolic pathways are of importance to wine chemistry:

- Sulfur-containing amino acids, cysteine, and methionine. In this pathway, the C–S bond is cleaved to yield an α-keto acid, NH_4^+, and the malodorous H_2S or CH_3SH, respectively (Chapter 10).
- Arginine can be metabolized to ornithine (a polynitrogenous α-amino acid) and urea. Under low nitrogen conditions, the urea will subsequently be used by yeast as a nitrogen source. However, in nitrogen-rich musts, the urea may accumulate and eventually form the possible human carcinogen, ethyl carbamate (Chapter 5).

22.3.5 Acetate ester formation

The formation of acetate esters is distinct from the formation of ethyl esters, which result from fatty acid metabolism (Chapter 22.2). Acetate esters are formed from higher alcohols that can serve as substrates for enzymatic acetylation by acetyl-CoA. Analogous acetylation reactions can occur for other alcohols, such as grape-derived 3-mercaptohexanol (Chapter 10) and 1-hexanol (Chapter 6). This reaction is catalyzed by acetyltransferase enzymes, particularly Atf1p (Figure 22.3.4). As discussed earlier (Chapter 7), acetate esters typically have fruity or floral aromas and are important contributors to the fermentation aromas of young wines.

The percentage of alcohol substrate that undergoes acetylation is low, and experiments in which 1-hexanol, 1-octanol, and other alcohols were spiked into wines showed that 0.2-–1.0% were converted to their corresponding acetate esters on a molar basis [17]. Most data indicate that acetate ester production during fermentation is more highly controlled by *ATF1* expression than by alcohol substrate availability [18], perhaps because the conversion efficiency is so low. This observation is in contrast to ethyl ester formation, for which fatty acid substrate availability appears to be more important (Chapter 22.2).

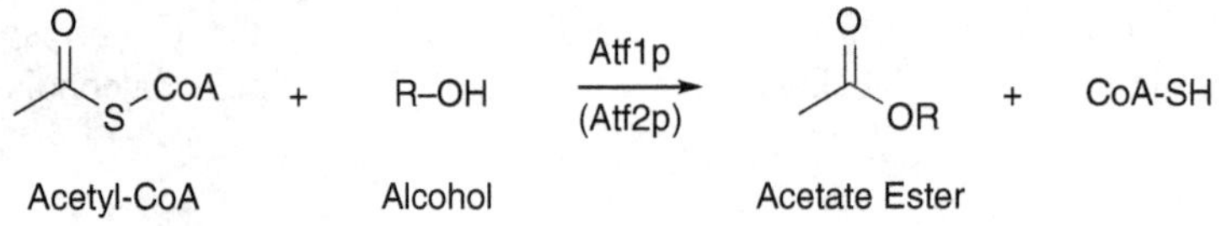

Figure 22.3.4 Reaction depicting formation of acetate esters from acetyl-CoA and higher alcohols by yeast via acetyltransferase enzymes (Atf1p and Atf2p)

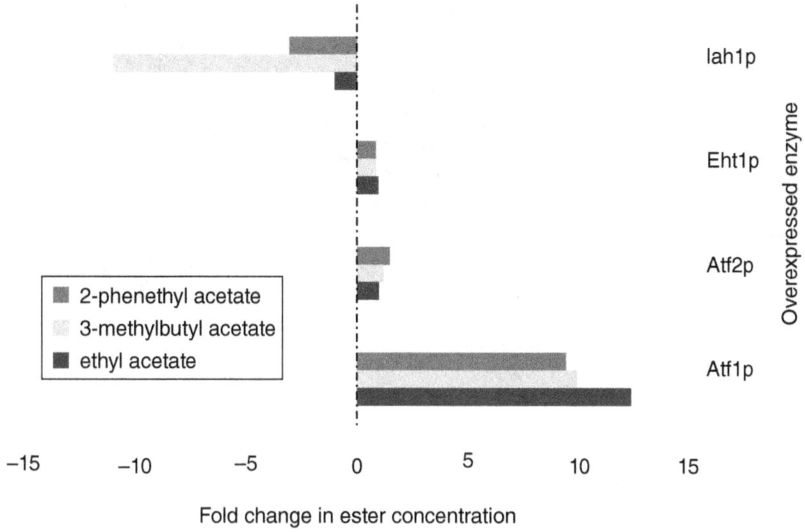

Figure 22.3.5 Effects of overexpression of genes involved in ester metabolism on acetate ester formation by transformed yeast. The largest increases are seen with Atf1p overexpression due to its role in catalyzing the acetylation of corresponding higher alcohols (Figure 22.3.4). Iah1p results in a decrease in acetate esters, likely due to esterase activity. Data from Lilly et al. [20]

Several approaches are available to winemakers to manipulate acetate ester production. Commercial yeast strains differ in acetate ester production, even under identical fermentation conditions, by an order of magnitude [19]. This variation is plausibly due to variation in Atf1p activity, as yeast genetically modified to overexpress *ATF1* show comparable order of magnitude changes in acetate ester formation (Figure 22.3.5) [20]. Overexpression of a different acetyltransferase enzyme (Atf2p) or an enzyme associated with ethyl ester formation (Eht1p, Chapter 22.2) resulted in minor increases in acetate ester concentrations. Acetate ester accumulation was decreased in strains expressing high esterase activity, for example, high activity of Iah1p (Figure 22.3.5).

Fermentation conditions may also affect *ATF1* expression. Most notably, aeration and unsaturated fatty acid additions are well known to decrease acetate ester formation [14], and these conditions can repress expression by at least four-fold [21]. High temperatures can also decrease acetate ester formation. Increasing the fermentation temperature from 20 °C to 30 °C reportedly lowers total isoamyl acetate production by 30% [22], and final wine concentrations are decreased by half due to additional losses resulting from increased volatilization at higher temperatures (Chapter 22.1.1). These factors also decrease fatty acid ethyl ester formation (Chapter 22.2), and in part explain the common winemaking practice of clarifying musts, minimizing oxygen exposure, and using cool fermentation temperatures when making fruity white wines.

22.3.6 YAN in the winery – requirements, approaches, and consequences

The effects of lower YAN concentrations on fermentation outcomes are summarized in Table 22.3.1. Typically, lower YAN results in slower fermentation kinetics and higher final sugar concentrations (Figure 22.3.6). While this and many of the other consequences of low YAN (e.g., greater H_2S and fusel alcohol production) are undesirable, excessively high YAN in must can result in higher YAN at the end of fermentation

Table 22.3.1 Effects of lower YAN on fermentations, summarized from Reference [1]

Outcome	Effect of lower YAN	Cross reference
Hydrogen sulfide production	Increase	Chapter 22.4
Release of thiols via β-lyase activity	Increase	Chapter 23.2
Fermentation rate and biomass production	Decrease	[26]
Higher alcohol production	Increase	This chapter
Acetic acid production	Decrease	Chapter 22.1
Ester production	Decrease	This chapter and Chapter 22.2
Ethyl carbamate or biogenic amine potential	Decrease	Chapter 5
Risk of microbial spoilage post-fermentation	Decrease	[27]

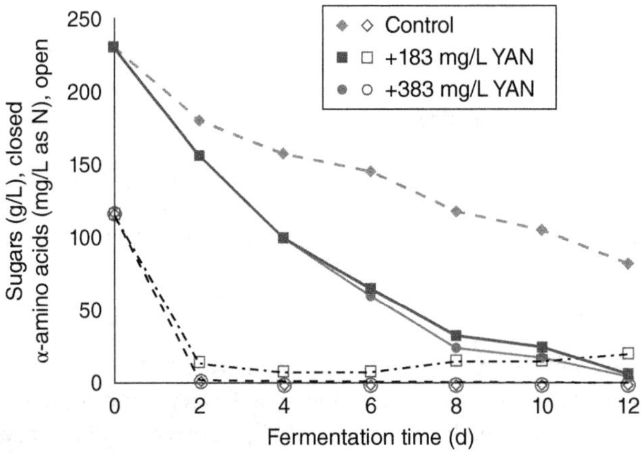

Figure 22.3.6 Effects of DAP supplementation on residual α-amino acid nitrogen and fermentation kinetics of a synthetic wine-like media. Fermentations were carried out with AWRI 796 yeast. Closed markers refer to sugar concentrations and open markers refer to α-amino acid nitrogen. The treatments were: control (diamond, YAN = 117 mg/L N, no DAP addition), + DAP, 183 mg/L as N (circle, total YAN = 300 mg/L), + DAP, 383 mg/L as N (diamond, total YAN = 500 mg/L). Data from Vilanova et al. [16]

(Figure 22.3.6), leading to challenges after alcoholic fermentation is complete (e.g., decreased microbial stability, greater potential for ethyl carbamate or biogenic amine formation). However, because the consequences of insufficient YAN are usually more immediately apparent and problematic to the winemaker (e.g., stuck or sluggish fermentations, off-aroma formation), it is more common to focus on the minimum necessary amount of YAN. Typical recommendations for YAN are in the order of 150–200 mg/L as N, but targets will vary depending on the criteria used. Concentrations as low as 70–140 mg/L are reportedly necessary to avoid any detectable residual nitrogen, while concentrations of 267 mg/L are recommended to avoid residual sugar at the end of fermentation [1].

YAN is typically increased by addition of NH_4^+, generally in the form of diammonium hydrogen phosphate (diammonium phosphate, DAP) [23]. More complex sources containing both amino acids and DAP derived from yeast autolysates are available commercially, although the concentration of amino acids in these preparations is usually less than 10 mg/L and thus likely to be of negligible importance [24]. These additions may be done prior to or after commencement of fermentation, but are recommended to take place in the first half of fermentation since the ability of yeast to take up nitrogen via permeases diminishes in the presence of ethanol [25].

References

1. Bell, S.J. and Henschke, P.A. (2005) Implications of nitrogen nutrition for grapes, fermentation and wine. *Australian Journal of Grape and Wine Research*, **11** (3), 242–295.
2. Huang, Z. and Ough, C.S. (1991) Amino-acid profiles of commercial grape juices and wines. *American Journal of Enology and Viticulture*, **42** (3), 261–267.
3. Theron, L.W. and Divol, B. (2014) Microbial aspartic proteases: current and potential applications in industry. *Applied Microbiology and Biotechnology*, **98** (21), 8853–8868.
4. Beltran, G., Novo, M., Rozes, N., *et al.* (2004) Nitrogen catabolite repression in *Saccharomyces cerevisiae* during wine fermentations. *FEMS Yeast Research*, **4** (6), 625–632.
5. Feldmann, H. (2010) *Yeast: molecular and cell biology*, Wiley-VCH, Weinheim.
6. Magasanik, B. and Kaiser, C.A. (2002) Nitrogen regulation in *Saccharomyces cerevisiae*. *Gene*, **290** (1–2), 1–18.
7. Cooper, T.C. (1982) Nitrogen metabolism in *Saccharomyces cerevisiae*, in *The molecular biology of the yeast Saccharomyces: metabolism and gene expression* (eds Strathern, J.N., Jones, E.W., Broach, J.R.), Cold Spring Harbor Laboratory Press, Cold Spring Harbor, NY, pp. 39–99.
8. Hazelwood, L.A., Daran, J.-M., van Maris, A.J.A., *et al.* (2008) The Ehrlich pathway for fusel alcohol production: a century of research on *Saccharomyces cerevisiae* metabolism. *Applied and Environmental Microbiology*, **74** (8), 2259–2266.
9. Ehrlich, F. (1907) Über die Bedingungen der Fuselölbildung und über ihren Zusammenhang mit dem Eiweissaufbau der Hefe. *Ber. Dtsch Chem. Ges.*, **40**, 1027–1047.
10. Vuralhan, Z., Morais, M.A., Tai, S.-L., *et al.* (2003) Identification and characterization of phenylpyruvate decarboxylase genes in *Saccharomyces cerevisiae*. *Applied and Environmental Microbiology*, **69** (8), 4534–4541.
11. Hernández-Orte, P., Ibarz, M.J., Cacho, J., Ferreira, V. (2005) Effect of the addition of ammonium and amino acids to musts of Airen variety on aromatic composition and sensory properties of the obtained wine. *Food Chemistry*, **89** (2), 163–174.
12. Dickinson, J.R., Harrison, S.J., Hewlins, M.J.E. (1998) An investigation of the metabolism of valine to isobutyl alcohol in *Saccharomyces cerevisiae*. *Journal of Biological Chemistry*, **273** (40), 25751–25756.
13. Hernández-Orte, P., Cacho, J.F., Ferreira, V. (2002) Relationship between varietal amino acid profile of grapes and wine aromatic composition. Experiments with model solutions and chemometric study. *Journal of Agricultural and Food Chemistry*, **50** (10), 2891–2899.
14. Malcorps, P., Cheval, J., Jamil, S., Dufour, J. (1991) A new model for the regulation of ester synthesis by alcohol acetyltransferase in *Saccharomyces cerevisiae* during fermentation. *Journal of the American Society of Brewing Chemists*, **49** (2), 47–53.
15. Nisbet, M.A., Tobias, H.J., Brenna, J.T., *et al.* (2014) Quantifying the contribution of grape hexoses to wine volatiles by high-precision [$U^{13}C$]-glucose tracer studies. *Journal of Agricultural and Food Chemistry*, **62** (28), 6820–6827.
16. Vilanova, M., Ugliano, M., Varela, C., et al. (2007) Assimilable nitrogen utilisation and production of volatile and non-volatile compounds in chemically defined medium by *Saccharomyces cerevisiae* wine yeasts. *Applied Microbiology and Biotechnology*, **77** (1), 145-157.
17. Dennis, E.G., Keyzers, R.A., Kalua, C.M., *et al.* (2012) Grape contribution to wine aroma: production of hexyl acetate, octyl acetate, and benzyl acetate during yeast fermentation is dependent upon precursors in the must. *Journal of Agricultural and Food Chemistry*, **60** (10), 2638–2646.
18. Saerens, S.M.G., Delvaux, F.R., Verstrepen, K.J., Thevelein, J.M. (2010) Production and biological function of volatile esters in *Saccharomyces cerevisiae*. *Microbial Biotechnology*, **3** (2), 165–177.
19. Steensels, J., Meersman, E., Snoek, T., *et al.* (2014) Large-scale selection and breeding to generate industrial yeasts with superior aroma production. *Applied and Environmental Microbiology*, **80** (22), 6965–6975.
20. Lilly, M., Bauer, F.F., Lambrechts, M.G., *et al.* (2006) The effect of increased yeast alcohol acetyltransferase and esterase activity on the flavour profiles of wine and distillates. *Yeast*, **23** (9), 641–659.
21. Fujii, T., Kobayashi, O., Yoshimoto, H., *et al.* (1997) Effect of aeration and unsaturated fatty acids on expression of the *Saccharomyces cerevisiae* alcohol acetyltransferase gene. *Applied and Environmental Microbiology*, **63** (3), 910–915.

22. Morakul, S., Mouret, J.-R., Nicolle, P., *et al.* (2013) A dynamic analysis of higher alcohol and ester release during winemaking fermentations. *Food and Bioprocess Technology*, **6** (3), 818–827.
23. Boulton, R.B., Singleton, V.L., Bisson, L.F., Kunkee, R.E. (1999) *Principles and practices of winemaking*, Springer, New York.
24. Stewart, A.C. and Butzke, C.E. (2011) Assessment of yeast nutrient supplements and residual nitrogen in wine. *American Journal of Enology and Viticulture*, **62** (3), 390A–391A.
25. Sablayrolles, J.-M., Dubois, C., Manginot, C., *et al.* (1996) Effectiveness of combined ammoniacal nitrogen and oxygen additions for completion of sluggish and stuck wine fermentations. *Journal of Fermentation and Bioengineering*, **82** (4), 377–381.
26. Bisson, L.F. (1999) Stuck and sluggish fermentations. *American Journal of Enology and Viticulture*, **50** (1), 107–119.
27. Fugelsang, K.C. and Edwards, C.G. (2007) *Wine microbiology – practical applications and procedures*, Springer, New York.

22.4
Sulfur Metabolism

22.4.1 Introduction

Yeasts require sulfur for growth due to its role in sulfur-containing amino acids, peptides, and proteins [1–3]. Unlike nitrogen, sulfur availability is rarely limiting to fermentations, and assimilable sources include inorganic species such as elemental sulfur, sulfate, and sulfite, and organic sources such as amino acids and peptides. However, grapes generally contain low amounts of organic sulfur compounds [4] such as cysteine and methionine, and the principal source of sulfur available to yeasts during winemaking is sulfate (SO_4^{2-}) [2], in which sulfur is in its highest oxidation state (+6). Sulfate is naturally abundant in grape juices or musts and can be found at concentrations up to several hundred mg/L[1] [5,6]. Several critical sulfur compounds affecting wine aroma and quality are produced as yeast secondary metabolites during fermentation (Chapter 10), and of primary interest in this regard is the formation of H_2S as a byproduct of sulfur amino acid metabolism. Fortunately, while total H_2S production for the duration of fermentation can be in the order of several hundred μg/L, much of this is entrained in CO_2 and lost due to volatilization such that only a fraction typically remains in the finished wine (e.g., 1–20 μg/L). Nonetheless, formation of H_2S is a malodorous nuisance to the winemaker during fermentation, and may lead to more challenging problems through the production of related volatile sulfur compounds that are not as easily purged from wine. Knowledge of formation of sulfur compounds by yeast is therefore indispensible and begins with an understanding of the origins of H_2S.

22.4.2 Sulfide production and assimilation

Biosynthesis of S-containing amino acids requires formation of sulfide (S^{2-}), the most reduced form of sulfur (oxidation state –2). To form sulfide from sulfate (the major sulfur source), yeast employ the sulfate reduction sequence (SRS) pathway (shown in Figure 22.4.1), which uses several enzymes for the uptake and activation

[1] Higher concentrations may result from oxidation of bisulfite added to juice or must, thereby increasing the natural sulfate levels.

Understanding Wine Chemistry, First Edition. Andrew L. Waterhouse, Gavin L. Sacks, and David W. Jeffery.

of sulfate, followed by its reduction to sulfite and then sulfide. Activation of sulfate by adenylation markedly increases its reduction potential, making its reduction to sulfite more thermodynamically favorable [7]. Following the reduction sequence, additional enzymatic steps see the incorporation of sulfide into cysteine and methionine via coupling with *O*-acetylhomoserine (O-AHS) [1,8]. Notably, while sulfite (SO_3^{2-}) is produced as an intermediate in the SRS pathway, extracellular sulfite (i.e., from the addition of bisulfite during winemaking) can also be utilized after diffusing into the cell (as SO_2, see below). This sulfur assimilation pathway can be summarized as follows:

- Extracellular sulfate is transported into the cell by sulfate permease.
- Sulfate is adenylated by ATP sulfurylase (1), forming adenosine-5′-phosphosulfate (APS) and pyrophosphate (PP_i).
- APS is phosphorylated to 3′-phosphoadenosyl-5′-phosphosulfate (PAPS) by APS kinase (2).
- PAPS is reduced to sulfite by PAPS reductase (3) and NADPH.
- If not excreted, sulfite is reduced to sulfide by sulfite reductase (4) and NADPH prior to its assimilation into S-containing amino acids or dissimilation as waste product H_2S.
- Further steps by sulfhydrylase, synthase, lyase, and methyltransferase enzymes produce homocysteine, which leads to cysteine and methionine.

Generally, factors that result in accumulation of sulfide but do not facilitate its incorporation into S-containing amino acids will result in greater accumulation of H_2S in the cell and diffusion into the fermenting must. The major sources of variation in H_2S formation (yeast genetics, nitrogen availability, etc.) are discussed in more detail below.

Yeast strain influences both sulfite and sulfide formation, depending on environmental and fermentation conditions. Most strains produce 10–30 mg/L of sulfite (low sulfite producers) although some can yield more than 100 mg/L (high sulfite producers) [3]. Many of these differences can be rationalized by differences in activities of enzymes depicted in Figure 22.4.1, meaning variations (or defects) in sulfate uptake, activation, and reduction can be responsible [1,9]. Greater production of sulfite can be attributed to increased sulfate permease and ATP sulfurylase activity or lack of repression (or feedback inhibition) by methionine or sulfite, and low sulfite reductase activity or affinity for sulfite. Conversely, low sulfite production is associated with increased activity of sulfite reductase and sulfhydrylases in the presence of sulfate and sulfite.

Similarly, genetic variations in the SRS pathway among yeasts can affect the amount of H_2S produced or accumulated. Strains can be categorized as non-producing, or low, medium, and high H_2S producers. Variability in production is similar for wild (including non-*Saccharomyces*) and commercial strains, and genetic variation can provide targets to microbiologists interested in identifying new yeast strains with low H_2S production. Sulfite reductase activity is of particular importance to controlling H_2S [10], because if sulfite accumulates[2] it can be diverted away from sulfide production. A number of surveys that assess the production of H_2S by different yeast strains (commercial or wild) show that lower sulfite reductase activity is correlated with lower H_2S at the end of fermentation [11–13] – and presumably higher total SO_2 as a consequence, although this is not always measured.

Media composition will also affect the SRS pathway – in particular, it is well known that low yeast assimilable nitrogen (YAN) is correlated with higher H_2S production (see Section 22.4.3). H_2S formation is also

[2]Yeast-mediated increases in sulfite concentration in wine may have other effects, such as inhibiting malolactic fermentation and contributing to acetaldehyde production, and high amounts produced by yeast should be factored in with other sulfite additions to stay within the legal limits for sulfite (and sulfate).

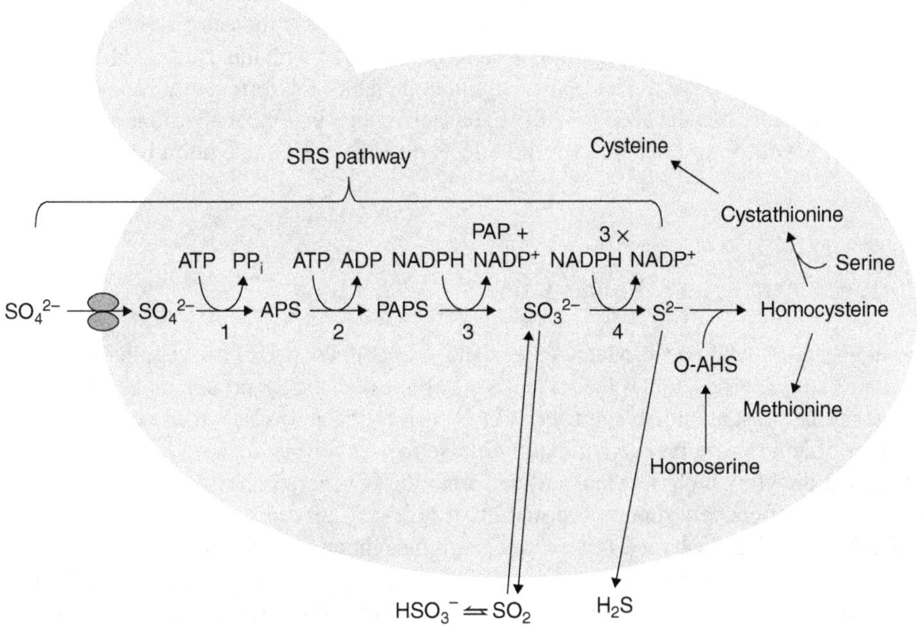

***Figure 22.4.1** Representation of the pathway for assimilation of sulfate into sulfur amino acids by yeast. After uptake, sulfate is adenylated with adenosine-5′-triphosphate (ATP) by the action of ATP sulfurylase (1), forming adenosine-5′-phosphosulfate (APS) and pyrophosphate (PP_i); APS is phosphorylated to 3′-phosphoadenosyl-5′-phosphosulfate (PAPS) by APS kinase (2); PAPS is reduced to sulfite by PAPS reductase (3) and NADPH, releasing phosphoadenosine phosphate (PAP); sulfite can be excreted, or reduced to sulfide by sulfite reductase (4) and NADPH; sulfide can be assimilated into S-containing amino acids or dissimilated as waste product H_2S*

reportedly higher in filtered juices as compared to synthetic media, possibly due to added stressors (e.g., from phenolics) or factors other than nitrogen deficiency (e.g., lack of methionine [11], see Section 22.4.3). The strain-dependent interaction of the SRS pathway with must nutrient status and fermentation conditions makes predicting H_2S production challenging in real juices and musts. As an example, strains that produced high amounts of H_2S in synthetic media also produced high amounts in Syrah juice, but strains that produced low levels of H_2S in synthetic media did not necessarily produce low H_2S in real juice [13].

A major pathway to H_2S production involves energetically taxing sulfate reduction to sulfite and ultimately sulfide during fermentation. However, SO_2 can readily diffuse into the yeast cell to yield intracellular sulfite (and bisulfite) [14], which is a better precursor to H_2S than sulfate[3] [15], especially under nitrogen-limiting conditions [4]. With the onset of nitrogen starvation, enzymes associated with sulfate uptake and initial stages of the SRS will be inhibited, but sulfite reduction to sulfide remains unaffected. Although sulfite reductase

[3] This does not imply that the modest amounts of bisulfite typically used during winemaking are important contributors to H_2S formation. Rather, it should emphasize the underlying biochemical reactions and their implications, as elaborated in the remainder of the paragraph with respect to bisulfite. Elemental sulfur from vineyard applications can also lead to formation of H_2S (and other volatile sulfur compounds) by non-enzymatic means during fermentation.

has relatively short-lived activity and cold-lability when extracted from yeast [16], reductase activity has been suggested to continue for several weeks after fermentation (presumably for as long as viable cells still exist), leading to continued formation of H_2S. Caution is therefore required with the addition of bisulfite (or SO_2) to wines with active yeast cells or yeast lees, especially in large tanks [2]; fortunately, the presence of lees will consume oxygen and decreases the need for sulfite addition at this time. Late additions are particularly problematic because they will occur after the purging effects on H_2S by CO_2 evolution have largely ceased (see Section 22.4.4).[4]

22.4.3 Nitrogen sources and H_2S formation

The S-containing amino acid end products (especially methionine) perform a regulatory role in the SRS pathway, meaning sulfide production is linked to the metabolic demand for protein biosynthesis (and S-amino acids). In most musts, yeast assimilable nitrogen (YAN) is limiting and sulfate is in excess. This results in an insufficient pool of amino acid precursors required to sequester sulfide, leading to overproduction of H_2S, particularly during the growth phase when sulfide formation is generally at its greatest [4,17]. Ammonium supplementation may effectively suppress formation of H_2S [18], as can a number of amino acids, particularly those that foster high growth rates (e.g., serine, arginine, glutamine, and asparagine) or act as precursors in the formation of cysteine and methionine (i.e., serine and aspartate) [4] (Figure 22.4.2). Differences in yeast strain are apparent in the study of Jiranek *et al.* [4], but both strains showed a lack of suppressive effect from proline, threonine, and cysteine, with the last of these three increasing production of H_2S. These outcomes can be rationalized based on the following:

- Proline is not a source of yeast assimilable nitrogen (Chapter 22.3).
- Threonine, which arises from aspartate via homoserine as an intermediate, inhibits the biosynthesis of homoserine, which is a key precursor to cysteine and methionine [19].
- Even though it has a regulatory role in the SRS pathway, cysteine can be catabolized to yield pyruvate, ammonium, and H_2S when nutrients are deficient [20].

Deficiencies of other nutrients involved in production of amino acid precursors (e.g., B group vitamins, pyroxidine and pantothenic acid), while less common,[5] can limit the production of methionine and lead to accumulation of H_2S [3,15].

Yeast catabolism of cysteine can potentially contribute to H_2S formation through β-lyase activity (Chapter 23.2), although cysteine is an unlikely source of H_2S under normal winemaking conditions due to its low abundance. However, degradation of proteins and glutathione (GSH, Chapter 10) may provide a pool of cysteine (or cystine, the oxidized dimeric form of cysteine), and lead to H_2S production [2,18,21]. Processes that minimize the presence or degradation of proteins in must (i.e., clarification, pre-fermentation bentonite treatment, avoidance of proteolytic activity) have been shown to reduce the amount of H_2S formed. Analogously, addition of GSH to a must or during yeast rehydration can lead to an increase in H_2S production. Interestingly, the uptake of GSH by yeast during rehydration seemed to be important, rather than uptake once fermentation starts [21].

---

[4] There is also the risk of inducing acetaldehyde accumulation by adding bisulfite during fermentation.

[5] Deficiencies such as these, particularly of thiamine (vitamin B1), are more common in the production of other fermented fruit beverages such as cider and perry.

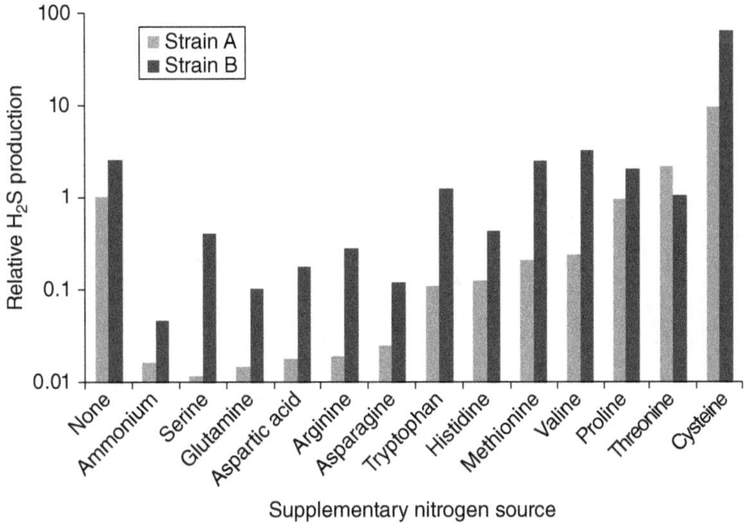

Figure 22.4.2 Relative production of H_2S (log scale, normalized to Strain A, no addition treatment) over 6 h in aliquots of fermenting chemically defined grape juice medium, initially containing sulfite (260 μM) and ammonium (8.3 mM), supplemented with ammonium or amino acids (total equivalent to 14.3 mM of nitrogen) 1 h before the predicted depletion of initial ammonium. Selected data from Reference [4]

22.4.4 Timing of formation and residual H_2S

Lack of must nutrients or their exhaustion during fermentation are commonly recognized as factors that lead to H_2S production. Despite the widespread practice of supplementing musts with assimilable nitrogen in the form of DAP, often prior to the onset of fermentation, there can be mixed results in the formation of H_2S. In some instances addition of DAP may increase overall H_2S production or residual concentration, depending on yeast strain and juice/must composition [17,22] (Figure 22.4.3). Because H_2S is highly volatile, the timing of its production, rather than the total amount produced, is important in determining how much might remain in the finished wine [23]. Factors affecting both total H_2S production and final H_2S concentration can be generalized as follows:

- Production is low to moderate during early stages of yeast growth and may continue throughout fermentation. Typically these conditions do not respond to nitrogen or nutrient supplementation, and may leave residual H_2S in wine.
- Production is at a maximum during the early–mid phase when yeast is actively growing and YAN becomes depleted – concurrent CO_2 evolution at this stage results in volatilization of most of this H_2S, although low fermentation vigor can leave residual H_2S in wine. DAP and nutrient supplementation can be beneficial for nitrogen-responsive strains in low YAN conditions.
- Production is usually low late in fermentation when yeast growth has ceased. However, there is also minimal purging due to decreased CO_2 evolution, which can result in increased risk of residual H_2S in wine. DAP addition tends to have no effect at this time.

In summary, higher total amounts of H_2S produced during fermentation do not necessarily equate to higher amounts in finished wine, since production after the midpoint of fermentation may be of greatest importance to residual H_2S in wine.

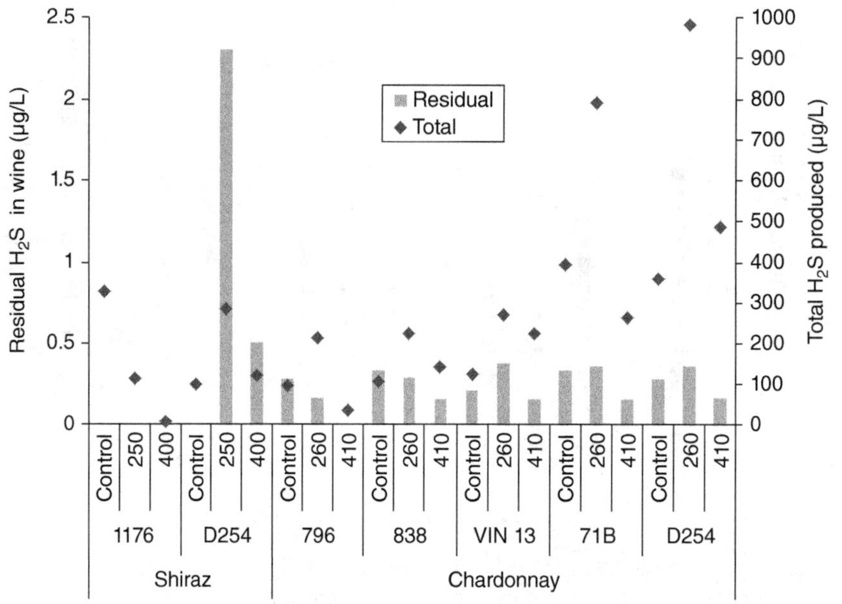

Figure 22.4.3 Variable influence of yeast strain and nitrogen supplementation (total YAN indicated, that is, 250/400 mg/L and 260/410 mg/L)) on the total production and final wine concentrations of H_2S in Shiraz (30 kg) and Chardonnay (200 mL) fermentations (note the vastly different y-scales for total and residual H_2S). Unsupplemented control treatments had 100 and 110 mg/L of YAN for Shiraz and Chardonnay fermentations, respectively. Data derived from References [17] and [22]

References

1. Rauhut, D. (1993) Yeasts – production of sulfur compounds, in *Wine microbiology and biotechnology* (ed. Fleet, G.H.), Harwood Academic Publishers, Chur, Switzerland, pp. 183–224.
2. Rauhut, D. (2009) Usage and formation of sulphur compounds, in *Biology of microorganisms on grapes, in must and in wine* (eds König, H., Unden, G., Fröhlich, J.), Springer-Verlag, Berlin and Heidelberg, pp. 181–207.
3. Eschenbruch, R. (1974) Sulfite and sulfide formation during winemaking – a review. *American Journal of Enology and Viticulture*, **25** (3), 157–161.
4. Jiranek, V., Langridge, P., Henschke, P.A. (1995) Regulation of hydrogen sulfide liberation in wine-producing *Saccharomyces cerevisiae* strains by assimilable nitrogen. *Applied and Environmental Microbiology*, **61** (2), 461–467.
5. Ough, C.S. and Amerine, M.A. (1988) Other constituents, in *Methods for analysis of musts and wines*, 2nd edn, John Wiley & Sons, Inc., New York, pp. 264–301.
6. Rankine, B.C. (2004) The composition of wines, in *Making good wine*, Pan Macmillan Australia, Sydney, NSW, pp. 259–265.
7. Yu, Z., Lemongello, D., Segel, I.H., Fisher, A.J. (2008) Crystal structure of *Saccharomyces cerevisiae* 3-phosphoadenosine-5-phosphosulfate reductase complexed with adenosine 3,5-bisphosphate. *Biochemistry*, **47** (48), 12777–12786.
8. Thomas, D. and Surdin-Kerjan, Y. (1997) Metabolism of sulfur amino acids in *Saccharomyces cerevisiae*. *Microbiology and Molecular Biology Reviews*, **61** (4), 503–532.
9. Pretorius, I.S. (2000) Tailoring wine yeast for the new millennium: novel approaches to the ancient art of winemaking. *Yeast*, **16** (8), 675–729.

10. Cordente, A.G., Heinrich, A., Pretorius, I.S., Swiegers, J.H. (2009) Isolation of sulfite reductase variants of a commercial wine yeast with significantly reduced hydrogen sulfide production. *FEMS Yeast Research*, **9** (3), 446–459.
11. Spiropoulos, A., Tanaka, J., Flerianos, I., Bisson, L.F. (2000) Characterization of hydrogen sulfide formation in commercial and natural wine isolates of *Saccharomyces*. *American Journal of Enology and Viticulture*, **51** (3), 233–248.
12. Mendes-Ferreira, A., Mendes-Faia, A., Leão, C. (2002) Survey of hydrogen sulphide production by wine yeasts. *Journal of Food Protection*, **65** (6), 1033–1037.
13. Kumar, G.R., Ramakrishnan, V., Bisson, L.F. (2010) Survey of hydrogen sulfide production in wine strains of *Saccharomyces cerevisiae*. *American Journal of Enology and Viticulture*, **61** (3), 365–371.
14. Divol, B., du Toit, M., Duckitt, E. (2012) Surviving in the presence of sulphur dioxide: strategies developed by wine yeasts. *Applied Microbiology and Biotechnology*, **95** (3), 601–613.
15. Wainwright, T. (1971) Production of H_2S by yeasts: role of nutrients. *Journal of Applied Bacteriology*, **34** (1), 161–171.
16. Jiranek, V., Langridge, P., Henschke, P.A. (1996) Determination of sulphite reductase activity and its response to assimilable nitrogen status in a commercial *Saccharomyces cerevisiae* wine yeast. *Journal of Applied Bacteriology*, **81** (3), 329–336.
17. Ugliano, M., Kolouchova, R., Henschke, P. (2011) Occurrence of hydrogen sulfide in wine and in fermentation: influence of yeast strain and supplementation of yeast available nitrogen. *Journal of Industrial Microbiology and Biotechnology*, **38** (3), 423–429.
18. Vos, P.J.A. and Gray, R.S. (1979) The origin and control of hydrogen sulfide during fermentation of grape must. *American Journal of Enology and Viticulture*, **30** (3), 187–197.
19. Jones, E.W. and Fink, G.R. (1982) Regulation of amino acid and nucleotide biosynthesis in yeast, in *The molecular biology of the yeast Saccharomyces: metabolism and gene expression* (eds Strathern, J.N., Jones, E.W., Broach, J.R.), Cold Spring Harbor Laboratory Press, Cold Spring Harbor, pp. 181–299.
20. Tokuyama, T., Kuraishi, H., Aida, K.O., Uemura, T. (1973) Hydrogen sulfide evolution due to a pantothenic acid deficiency in the yeast requiring this vitamin, with special reference to the effect of adenosine triphosphate on yeast cysteine desulfhydrase. *The Journal of General and Applied Microbiology*, **19** (6), 439–466.
21. Winter, G., Henschke, P.A., Higgins, V.J., *et al.* (2011) Effects of rehydration nutrients on H_2S metabolism and formation of volatile sulfur compounds by the wine yeast VL3. *AMB Express*, **1** (36), 1–11.
22. Ugliano, M., Fedrizzi, B., Siebert, T., *et al.* (2009) Effect of nitrogen supplementation and *Saccharomyces* species on hydrogen sulfide and other volatile sulfur compounds in Shiraz fermentation and wine. *Journal of Agricultural and Food Chemistry*, **57** (11), 4948–4955.
23. Ugliano, M., Winter, G., Coulter, A.D., Henschke, P.A. (2009) Practical management of hydrogen sulfide during fermentation – an update. *The Australian and New Zealand Grapegrower and Winemaker*, **545a**, 30–38.

22.5

Bacterial Fermentation Products

22.5.1 Introduction

While yeast is necessary for alcoholic fermentation during wine production, several types of bacteria may also affect wine composition. Although the number of bacterial species on Earth may number in the millions, only a small percentage can grow under conditions of high osmotic stress (juice), high alcohol (wine), and/or low pH (both grape juice and wine). Those with the best established role in wine production are lactic acid bacteria (LAB) and two members of the acetic acid bacteria (AAB) family: *Acetobacter* and *Gluconobacter*. The changes imparted by these microflora can be desirable (and encouraged) in some circumstances, as is frequently the case for *Oenococcus oeni* (an LAB) in conducting malolactic fermentation (MLF). However, modifications induced by AAB as well as some LAB are often associated with spoilage [1–3]. Because some metabolic transformations by these bacteria are analogous to those already described for yeast, for example, LAB are capable of hydrolyzing glycosides and reducing aldehydes to alcohols, the focus of this chapter will be on transformations that are associated most strongly with bacteria and have not been addressed elsewhere, as follows:

- Wine deacidification due to MLF (Chapter 3)
- Production of "lactic" volatile compounds such as diacetyl and acetoin (Chapter 9)
- Production of spoilage compounds such as mannitol and β-glucans (Chapter 2), acetic acid (Chapter 3), biogenic amines and imines (Chapter 5), and 2-ethoxyhexa-3,5-diene (Chapter 18).

22.5.2 Lactic acid bacteria

22.5.2.1 Malolactic fermentation

As the name implies, MLF involves the conversion of the diprotic grape-derived L-malic acid into the monoprotic L-lactic acid due to LAB metabolism, although strictly speaking this is an enzymatic decarboxylation

Understanding Wine Chemistry, First Edition. Andrew L. Waterhouse, Gavin L. Sacks, and David W. Jeffery.

Table 22.5.1 Mean values for pH and selected MLF-related compounds in wines from different grape varieties inoculated with two strains of Oenococcus oeni. Data from Reference [6]

Wine composition	Grape variety								
	Syrah			Cabernet Sauvignon			Merlot		
	Pre-MLF	Strain 1[a]	Strain 2[b]	Pre-MLF	Strain 1[a]	Strain 2[b]	Pre-MLF	Strain 1[a]	Strain 2[b]
pH	3.71	4.01	4.06	3.21	3.27	3.24	3.13	3.17	3.17
Non-volatiles (g/L)									
Titratable acidity[c]	6.74	3.90	3.86	6.17	5.21	5.56	5.66	5.18	5.19
Malic acid	3.78	0.03	0.10	1.41	0.09	0.11	0.93	0.08	0.11
Lactic acid	0.07	2.04	2.19	0.04	1.00	0.88	0.02	0.62	0.56
Citric acid	0.38	0.02	0.10	0.32	0.04	0.19	0.18	0.04	0.12
Volatiles (mg/L)									
Diaectyl	6.01	10.88	4.59	5.06	13.21	6.95	3.70	6.92	3.66
Acetoin	1.29	3.26	0.45	1.34	15.39	4.83	1.44	5.29	1.30
Volatile acidity[d]	0.21	0.37	0.38	0.26	0.36	0.31	0.30	0.35	0.36
Acetaldehyde	22.06	6.46	4.20	25.69	8.27	7.87	20.28	11.14	4.87
Ethyl lactate	3.27	48.14	52.34	7.13	48.00	40.94	5.76	28.06	28.57

[a] Commercial strain *O. oeni* PN4.
[b] Indigenous strain *O. oeni* C22L9.
[c] Expressed as tartaric acid.
[d] Expressed as acetic acid (g/L).

rather than fermentation.[1] *Oenococcus oeni* (formerly *Leuconostoc oenos*) is the LAB species most adapted to wine conditions (low pH, high ethanol) and is the dominant LAB present at the end of alcoholic fermentation [4,5]. Strains of *O. oeni* are therefore preferred for conducting MLF, which results in a decrease in wine titratable acidity (TA), with deacidification of approximately 1–3 g/L as tartaric acid equivalents[2] and an increase in pH of around 0.1–0.3 units[3] (Table 22.5.1), as well as a decrease in sourness (Chapter 3). In spite of the pH increase, the removal of malic acid substrate and other nutrients can improve the microbial stability of the resulting wine. MLF is desired for some white wines (often to affect flavor) and most red wines (for flavor and/or stability). Usually, this is achieved through inoculation with a selected commercial strain of *O. oeni* after alcoholic fermentation is complete, although spontaneous or inoculated MLF coincident with alcoholic fermentation can also occur.

---

[1] LAB do conduct fermentation in the true sense of the term as outlined further below, but the conversion of malic to lactic acid is called a fermentation due to production of CO_2, which is analogous to alcoholic fermentation by yeast.

[2] Complete malolactic conversion of 1 g/L of malic acid (approx. 15 meq/L) will yield 0.67 g/L of lactic acid (approximately 7.5 meq/L), resulting in a TA decrease of about 7.5 meq/L (or 0.56 g/L as tartaric acid equivalents). The variation in the change in TA and pH during MLF is thus largely due to the large range of malic acid concentrations found in grapes (2–6 g/L).

[3] Deacidification by MLF is especially desirable in cool climate regions where grape malic acid levels are naturally high, for example, TA > 8 g/L as tartaric acid equivalents. The increase in pH due to MLF tends to be smaller than for a comparable decrease in TA achieved through deacidification with carbonate salts, which can be advantageous for preventing microbial spoilage. In warmer regions where malic acid levels are much lower and wines already possess low TA and high pH, MLF may still be undertaken to achieve other organoleptic changes or to prevent spoilage by wild LAB strains. In this case, the deacidifying effect of MLF is undesirable but can be counteracted by acidification with tartaric acid to adjust the final pH and TA of wine to fall within appropriate parameters (Chapter 3).

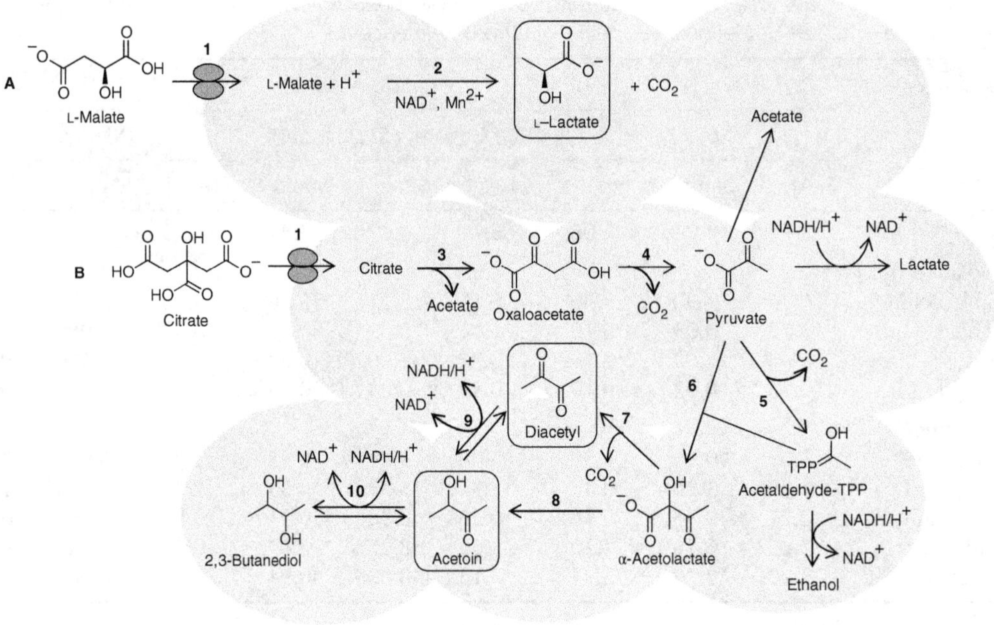

Figure 22.5.1 Formation of L-lactic acid from L-malic acid metabolism (pathway A) and MLF-associated volatile compounds, diacetyl and acetoin, from citric acid metabolism (pathway B) by Oenococcus oeni. Other metabolites arising from citrate, via pyruvate, are also shown. Numbers for the steps refer to the following: 1, carboxylate transporter; 2, malolactic enzyme; 3, citrate lyase; 4, oxaloacetate decarboxylase; 5, pyruvate decarboxylase; 6, α-acetolactate synthase, 7, oxidative decarboxylation (nonenzymatic), 8, α-acetolactate decarboxylase, 9, diacetyl reductase, 10, acetoin reductase

L-Malic acid is transported into the cell as monovalent malate HM^- [7] where the decarboxylation reaction (Figure 22.5.1, pathway A) is catalyzed by the malolactic enzyme (L-malate:NAD^+ carboxylase) in the presence of NAD^+ and Mn^{2+} as cofactors [4, 8]. The resulting L-lactic acid and CO_2 are excreted from the cell by diffusion. Unlike the reactions involved in glycolysis (Chapter 22.1), the enzymatic decarboxylation of malic to lactic acid does not directly yield ATP and the process is energetically unfavorable. However, this reaction consumes a proton, increasing intracellular pH and resulting in a proton gradient (i.e., proton-motive force, PMF) across the cell membrane. The PMF facilitates malate transport and, in combination with cell membrane ATPases, can be used to generate energy in the form of ATP [5].

22.5.2.2 Metabolism of citric acid by LAB

Another important outcome of MLF is the metabolism of citric acid to form diacetyl, acetoin, and 2,3-butanediol (Figure 22.5.1, pathway B) [4, 5, 9, 10], of which the "buttery" smelling diacetyl has the lowest sensory threshold and the greatest impact on wine flavor (Chapter 9) [2]. As with malate metabolism, citrate is transported into the cell as a singly charged anion and cleaved to yield acetate and oxaloactetate. Although alternate pathways exist, the majority of oxaloacetate will eventually be transformed into α-acetolactate via reactions that produce pyruvate and acetaldehyde-thiamine pyrophosphate (TPP). Under semi-aerobic conditions, α-acetolactate can spontaneously react with O_2 to yield diacetyl (Figures 22.5.1 and 22.5.2).

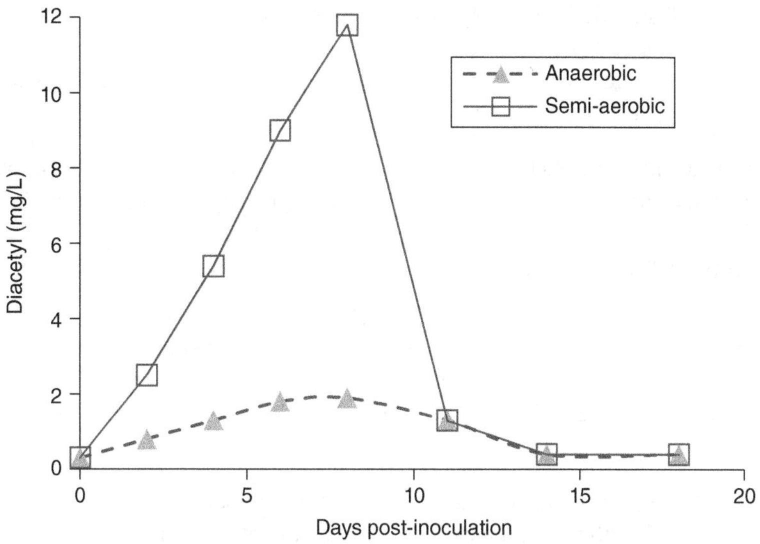

Figure 22.5.2 *Diacetyl concentrations (mg/L) during MLF as a function of time after inoculation, showing that semi-aerobic conditions promote greater formation of diacetyl than anaerobic conditions. In both situations, diacetyl will eventually be reduced to acetoin and 2,3-butanediol through LAB metabolism. Data from Reference [12]*

However, under anaerobic conditions typical to wine production, α-acetolactate can be sequentially converted to less odorous acetoin and then reduced to 2,3-butanediol. This process helps to maintain redox balance by regenerating NAD^+ from NADH, and is analogous to the role of acetaldehyde reduction to ethanol during alcoholic fermentation (Chapter 22.1). After reaching a maximum, diacetyl can be enzymatically reduced by LAB to acetoin, and subsequently 2,3-butanediol. Factors that affect diacetyl production[4] and stability in wine are reviewed elsewhere [11], but major effects include the following [4, 12]:

- Citrate is almost fully consumed after malate degradation is complete[5] and higher initial concentrations of citric acid in wine will result in greater diacetyl (and acetate) production.
- Semi-aerobic conditions will favor the accumulation of diacetyl through its non-enzymatic formation.
- Differences among bacterial strains can alter metabolite profiles; for example, higher diacetyl reductase activity in a strain can result in greater conversion of diacetyl to acetoin. Additionally, storage in the presence of LAB or yeast lees will result in lower diacetyl concentrations due to the presence of reductases.

[4] Winemakers may aim to increase diacetyl concentrations by fortifying wine with citric acid, whereas the higher concentrations of citric acid present in other fermented fruit beverages (e.g., some pears used for perry) mean that MLF should be strictly controlled or avoided altogether.

[5] There are two related consequences worthy of attention. Firstly, citric acid may not be fully metabolized if a wine is sulfited once MLF is deemed complete based on the disappearance of malic acid. Secondly, diacetyl concentrations peak roughly when all malic acid has been metabolized. These scenarios can therefore provide a control point for optimizing diacetyl concentration (and buttery characteristics) in finished wine, by allowing complete citric acid utilization and then waiting for adequate enzymatic reduction to acetoin/2,3-butanediol prior to adding sulfite post-MLF.

- Conditions that favor faster MLF (due to higher temperature and pH) tend to produce greater amounts of acetic acid at the expense of diacetyl/acetoin.
- Diacetyl can be partially and reversibly bound by SO_2 (Chapters 9 and 17), decreasing its volatility and masking the intensity of its buttery aroma.

Metabolism of citric acid will also yield lactate, ethanol, ethyl lactate (from esterification of lactate), and at least one molar equivalent of acetic acid; hence volatile acidity (VA) typically increases during MLF (Table 22.5.1).

22.5.2.3 Fermentation of sugars by LAB

L-Malic acid and citric acid cannot be used as sole carbon sources for growth of LAB [13], and biomass production requires fermentable sugars or amino acids [5]. Co-metabolism of glucose and citric acid is energetically favorable for *O. oeni* and leads to an increased growth rate and enhanced biomass production [10]. LAB can be distinguished by their ability to ferment sugars and are classified as homofermentative or heterofermentative [1, 14]:

- Homofermentative LAB produce lactic acid[6] as the primary metabolite by reduction of pyruvate formed through glycolysis (EMP pathway, Chapter 22.1). In wine, homofermentative LAB are typically associated with lactic spoilage, and include *Pediococcus* spp. and some *Lactobacillus* spp.
- Heterofermentative LAB such as *O. oeni* and a number of *Lactobacillus* spp. metabolize hexoses and pentoses (and other carbohydrates) using the pentose-phosphate (phosphoketolase) pathway (Figure 22.5.3) to produce not only D-lactic acid (in addition to L- and DL-lactic acid in the case of some *Lactobacillus* spp.), but also acetic acid, ethanol, CO_2, and other products such as glycerol and erythritol [15, 16].
- Heterofermentative *Lactobacilli* can be subdivided into strict (e.g., *L. brevis* and *L. hilgardii*) and facultative heterofermenters (e.g., *L. casei* and *L. plantarum*) based on their metabolism of hexoses; facultative heterofermenters produce only lactic acid by the EMP pathway and the strict heterofermenters produce the array of products mentioned above. In all cases, heterofermentative LAB ferment pentoses to lactic acid and acetic acid (or ethanol under reductive conditions) as the main products by the phosphoketolase pathway. Since pyruvate is an intermediate in the pathway, volatile compounds such as diacetyl and acetoin can also be formed (see Figure 22.5.1).

22.5.2.4 Effects of LAB on wine composition and stability

Similar to yeast, LAB have the ability to reduce aldehydes and other carbonyls to their corresponding alcohols (Chapter 22.1). As a result, concentrations of several key SO_2 binders, particularly acetaldehyde, will decrease during MLF (see Table 22.5.1). The concentrations of other volatile compounds can also be altered during MLF due to the production of glycosidases (Chapter 23.1) and esterases (Chapter 22.2). The presence of residual sugars in a wine (whether intentional or not) can encourage the growth of spoilage organisms and the formation of unwanted metabolites (see Section 22.5.3 below). Wine pH less than 3.5 favors the presence of *O. oeni* (with its often desirable organoleptic consequences) and helps to eliminate the potential for spoilage by other bacteria [17]; however, if the pH is higher than 3.5, *Pediococcus* spp. can flourish and undertake MLF. Even after the addition of SO_2, some LAB (particularly *Lactobacillus* spp. and *Pediococcus* spp.) can remain viable post-MLF, producing undesirable organoleptic changes over time. In general, LAB are much more susceptible to SO_2 than yeasts, and sulfiting of wine is usually avoided if MLF is to be encouraged.

[6] Depending on the strain, LAB can produce D-, L-, or DL-lactic acid by fermentation due to differences in the stereospecificity of lactate dehydrogenase enzymes, which reduce the ketone group of pyruvate to produce the stereocenter in lactic acid. In contrast, decarboxylation of L-malic acid yields exclusively L-lactic acid, as the stereocenter is unaffected by the malolactic transformation.

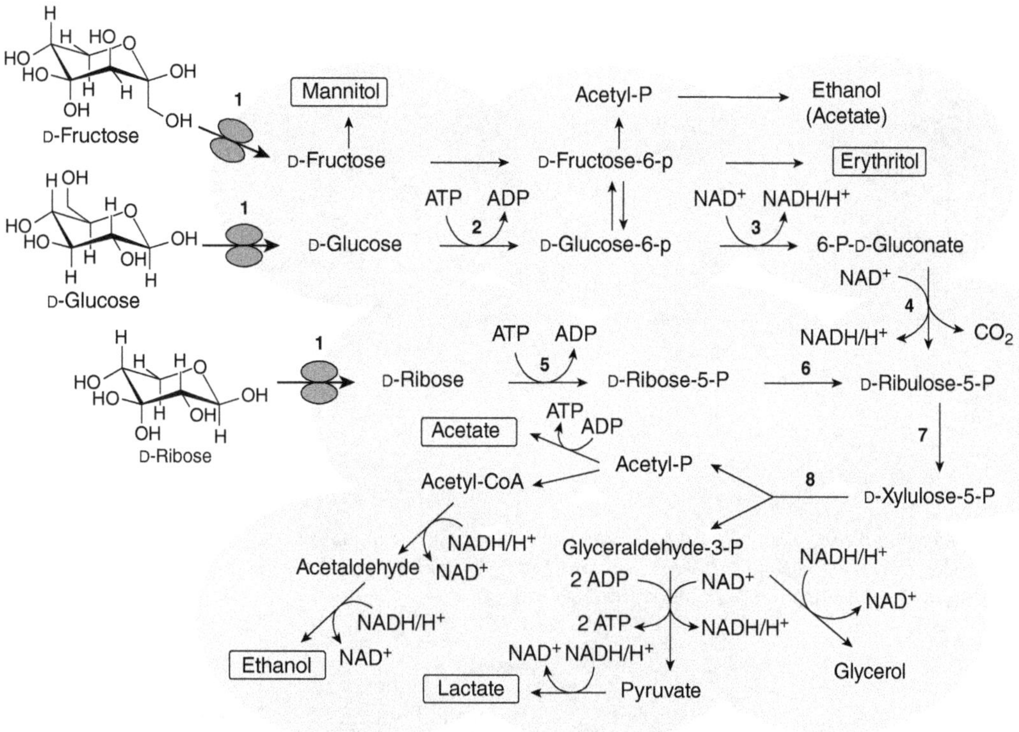

Figure 22.5.3 Heterofermentative pathways for LAB based on hexose and pentose sugars (shown in the β-pyranose conformation), yielding a range of metabolites. Numbers for the steps refer to the following: 1, sugar transporter; 2, hexokinase; 3, glucose-6-phosphate dehydrogenase; 4, 6-phosphogluconate dehydrogenase; 5, ribokinase; 6, ribose 5-phosphate ketol-isomerase; 7, ribulose-5-phosphate 3-epimerase; 8, phosphoketolase

The changes in major indigenous LAB species have been mapped from the grapes/must through to the end of MLF, revealing that most *Lactobacillus* spp., *O. oeni*, and *Pedioccocus* spp. cells present in the must do not survive alcoholic fermentation, which is attributed to increases in ethanol, SO_2, and other yeast metabolites. Assuming no LAB inoculation occurs, the few surviving cells of *O. oeni* develop and conduct MLF [1, 4, 17]. Ultimately, apart from the use of SO_2 and control of wine pH, sterile filtration (Chapter 26.3) is necessary to remove these bacteria and render a wine microbiologically stable.

22.5.3 Spoilage of wine by bacteria

Bacterial spoilage is defined as the unintended growth of bacteria and the subsequent production (or accumulation to objectionable levels) of compounds that negatively impact on wine quality due to undesirable organoleptic and/or health effects [1–4, 18]. Several compounds formed during LAB spoilage have been described:

- An increase of acetic acid and diacetyl, as described above.
- Formation of mannitol from D-fructose (Figure 22.5.3), which causes a viscous texture, sweet taste, and irritating finish (Chapter 2).

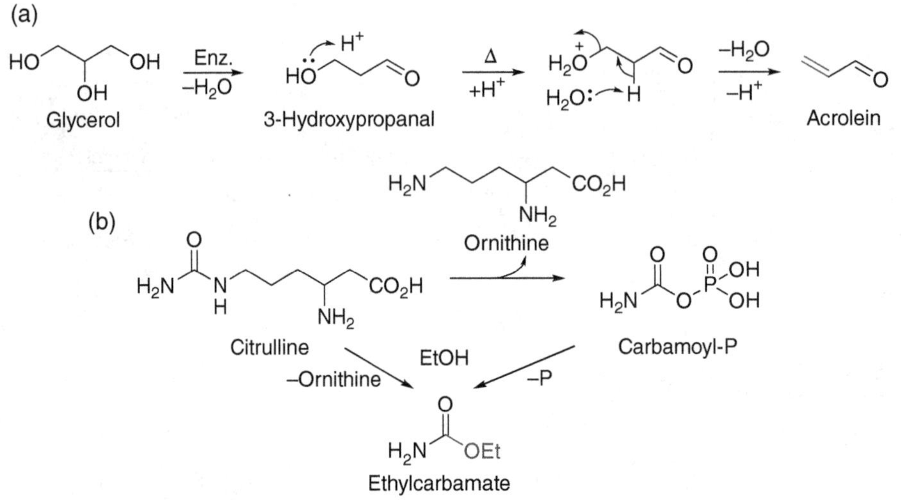

Figure 22.5.4 Formation of (a) acrolein under acidic wine conditions from glycerol produced by heterofermentative LAB, where Enz. refers to glycerol hydro-lyase and Δ (i.e., heat) indicates an impact from temperature, and (b) ethyl carbamate from citrulline or carbamyl-P by ethanolysis

- Metabolism of glycerol (*amertune*) to acrolein (i.e., 2-propenal, Figure 22.5.4). In addition to being toxic, acrolein can induce bitterness following reaction with phenolics, a more likely issue in red wine.
- *Ropiness*, where the production of exopolysaccharides (β-D-glucan from fermentation of residual glucose) by *Pediococcus* spp. leads to an abnormally viscous wine texture.
- Heterofermentative LAB can produce a *mousy* off-flavor due to heterocyclic imines such as 2-acetyl-3,4,5,6-tetrahydropyridine (Chapter 5) from ornithine or lysine metabolism, in conjunction with acetyl-CoA production from acetyl-P during fermentation of sugars (Figure 22.5.3).
- Biogenic amines such as histamine and putrescine can arise enzymatically from decarboxylation of the corresponding amino acids due primarily to the presence of *Pediococcus* spp. and *Lactobacillus* spp. (Chapter 5).
- Metabolism of arginine produces citrulline and carbamoyl-P through the arginine deaminase pathway, and these intermediates can react with ethanol to yield carcinogenic ethyl carbamate (Figure 22.5.4), although production of urea by yeast is the main source of this compound (Chapter 5).

LAB also possess cinnamoyl decarboxylase activity, which can provide a source of 4-vinylphenol/4-vinylguaiacol from hydroxycinnamic acids present in wine, thereby increasing the pool of compounds that can be metabolized by *Brettanomyces* (Chapters 12 and 23.3). Furthermore, the presence of reductases means care should be taken when employing sorbic acid (or potassium sorbate) as a preservative to prevent refermentation in sweet wines. Enzymatic reduction of the acid functional group to a primary alcohol by LAB provides an activated dienol, which can undergo rearrangement to produce 2-ethoxyhexa-3,5-diene, a compound responsible for *geranium taint* (Chapter 18).

AAB in the form of *Acetobacter* spp. and *Gluconobacter* spp. only serve as spoilage bacteria in wine, in contrast to the potentially desirable role that LAB can play. The presence of AAB contributes to production of acetaldehyde, acetic acid, and ethyl acetate due to oxidation of ethanol, leading to increases in VA and vinegary characteristics in wine (Chapter 3). The risks from *Acetobacter* spp. tend to arise post-fermentation, during extended maturation in barrels, storage in tank, or post-bottling in conjunction with exposure to air. Rotting grapes present higher populations of AAB and a greater potential for spoilage before fermentation starts. *Gluconobacter* is typically found on grapes and in musts but does not withstand alcoholic fermentation; thus, its primary role in wine spoilage is from damaged grapes, where it generates high concentrations of both gluconic acid (Chapter 2) and acetic acid.

22.5.3.1 Protecting against spoilage

Disease in the vineyard and operations in the winery will affect the presence, viability, and growth of microorganisms. Aside from controlling these aspects, a number of other compositional and winemaking parameters influence bacterial activity [2, 18]. Some have been mentioned in the last paragraph of Section 22.5.2 above with respect to LAB, namely pH control (by addition of tartaric acid or other permitted acids), maintaining effective levels of SO_2 (at each stage of the winemaking process), and sterile filtration (especially prior to bottling). Other precautionary measures include the following:

- Minimizing residual sugar and nitrogen requires healthy fermentations, fermenting to dryness to avoid the risk of refermentation, and avoiding the overuse of diammonium phosphate.
- Pasteurization, high pressure, or ultrasonic treatments are used to reduce the load of viable bacteria.
- Appropriate temperature control, such as cooler temperatures, for example, 15 °C, reduces the rate of bacterial growth.
- Inoculating with a commercial LAB strain with known characteristics; spontaneous MLF by indigenous LAB can produce the range of undesirable results mentioned above.
- The use of preservatives other than SO_2, for example, dimethyldicarbonate, depending on local regulations (Chapter 27).
- Limiting exposure to oxygen by eliminating ullage in tanks and using inert gas coverage for wine transfers (Chapter 19).
- Using clean wine for topping up vessels.
- Routine quality control, such as testing of filter integrity and microbial populations.

References

1. Costantini, A., García-Moruno, E., Moreno-Arribas, M.V. (2009) Biochemical transformations produced by malolactic fermentation, in *Wine Chemistry and Biochemistry* (eds Moreno-Arribas, M.V. and Polo, M.C.), Springer, New York, pp. 27–57.
2. Bartowsky, E.J. (2009) Bacterial spoilage of wine and approaches to minimize it. *Letters in Applied Microbiology*, **48** (2), 149–156.
3. Bartowsky, E.J. and Pretorius, I.S. (2009) Microbial formation and modification of flavor and off-flavor compounds in wine, in *Biology of microorganisms on grapes, in must and in wine*, (eds König, H., Unden, G., Fröhlich, J.), Springer-Verlag, Berlin and Heidelberg, pp. 209–231.
4. Lonvaud-Funel, A. (1999) Lactic acid bacteria in the quality improvement and depreciation of wine. *Antonie van Leeuwenhoek*, **76** (1–4), 317–331.
5. Versari, A., Parpinello, G.P., Cattaneo, M. (1999) *Leuconostoc oenos* and malolactic fermentation in wine: a review. *Journal of Industrial Microbiology and Biotechnology*, **23** (6), 447–455.
6. Ruiz, P., Izquierdo, P.M., Seseña, S., *et al.* (2012) Malolactic fermentation and secondary metabolite production by *Oenoccocus oeni* strains in low pH wines. *Journal of Food Science*, **77** (10), M579–M585.
7. Salema, M., Poolman, B., Lolkema, J.S., *et al.* (1994) Uniport of monoanionic L-malate in membrane vesicles from *Leuconostoc oenos*. *European Journal of Biochemistry*, **225** (1), 289–295.

8. Radler, F. (1986) Microbial biochemistry. *Experientia*, **42** (8), 884–893.
9. Swiegers, J.H., Bartowsky, E.J., Henschke, P.A., Pretorius, I.S. (2005) Yeast and bacterial modulation of wine aroma and flavour. *Australian Journal of Grape and Wine Research*, **11** (2), 139–173.
10. Ramos, A. and Santos, H. (1996) Citrate and sugar cofermentation in *Leuconostoc oenos*, a ^{13}C nuclear magnetic resonance study. *Applied and Environmental Microbiology*, **62** (7), 2577–2585.
11. Bartowsky, E.J. and Henschke, P.A. (2004) The "buttery" attribute of wine – diacetyl – desirability, spoilage and beyond. *International Journal of Food Microbiology*, **96** (3), 235–252.
12. Nielsen, J.C. and Richelieu, M. (1999) Control of flavor development in wine during and after malolactic fermentation by *Oenococcus oeni*. *Applied and Environmental Microbiology*, **65** (2), 740–745.
13. Liu, S.Q., Davis, C.R., Brooks, J.D. (1995) Growth and metabolism of selected lactic acid bacteria in synthetic wine. *American Journal of Enology and Viticulture*, **46** (2), 166–174.
14. Zúñiga, M., Pardo, I., Ferrer, S. (1993) An improved medium for distinguishing between homofermentative and heterofermentative lactic acid bacteria. *International Journal of Food Microbiology*, **18** (1), 37–42.
15. Unden, G. and Zaunmüller, T. (2009) Metabolism of sugars and organic acids by lactic acid bacteria from wine and must, in *Biology of microorganisms on grapes, in must and in wine* (eds König, H., Unden, G., Fröhlich, J.), Springer-Verlag, Berlin and Heidelberg, pp. 135–147.
16. Kandler, O. (1983) Carbohydrate metabolism in lactic acid bacteria. *Antonie van Leeuwenhoek*, **49** (3), 209–224.
17. Wibowo, D., Eschenbruch, R., Davis, C.R., *et al.* (1985) Occurrence and growth of lactic acid bacteria in wine: a review. *American Journal of Enology and Viticulture*, **36** (4), 302–313.
18. du Toit, M. and Pretorius, I.S. (2000) Microbial spoilage and preservation of wine: using weapons from nature's own arsenal – a review. *South African Journal for Enology and Viticulture*, **21** (Special Issue), 74–96.

23

Grape-Derived Aroma Precursors

In contrast to the previous chapters, which consider wine compounds common to all alcoholic beverages, these chapters detail the (bio)chemical reactions that transform grape-derived precursors into (varietal) secondary and tertiary aroma compounds. The subsequent sections describe:

- Biosynthesis of glycosides in the berry, their extraction, and hydrolysis during fermentation and aging
- Formation of *S*-conjugates (thiol precursors) pre- and post-harvest, cleavage during fermentation, the concept of sulfur mass balance and alternative pathways to thiol formation
- Conversion of other grape components – polyunsaturated fatty acids (C_6 volatiles), hydroxycinnamic acids (*Brettanomyces* and volatile phenols), and *S*-methylmethionine (dimethyl sulfide).

Understanding Wine Chemistry, First Edition. Andrew L. Waterhouse, Gavin L. Sacks, and David W. Jeffery.

23.1

Glycosidic Precursors to Wine Odorants

23.1.1 Introduction

Glycosides are plant secondary metabolites consisting of a non-sugar component, called an *aglycone*, attached to one or more sugars (Figure 23.1.1).[1] Glycosides are ubiquitous in the plant kingdom and found in all major plant organs, including fruit, leaf, seed, flower, bark, and root [1, 2].[2] Most aglycones are non-polar or semi-polar – glycosylation increases their water solubility and lowers reactivity, which in turn facilitates the transport, accumulation and storage, and detoxification of these compounds [1–3]. Glycoside metabolism in plants involves glycoside transferases and hydrolases to catalyze the glycosylation and hydrolysis of aglycones [3, 4]. Several glycosides of non-volatile phenolic aglycones are discussed earlier in this book, including anthocyanins and flavonols (Chapter 11). However, the focus of this chapter is on glycosides of *volatile* aglycones, including aliphatic alcohols (e.g., C_6 compounds), shikimate derivatives (e.g., benzyl alcohol, phenols, vanillin) and mevalonate/deoxyxylulose phosphate (DXP) derivatives (e.g., monoterpenoids, C_{13}-norisoprenoids) (Figure 23.1.2), which can serve as precursors of odorous compounds following hydrolysis [5]. These glycosides are often referred to as *bound* compounds, which can be converted to *free* odorants during fermentation and storage. Several factors are important in determining the eventual concentrations of free aroma compounds arising from glycosidic precursors, including the amounts of precursors present in juice or must, enzymatic activity during fermentation, and pH/temperature during storage.

[1]That is, an aglycone linked by its functional group to a sugar (usually an alcohol, giving an *O*-glycoside) through a glycosidic bond involving the anomeric hydroxyl. The position of the anomeric hydroxyl can be axial (i.e., makes a large (90°) angle relative to a plane passing closest to the majority of ring atoms) or equatorial (makes a small angle, as seen in Figure 23.1.1); in many cases the β-anomer has the hydroxyl equatorial and the α-anomer has it axial. However, the naming of α- and β-anomers, which is beyond the scope of this chapter, does not simply derive from the anomeric hydroxyl being axial or equatorial in the cyclic form of a sugar.

[2]A complete account of all economically important plant-derived glycosides is well outside the scope of this book, but such a list would include salicin from willow bark, which is a glucoside of salicylic acid and a close relative of aspirin; indican, the glucoside of indigo dye; and naringin, a key bitter component of grapefruit juice.

Understanding Wine Chemistry, First Edition. Andrew L. Waterhouse, Gavin L. Sacks, and David W. Jeffery.

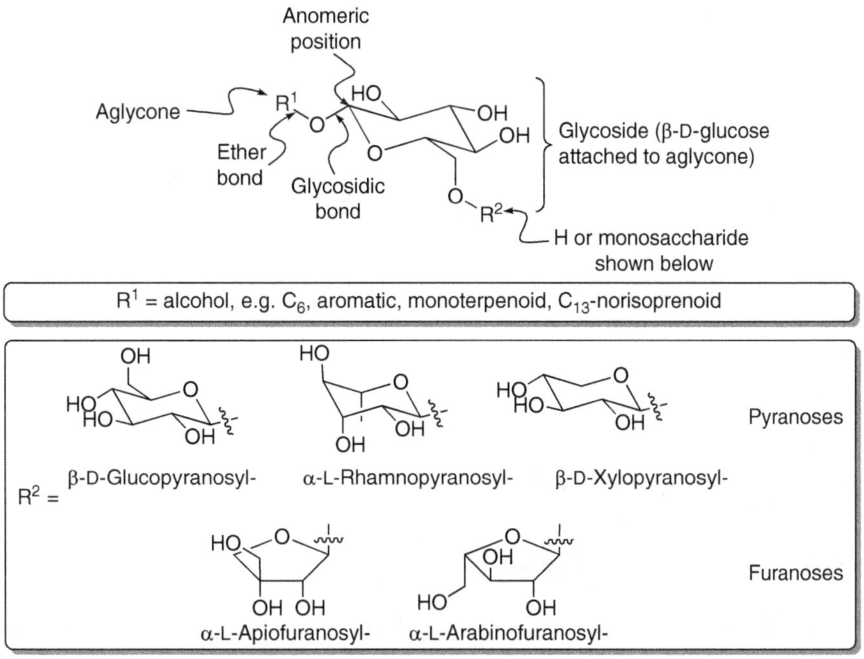

Figure 23.1.1 *Structure of a generic glycoside, showing the aglycone (R^1) attached at the anomeric hydroxyl of* D-*glucose (β-anomer shown; hence β-*D-*glucoside). The bond between aglycone and oxygen is known as the ether bond, whereas the one between the oxygen and sugar moiety is the glycosidic bond. Different sugars (R^2) can be attached to the 6-position hydroxyl of glucose, usually through their own anomeric position as indicated (∿∿). Note for the β-anomer of* D-*glucose the hydroxyls are all in the equatorial position and the hydrogens, although not drawn, are all axial*

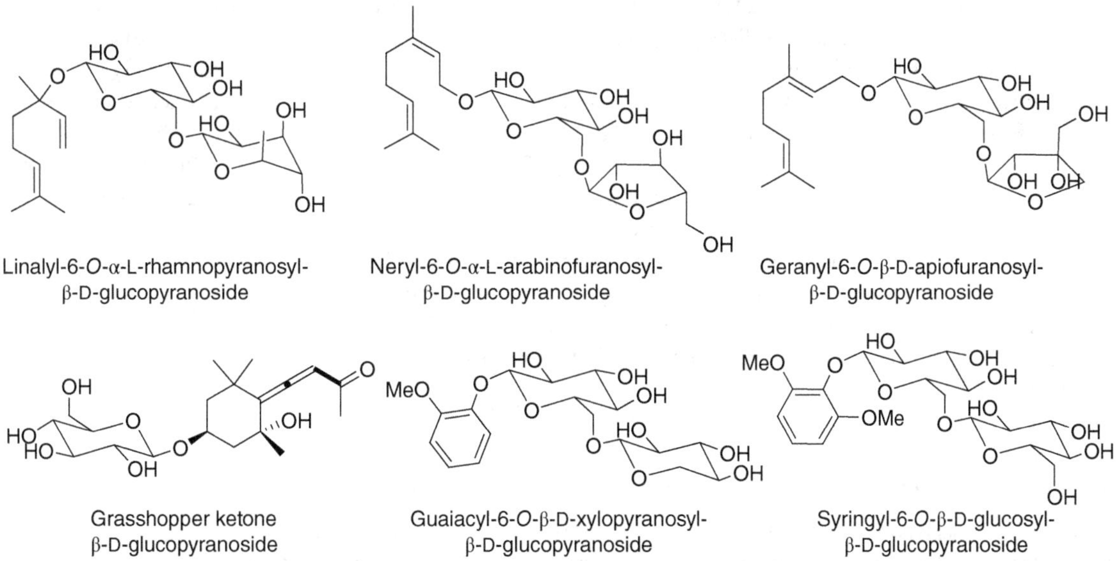

Figure 23.1.2 *Examples of the types of glycosides of volatile aglycones (monoterpenoid, C_{13}-norisoprenoid, phenolic) identified in grapes and wines*[3]

[3] A rhamnosyl-glucoside is commonly called a rutinoside and a glucosyl-glucoside can be termed a gentiobioside (or a sophoroside if the linkage is 2-*O*- rather than 6-*O*-).

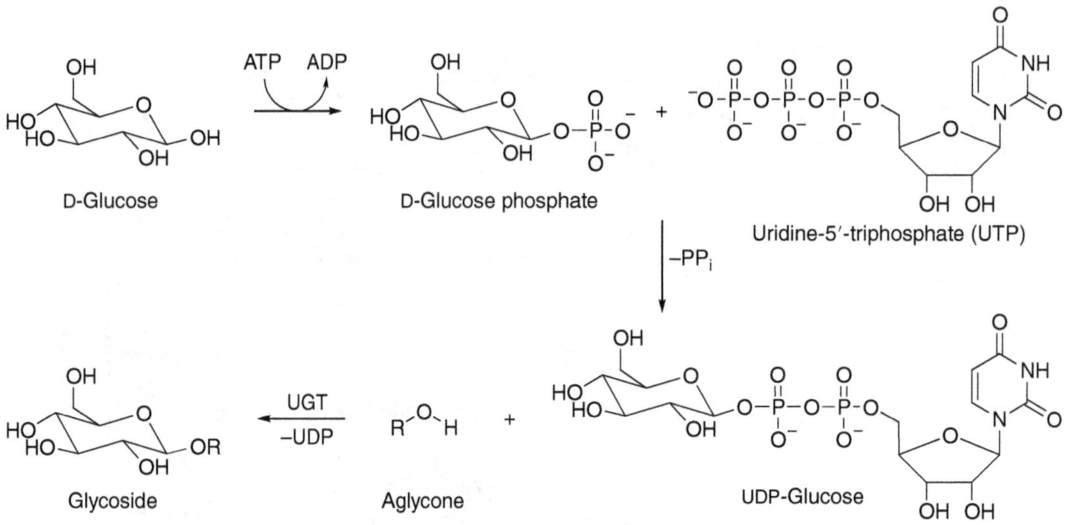

Figure 23.1.3 Putative grape berry glycosylation reaction showing formation and transfer of an activated glycosyl residue (e.g., UDP-glucose) to an aglycone acceptor under the action of UGT[4]

23.1.2 Formation of glycosidic aroma precursors in grape berries

Glycosides in wine originate from the grape berry, where they accumulate during ripening, and appear to be well correlated with the concentration of their corresponding aglycone(s). The aglycone is always attached directly to β-D-glucose, yielding a monosaccharide (i.e., an *O*-β-D-glucoside, Figure 23.1.1, R^2=H), but the glucose can be further substituted by other sugars (α-L-arabinofuranose, α-L-rhamnopyranose, β-D-xylopyranose, β-D-apiofuranose, β-D-glucose, Figure 23.1.2) to give the corresponding disaccharides [1, 6]. The mechanisms underpinning glycosylation of volatiles in grape berries have not been fully elucidated but their formation putatively involves the reaction of an alcohol with a sugar. This is an energetically unfavorable reaction, and requires activated nucleotide sugars and the presence of uridine diphosphate (UDP) glycosyltransferase enzymes (UGTs, Figure 23.1.3) [7]. The aglycone components are often biosynthesized in the grape berry, either directly in the case of monoterpenoids (Chapter 8, and polyhydroxylated variants outlined below) and higher alcohols (Chapter 6), or as a result of the breakdown of carotenoids (Chapter 8), as described in more detail below. Grapes can also produce trace concentrations of volatile phenol glycosides, but concentrations capable of releasing suprathreshold amounts of aglycones are typically observed only following contamination of grapevines by wildfire smoke (Chapter 12). The major classes of glycosidic precursors to wine aroma compounds, and factors determining their levels in grapes, are summarized in Table 23.1.1.

23.1.2.1 Carotenoid-derived precursors

Of the various aglycone classes, formation of C_{13}-norisoprenoid precursors deserves special discussion due to their genesis from multiple precursors and intermediates. As mentioned in Chapter 8, the key odorous

[4] Conjugation can occur with either retention of configuration at the anomeric position of the sugar (as shown here) or inversion.

Table 23.1.1 *Classes of wine aroma compounds bound as glycoconjugates in grape berries, examples of the volatiles formed from each, and factors affecting berry glycoside concentrations*

Aglycone class	Key volatiles formed	Major factor(s) determining concentration
Monoterpenoids[a]	Linalool, geraniol	Cultivar: higher in Muscat-type grapes Maturity: accumulation commences ~4 weeks after veraison and continues through ripening
C_{13}-Norisoprenoids[a]	TDN, β-damascenone	Cultivar (for TDN): higher in Riesling Growing conditions (for TDN): more cluster exposure to sun, lower N and water. Maturity: accumulation commences 1–3 weeks after veraison and continues through ripening
Volatile phenols	Guaiacol, 4-methylguaiacol, vanillin, syringol	Growing conditions: exposure of grapes to smoke around veraison, or upon application of oak extracts to vines
Higher alcohols	Hexanol	Unknown

[a] May also be formed by rearrangement of odorless polyols, as described below.

C_{13}-norisoprenoids are derived from precursors linked to carotenoid degradation [8, 9]. Carotenoids are widely distributed, naturally occurring pigments present in mature grape berries at total concentrations around 0.4–2.5 mg/kg, primarily as lutein and β-carotene. Carotenoids have key roles in photosynthesis (including in green berries) and concentrations in unripe grapes are often 3-fold higher than at maturity [10, 11]. Carotenoid degradation will commence 1 week pre-veraison, and C_{13}-norisoprenoid precursors will begin to accumulate about 1–2 weeks after veraison [12]. While oxidative degradation could potentially occur spontaneously, several carotenoid cleavage dioxygenases (CCD) have been identified in grapes [13], and at least one (encoded by *VvCCD1*) is demonstrated to produce C_{13}-norisoprenoids from carotenoids [12]. Exemplary pathways for formation of two odor-active C_{13}-norisoprenoids from two different carotenoids involve:

- Direct formation in grapes from oxidative cleavage of carotenoids, as appears to happen with β-ionone (Chapter 8).
- Formation of non-odorous C_{13}-norisoprenoid intermediates, which can undergo acid-catalyzed rearrangements during storage; for example, allenic triol can serve as a precursor to β-damascenone during acidic storage (Figure 23.1.4a). These intermediates often contain multiple OH groups (i.e., they are polyols) and may exist in grapes as glycosides. This pathway appears to occur for TDN, vitispirane, actinidols, and some β-damascenone precursors, and will require an initial hydrolysis step to release the glycoside, as described further below.

23.1.2.2 Polyhydroxylated monoterpenoids

Free and glycosidically bound monoterpenoid diols (i.e., diendiols related to linalool, geraniol, nerol, and citronellol) have been identified in grape juices [14–18]. Odorless polyhydroxylated variants are produced enzymatically, and glycosylated as outlined above. The pool of bound monoterpenoids fluctuates but generally increases with grape berry ripening, and exceeds the concentration of free monoterpenoids in

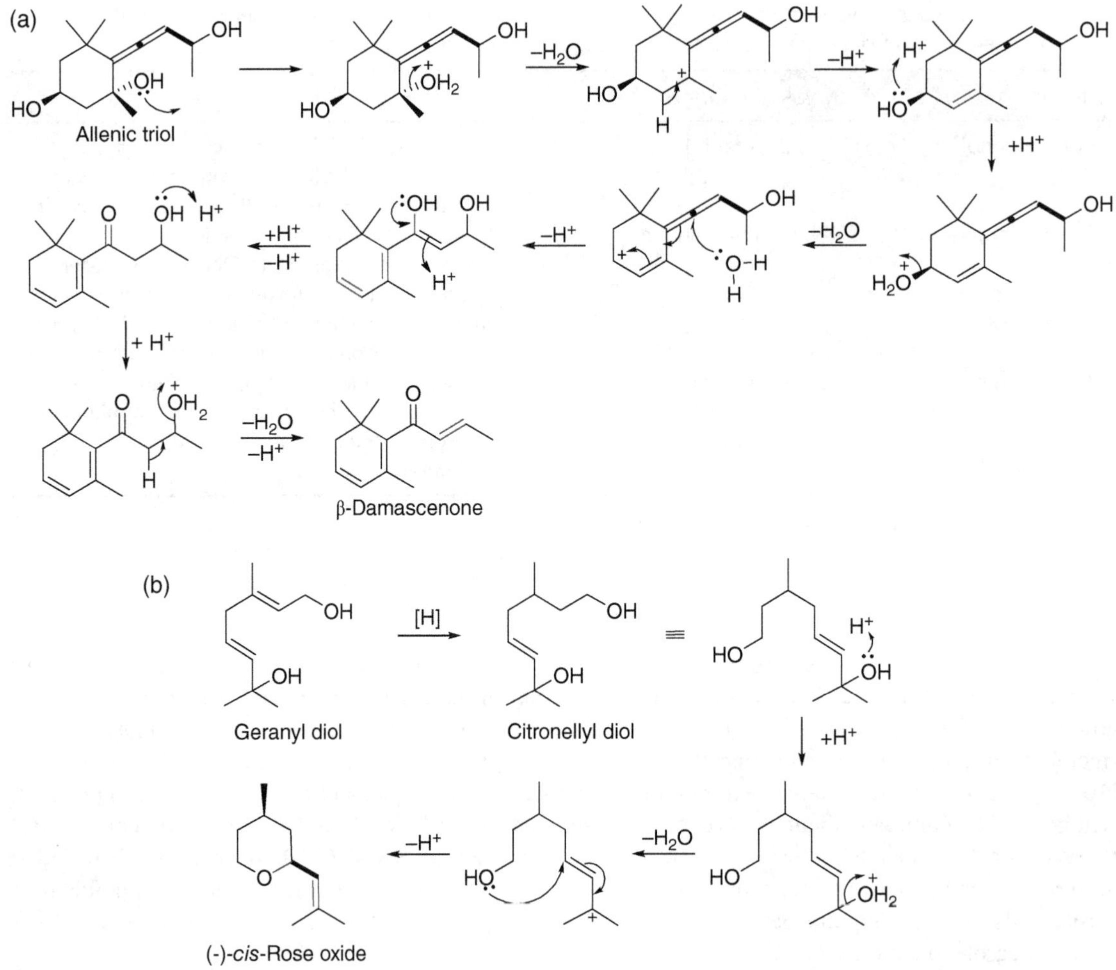

Figure 23.1.4 Acid-catalyzed rearrangements of non-odorous precursors leading to formation of (a) β-damascenone from carotenoid-derived allenic triol (earlier polyol intermediates in this rearrangement sequence may also be accumulated as glycosidic precursors in grapes, where glucose is conjugated to one of the hydroxyl groups) and (b) (−)-cis-rose oxide from citronellyl diol, which can form from geranyl diol after enzymatic reduction ([H]) of a double bond

mature berries. The main exception to this relates to a linalyl diendiol ((*E*)-3,7-dimethylocta-1,5-dien-3,7-diol), which was found to increase rapidly during ripening and eventually exceed the concentration of all other free monoterpenoids [19]. As with aroma precursors, such as C_{13}-norisoprenoids described above, acid-catalyzed hydrolysis of glycosides and rearrangements lead to a range of wine aroma compounds, such as hotrienol, *cis*-rose oxide (and linalool or nerol oxides [14]), and wine lactone (Chapters 8 and 25). An example showing the formation of *cis*-rose oxide from citronellyl diol (which can arise from geranyl diol) [20] is shown in Figure 23.1.4b.

Like other grape-derived aroma precursors such as varietal thiol conjugates (Chapter 23.2), glycosides are distributed between the pulp (including juice) and skin of grape berries to varying extents depending on the aglycone, with a slight predominance in favor of skin [15, 21, 22]. From the analysis of volatiles released during hydrolysis experiments, it can be inferred that juice and must concentrations of different glycosides are in the order of tens to thousands of μg/L [23–25].

23.1.3 Glycosidic aroma precursors – extraction

Because glycosidic aroma precursors are proportionally higher in the skins, considerable research has explored strategies to increase glycoside extraction in wines that are pressed prior to fermentation, that is, most white wines. Most commonly, this is achieved through skin contact, often in conjunction with enzyme preparations:

- Pre-fermentation skin contact facilitates extraction of glycosidic precursors. Macerating crushed grapes for 4–12 hours at 10 °C [26] or 6–23 hours at 15–18 °C prior to pressing and fermenting [27–31], led to significant increases (up to 2-fold in total for a compound class) for a range of volatile compounds arising from glycosides, including monoterpenoids, benzenoids, and C_{13}-norisoprenoids.
- The use of commercial pectolytic enzymes (i.e., pectinases, also used for juice clarification, Chapter 21) can aid in pre-fermentative extraction of aroma precursors from skins of white and red grape varieties [32]. Treatment of grapes [33] and macerating musts [34] with pectinases enhanced glycoside concentrations in a similar manner to skin contact alone.
- Conversely, the addition of pectinase during clarification of juice has also been found to significantly lower the concentration of glycosides [35]. This may be because many pectinase preparations can possess glycosidase side-activity (Table 23.1.2), which hydrolyses glycosides and releases bound aroma compounds, as discussed below. This aspect of pectinase use will affect wine aroma release from glycosides rather than extraction of the glycosides from cells.

Table 23.1.2 Enzymatic activities (nkat/mg)[5] of some commercial pectinase/hemicellulase preparations that have been used in winemaking. Data from Reference [38]

Enzyme preparation	β-Glucosidase	α-Arabinosidase	α-Rhamnosidase	β-Apiosidase
AR 2000	5.7	14.7	0.3	1.08
Cellulase A	6.1	0.6	0.07	–
Hemicellulase	7.1	7.0	0.9	–
Novoferm 12	8.4	0.5	0.05	0.15
Pectinase 263	7.2	1.4	0.3	0.2
Pectinol D5S	0.5	0.7	–	–
Pectinol VR	0.2	0.1	–	–
Pektolase 3PA	1.5	3.8	0.04	0.3
Rohament CW	3.3	0.7	0.4	–
Ultrazym 100	0.5	0.1	–	0.03

[5]The SI unit for catalytic activity in mol/s is designated a katal (abbreviated kat); nkat represents nanokatals, which are expressed per mg of enzyme preparation.

23.1.4 Hydrolysis of glycosidic aroma precursors – mechanisms

Being non-volatile, glycosidic aroma precursors need to be hydrolyzed to release the volatile aglycone. Hydrolysis can be acid- or enzyme-catalyzed [6, 36], and the resulting products will differ depending on the mode of hydrolysis [37] and structure of glycosides and aglycones [1].

23.1.4.1 Acid-catalyzed hydrolysis

Glycosides undergo slow acid hydrolysis during wine storage [3, 36, 38] and hydrolytic studies have been conducted using mild (e.g., wine-like pH, 50 °C, several weeks) or harsh (e.g., pH 1, 100 °C, 1 hour) acidic conditions. The former is more suited to examination of compositional changes during wine aging while the latter "forcing conditions" are more useful for evaluating the maximum pool of odorants that could be released from grape glycosidic precursors. Acid hydrolysis can result in cleavage of either the *ether* or *glycosidic* bonds (see Figure 23.1.1), depending on the structure of the aglycone, and released volatile compounds will often undergo further acid-catalyzed rearrangements (Figures 23.1.4 and 23.1.5); higher temperatures and lower pH will lead to greater release and rearrangement of aglycones [37, 39, 40]. Glycosylation stabilizes an aglycone, such that free geraniol rearranges about an order of magnitude faster than geranyl glucoside to form related odorants (e.g., linalool). Glycosylation may also change the relative amounts of products formed following hydrolysis; for example, 50% more β-damascenone is produced relative to its odorless counterpart 3-hydroxy-β-damascone from a glycosidic precursor, as opposed to from the corresponding precursor aglycone [1, 39].

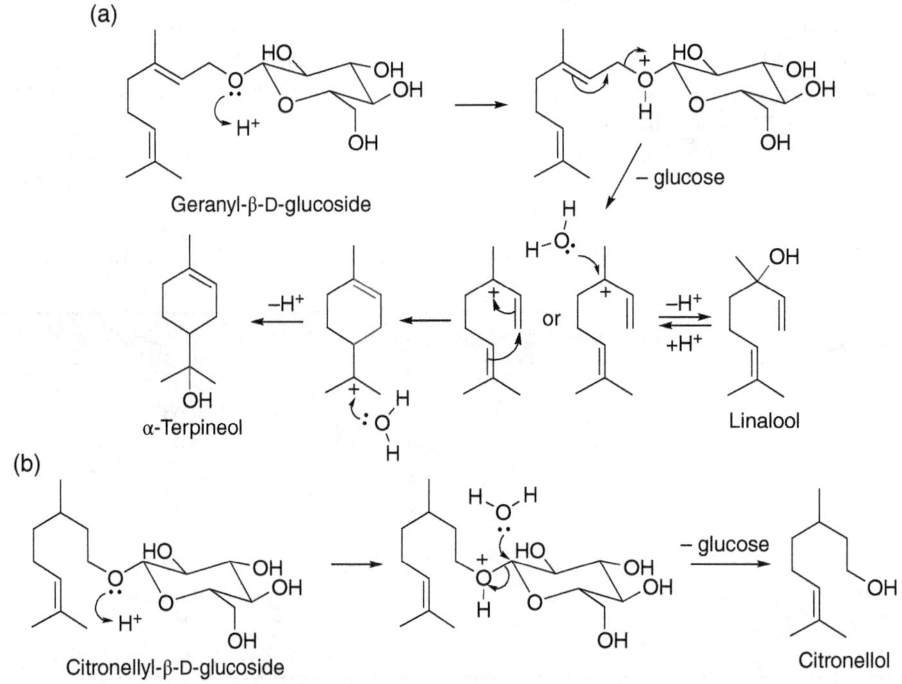

Figure 23.1.5 Acid-catalyzed glycoside hydrolysis of (a) an activated alcohol (i.e., allylic-stabilized carbocation intermediate) such as geraniol (ether bond cleavage) and subsequent reaction of liberated carbocation to form new compounds including linalool and α-terpineol, and (b) unactivated alcohol citronellol (glycosidic bond cleavage) to produce the volatile aglycone, which is similar to the mechanism of enzymatic hydrolysis [4]

23.1.4.2 Enzyme-catalyzed hydrolysis

Aglycone liberation by enzymes (glycosidases) can be single step or sequential [1, 3, 6, 36]. Simple glucosides can be cleaved by the action of β-glucosidase (single step) whereas disaccharides incorporating a sugar other than glucose often require specific enzymes (exoglycosidases, e.g., α-arabinosidase, etc., see Table 23.1.2) to hydrolyze the intersugar linkage, cleaving the terminal sugar prior to the glucose-aglycone bond being hydrolyzed (sequential). There are, however, diglycosidases (endo-β-glucosidases) that can cleave the disaccharide directly in a single step. Grapes contain endogenous glycosidases, with β-glucosidase being the most dominant, but these are essentially inactive at juice pH and inhibited by glucose. Given the abundance of glycosides in grapes and their relatively slow release during fermentation, alternative sources of glycosidases have been evaluated for their potential to liberate aglycones and improve wine aroma. Yeast and bacteria have glycosidase activity during fermentation (Section 23.1.5 and [40]), but commercial preparations typically involve exogenous glycosidases from *Aspergillus niger*, which have greater pH, temperature, glucose, and ethanol tolerance. Ideally, an enzyme preparation will have a broad range of activities to sequentially hydrolyze different disaccharide classes. However, commercial preparations show considerable variation in their specificity towards different glycosides (see Table 23.1.2) [3, 36, 38], which means that different preparations can lead to different sensory profiles. Glycosides can also be released enzymatically by oral microflora [41] (e.g., volatile phenol glycosides, Chapter 12), which can result in a more intense "aftertaste" following consumption of the wine.

The glycosidase side activity of pectinase preparations have been frequently shown to increase free volatile concentrations (Table 23.1.3). However, this glycosidase activity is not necessarily desirable [3, 38, 42]. In the case of red winemaking, β-glucosidase activity towards anthocyanin glucosides should be minimal, due to the important stabilizing influence of glucosylation on anthocyanins (Chapter 16). Similarly, enzyme preparations with cinnamate esterase activity are avoided due to their hydrolytic effect on hydroxycinnamoyl

Table 23.1.3 Effect of exogenous enzyme treatment on the concentrations (μg/L) of classes of aroma compounds in wines of selected grape varieties

Variety	Monoterpenes		Norisoprenoids		Benzenoids[a]		Study
	Control	Treated	Control	Treated	Control	Treated	
Muscat	4384	7718	9	542	–[b]	–	[6]
	1520	2010	–	–	24839	28213	[43]
Riesling	2418	3119	ND[c]	407	–	–	[6]
Sauvignon Blanc	53	198	ND	141	–	–	[6]
	17	26	–	–	19410	22570	[44]
Shiraz	145	1138	ND	847	–	–	[6]
Traminer	72	336	–	–	15691	25606	[44]
Palomino	43	55	–	–	17686	20288	[44]
Chardonnay	45	79	–	–	24438	30053	[43]
	7	9	–	–	10574	10218	[45]
Emir	14	25	48	74	1226	1874	[29]
Airén	56	93	–	–	26128	32045	[43]
	ND	ND	–	–	12217	13640	[45]
Macabeo	52	85	–	–	30264	37665	[43]
	8	9	–	–	8687	9498	[45]

[a] Concentrations predominantly result from 2-phenylethanol in the tens of mg/L range.

[b] –, not reported.

[c] ND, not detected.

tartrate esters (Chapters 13 and 23.3); release of the free hydroxycinnamic acids provides a source of precursors for "Brett" off-flavor formation (Chapters 12 and 23.3). The presence of other esterases is also cause for concern, considering the significance of volatile esters to wine aroma (Chapter 7). As an example, a wine treated with AR 2000 for 15 and 30 days revealed a dramatic decrease in several important "fruity" acetate esters. There was also a several-fold increase in diethyl succinate and diethyl malate, which may not be so consequential but is indicative of the catalytic role of esterases in achieving equilibrium concentrations of esters, through concurrent esterification and hydrolysis reactions (Chapter 7). An alternative enzyme with glycosidase activity, Novarom G, showed no such effect on esters [42].

23.1.4.3 Comparison of hydrolysis by enzyme and acid

As outlined above, differences are noted depending on the nature of the glycosides and mode of hydrolysis. To summarize:

- Acid hydrolysis may result in cleavage at either the ether or glycosidic bonds, while enzymatic hydrolysis results in cleavage at the glycosidic group (see Figure 23.1.5).
- Acid hydrolysis often results in additional acid-catalyzed rearrangements and, by comparison, enzymatic hydrolysis usually results in minimal changes to the aglycone.
- Acid hydrolysis is less sensitive to the sugar group, whereas in enzymatic hydrolysis disaccharide glycoside precursors are usually hydrolyzed more slowly because of the need for multiple selective enzymes.

During winemaking and storage both acid and enzymatic hydrolysis processes will occur, concurrently at some points. Nonetheless, much of the knowledge of aroma precursors has come from studies selectively employing either acid or enzyme hydrolysis of extracts, which are useful approaches for separately assessing the effects of these factors on the release of glycosides. The study of Loscos *et al.* [37] reinforces earlier research and serves as a useful example for comparing *natural* and *accelerated* hydrolysis techniques. Using precursor extracts from different grape varieties, it was shown that alcoholic fermentation tended to release the lowest amount of volatile compounds, enzymatic hydrolysis was very efficient yet the most different to other treatments, and harsh acid hydrolysis (with intermediate release of volatiles) better represented the aroma potential of grapes due to mimicking the acid-catalyzed reactions and rearrangements that occur in wine over time. Generalizing the results:

- Enzymatic hydrolysis was effective at increasing free volatile benzenoid and monoterpenoids. However, the increase in benzenoids was in part due to vinylphenol formation due to cinnamate esterase side activity, as mentioned above and in Chapter 23.3.
- Harsh acid hydrolysis contributed most to C_{13}-norisoprenoids and monoterpenoids.
- Alcoholic fermentation resulted in relatively high formation of benzenoids (especially 2-phenylethanol through amino acid metabolism rather than glycoside release) and lactones (Table 23.1.4).

Overall, the concentrations of volatiles released by harsh acid hydrolysis were better correlated with volatile concentrations arising through alcoholic fermentation than with enzymatic hydrolysis. However, harsh hydrolytic treatment may lead to unrealistic outcomes, as highlighted with studies on the formation of β-damascenone from precursors that would produce negligible amounts of β-damascenone under normal storage conditions [46]. Importantly from a winemaking perspective, due to the low level of hydrolysis, a sizeable pool of precursors remains unaffected by alcoholic fermentation that can then play a role during aging. Additionally, these glycosides can affect aftertaste through in-mouth release by oral microflora (Chapter 12).

Table 23.1.4 Aroma compounds released (sum of relative peak areas) from precursor extracts of a range of grape varieties by hydrolysis under different conditions. Data from Reference [37]

Hydrolytic Treament	Compounds Released[a]	Control[b]	Verdejo	Tempranillo	Chardonnay	Cab Sauv	Merlot	Muscat	Grenache
Fermentation (synthetic must, 200 g/L glucose, Stellevin NT 116, 54 days)	Monoterpenoids	27	35	32	54	31	47	266	36
	Norisoprenoids[c]	1	13	7	13	11	9	9	7
	Benzenoids[d]	13	269	282	189	231	283	184	624
Acid (pH 2.5 citrate buffer, 100 °C, 1h)	Monoterpenoids	2	22	24	81	30	66	1325	95
	Norisoprenoids[c]	0.1	187	174	276	192	214	160	312
	Benzenoids[d]	1	214	248	250	265	306	130	771
Enzyme (AR 2000, pH 5 citrate/ phosphate buffer, 40 °C, 16 h)	Monoterpenoids	1	96	218	137	105	97	963	162
	Norisoprenoids[c]	0.1	26	19	44	21	13	21	30
	Benzenoids[d]	30	7022	9941	5736	10373	5242	2720	11543

[a]Total amount determined from the sum of relative areas for a given class of compounds.
[b]Treatment performed without addition of precursor extract.
[c]3-Oxo-α-ionol not included.
[d]Selected compounds including vinylphenols, vanillin derivatives and benzene derivatives.

23.1.5 Hydrolysis of glycosides under fermentation and aging conditions

Most yeast and lactic acid bacteria (LAB) strains express glycosidases, and further hydrolysis or rearrangements of aglycones can occur under the acidic conditions of wine storage. The glycoside pool thus provides winemakers with tools to manipulate wine aroma and flavor because relative enzyme activity and substrate specificity will vary among strains and fermentation conditions [47–49]. In fact, many yeast suppliers will advertise yeast strains with glycosidase activity appropriate for a particular wine style.

23.1.5.1 Microbial strain and lees effects

The ability of yeast strains to release monoterpenes from glycosides has been demonstrated by supplementing a model juice media with glycoside precursor fraction containing no free monoterpenes (Figure 23.1.6a). The resulting model wines had 2–5-fold higher concentrations of monoterpenes than the control. However, interpreting the effects of either aging or microbial action on glycosides is often complicated by additional changes that may occur to odorants during fermentation or storage. In fermentations of real juices, monoterpenes may appear to decrease, quite substantially in some cases (e.g., geraniol, Figure 23.1.6b). This is likely because their release from glycosides can be less than the losses of free monoterpenes present in juice, as a result of lees binding, volatilization, yeast metabolism, or chemical reactions.

Apart from hydrolyzing glycosides to varying amounts during model fermentations (e.g., up to 10-fold variation for release of some monoterpenoids), some yeast strains produce higher concentrations of vinylphenols (attributed to esterase activity on hydroxycinnamic acids,[6] Chapter 23.3) or lactones from precursor fractions [50, 52, 53]. Not

[6]Most extraction techniques for isolating glycosides in research studies are rather crude. Typically, reversed-phase (C-18) sorbents are used to isolate a methanol soluble fraction. While this will isolate glycosides, many other semi-polar compounds (including phenolics like hydroxycinnamic acids) will also be extracted.

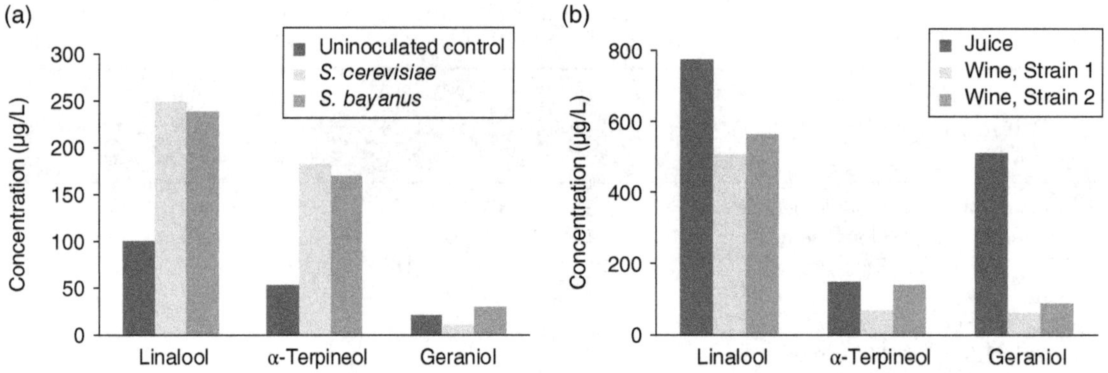

Figure 23.1.6 Effect of yeast strain on monoterpene concentrations in (a) a model juice system supplemented with a grape glycoside fraction, showing that yeast glycosidase activity during fermentation can result in an increase in free monoterpenes (data from Reference [50]), and (b) fermentations of real juices, highlighting an apparent decrease in monoterpenes due to losses of free monoterpenes present in the juice (data from Reference [51])

Table 23.1.5 Concentrations (µg/L) of free and bound aroma compounds in Riesling juice and wines made with different yeast strains. Data from Reference [54]

Compound	Juice		PDM		D47		Fermiblanc		VL1		Native[a]	
	Bound	Free	Bound	Free	Bound	Free	Bound	Free	Bound	Free	Bound	Free
Linalool	70	32	55	23	59	40	42	38	63	43	48	34
Nerol	20	11	16	<1	14	1	5	2	7	1	7	<1
Geraniol	42	22	28	2	29	2	20	2	18	3	18	4
α-Tepineol	114	8	91	32	84	35	62	24	80	37	60	50
Benzyl alcohol	16	8	14	5	14	6	11	7	13	9	11	1
2-Phenylethanol	160	33	149	88	135	101	89	122	130	107	98	139
Total Monoterpenoids	246	73	190	57	186	78	129	66	168	200	133	88
Total Benzenoids	176	41	163	93	149	107	100	129	143	116	109	140

[a] Uninoculated fermentation.

surprisingly, the sensory effects of different strains were easily noted in the model wines but are less obvious when fermenting real Muscat or Riesling grape juices [51, 54]. Nonetheless, the composition of free and bound monoterpenoids and benzenoids differs between yeast strains and the initial juice [54], as shown for Riesling in Table 23.1.5. These data reinforce the notion that winemakers have a level of control on wine composition and aroma through the choice of commercial yeast strain (or by conducting uninoculated fermentation), and also indicate that substantial quantities of residual glycosides remain in the new wine that could be released during aging (Chapter 25).

Similar to the effects of yeast strain, lactic acid bacterial (LAB) strains used for malolactic fermentation (MLF) can also affect glycoside and aglycone concentrations in wine. However, declines in glycosides are not necessarily mirrored with increases in the respective concentrations of free volatiles, likely as a result of other transformations or binding of aglycones to polysaccharides produced by LAB [55]. A study of LAB strains conducting MLF in model wine (pH 3.4) containing Muscat precursor extract showed a decrease in monotepenoid glycosides compared to the control (Figure 23.1.7a), and a related increase in the corresponding aglycones or their rearrangement products (Figure 23.1.7b) [56]. Notably, there was significant glycoside

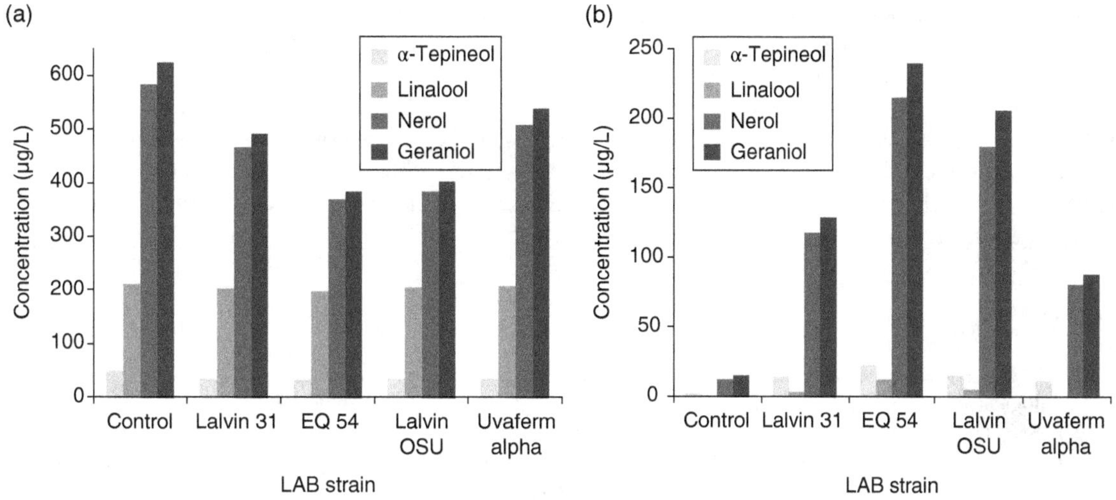

Figure 23.1.7 *Effects of LAB strain in a model wine system supplemented with a grape glycoside fraction relative to uninoculated control post-MLF, showing (a) a decrease in monoterpene glycoside concentrations and (b) an increase in free monoterpenoids. Data from Reference [56]*

hydrolysis due to MLF, and variable release according to the nature of the aglycone; up to 7% linalool glycoside was hydrolyzed in contrast to 34–38% of α-terpineol, nerol, and geraniol precursors. However, hydrolysis of glycosides usually exceeds the amount of volatiles formed [57]. Glycosidase activity decreases at low pH and several strains show a significant decrease in precursor hydrolysis (around 20–70%) for model wine systems [56]. Similar to yeast, model glycoside systems treated with different LAB strains can be differentiated sensorially, but these changes are less apparent with real wines [58].

Storage on yeast lees can either increase or decrease the concentration of compounds derived from grape precursors, depending on the grape variety (which would affect precursor pool size and pH) and yeast strain (which would affect activity of glycosidases and other enzymes). Maintaining wines on lees for 20 days after alcoholic fermentation showed an increase (around 1.5–3-fold) in some isoprenoids and lactones in the case of Airen wine, but a decrease (around 1.5–2-fold) in Macabeo wine, except for β-ionone, which increased slightly [59]. Some variation was also seen with wine made from sterile Macabeo juice supplemented with grape precursor extract and stored on lees from different yeast strains for 3 and 9 months. A range of isoprenoids (e.g., linalool, α-terpineol, Riesling acetal, vitispiranes, TDN) increased by up to 2-fold whereas other compounds (some isoprenoids along with benzenoids and lactones) decreased by up to 3-fold [60]. Overall, storage time seemed to be more important than yeast strain, but lees were determined to be taking an active role in the observed changes (i.e., changes were beyond what could be explained by acid hydrolysis alone).

23.1.5.2 Aging effects

Glycoside hydrolysis occurs prior to and during fermentation, but a latent pool of aroma compounds in the form of grape-derived glycosides still remains in finished wine that can continue to transform under abiotic conditions (Chapter 25) [61]. Most of these reactions are acid-catalyzed, and thus will occur faster at lower pH and higher temperature. However, the concentration and compositional distribution of free volatiles that can be released under real conditions is generally less than that released from the total bound pool during accelerated aging trials (as described above, e.g., 45 °C for several weeks) [39]. For example, many glycosylated β-damascenone precursors can be converted to free β-damascenone at high temperatures such as those encountered during cooking,[7] but release of β-damascenone is negligible under typical wine storage conditions [46]. Presumably, this indicates that hydrolysis of several β-damascenone precursors involves high activation energies, although specific data on this topic are unavailable. Furthermore, glycoside hydrolysis during aging occurs concurrently with degradation reactions involving odorant aglycones (Chapter 25). For many key odorants (e.g., geraniol, linalool), the rate of release from glycosides is slower than the degradation reactions, and an increase in free volatiles would only be observed in cases where the pool of bound compounds is much higher in concentration than that of the free volatiles. However, other compounds (particularly TDN and *cis*-rose oxide, Chapter 8) appear to be very stable once formed and can therefore accumulate during storage.

References

1. Winterhalter, P. and Skouroumounis, G. (1997) Glycoconjugated aroma compounds: occurrence, role and biotechnological transformation, in *Advances in biochemical engineering/biotechnology: biotechnology of aroma compounds* (ed. Berger, R.G.), Springer-Verlag, Berlin, pp. 73–105.
2. Hjelmeland, A.K. and Ebeler, S.E. (2015) Glycosidically bound volatile aroma compounds in grapes and wine: a review. *American Journal of Enology and Viticulture*, **66** (1), 1–11.
3. Sarry, J.-E. and Günata, Z. (2004) Plant and microbial glycoside hydrolases: volatile release from glycosidic aroma precursors. *Food Chemistry*, **87** (4), 509–521.
4. Sinnott, M.L. (1990) Catalytic mechanism of enzymic glycosyl transfer. *Chemical Reviews*, **90** (7), 1171–1202.
5. Baumes, R. (2009) Wine aroma precursors, in *Wine chemistry and biochemistry* (eds Moreno-Arribas, M.V. and Polo, M.C.), Springer, New York, pp. 251–274.
6. Gunata, Z., Dugelay, I., Sapis, J.C., *et al.* (eds) (1993) Role of enzymes in the use of the flavour potential from grape glycosides in winemaking in *Progress in flavour precursor studies: analysis, generation, biotechnology*, September 30–October 2, 1992, Wurzburg, Germany, Allured Publishing Corporation, Carol Stream, IL.
7. Bowles, D. and Lim, E.-K. (2010) Glycosyltransferases of small molecules: their roles in plant biology, in *Encyclopedia of life sciences (ELS)*, John Wiley & Sons, Ltd, Chichester, UK, pp. 1–10.
8. Baumes, R., Wirth, J., Bureau, S., *et al.* (2002) Biogeneration of C_{13}-norisoprenoid compounds: experiments supportive for an apo-carotenoid pathway in grapevines. *Analytica Chimica Acta*, **458** (1), 3–14.
9. Mendes-Pinto, M.M. (2009) Carotenoid breakdown products the-norisoprenoids-in wine aroma. *Archives of Biochemistry and Biophysics*, **483** (2), 236–245.
10. Razungles, A.J., Baumes, R.L., Dufour, C., *et al.* (1998) Effect of sun exposure on carotenoids and C-13-norisoprenoid glycosides in Syrah berries (*Vitis vinifera* L.). *Sciences Des Aliments*, **18** (4), 361–373.
11. Oliveira, C., Barbosa, A., Ferreira, A.C.S., *et al.* (2006) Carotenoid profile in grapes related to aromatic compounds in wines from Douro region. *Journal of Food Science*, **71** (1), S1–S7.
12. Mathieu, S., Terrier, N., Procureur, J., *et al.* (2005) A carotenoid cleavage dioxygenase from *Vitis vinifera* L.: functional characterization and expression during grape berry development in relation to C_{13}-norisoprenoid accumulation. *Journal of Experimental Botany*, **56** (420), 2721–2731.
13. Young, P., Lashbrooke, J., Alexandersson, E., *et al.* (2012) The genes and enzymes of the carotenoid metabolic pathway in *Vitis vinifera* L. *BMC Genomics*, **13** (1), 243.
14. Williams, P.J., Strauss, C.R., Wilson, B. (1980) Hydroxylated linalool derivatives as precursors of volatile monoterpenes of muscat grapes. *Journal of Agricultural and Food Chemistry*, **28** (4), 766–771.
15. Wilson, B., Strauss, C.R., Williams, P.J. (1986) The distribution of free and glycosidically-bound monoterpenes among skin, juice, and pulp fractions of some white grape varieties. *American Journal of Enology and Viticulture*, **37** (2), 107–111.

[7] This rapid release of β-damascenone precursors can be demonstrated in the home kitchen during cooking of apple sauce or apple pie, which yields the quintessential "cooked apple" aroma of these products.

16. Strauss, C.R., Wilson, B., Williams, P.J. (1988) Novel monoterpene diols and diol glycosides in *Vitis vinifera* grapes. *Journal of Agricultural and Food Chemistry*, **36** (3), 569–573.
17. Luan, F., Hampel, D., Mosandl, A., Wüst, M. (2004) Enantioselective analysis of free and glycosidically bound monoterpene polyols in *Vitis vinifera* L. cvs. Morio Muscat and Muscat Ottonel: evidence for an oxidative monoterpene metabolism in grapes. *Journal of Agricultural and Food Chemistry*, **52** (7), 2036–2041.
18. Luan, F., Mosandl, A., Münch, A., Wüst, M. (2005) Metabolism of geraniol in grape berry mesocarp of *Vitis vinifera* L. cv. Scheurebe: demonstration of stereoselective reduction, *E*/*Z*-isomerization, oxidation and glycosylation. *Phytochemistry*, **66** (3), 295–303.
19. Wilson, B., Strauss, C.R., Williams, P.J. (1984) Changes in free and glycosidically bound monoterpenes in developing muscat grapes. *Journal of Agricultural and Food Chemistry*, **32** (4), 919–924.
20. Koslitz, S., Renaud, L., Kohler, M., Wust, M. (2008) Stereoselective formation of the varietal aroma compound rose oxide during alcoholic fermentation. *Journal of Agricultural and Food Chemistry*, **56** (4), 1371–1375.
21. Gunata, Y.Z., Bayonove, C.L., Baumes, R.L., Cordonnier, R.E. (1985) The aroma of grapes. Localisation and evolution of free and bound fractions of some grape aroma components c.v. Muscat during first development and maturation. *Journal of the Science of Food and Agriculture*, **36** (9), 857–862.
22. Gomez, E., Martinez, A., Laencina, J. (1994) Localization of free and bound aromatic compounds among skin, juice and pulp fractions of some grape varieties. *Vitis*, **33**, 1–4.
23. Gunata, Y.Z., Bayonove, C.L., Baumes, R.L., Cordonnier, R.E. (1985) The aroma of grapes I. Extraction and determination of free and glycosidically bound fractions of some grape aroma components. *Journal of Chromatography A*, **331**, 83–90.
24. Schneider, R., Razungles, A., Augier, C., Baumes, R. (2001) Monoterpenic and norisoprenoidic glycoconjugates of *Vitis vinifera* L. cv. Melon B. as precursors of odorants in Muscadet wines. *Journal of Chromatography A*, **936** (1–2), 145–157.
25. Sefton, M.A., Francis, I.L., Williams, P.J. (1993) The volatile composition of Chardonnay juices: a study by flavor precursor analysis. *American Journal of Enology and Viticulture*, **44** (4), 359–370.
26. Rodrıguez-Bencomo, J.J., Méndez-Siverio, J.J., Pérez-Trujillo, J.P., Cacho, J. (2008) Effect of skin contact on bound aroma and free volatiles of Listán blanco wine. *Food Chemistry*, **110** (1), 214–225.
27. Cabaroglu, T., Canbas, A., Baumes, R., *et al.* (1997) Aroma composition of a white wine of *Vitis vinifera* L. cv. Emir as affected by skin contact. *Journal of Food Science*, **62** (4), 680–683.
28. Cabaroglu, T. and Canbas, A. (2002) The effect of skin contact on the aromatic composition of the white wine of *Vitis vinifera* L. cv. Muscat of Alexandria grown in Southern Anatolia. *Acta Alimentaria*, **31** (1), 45–55.
29. Cabaroglu, T., Selli, S., Canbas, A., *et al.* (2003) Wine flavor enhancement through the use of exogenous fungal glycosidases. *Enzyme and Microbial Technology*, **33** (5), 581–587.
30. Selli, S., Canbas, A., Cabaroglu, T., *et al.* (2006) Aroma components of cv. Muscat of Bornova wines and influence of skin contact treatment. *Food Chemistry*, **94** (3), 319–326.
31. Sánchez-Palomo, E., Pérez-Coello, M.S., Díaz-Maroto, M.C., *et al.* (2006) Contribution of free and glycosidically-bound volatile compounds to the aroma of muscat "a petit grains" wines and effect of skin contact. *Food Chemistry*, **95** (2), 279–289.
32. Ugliano, M. (2009) Enzymes in winemaking, in *Wine chemistry and biochemistry* (eds Moreno-Arribas, M.V. and Polo, M.C.), Springer, New York, pp. 103–126.
33. Itu, N.L., Rapeanu, G., Hopulele, T. (2011) Effect of maceration enzymes addition on the aromatic white winemaking. *Annals of the University "Dunarea de Jos" of Galati – Fascicle VI: Food Technology*, **35** (1), 77–91.
34. Gil, J.V. and Valles, S. (2001) Effect of macerating enzymes on red wine aroma at laboratory scale: exogenous addition or expression by transgenic wine yeasts. *Journal of Agricultural and Food Chemistry*, **49** (11), 5515–5523.
35. Moio, L., Ugliano, M., Gambuti, A., *et al.* (2004) Influence of clarification treatment on concentrations of selected free varietal aroma compounds and glycoconjugates in Falanghina (*Vitis vinifera* L.) must and wine. *American Journal of Enology and Viticulture*, **55** (1), 7–12.
36. Pogorzelski, E. and Wilkowska, A. (2007) Flavour enhancement through the enzymatic hydrolysis of glycosidic aroma precursors in juices and wine beverages: a review. *Flavour and Fragrance Journal*, **22** (4), 251–254.
37. Loscos, N., Hernandez-Orte, P., Cacho, J., Ferreira, V. (2009) Comparison of the suitability of different hydrolytic strategies to predict aroma potential of different grape varieties. *Journal of Agricultural and Food Chemistry*, **57** (6), 2468–2480.

38. Gunata, Z. (2003) Flavor enhancement in fruit juices and derived beverages by exogenous glycosidases and consequences of the use of enzyme preparations, in *Handbook of food enzymology* (eds Whitaker, J.R., Voragen, A.G.J., Wong, D.W.S.), Marcel Dekker Inc., New York.
39. Skouroumounis, G.K. and Sefton, M.A. (2000) Acid-catalyzed hydrolysis of alcohols and their β-D-glucopyranosides. *Journal of Agricultural and Food Chemistry*, **48** (6), 2033–2039.
40. Maicas, S. and Mateo, J.J. (2005) Hydrolysis of terpenyl glycosides in grape juice and other fruit juices: a review. *Applied Microbiology and Biotechnology*, **67** (3), 322–335.
41. Hemingway, K.M., Alston, M.J., Chappell, C.G., Taylor, A.J. (1999) Carbohydrate-flavour conjugates in wine. *Carbohydrate Polymers*, **38** (3), 283–286.
42. Tamborra, P., Martino, N., Esti, M. (2004) Laboratory tests on glycosidase preparations in wine. *Analytica Chimica Acta*, **513** (1), 299–303.
43. Castro Vázquez, L., Pérez-Coello, M.S., Cabezudo, M.D. (2002) Effects of enzyme treatment and skin extraction on varietal volatiles in Spanish wines made from Chardonnay, Muscat, Airén, and Macabeo grapes. *Analytica Chimica Acta*, **458** (1), 39–44.
44. Valcarcel, M.C. and Palacios, V. (2008) Influence of "Novarom G" pectinase β-glycosidase enzyme on the wine aroma of four white varieties. *Food Science and Technology International*, **14** (5), 95–102.
45. Sánchez-Palomo, E., Hidalgo, M.C.D.a.-M., González-Viñas, M.Á., Pérez-Coello, M.S. (2005) Aroma enhancement in wines from different grape varieties using exogenous glycosidases. *Food Chemistry*, **92** (4), 627–635.
46. Sefton, M.A., Skouroumounis, G.K., Elsey, G.M., Taylor, D.K. (2011) Occurrence, sensory impact, formation, and fate of damascenone in grapes, wines, and other foods and beverages. *Journal of Agricultural and Food Chemistry*, **59** (18), 9717–9746.
47. McMahon, H., Zoecklein, B.W., Fugelsang, K., Jasinski, Y. (1999) Quantification of glycosidase activities in selected yeasts and lactic acid bacteria. *Journal of Industrial Microbiology and Biotechnology*, **23** (3), 198–203.
48. Mansfield, A.K., Zoecklein, B.W., Whiton, R.S. (2002) Quantification of glycosidase activity in selected strains of *Brettanomyces bruxellensis* and *Oenococcus oeni*. *American Journal of Enology and Viticulture*, **53** (4), 303–307.
49. Grimaldi, A., Bartowsky, E., Jiranek, V. (2005) A survey of glycosidase activities of commercial wine strains of *Oenococcus oeni*. *International Journal of Food Microbiology*, **105** (2), 233–244.
50. Ugliano, M., Bartowsky, E.J., McCarthy, J., *et al.* (2006) Hydrolysis and transformation of grape glycosidically bound volatile compounds during fermentation with three *Saccharomyces* yeast strains. *Journal of Agricultural and Food Chemistry*, **54** (17), 6322–6331.
51. Delcroix, A., Günata, Z., Sapis, J.-C., *et al.* (1994) Glycosidase activities of three enological yeast strains during winemaking: effect on the terpenol content of Muscat wine. *American Journal of Enology and Viticulture*, **45** (3), 291–296.
52. Fernandez-Gonzalez, M., Di Stefano, R., Briones, A. (2003) Hydrolysis and transformation of terpene glycosides from muscat must by different yeast species. *Food Microbiology*, **20** (1), 35–41.
53. Loscos, N., Hernandez-Orte, P., Cacho, J., Ferreira, V. (2007) Release and formation of varietal aroma compounds during alcoholic fermentation from nonfloral grape odorless flavor precursors fractions. *Journal of Agricultural and Food Chemistry*, **55** (16), 6674–6684.
54. Zoecklein, B.W., Marcy, J.E., Williams, J.M., Jasinski, Y. (1997) Effect of native yeasts and selected strains of *Saccharomyces cerevisiae* on glycosyl glucose, potential volatile terpenes, and selected aglycones of white Riesling (*Vitis vinifera* L.) wines. *Journal of Food Composition and Analysis*, **10** (1), 55–65.
55. Boido, E., Lloret, A., Medina, K., *et al.* (2002) Effect of β-glycosidase activity of *Oenococcus oeni* on the glycosylated flavor precursors of Tannat wine during malolactic fermentation. *Journal of Agricultural and Food Chemistry*, **50** (8), 2344–2349.
56. Ugliano, M., Genovese, A., Moio, L. (2003) Hydrolysis of wine aroma precursors during malolactic fermentation with four commercial starter cultures of *Oenococcus oeni*. *Journal of Agricultural and Food Chemistry*, **51** (17), 5073–5078.
57. Ugliano, M. and Moio, L. (2006) The influence of malolactic fermentation and *Oenococcus oen*i strain on glycosidic aroma precursors and related volatile compounds of red wine. *Journal of the Science of Food and Agriculture*, **86** (14), 2468–2476.
58. Hernandez-Orte, P., Cersosimo, M., Loscos, N., *et al.* (2009) Aroma development from non-floral grape precursors by wine lactic acid bacteria. *Food Research International*, **42** (7), 773–781.

59. Bueno, J., Peinado, R., Medina, M., Moreno, J. (2006) Effect of a short contact time with lees on volatile composition of Airen and Macabeo wines. *Biotechnology Letters*, **28** (13), 1007–1011.
60. Loscos, N., Hernandez-Orte, P., Cacho, J., Ferreira, V. (2009) Fate of grape flavor precursors during storage on yeast lees. *Journal of Agricultural and Food Chemistry*, **57** (12), 5468–5479.
61. Gunata, Y.Z., Bayonove, C.L., Baumes, R.L., Cordonnier, R.E. (1986) Stability of free and bound fractions of some aroma components of grapes cv. Muscat during the wine processing: preliminary results. *American Journal of Enology and Viticulture*, **37** (2), 112–114.

23.2

S-Conjugates

23.2.1 Introduction

Varietal thiols are potent impact compounds contributing pleasant tropical and citrus aromas to several varietal wines, most notably Sauvignon Blanc (Chapter 10), but these compounds are largely absent from grape juice. The majority of these thiols appear to be released by yeast enzyme activity cleaving the C–S bonds of non-volatile sulfur-containing precursors. Early studies identified conjugates of the sulfur-containing amino acid cysteine (Cys) and the tripeptide glutathione (GSH) as precursors capable of releasing key thiols 3-mercaptohexan-1-ol (3-MH) and 4-mercapto-4-methylpentan-2-one (4-MMP)[1] during fermentation (Figure 23.2.1) [1–4]. These thiol precursors[2] are found in grape skin and pulp at tens to hundreds of μg/L, with grape botrytis infection and ripeness, grape pressing, skin contact, and other processing operations greatly affecting their concentrations in juice (see Table 23.2.1 for some examples). Identical or related *S*-conjugates have been found in other natural matrices such as passion fruit, onion, bell pepper, and even human sweat, where they can serve as precursors to potent odorous thiols [5]. Research surrounding varietal thiols and their precursors is still very active, and complete explanations are not yet available. As a result this chapter includes more speculation based on available data than most other chapters.

[1]The prefix "mercapto" to designate a thiol functionality is obsolete and should be replaced with "sulfanyl" according to IUPAC nomenclature. However, for historical purposes mercapto is still widely used in the literature to name these compounds and create their abbreviations. Newly identified compounds often have the correct sulfanyl prefix assigned, but for consistency these too are referred to as mercapto in this text; for example, 2-methyl-3-sulfanylbutan-1-ol becomes 3-mercapto-2-methylbutan-1-ol (3-MMB) as in Figure 23.2.1. Some authors prefer to use the modern nomenclature for all thiols, so it is common to see 3-MH termed 3-SH based on abbreviation of 3-sulfanylhexan-1-ol, for example.

[2]Chiral thiols such as 3-MH are present in wine as pairs of enantiomers, which correspond to the diastereomers of the respective thiol precursors.

Understanding Wine Chemistry, First Edition. Andrew L. Waterhouse, Gavin L. Sacks, and David W. Jeffery.

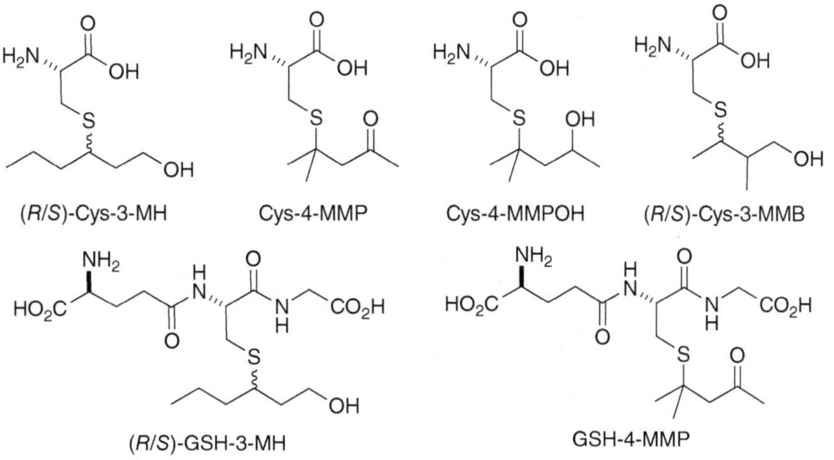

Figure 23.2.1 Varietal thiol-related cysteine and glutathione conjugates identified in grape juice or wine. MH = mercaptohexan-1-ol, MMP = mercapto-4-methylpentan-2-one, MMPOH = mercapto-4-methylpentan-2-ol, and MMB = mercapto-2-methylbutan-1-ol. (R/S) refers to the alkyl chain stereocenter and designates that these chiral compounds are present as pairs of diastereomers. This designation is often omitted but the presence of diastereomers should still be assumed

23.2.2 Formation of *S*-conjugate precursors in berries and juice

S-conjugate precursors of both 3-MH and 4-MMP are well characterized in grapes. In contrast, *O*-acetates such as 3-mercaptohexyl acetate (3-MHA) do not have a direct grape-derived precursor; rather, 3-MHA arises from acetylation of 3-MH liberated during fermentation (see Section 23.2.3). Formation of both Cys- and GSH-conjugates in grapes likely results from conjugation of GSH to electrophilic α,β-unsaturated alkenals (e.g., (*E*)-2-hexenal to yield 3-MH precursors, Figure 23.2.2) or alkenones (e.g., 4-methylpent-3-en-2-one [mesityl oxide], giving 4-MMP precursors) [5, 12, 13].[3] GSH conjugation reactions catalyzed by glutathione *S*-transferases (GSTs) are well-known detoxification mechanisms in the plant kingdom [14, 15], as well as among animals.[4] The resulting adducts typically show lower reactivity and greater solubility than their precursors. The same role of GSTs in grapevines appears to at least partially explain the presence of *S*-conjugates in grapes [16]. Additional *S*-conjugates may be formed post-harvest during grape processing or fermentation [3].

Metabolism of GSH-conjugates is well studied in other plants [17], and it appears that some of these metabolites also form in grape berries. The tripeptide GSH is enzymatically hydrolyzed to first yield the dipeptide conjugate (Cysgly) followed by the Cys conjugate [18]. There is strong evidence that these steps also occur in grapes (or juice), as these conjugates have been detected for 3-MH [19, 20]. Intermediates necessarily undergo other enzymatic transformations, such as reduction of an aldehyde to yield the corresponding alcohol (Figure 23.2.2).

[3] Such carbonyl compounds are typically formed due to biological lipid oxidation. For example, (*E*)-2-hexenal arises through oxidative degradation of linolenic acid by enzymes in the grape berry (Chapter 6). Similarly, although mesityl oxide is yet to be identified in grapes, its presence is known in other foodstuffs, and it is also likely to arise from the oxidative degradation of fatty acids.

[4] Many drugs are partially or fully metabolized in humans through GSTs, including the active ingredient in Tylenol (acetaminophen). The resulting GSH conjugates are more readily excreted.

Table 23.2.1 *Viticultural and pre-fermentation processing effects on Cys-3-MH and GSH-3-MH concentrations in Sauvignon Blanc berries and juices*

Harvest method	Treatment	Precursor concentration, sum of diastereomers (μg/L)		Interpretation	Study
		Cys-3MH	GSH-3MH[a]		
Hand	–[b]	11.7–11.8	73.0–77.8	Hand harvesting tends to yield lower concentrations of precursors due to minimizing both berry damage and precursor extraction	[6]
	–[c]	7.6–57.0	33–224		[7]
	Separated skin[d]	30	32	Although variable for different thiol precursors, greater amounts are mostly found in the skins of berries	[8]
	Separated pulp[d]	8	24		
	Botrytis 0%	100	NR	Botrytis infection stimulates production of precursors by grape berries, in line with a berry damage and detoxification mechanism	[9]
	Botrytis 100%, not desiccated	2188	NR		
	Botrytis 100%, desiccated	4486	NR		
	As above, picked 1 week later	3449	NR		
	Veraison[e]	0.1–1.5	4.8–13.2	Precursor concentrations increase as grapes ripen, in line with a berry damage and detoxification mechanism	[10]
	Two weeks before harvest[e]	2.9–4.9	18.8–41.8		
	Harvest[e]	27.4–49.7	144.3–225.6		
Machine	No SO_2	38.1	253.7	Precursor concentrations increase due to transportation of grapes (as a result of berry damage, enzymatic reactions, and maceration), and decrease when high levels of SO_2 are added during harvest (due to an inhibitory effect on enzymatic reactions)	[6]
	50 mg/L SO_2	26.2	214.8		
	500 mg/L SO_2	9.46	114.0		
	Transport, no SO_2 addition[f]	269.0	507.8		
	Transport, 50 mg/L SO_2[f]	213.2	440.3		
	Transport, 500 mg/L SO_2[f]	92.9	252.1		
	Free run juice[c]	9.6–39.5	33–322	Precursor concentrations increase with skin contact and pressing at higher pressures, in accord with their localization within berry skins	[11]
	Pressed to 1 bar[c]	16.5–111	200–541		
	Free run juice[g]	14–18	NR		
	Skin contact 1 hour[g]	18–22	NR		
	Pressed to 1.2–1.4 atm[g]	31–74	NR		

[a] NR, not reported.
[b] Range for two different locations within a single vineyard.
[c] Range for five different juices from NZ.
[d] Mean result for three different regions of France. Units are μg/kg of fresh skin or pulp.
[e] Five different clones co-located in one vineyard in the Adelaide Hills, Australia. Units are μg/kg of fresh berries.
[f] Transported 800 km in 12 h from SA to NSW, Australia.
[g] Range for three different juices from NZ.

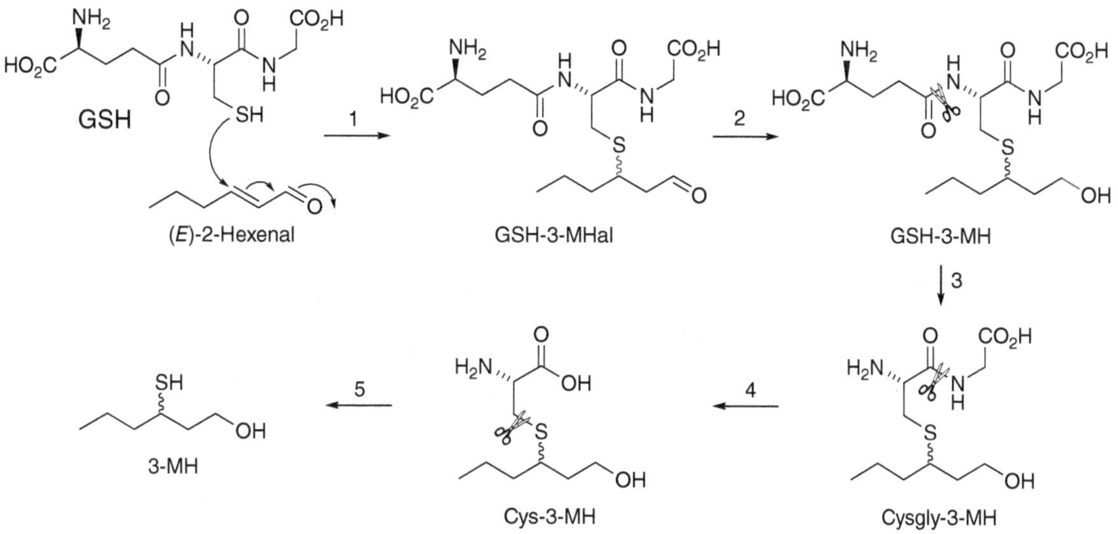

Figure 23.2.2 Formation of thiol precursor GSH-3-MH from nucleophilic GSH and electrophilic (E)-2-hexenal, and degradation to yield other precursors and varietal aroma compound 3-MH in wine. Steps: 1 = conjugation by glutathione S-transferase (possibly non-enzymatic); 2 = enzymatic carbonyl reduction; 3 = removal of glutamyl residue by γ-glutamyltranspeptidase; 4 = removal of glycine residue by peptidases; 5 = release of varietal thiol by carbon–sulfur lyase. Note that the precise order of carbonyl reduction relative to cleavage of amino acids is unknown.

The concentrations of GSH-3-MH and Cys-3-MH are strongly affected by both growing and processing conditions (e.g., see Reference [2]), as outlined in Table 23.2.1. In brief:

- Hand harvesting leads to lower precursor concentrations compared to mechanical harvesting.
- Precursor concentrations increase as a result of disease, grape maturity, and transportation of harvested fruit.
- Precursors are found in grape skins and their concentrations increase with skin contact and pressure used during pressing to juice.

23.2.3 Conversion of *S*-conjugate precursors during fermentation

Unlike glycosylated precursors, which can liberate key odorants through both enzymatic and non-enzymatic pathways, *S*-conjugates appear to only release volatile thiols through enzymatic mechanisms, primarily during fermentation [2, 21] (Figures 23.2.2 and 23.2.3). This release requires two steps:

1. *S*-conjugate precursor uptake into the cell.
2. Precursor cleavage by carbon–sulfur (C-S) lyases (i.e. β-lyase enzymes).

Historically, aspects of C-S lyase activity were determined before those related to uptake of precursors by transporters; however, since uptake by yeast has to occur before thiol release, this phenomenon is discussed first.

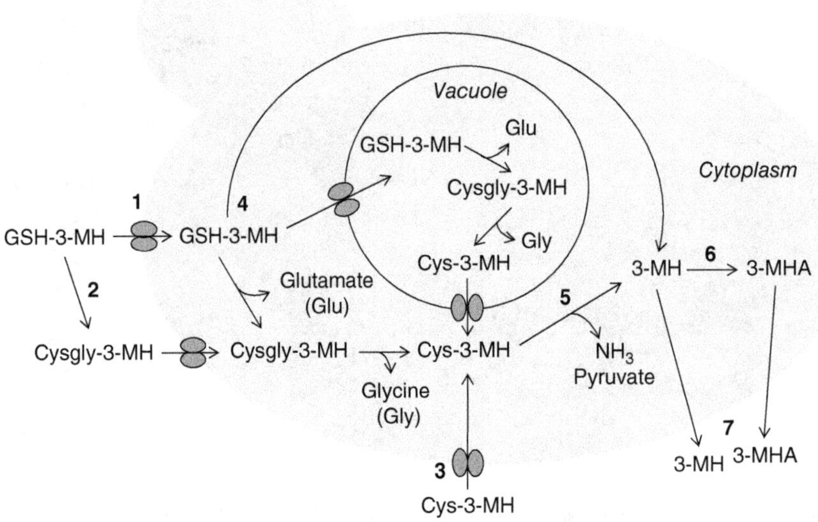

*Figure 23.2.3 Representation of potential pathways for 3-MH liberation from grape-derived GSH-3-MH, Cysgly-3-MH, and Cys-3-MH present in juice, and conversion to 3-MHA during fermentation based on information compiled in Reference [24]. Bolded numbers indicate metabolic processes as follows: **1**, GSH transporter encoded by OPT1 enables uptake of GSH-3-MH; **2**, γ-glutamyltranspeptidase cleavage of glutamate and transport of Cysgly-3-MH; **3**, general amino acid transporter encoded by GAP1 for uptake of Cys-3-MH; **4**, metabolism of GSH-3-MH to 3-MH either directly or step-wise via other precursors; **5**, cleavage of 3-MH from Cys-3-MH by carbon–sulfur lyase; **6**, acetylation of 3-MH by alcohol acetyltransferase (AAT) to afford 3-MHA; **7**, unknown mechanism leading to volatile thiol release into wine*

23.2.3.1 Precursor uptake

As discussed later, the primary reason for microbial catabolism of *S*-conjugates appears to be that they can serve as a source of yeast assimilable nitrogen (YAN). These conjugates are not a preferred nitrogen source, however, and their uptake decreases under conditions that induce nitrogen catabolite repression (NCR), that is, when a preferred nitrogen source is present (see Chapter 22.3 for a full discussion of NCR). For example, fermentations performed in the presence of excess diammonium phosphate (DAP), a more preferred source of YAN, led to lower consumption of Cys-3-MH. In contrast, addition of urea, a less preferred nitrogen source, had no effect on Cys-3-MH consumption [22]. In comparison, oxygen, vitamins, and sugars had negligible effects on thiol formation.

Yeast transporters and permeases are linked to precursor uptake (Figure 23.2.3). Mutant yeast strains that express the general amino acid permease (*GAP1*) gene, even under NCR conditions (Chapter 22.3), also produced more 3-MH from Cys-3MH than the parent strains in a synthetic medium, indicating that Gap1p is likely critical for Cys-precursor uptake [22]. Analogous work indicated that the oligopeptide transporter gene *OPT1*, which encodes for the main GSH transporter Opt1p, is critical for GSH-3-MH uptake [23]. Deletion of *OPT1* in a strain of *S. cerevisiae* not only limited GSH uptake but also led to a 2-fold decrease in 3-MH (and 3-MHA) production during fermentation of Sauvignon Blanc juice in comparison to the wild-type strain. The authors hypothesized that half of the 3-MH arose from a precursor entering via Opt1p (i.e., the precursor resembled GSH and was likely to be GSH-3-MH) or that GSH was possibly an activator of intracellular 3-MH release from *S*-conjugate precursors, so the limited uptake of GSH without Opt1p resulted in a lower 3-MH concentration [23].

23.2.3.2 Precursor cleavage

Volatile thiols can be liberated from *S*-conjugates as a result of C-S lyase activity. This was first demonstrated with bacterial β-lyase cleavage of synthetic cysteine conjugates and conjugates isolated from Sauvignon Blanc grape juice; the work also showed thiol release from fermentations with *S. cerevisiae* [25].[5] This spurred further interest in using yeast strain selection to optimize release of varietal thiols during winemaking (Chapter 29).

Cleavage of *S*-conjugate precursors by yeast has been investigated through studies of *S. cerevisiae* mutants in which putative β-lyase genes were either deleted or overexpressed, and allowed to ferment synthetic media containing *S*-conjugate precursors. Several candidate genes have been shown to affect 4-MMP release from Cys-4-MMP in synthetic media [26]. Summarizing the key points regarding these β-lyases:

- *IRC7* (called *METC* at times) appears to be the major β-lyase gene responsible for liberation of 4-MMP (in particular) and 3-MH from cysteine conjugates during fermentation in synthetic media with VL3 strains, as seen with the decrease of 4-MMP and 3-MH of around 95 and 40%, respectively, for a deletion strain [27]. In support of this result, *IRC7* overexpression resulted in a 100-fold increase in 4-MMP and doubling of 3-MH (and 3-MHA) using Sauvignon Blanc juice [28].
- However, substrate preferences will vary among β-lyases, which provides an explanation for how gene expression differences among yeast strains could result in different thiol profiles. For example, another β-lyase gene (*STR3*) with primary activity towards cystathionine was also identified in *S. cerevisiae*; the corresponding enzyme (Str3p) was purified and shown to have β-lyase side activity towards Cys-4-MMP and Cys-3-MH. Str3p showed a preference for Cys-3-MH over Cys-4-MMP at low concentrations (0.25 mM) although at higher precursor concentrations (2 mM) Cys-4-MMP was preferred [29]. Overexpression of *STR3* in commercial wine yeast VIN13 and fermentation of Sauvignon Blanc juice revealed increases in 3-MH in the order of 25%.
- *IRC7* expression, and presumably the expression of other β-lyase genes, is influenced by NCR transcriptional regulation, and shows stereoselectivity in the production of 3-MH enantiomers from Cys-3-MH diastereomers [27].
- Genetic modification of a commercial yeast strain, VIN13, through insertion and overexpression of an *E. coli* tryptophanase gene (*tna*A) has been shown to increase thiol release during model fermentations by up to 25 times compared to the parent strain [30].

These observations also likely explain variation in vendors' descriptions of yeast strains – several commercial strains are advertised as being ideal for producing Sauvignon Blanc wines in a fruity style, presumably because of either high β-lyase activity, high uptake efficiency, or a combination of both (although the relative importance of each is unknown). Ultimately, in light of the low conversion efficiency of precursors (see Section 23.2.4) there is great potential for increasing relative or total thiol release through yeast strain selection (Chapter 29).

23.2.3.3 Acetylation by yeast

An important outcome of fermentation and varietal thiol release is the conversion of 3-MH into its *O*-acetylated derivative 3-MHA, which has similar aroma qualities but over a 10-fold lower sensory threshold (Chapter 10). In general, a small proportion of 3-MHA may be produced (in the order of 10–20% of 3-MH

[5] The "β" in β-lyase refers to the fact that *S*-conjugates undergo β-elimination; that is, the leaving group (thiol) is lost from the β-position of the amino acid (and H^+ is lost from the α-position). The preferred substrates for bacterial β-lyases, which are expressed to increase available nitrogen, are often free amino acids (cysteine, tryptophan, etc.). Microbial β-lyase activity in the oral cavity can generate H_2S from cysteine, and is an often-cited cause of halitosis. In the same way, thiols such as 3-MH can also be released from cysteine conjugates due to oral microflora.

concentration), but the impact on overall wine aroma can be substantial, particularly in young wines [31]. As with release of thiol precursors, the extent of 3-MHA production is dictated by genetic differences among yeast strains. The main enzyme catalyzing this transformation has been identified as alcohol acetyltransferase Atf1p [32], which is also involved in formation of other acetate esters (Chapter 22.3) under the control of *S. cerevisiae ATF1* gene [33]. Overexpression of *ATF1* in commercial yeast VIN13 led to around a 7-fold increase in 3-MHA for model fermentations spiked with 3-MH (approximately 50% conversion), whereas overexpression of other genes involved in ester metabolism (*ATF2, EHT1, IAH1*) either produced the same amount of 3-MH as the parent strain or somewhat less in the case of *IAH1*, which encodes an esterase.

23.2.4 Mass balance and alternative pathways to volatile thiol formation

Although Cys- and GSH-conjugates can serve as thiol precursors during fermentation, the correlation of these *S*-conjugates in must to corresponding volatile thiol concentrations in wine is very poor, even under similar fermentation conditions [34]. Additionally, high precursor concentrations remain in finished wine (and precursors can form during winemaking) [10] and conjecture remains about the dominant contributors to thiols in wine. Put simply, there is currently a mass balance problem, in that the extent of conversion of known precursors does not explain the amounts of thiols formed during fermentation. For example, based on a molar conversion yield of <1% for labeled Cys-3-MH, the contribution of naturally present Cys-precursor (160 nM) to total 3-MH (42 nM) in Sauvignon Blanc was calculated at just 3% (or 1.26 nM) [23]; a similar result has been reported for GSH-3-MH [35]. Other reports have observed similarly low conversion efficiencies, with Cys-3-MH yielding roughly twice as much 3-MH (1% molar conversion of precursor) as GSH-3-MH (0.5% conversion) [24].

Accounting for the remaining >90% of 3-MH precursor is not a simple issue to resolve, and has been a subject of several recent investigations. The transporters discussed above have a role in providing the precursors intracellularly, and differences in uptake efficiency or regulation could explain some variation. Additionally, among other possibilities related to precursor reactivity/stability, GSH-3-MH may not be cleaved directly to yield 3-MH, but rather undergo degradation via several steps to yield Cys-3-MH (Figure 23.2.3), akin to GSH metabolism with enzymatic cleavage of glutamate and glycine residues [24]. Essentially this suggests that Cys-3-MH is more easily utilized by yeast, and precursor ratios could therefore play a role.

Other studies have investigated alternate pathways leading to volatile thiols or their precursors during fermentation. Potential biogenetic pathways involve conjugate addition of a thiol-containing compound such as H_2S, cysteine, or GSH to (*E*)-2-hexenal or mesityl oxide during processing or fermentation [6, 20, 36, 37], yielding 3-MH (and 3-MHA) and 4-MMP more directly. H_2S formed through fermentation does appear to yield 3-MH, although the contribution is very minor. Formation of additional cysteine and GSH conjugates once fermentation commences is not as well studied, but would presumably augment the initial precursor pool (see Section 23.2.2). Additionally, under certain conditions, yeast may convert (*E*)-2-hexen-1-ol (a C_6 alcohol, Chapter 6) into (*E*)-2-hexenal, yielding additional 3-MH and 3-MHA in wine [38], which may occur after the (*E*)-2-hexenal so produced combines with H_2S arising early during fermentation (Chapter 22.4). However, the contribution of H_2S and C_6 reaction products to the volatile thiol pool seems to be low. This is not an unexpected scenario, since the varietal thiol concentrations show stronger grape variety dependence (Chapter 10) than either C_6 compounds or yeast-derived H_2S.

Alternatively, other grape-derived conjugates may exist that are yet to be identified, and precursor intermediates such as GSH-3-MHal and Cysgly-3-MH (Figure 23.2.2) and similar conjugates could conceivably contribute to wine varietal thiol concentrations [6, 19]. It is also theoretically possible that other sources of (*E*)-2-hexenal could ultimately lead to 3-MH, such as the (*E*)-2-hexenal sulfonate adduct (Chapter 17) [39] or other unknown adducts. In the end, additional studies will be necessary to fully account for the presence of varietal thiols, particularly 3-MH and 3-MHA.

References

1. Roland, A., Schneider, R., Razungles, A., Cavelier, F. (2011) Varietal thiols in wine: discovery, analysis and applications. *Chemical Reviews*, **111** (11), 7355–7376.
2. Coetzee, C. and du Toit, W.J. (2012) A comprehensive review on Sauvignon blanc aroma with a focus on certain positive volatile thiols. *Food Research International*, **45** (1), 287–298.
3. Capone, D.L., Sefton, M.A., Jeffery, D.W. (2012) Analytical investigations of wine odorant 3-mercaptohexan-1-ol and its precursors, in *Flavor chemistry of wine and other alcoholic beverages*, American Chemical Society, pp. 15–35.
4. Peña-Gallego, A., Hernández-Orte, P., Cacho, J., Ferreira, V. (2012) *S*-Cysteinylated and *S*-glutathionylated thiol precursors in grapes. A review. *Food Chemistry*, **131** (1), 1–13.
5. Starkenmann, C., Troccaz, M., Howell, K. (2008) The role of cysteine and cysteine–*S* conjugates as odour precursors in the flavour and fragrance industry. *Flavour and Fragrance Journal*, **23** (6), 369–381.
6. Capone, D.L. and Jeffery, D.W. (2011) Effects of transporting and processing Sauvignon blanc grapes on 3-mercaptohexan-1-ol precursor concentrations. *Journal of Agricultural and Food Chemistry*, **59** (9), 4659–4667.
7. Allen, T., Herbst-Johnstone, M., Girault, M., *et al.* (2011) Influence of grape-harvesting steps on varietal thiol aromas in Sauvignon blanc wines. *Journal of Agricultural and Food Chemistry*, **59** (19), 10641–10650.
8. Roland, A., Schneider, R., Charrier, F., *et al.* (2011) Distribution of varietal thiol precursors in the skin and the pulp of Melon B. and Sauvignon Blanc grapes. *Food Chemistry*, **125** (1), 139–144.
9. Thibon, C., Shinkaruk, S., Jourdes, M., *et al.* (2010) Aromatic potential of botrytized white wine grapes: identification and quantification of new cysteine-*S*-conjugate flavor precursors. *Analytica Chimica Acta*, **660** (1–2), 190–196.
10. Capone, D.L., Sefton, M.A., Jeffery, D.W. (2011) Application of a modified method for 3-mercaptohexan-1-ol determination to investigate the relationship between free thiol and related conjugates in grape juice and wine. *Journal of Agricultural and Food Chemistry*, **59** (9), 4649–4658.
11. Maggu, M., Winz, R., Kilmartin, P.A., *et al.* (2007) Effect of skin contact and pressure on the composition of Sauvignon Blanc must. *Journal of Agricultural and Food Chemistry*, **55** (25), 10281–10288.
12. Peyrot des Gachons, C., Tominaga, T., Dubourdieu, D. (2002) Sulfur aroma precursor present in *S*-glutathione conjugate form: identification of *S*-3-(hexan-1-ol)-glutathione in must from *Vitis vinifera* L. cv. Sauvignon Blanc. *Journal of Agricultural and Food Chemistry*, **50** (14), 4076–4079.
13. Thibon, C., Cluzet, S., Merillon, J.M., *et al.* (2011) 3-Sulfanylhexanol precursor biogenesis in grapevine cells: the stimulating effect of *Botrytis cinerea*. *Journal of Agricultural and Food Chemistry*, **59** (4), 1344–1351.
14. Marrs, K.A. (1996) The functions and regulation of glutathione *S*-transferases in plants. *Annual Review of Plant Physiology and Plant Molecular Biology*, **47**, 127–158.
15. Dixon, D.P., Skipsey, M., Edwards, R. (2010) Roles for glutathione transferases in plant secondary metabolism. *Phytochemistry*, **71** (4), 338–350.
16. Kobayashi, H., Takase, H., Suzuki, Y., *et al.* (2011) Environmental stress enhances biosynthesis of flavor precursors, *S*-3-(hexan-1-ol)-glutathione and *S*-3-(hexan-1-ol)-L-cysteine, in grapevine through glutathione *S*-transferase activation. *Journal of Experimental Botany*, **62** (3), 1325–1336.
17. Schroder, P. (1997) *Fate of glutathione S-conjugates in plants. Degradation of the glutathione moiety*, Kluwer Academic Publishers, Dordrecht, The Netherlands.
18. Rennenberg, H. (1982) Glutathione metabolism and possible biological roles in higher plants. *Phytochemistry*, **21** (12), 2771–2781.
19. Capone, D.L., Pardon, K.H., Cordente, A.G., Jeffery, D.W. (2011) Identification and quantitation of 3-*S*-cysteinylglycinehexan-1-ol (Cysgly-3-MH) in Sauvignon blanc grape juice by HPLC-MS/MS. *Journal of Agricultural and Food Chemistry*, **59** (20), 11204–11210.
20. Capone, D.L., Black, C.A., Jeffery, D.W. (2012) Effects on 3-mercaptohexan-1-ol precursor concentrations from prolonged storage of Sauvignon Blanc grapes prior to crushing and pressing. *Journal of Agricultural and Food Chemistry*, **60** (13), 3515–3523.
21. Dufour, M., Zimmer, A., Thibon, C., Marullo, P. (2013) Enhancement of volatile thiol release of *Saccharomyces cerevisiae* strains using molecular breeding. *Applied Microbiology and Biotechnology*, **97** (13), 5893–5905.
22. Subileau, M., Schneider, R., Salmon, J.-M., Degryse, E. (2008) Nitrogen catabolite repression modulates the production of aromatic thiols characteristic of Sauvignon Blanc at the level of precursor transport. *FEMS Yeast Research*, **8** (5), 771–780.

23. Subileau, M., Schneider, R., Salmon, J.-M., Degryse, E. (2008) New insights on 3-mercaptohexanol (3MH) biogenesis in Sauvignon Blanc wines: Cys-3MH and (*E*)-hexen-2-al are not the major precursors. *Journal of Agricultural and Food Chemistry*, **56** (19), 9230–9235.
24. Winter, G., Van Der Westhuizen, T., Higgins, V.J., *et al.* (2011) Contribution of cysteine and glutathione conjugates to the formation of the volatile thiols 3-mercaptohexan-1-ol (3MH) and 3-mercaptohexyl acetate (3MHA) during fermentation by *Saccharomyces cerevisiae*. *Australian Journal of Grape and Wine Research*, **17** (2), 285–290.
25. Tominaga, T., Peyrot des Gachons, C., Dubourdieu, D. (1998) A new type of flavor precursors in *Vitis vinifera* L. cv. Sauvignon Blanc: *S*-cysteine conjugates. *Journal of Agricultural and Food Chemistry*, **46** (12), 5215–5219.
26. Howell, K.S., Klein, M., Swiegers, J.H., *et al.* (2005) Genetic determinants of volatile-thiol release by *Saccharomyces cerevisiae* during wine fermentation. *Applied and Environmental Microbiology*, **71** (9), 5420–5426.
27. Thibon, C., Marullo, P., Claisse, O., *et al.* (2008) Nitrogen catabolic repression controls the release of volatile thiols by *Saccharomyces cerevisiae* during wine fermentation. *FEMS Yeast Research*, **8** (7), 1076–1086.
28. Roncoroni, M., Santiago, M., Hooks, D.O., *et al.* (2011) The yeast *IRC7* gene encodes a β-lyase responsible for production of the varietal thiol 4-mercapto-4-methylpentan-2-one in wine. *Food Microbiology*, **28** (5), 926–935.
29. Holt, S., Cordente, A.G., Williams, S.J., *et al.* (2011) Engineering *Saccharomyces cerevisiae* to release 3-mercaptohexan-1-ol during fermentation through overexpression of an *S. cerevisiae* gene, *STR3*, for improvement of wine aroma. *Applied and Environmental Microbiology*, **77** (11), 3626–3632.
30. Swiegers, J.H., Capone, D.L., Pardon, K.H., *et al.* (2007) Engineering volatile thiol release in *Saccharomyces cerevisiae* for improved wine aroma. *Yeast*, **24** (7), 561–574.
31. Dubourdieu, D., Tominaga, T., Masneuf, I., *et al.* (2006) The role of yeasts in grape flavor development during fermentation: the example of Sauvignon blanc. *American Journal of Enology and Viticulture*, **57** (1), 81–88.
32. Swiegers, J.H., Willmott, R., Hill-Ling, A., *et al.* (2006) Modulation of volatile thiol and ester aromas by modified wine yeast, in *Developments in food science* (eds Bredie, W.L.P. and Petersen, M.A.), Elsevier, Amsterdam, The Netherlands, pp. 113–116.
33. Mason, A.B. and Dufour, J.-P. (2000) Alcohol acetyltransferases and the significance of ester synthesis in yeast. *Yeast*, **16** (14), 1287–1298.
34. Pinu, F.R., Jouanneau, S., Nicolau, L., *et al.* (2012) Concentrations of the volatile thiol 3-mercaptohexanol in Sauvignon blanc wines: no correlation with juice precursors. *American Journal of Enology and Viticulture*, **63** (3), 407–412.
35. Roland, A., Schneider, R., Guernevé, C.L., *et al.* (2010) Identification and quantification by LC-MS/MS of a new precursor of 3-mercaptohexan-1-ol (3MH) using stable isotope dilution assay: elements for understanding the 3MH production in wine. *Food Chemistry*, **121** (3), 847–855.
36. Schneider, R., Charrier, F., Razungles, A., Baumes, R. (2006) Evidence for an alternative biogenetic pathway leading to 3-mercaptohexanol and 4-mercapto-4-methylpentan-2-one in wines. *Analytica Chimica Acta*, **563** (1–2), 58–64.
37. Roland, A., Vialaret, J., Razungles, A., *et al.* (2010) Evolution of *S*-cysteinylated and *S*-glutathionylated thiol precursors during oxidation of Melon B. and Sauvignon blanc musts. *Journal of Agricultural and Food Chemistry*, **58** (7), 4406–4413.
38. Harsch, M.J., Benkwitz, F., Frost, A., *et al.* (2013) A new precursor of 3-mercaptohexan-1-ol in grape juice: thiol-forming potential and kinetics during early stages of must fermentation. *Journal of Agricultural and Food Chemistry*, **61** (15), 3703–3713.
39. Duhamel, N., Piano, F., Davidson, S.J., *et al.* (2015) Synthesis of alkyl sulfonic acid aldehydes and alcohols, putative precursors to important wine aroma thiols. *Tetrahedron Letters*, **56** (13), 1728–1731.

23.3

Conversion of Variety Specific Components, Other

23.3.1 Introduction

Along with glycosides (Chapter 23.1) and *S*-conjugates (Chapter 23.2), grape-derived polyunsaturated fatty acids (PUFAs), hydroxycinnamic acids (HCAs), and *S*-methylmethionine (SMM) can also serve as odorant precursors. As with glycosides and *S*-conjugates, volatiles formed from these precursors would not be formed through fermentation of a simple sugar + ammonium medium, and are strongly dependent on processing or storage conditions (e.g., maceration conditions, microbial populations, wine age) as well as precursor concentrations.

Several other odorants found in wine appear to have poorly defined relationships with non-odorous grape-derived precursors, but are not considered in detail in this chapter. For example, *ortho*-aminoacetophenone (*o*-AAP, "dirty dishrag, foxy" aroma) is correlated with the "atypical aging" off-flavor found in some wines (Chapter 5). Although *o*-AAP is plausibly formed via oxidation of plant hormone, indole acetic acid (IAA), the correlation between the two components is weak [1], and the proposed mechanism has not been demonstrated in wine or wine-like conditions. Similarly, indole and skatole in wine are likely to be related to tryptophan (Chapter 5), but it is unclear if this is connected to grape tryptophan content.

23.3.2 Polyunsaturated fatty acid precursors of C_6 compounds

The grassy-smelling C_6 alcohols – namely hexan-1-ol (hexanol), (*Z*)-3-hexen-1-ol (*cis*-3-hexen-1-ol), and (*E*)-2-hexen-1-ol (*trans*-2-hexenol) – were introduced earlier in this book (Chapter 6). These alcohols are present at negligible concentrations in most intact plant tissues including grapes,[1] but instead are formed from enzymatic oxidation of the major fatty acids in grape berries – α-linolenic and α-linoleic acids, both

[1] The fact that C_6 compounds are enzymatically formed from unsaturated fatty acids can be demonstrated by macerating plant material under highly denaturing conditions, that is, in cold methanol. Several food processing decisions affect final flavor by modifying the activity of lipid-degradation pathways; for example, the difference in aroma between green and black tea arises in part because the former is thermally treated following harvest, which inactivates lipoxygenase as well as other enzymes, for example, polyphenol oxidase.

Understanding Wine Chemistry, First Edition. Andrew L. Waterhouse, Gavin L. Sacks, and David W. Jeffery.

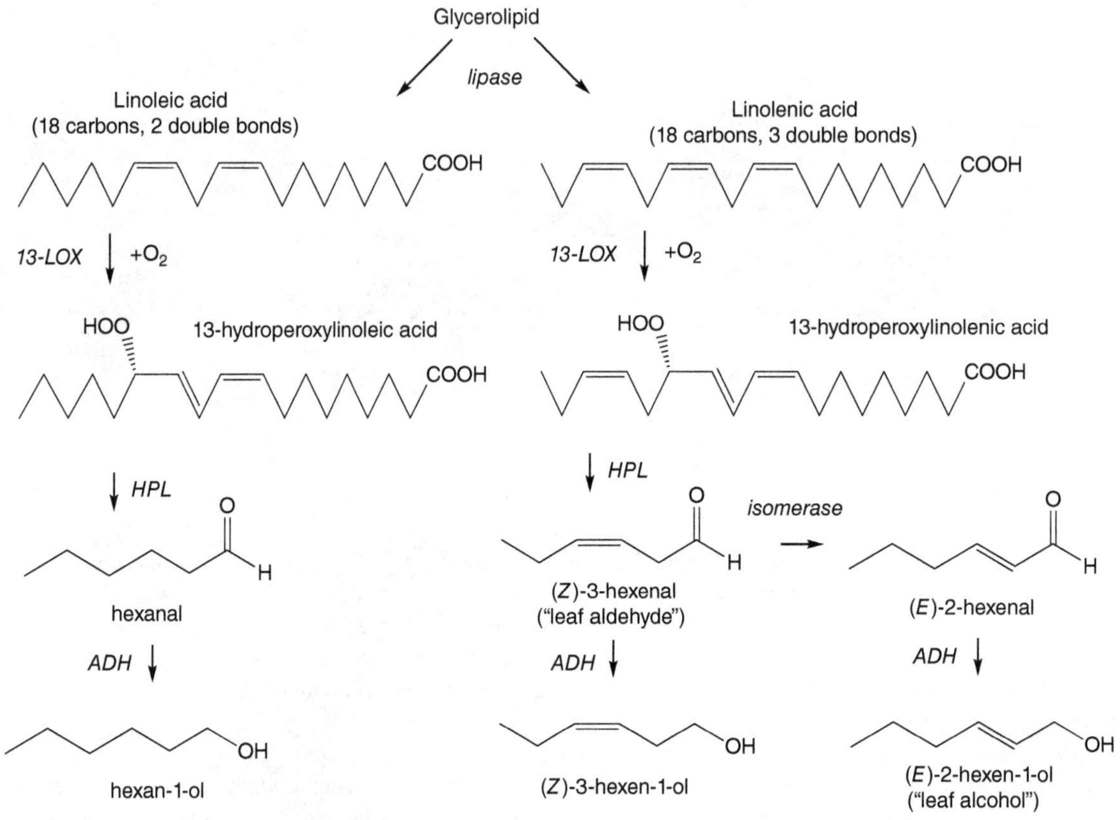

Figure 23.3.1 Enzymatic pathway for formation of C_6 alcohols and aldehydes in grapes and other plants. HPL = hydroperoxide lyase, LOX = lipoxgenase, ADH = alcohol dehydrogenase

PUFAs – following disruption of grape tissue. A key difference between PUFA precursors and other enzymatically transformed precursors discussed elsewhere in Chapter 23 is that PUFAs are enzymatically degraded prior to fermentation through grape enzymes released when berries are crushed, rather than during or after fermentation through microbial action.

The lipid oxidation pathway is outlined in Figure 23.3.1 [2]. In intact berry tissues, PUFAs are predominantly found as glycerolipids, where they have key roles in forming cell membranes (Chapter 22.2), or as triacylglycerides with roles as energy stores. Disruption of the plant tissue (i.e., by crushing grapes) brings the glycerolipids into contact with the critical enzymes in the lipid oxidation pathway. Firstly, a lipase enzyme hydrolyzes the glycerolipid, releasing free PUFAs. These can be stereospecifically converted to their corresponding 13-hydroperoxides in the presence of oxygen by lipoxygenase enzyme (13-LOX).[2] In the case of 13-hydroperoxylinolenic acid, the hydroperoxide lyase enzyme (HPL) then catalyzes the formation of (*Z*)-3-hexenal, which can subsequently be isomerized and/or reduced by additional enzymes to form other related

[2]Grapes and other plants also possess enzymes capable of forming 9-hydroperoxides, although generally 9-LOX and 9-HPL activity are much lower than their respective 13- analogues. One exception is Curcubitacae (e.g., cucumbers, melons, squash), where relatively high 9-LOX and 9-HPL activity yields superthreshold concentrations of C_9 aldehydes (and alcohols) like (*E,Z*)-2,6-nonadienal ("cucumber aldehyde") responsible for the characteristic odor of these plants.

C_6 compounds, including (*Z*)-3-hexen-1-ol, (*E*)-2-hexenal, and (*E*)-2-hexen-1-ol. An analogous pathway is responsible for the formation of hexanal and 1-hexanol from α-linoleic acid.

Lipid-derived oxidation products are responsible for the major detectable volatiles in crushed grapes – in one report, 16 of the 27 volatiles identified in a grape macerate were direct products or derivatives of unsaturated fatty acids [3]. Of these volatiles, the C_6 aldehydes, particularly (*E*)-2-hexenal (sensory threshold in water = 17 μg/L), hexanal (4.5 μg/L), and (*Z*)-3-hexenal (0.25 μg/L), are likely the major contributors to the green/grassy aromas of freshly crushed grapes. Blender macerated Cabernet Sauvignon berries reportedly have 8000 μg/kg of (*E*)-2-hexenal, 2400 μg/kg of hexanal, and 910 μg/kg of (*Z*)-3-hexenal, which translates into OAV > 470 for all three compounds [4]. In real musts, which hopefully undergo less thorough maceration, C_6 aldehyde concentrations following crushing are generally about 10% of the aforementioned values [5], but still above sensory threshold values. Considering these high OAVs, it is likely that the major retronasal aroma character perceived when chewing on grape skins is due to these "green" aldehydes, particularly for non-aromatic grape varieties.

23.3.2.1 Changes during fermentation

During fermentation, hexanol and (*Z*)-3-hexenol remain relatively stable or increase, while the C_6 aldehydes and 2-alkenols decrease by 95% or more to nearly undetectable concentrations within the first 24–48 hours of fermentation [6]. There are several reasons for this phenomenon:

- During fermentation, alkenals and 2-alkenols will be reduced to corresponding saturated alcohols (Figure 23.3.2). This is the major pathway accounting for loss of (*E*)-2-hexen-1-ol, hexanal, and (*E*)-2-hexenal [7, 8] (Chapter 22.1).
- (*E*)-2-Hexenal can react with grape-derived glutathione to form a precursor for 3-mercaptohexan-1-ol (Chapter 23.2).
- A small portion of hexanol (i.e., <1%), and possibly other lipid derived alcohols, may be enzymatically esterified to their corresponding acetate esters [7], as described earlier (Chapters 7 and 22.3).
- Other possible fates for C_6 compounds during fermentation include volatilization, binding to lees, or partial oxidation to hexanoic acid.

23.3.2.2 Factors affecting C_6 production and formation from precursors

Under controlled maceration conditions, C_6 aldehyde formation from grapes is reported to reach a maximum around veraison [9] and in some instances will decrease with increasing grape maturity – by a factor of two in the case of Spanish Macabeo [10]. As a result, higher concentrations of C_6 alcohols are often found in

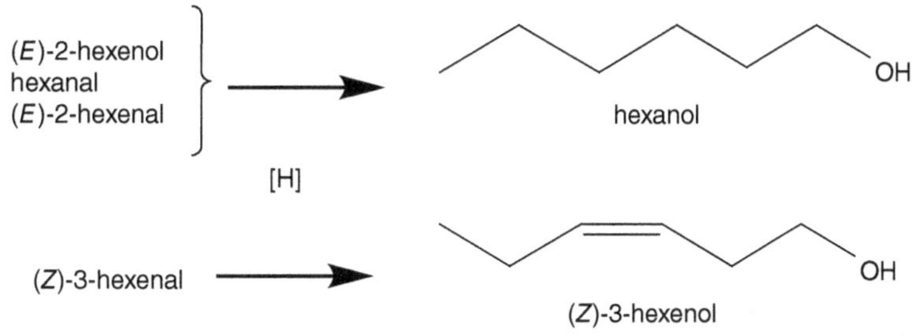

Figure 23.3.2 Fate of unstable C_6 compounds in the reductive fermentation environment. [H] represents a reducing equivalent, typically NAD(P)H

wines produced from early-harvest grapes [11]. PUFA content in grapes decreases by about a factor of two between veraison and maturity although the timing of this decline is not fully synchronized with the decrease in C_6 aldehyde production [10]. Surprisingly, LOX activity increases during berry ripening, suggesting that activity of other enzymes in the lipid oxidation pathway may be more important for explaining changes in C_6 production and relative concentrations [9, 10]. As a caveat, comparing quantitative results for grapes across studies is often challenging because of differences in maceration conditions. For example, in some reports whole berries are processed to a fine powder [9]; this approach is expected to yield higher C_6 compounds than typical wine maceration conditions, both because of the greater degree of tissue damage, especially to seeds (a rich source of PUFAs).

Because formation of C_6 compounds requires maceration, pre-fermentation conditions can have a profound effect on their eventual concentrations in wine. Factors affecting C_6 formation during winemaking have been studied [12–14] and include:

- Extent of maceration and solids contact time prior to alcoholic fermentation (e.g., extent of clarification, temperature, time, degree of crushing). This parameter is of particular importance to white wine production. For example, avoiding skin contact or clarifying white juice by settling results in lower PUFAs in must and lower concentrations of C_6 volatiles (e.g., 1-hexanol, Figure 23.3.3) in must and wine [15]. Similarly, mechanical harvesting, crushing, and hard pressing can result in up to 10-fold increases in final C_6 concentrations in wine as compared to hand-harvesting and hand-pressing [16] (Figure 23.3.4).
- Oxygen availability. The LOX enzyme requires O_2, and saturation of a must with air prior to fermentation ("hyperoxidation") results in a two-fold increase in hexanol in finished wines [17].
- Inhibition of enzyme activity, for example, through SO_2 addition [18] or thermovinification [19], will decease C_6 formation
- Presence of material other than grape berries (MOG). For example, on a weight-by-weight basis, macerated grape leaf produces 100-fold more (*Z*)-3-hexenal than grape berries, as well as higher concentrations of several other C_6 compounds [4].

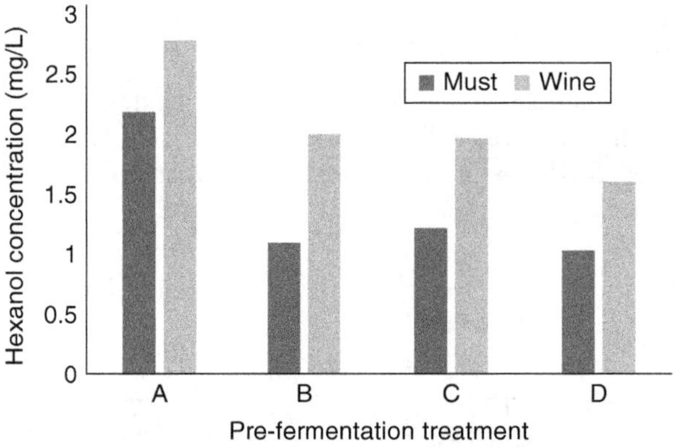

Figure 23.3.3 Effect of pre-fermentation clarification treatments on 1-hexanol, a C_6 alcohol, in must and in wine. A, control treatment; Airen white wine grapes, 16 h skin contact following crushing, no clarification following pressing. B, must clarified by settling for 24 h at 20 °C following pressing. C, must clarified by settling for 24 h at 15 °C following pressing. D, must clarified by settling with added pectinase enzymes for 24 h at 15 °C following pressing. Data from Reference [15]

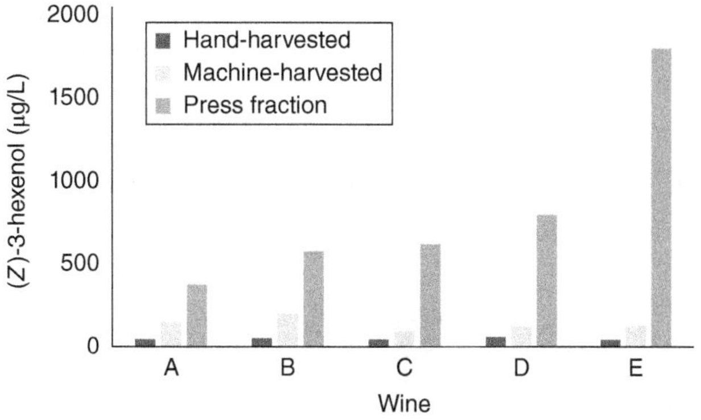

Figure 23.3.4 Effect of pre-fermentation grape handling on concentrations of (Z)-3-hexenol in five different wines (A–E). Wines were produced from Sauvignon Blanc grapes that were either (i) hand-harvested and hand-pressed, (ii) machine-harvested and hand-pressed, or (iii) machine-harvested, crushed, and mechanically pressed at 1 bar. Data from Reference [16]

23.3.3 Hydroxycinnamic acids, *Brettanomyces*, and volatile phenols

Wines possess modest concentrations of HCAs (~60 mg/L total), which in grapes exist almost exclusively as their tartaric acid esters (Chapter 13). HCAs are synthesized pre-veraison, and will decline by a factor of 2 on a weight-by-weight (but not per berry) basis during ripening due to the increase in berry size [20]. Concentrations of HCA tartrate esters will vary considerably among cultivars, as discussed in Chapter 13. Although HCAs (particularly caffeic acid) have important roles in oxidation reactions (Chapter 24), they are also of interest because of their ability to serve as precursors for odor-active volatile phenols. The conversion of HCA tartrate esters to volatile phenols is summarized in Figure 23.3.5.

23.3.3.1 Tartrate ester hydrolysis

Formation of volatile phenols requires initial hydrolysis of the tartrate ester group. Cinnamic acid esters undergo very slow hydrolysis at wine pH and at ambient temperature due to resonance stabilization of the carbonyl (Figure 23.3.5, top right) – estimated half-lives are on the order of decades for the hydrolysis of methyl and ethyl cinnamate at pH 3.4 and room temperature [21]. Average concentrations of caftaric acid in commercial Pinotage wines were inversely correlated with wine age and were 34% lower in wines from the 1998 vintage versus 2001 vintage (53 mg/L versus 79 mg/L) [22], which can be extrapolated to a half-life of ~7 years in wine. Ratios of esterified and free forms of HCA in wine can be as high as 9:1 even after ~15 years of storage [23]. Thus, acid hydrolysis of HCA tartrate esters is probably negligible for most wines and release of free HCAs through *cinnamic acid ester hydrolase* activity is of much greater importance.

- Some yeasts have modest *cinnamic acid ester hydrolase* (also called *cinnamic acid esterase*, CE, or *hydroxycinnamate ester hydrolase*, HCEH) activity. One report on Chardonnay showed a steady increase in *trans*-caffeic acid from undetectable to 11.6 mg/L caffeic acid equivalents (CAE) over the course of fermentation, and a corresponding decrease in its ester form (*trans*-caftaric acid) from a maximum of 41.8 mg/L CAE to 30.2 mg/L CAE by the end of fermentation [24], or a conversion of approximately 25% ignoring oxidative losses. Greater conversion (~75%) was observed for coutaric acid to coumaric acid.

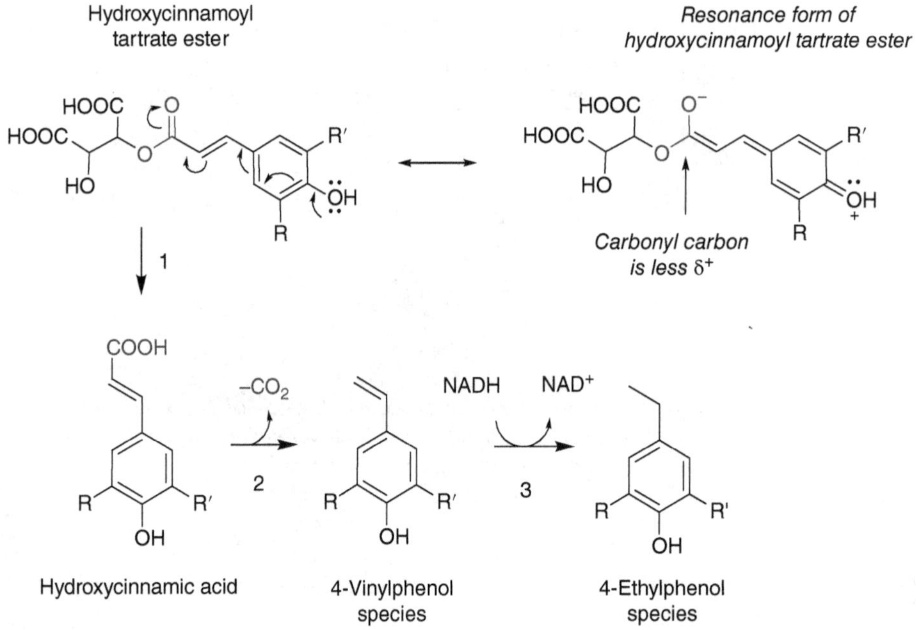

Figure 23.3.5 Reaction pathway to form odorous vinylphenols and ethylphenols from hydroxycinnamoyl (HCA) tartrate esters in grapes. The resonance stabilization of HCA tartrate esters (top right) decreases the rate of non-enzymatic acid hydrolysis as compared to simpler esters. 1, Hydroxycinnamoyl esterase (CE); certain lactic acid bacteria, pectinase preparations, and to a lesser extent yeast. 2, Hydroxyxcinnamate decarboxylase (HCD); S. cerevesiae, Brettanomyces and other yeasts. 3, Vinylphenol reductase (VPR); Brettanomyces

- Lactic acid bacteria possess widely variable CE activity – for the same wine, the VFO strain of *O. oeni* resulted in near-quantitative conversion of caftaric to caffeic acid, while other strains (Alpha, VP71) had no significant impact [25]. Thus, wines that undergo malolactic fermentation (e.g., many reds, Chapter 19) are expected to have greater concentrations of free HCAs.
- Finally, pectinase enzyme preparations added to increase juice yield (Chapters 21, 23.1, and 27) can demonstrate unintended CE side activity (Figure 23.3.6). Similar to LAB strains, the degree of CE activity varies considerably – in one study, degradation of coutaric acid ranged from negligible to >70% [26]. High CE activity is of concern to winemakers since it results in high free HCAs before fermentation, precursors for the malodorous vinyl- or ethylphenols, as described in the next sections. As a result, many commercial pectinases are selected to have (and advertised as possessing) low CE activity.

23.3.3.2 Vinylphenol formation

Free HCAs and their ethyl esters possess some astringent and bitter properties (Chapter 13). During fermentation, the *trans* forms of HCAs can be decarboxylated to yield volatile vinylphenols with "smoky" and "medicinal" odors (Chapter 12). In particular, ferulic acid can be converted to 4-vinylguaiacol (4-VG) and coumaric acid can be converted to 4-vinylphenol (4-VP). This transformation is carried out through *hydroxycinnamic decarboxylase* (HCD) enzymes, also referred to as *cinnamic acid decarboxylases* (Figure 23.3.5). Most *S. cerevesiae* wine yeast strains possess decarboxylase ability [27, 28], as do *Brettanomyces* spoilage yeasts (described in

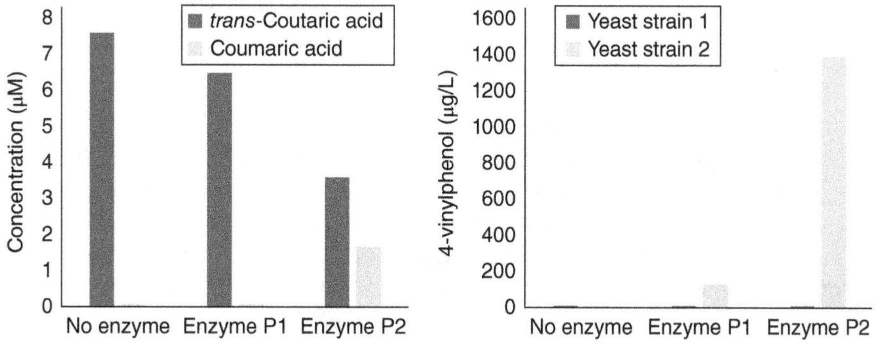

Figure 23.3.6 (Left) Demonstration of CE side activity in two commercial pectinase enzymes, P1 and P2, as evidenced by a decrease in coutaric acid and an increase in coumaric acid in Muscat of Frontignan juice samples incubated for 20 h at 20 °C. (Right) 4-Vinylphenol (4-VP) concentration in wines following fermentation of control and enzyme-treated juices. The highest 4-VP accumulation required both high free coumaric acid (from enzyme preparation P2) and a yeast with high cinnamic acid decarboxylase activity (Strain 2). Data from Reference [26]

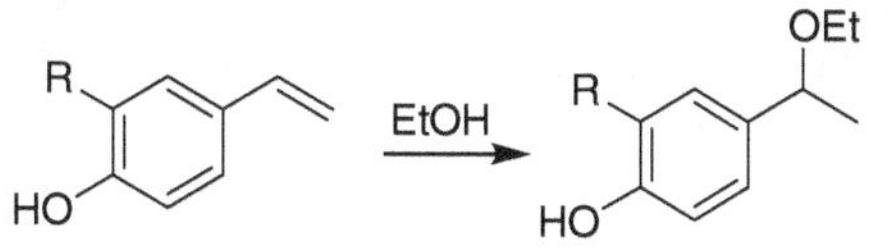

Figure 23.3.7 Reaction of vinylphenol (left) to form a 4-(1-ethoxyethyl)phenol [26]

more detail below) [29].[3] Disruption of either one of two genes (*FDC1* or *PAD1*) results in a loss of HCD activity in yeast [27]. Cinnamic acid derivatives have antimicrobial activities, and the function of the decarboxylase genes in yeasts is apparently to protect them against these common plant secondary metabolites [30].

Although most wine yeasts possess decarboxylase activity, the extent of HCA decarboxylation can vary considerably among strains – from <10% to 75% in a model juice system supplemented with 12 mg/L coumaric acid and fermented with one of three different strains [31]. As a result, vinylphenol concentrations can differ by an order of magnitude or more as a function of yeast strain, although initial concentrations of free HCAs will play an equally critical role (Figure 23.3.6 [26]).

As a caveat, interpreting differences among vinylphenol concentrations in real wine fermentations is often not straightforward, since formation will depend on not only initial hydroxycinnamate concentrations but also the rate of release of free HCAs from tartrate esters, and the reactivity of vinylphenols. 4-VG and 4-VP decrease in aged wines due to formation of 1-ethoxyethanol adducts (Figure 23.3.7, [26]), and the half-life of vinylphenol in white wines is approximately 6 months at 16–18 °C. Vinylphenols may decrease at faster rates in red wines due to their reactivity with anthocyanins to yield vinylphenolic pyranoanthocyanins (Chapter 16), or through further metabolism by *Brettanomyces*, as described next.

[3] In contrast to wine yeasts, most commercial brewing yeasts lack hydroxycinnamate decarboxylase activity because the clove/phenolic aroma of vinylphenols is considered undesirable in many beer styles. Yeasts with hydroxycinnamate decarboxylase activity are referred to as "POF (+)" (phenolic off-flavor positive) and are usually avoided. Some exceptions exist to this statement: many German wheat beers derive their characteristic flavor from POF (+) *Saccharomyces* yeasts, and some Belgian (or Belgian-style) ales are produced with *Brettanomyces* strains that will generate ethylphenols as well as vinylphenols.

23.3.3.3 Ethylphenol formation

Although 4-VG and 4-VP are detectable in many wines, particularly whites, they rarely appear to contribute to wine aroma due to their high reported sensory thresholds (770 μg/L) [32]. However, these vinylphenols can be enzymatically reduced to their more potent alkyl analogs – 4-ethylguaiacol (4-EG) and 4-ethylphenol (4-EP) – by the action of *vinylphenol reductase* (VPR) enzymes. In contrast to CE and HCD activity, which are found in a wide range of wine microorganisms, VPR activity is found almost exclusively in *Brettanomyces* yeasts [33], and a VPR enzyme from *Brettanomyces* has recently been purified and the protein sequenced [34]. Although *Brettanomyces* metabolism can result in several changes to wine organoleptic properties, including an increase in acetic acid and a loss of anthocyanins due to glycosidases [35], the most characteristic aspect of "Brett" flavor is an increase in "phenolic" or "medicinal" odor arising from 4-EP and related volatile phenols (Chapter 12) [36]. This formation of ethylphenols, and associated aromas, are generally (but not universally) considered undesirable [35, 44].

Brettanomyces grows at a much slower rate than *Saccharomyces* and has a negligible presence during alcoholic fermentation, in large part because alcoholic fermentation by "Brett" is slowed under anaerobic conditions (due to the "negative Pasteur effect") [37]. However, *Brettanomyces* can remain viable even in the harsh conditions of a finished wine (i.e., high ethanol, low pH, low O_2, low nutrients, SO_2) [38, 39], and requires minimal amounts of sugars to sustain growth (<1 g/L) [40]. Finally, in contrast to *Saccharomyces*, some *Brettanomyces* can use ethanol as a carbon source with minimal oxygen present, and most strains have the ability to utilize cellobiose, a $\beta(1 \rightarrow 4)$ disaccharide of two glucose molecules and a byproduct of cellulose degradation, that is, from toasting oak [39]. Viable 4-EP-producing *Brettanomyces* have been found several mm deep in the interior of used oak barrels [41], and thus can withstand many common surface cleaning and sanitation protocols [42]. Oak storage also introduces more oxygen, which stimulates *Brettanomyces* growth. As a consequence, barrels in wineries often serve as sites of *Brettanomyces* growth as well as reservoirs for further contamination [35, 42].

Almost all of ethylphenol production occurs towards the end of the growth phase and start of the stationary phase, and production is reported to cease when suitable carbon sources (e.g., glucose) are no longer detectable [43]. *Brettanomyces* can convert free vinylphenols to ethylphenols, but the majority of ethylphenol production occurs via the two-step enzyme-catalyzed processes of HCA → vinylphenol → ethylphenol [45]. Conversion of HCA precursors can be highly efficient under model conditions – one strain was able to convert 90% of coumaric acid to 4-EP in a model wine containing 2% glucose [43]. However, this conversion efficiency has a strong strain dependence, with variation in production of ethylphenols across 37 *Brettanomyces* strains ranging from undetectable to >2500 μg/L of both 4-EP and 4-EG using the same wine substrate [36] (Figure 23.3.8). Interestingly, 4-EP and 4-EG appear to be highly correlated, indicating that there is relatively little variation in VPR selectivity towards the main vinylphenols, and their relative concentrations ultimately appear to be related to that of the precursor HCAs (Chapter 12). Caffeic acid can also be reduced by *Brettanomyces* to form 4-ethylcatechol (4-EC) [45], although this conversion has not been as well studied as for 4-EP and 4-EG due to its lower efficiency, the higher olfactory threshold of 4-EC, and challenges associated with analysis of 4-EC.

In summary, factors that will affect final ethylphenol species concentrations include:

- Initial hydroxycinnamate ester concentration, which can be strongly affected by cultivar and winemaking conditions (Chapter 13).
- Extent of conversion of hydroxycinnamate esters to free HCAs, for example, by microbial enzymatic activity.
- Stability of HCAs during fermentation and storage and prior to *Brettanomyces* growth.
- Most importantly, the introduction and growth of *Brettanomyces* strains with VPR activity.

The physiological conditions that promote *Brettanomyces* growth include lower SO_2, higher O_2, and avoidance of filtration or other stabilization approaches [35]. These are also conditions that are more widely encountered in red winemaking and, along with the more widespread use of extended oak aging, can rationalize the more frequent appearance of the "Brett" character in red wines.

23.3.4 *S*-methylmethionine and dimethyl sulfide

Dimethyl sulfide (DMS, "truffle", "canned corn") can contribute both to the aroma of aged wine (Chapter 25) and, at high concentrations, to sulfurous off-aromas (Chapter 10). DMS is generally present at low µg/L concentrations in young wines (<1 year old), but will be formed during storage from grape-derived precursors [46]. The most important of these precursors appears to be *S*-methylmethionine (SMM) [47], which can form DMS upon hydrolysis. At pH values above 5 (as is found in beer wort and many vegetables), this hydrolysis appears to be base-catalyzed and will increase with increasing pH [48], but at pH < 5 the release appears to proceed through nucleophilic substitution by water [49], and no pH dependence is expected (Figure 23.3.9).

SMM is widespread in the plant kingdom, where it can be formed by methylation of methionine. The metabolic purpose of SMM is somewhat unclear but its primary role seems to be as a convenient means to transport methionine from its place of origin (vegetative tissues like leaves) into sinks (like seeds) [50]. The ability of SMM to form DMS during the thermal processing and storage of plant-based foods and beverages

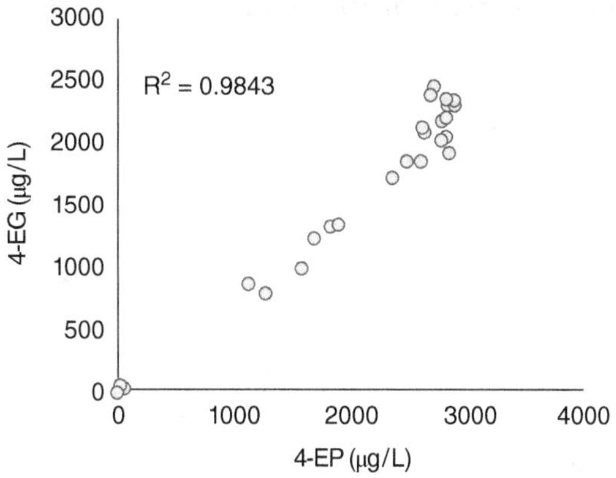

Figure 23.3.8 Correlation of 4-EG and 4-EP production for 37 Brettanomyces strains inoculated into an off-dry Grenache rosé wine. Note that 13 of the strains (37%) produced negligible amounts (<150 µg/L total) of ethylphenols, whereas around half of the strains were high producers (>2000 g/L total). Data from Conterno et al. [39]

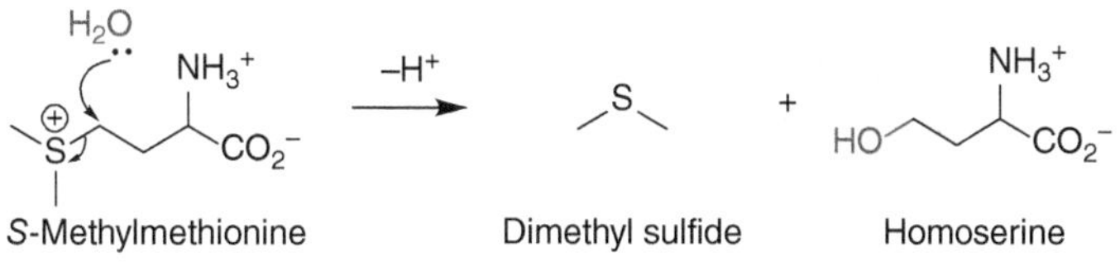

Figure 23.3.9 Hydrolytic formation of dimethyl sulfide (DMS) from S-methylmethionine (SMM) precursor to yield DMS [49]

has been well studied [51]. Higher concentrations of SMM are typically observed in vegetative tissues rather than fleshy fruits, with SMM concentrations in cabbage and asparagus in the range of 0.1–0.2% w/w [45], or about 1000-fold higher than what is typically observed in grapes [52] based on "potential DMS" measurements (see the next paragraph). The typical concentrations of DMS in these vegetables following cooking (10–20 mg/kg) are also two to three orders of magnitude higher than the highest concentrations reported in wine (Chapter 10).[4]

SMM can be quantified in grape juice [53], but more commonly "potential DMS" (pDMS) is determined by heating a juice or wine sample under alkaline conditions and quantifying the DMS formed [47]. One study has reported a wide range of pDMS in grapes of different varieties and growing locations, from low μg/L to > 4000 μg/L [53]. A particularly high concentration of pDMS was observed in a late harvest Petit Manseng and higher DMS concentrations are correlated with later harvest times [11], but the factors that control SMM in grapes are generally not well studied. Regardless, the pDMS concentration of grape juice is an excellent predictor of wine DMS following storage under controlled conditions [52]. From a limited number of Syrah grape samples, total wine DMS (pDMS + free) was about 70% of grape pDMS, indicating that SMM is mostly stable during fermentation. However, some yeast can reportedly use SMM as a nitrogen source during beer fermentations [48], and variation among wine yeasts has not been characterized.

Yeast do not appear to enzymatically hydrolyze SMM to DMS during fermentation. As mentioned above, the primary reaction will involve water as a nucleophile at wine pH and therefore will not show strong pH dependence. However, storage time and temperature will be critically important. The hydrolytic release of DMS from SMM has a high activation energy barrier, 186 kJ/mol [54] (Chapter 25). While SMM hydrolysis can occur rapidly around the boiling point of water (half-life = 30 min in beer wort, 98.5 °C), the release will be much slower under typical wine storage temperatures (half-life ~5 years, Figure 23.3.10).

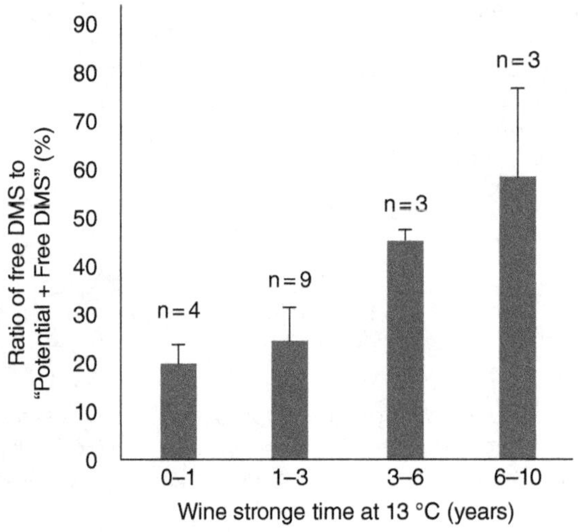

Figure 23.3.10 Mean percentage of DMS in free form as compared to the total DMS pool (pDMS + free DMS) as a function of wine age for Grenache and Syrah wines. Values over top of bars indicate number (n) of wines in an age category. Error bars represent standard deviations. Data from Reference [52]

[4] The high concentrations of DMS produced during the cooking of cabbage, asparagus, and many other vegetables in comparison to fruits can be traced not only to the high SMM concentrations of many vegetables, but also to their relatively high pH. The fact that SMM hydrolysis proceeds faster at alkaline pH explains why many cabbage recipes recommend the addition of acid during cooking to diminish strong aromas.

References

1. Linsenmeier, A., Rauhut, D., Sponholz, W.R. (2010) Ageing and flavour deterioration in wine, in *Managing wine quality, Vol. 2, Oenology and wine quality* (ed. Reynolds, A.G.), Woodhead Publishing and CRC Press, Oxford and Boca Raton, FL.
2. Matsui, K. (2006) Green leaf volatiles: hydroperoxide lyase pathway of oxylipin metabolism. *Current Opinion in Plant Biology*, **9** (3), 274–280.
3. Canuti, V., Conversano, M., Calzi, M.L., *et al.* (2009) Headspace solid-phase microextraction–gas chromatography–mass spectrometry for profiling free volatile compounds in Cabernet Sauvignon grapes and wines. *Journal of Chromatography A*, **1216** (15), 3012–3022.
4. Hashizume, K. and Samuta, T. (1997) Green odorants of grape cluster stem and their ability to cause a wine stemmy flavor. *Journal of Agricultural and Food Chemistry*, **45** (4), 1333–1337.
5. Iyer, M.M., Sacks, G.L., Padilla-Zakour, O.I. (2010) Impact of harvesting and processing conditions on green leaf volatile development and phenolics in Concord grape juice. *Journal of Food Science*, **75** (3), C297–C304.
6. Harsch, M.J., Benkwitz, F., Frost, A., *et al.* (2013) New precursor of 3-mercaptohexan-1-ol in grape juice: thiol-forming potential and kinetics during early stages of must fermentation. *Journal of Agricultural and Food Chemistry*, **61** (15), 3703–3713.
7. Dennis, E.G., Keyzers, R.A., Kalua, C.M., *et al.* (2012) Grape contribution to wine aroma: production of hexyl acetate, octyl acetate, and benzyl acetate during yeast fermentation is dependent upon precursors in the must. *Journal of Agricultural and Food Chemistry*, **60** (10), 2638–2646.
8. Herraiz, T., Herraiz, M., Reglero, G., *et al.* (1990) Changes in the composition of alcohols and aldehydes of C6 chain length during the alcoholic fermentation of grape must. *Journal of Agricultural and Food Chemistry*, **38** (4), 969–972.
9. Kalua, C.M. and Boss, P.K. (2010) Comparison of major volatile compounds from Riesling and Cabernet Sauvignon grapes (*Vitis vinifera* L.) from fruitset to harvest. *Australian Journal of Grape and Wine Research*, **16** (2), 337–348.
10. Iglesias, J.L.M., Dabila, F.H., Marino, J.I.M., *et al.* (1991) Biochemical aspects of the lipids in *Vitis vinifera* grapes (Macabeo variety). 1. Linoleic and linolenic acids as aromatic precursors. *Nahrung-Food*, **35** (7), 705–710.
11. Bindon, K., Varela, C., Kennedy, J., *et al.* (2013) Relationships between harvest time and wine composition in *Vitis vinifera* L. cv. Cabernet Sauvignon. 1. Grape and wine chemistry. *Food Chemistry*, **138** (2–3), 1696–1705.
12. Ramey, D., Bertrand, A., Ough, C.S., *et al.* (1986) Effects of skin contact temperature on Chardonnay must and wine composition. *American Journal of Enology and Viticulture*, **37** (2), 99–106.
13. Joslin, W.S. and Ough, C.S. (1978) Cause and fate of certain C_6 compounds formed enzymatically in macerated grape leaves during harvest and wine fermentation. *American Journal of Enology and Viticulture*, **29** (1), 11–17.
14. Capone, D.L., Black, C.A., Jeffery, D.W. (2012) Effects on 3-mercaptohexan-1-ol precursor concentrations from prolonged storage of Sauvignon Blanc grapes prior to crushing and pressing *Journal of Agricultural and Food Chemistry*, **60** (13), 3515–3523.
15. Ferreira, B., Hory, C., Bard, M.H., *et al.* (1995) Effects of skin contact and settling on the level of the C18:2, C18:3 fatty-acids and C6 compounds in Burgundy Chardonnay musts and wines. *Food Quality and Preference*, **6** (1), 35–41.
16. Herbst-Johnstone, M., Araujo, L., Allen, T., *et al.* (2012) Effects of mechanical harvesting on "Sauvignon blanc" aroma, in *1st International Workshop on Vineyard mechanization and grape and wine quality* (ed. Poni, S.), ISHS Acta Horticulturae, pp. 179–186.
17. Cejudo-Bastante, M.J., Castro-Vázquez, L., Hermosín-Gutiérrez, I., Pérez-Coello, M.S. (2011) Combined effects of prefermentative skin maceration and oxygen addition of must on color-related phenolics, volatile composition, and sensory characteristics of Airén white wine. *Journal of Agricultural and Food Chemistry*, **59** (22), 12171–12182.
18. Makhotkina, O., Herbst-Johnstone, M., Logan, G., *et al.* (2013) Influence of sulfur dioxide additions at harvest on polyphenols, C_6-compounds, and varietal thiols in Sauvignon blanc. *American Journal of Enology and Viticulture*, **64** (2), 203–213.
19. Fischer, U., Strasser, M., Gutzler, K. (2000) Impact of fermentation technology on the phenolic and volatile composition of German red wines. *International Journal of Food Science and Technology*, **35** (1), 81–94.
20. Ong, B.Y. and Nagel, C.W. (1978) Hydroxycinnamic acid–tartaric acid ester content in mature grapes and during the maturation of white Riesling grapes. *American Journal of Enology and Viticulture*, **29** (4), 277–281.

21. Rayne, S. and Forest, K. (2011) Estimated carboxylic acid ester hydrolysis rate constants for food and beverage aroma compounds. *Nature Precedings*. doi: 10.1038/npre.2011.6471.1.
22. Schwarz, M., Hofmann, G., Winterhalter, P. (2004) Investigations on anthocyanins in wines from *Vitis vinifera* cv. Pinotage: factors influencing the formation of Pinotin A and its correlation with wine age. *Journal of Agricultural and Food Chemistry*, **52** (3), 498–504.
23. Cheynier, V.F., Trousdale, E.K., Singleton, V.L., *et al.* (1986) Characterization of 2-*S*-glutathionyl caftaric acid and its hydrolysis in relation to grape wines. *Journal of Agricultural and Food Chemistry*, **34** (2), 217–221.
24. Somers, T.C., Vérette, E., Pocock, K.F. (1987) Hydroxycinnamate esters of *Vitis vinifera*: changes during white vinification, and effects of exogenous enzymic hydrolysis. *Journal of the Science of Food and Agriculture*, **40** (1), 67–78.
25. Burns, T.R. and Osborne, J.P. (2013) Impact of malolactic fermentation on the color and color stability of Pinot noir and Merlot wine. *American Journal of Enology and Viticulture*, **64** (3), 370–377.
26. Dugelay, I., Gunata, Z., Sapis, J.C., *et al.* (1993) Role of cinnamoyl esterase activities from enzyme preparations on the formation of volatile phenols during winemaking. *Journal of Agricultural and Food Chemistry*, **41** (11), 2092–2096.
27. Mukai, N., Masaki, K., Fujii, T., *et al.* (2010) PAD1 and FDC1 are essential for the decarboxylation of phenylacrylic acids in *Saccharomyces cerevisiae*. *Journal of Bioscience and Bioengineering*, **109** (6), 564–569.
28. Chatonnet, P., Dubourdieu, D., Boidron, J.-n., Lavigne, V. (1993) Synthesis of volatile phenols by *Saccharomyces cerevisiae* in wines. *Journal of the Science of Food and Agriculture*, **62** (2), 191–202.
29. Edlin, D.A.N., Narbad, A., Gasson, M.J., *et al.* (1998) Purification and characterization of hydroxycinnamate decarboxylase from *Brettanomyces anomalus*. *Enzyme and Microbial Technology*, **22** (4), 232–239.
30. Clausen, M., Lamb, C.J., Megnet, R., Doerner, P.W. (1994) PAD1 encodes phenylacrylic acid decarboxylase which confers resistance to cinnamic acid in *Saccharomyces cerevisiae*. *Gene*, **142** (1), 107–112.
31. Benito, S., Palomero, F., Morata, A., *et al.* (2009) Minimization of ethylphenol precursors in red wines via the formation of pyranoanthocyanins by selected yeasts. *International Journal of Food Microbiology*, **132** (2–3), 145–152.
32. Nikfardjam, M.P., May, B., Tschiersch, C. (2009) Analysis of 4-vinylphenol and 4-vinylguaiacol in wines from the Wuerttemberg region (Germany). *Mitteilungen Klosterneuburg, Rebe und Wein, Obstbau und Früchteverwertung*, **59** (2), 84–89.
33. Chatonnet, P., Viala, C., Dubourdieu, D. (1997) Influence of polyphenolic components of red wines on the microbial synthesis of volatile phenols. *American Journal of Enology and Viticulture*, **48** (4), 443–448.
34. Granato, T., Romano, D., Vigentini, I., *et al.* (2015) New insights on the features of the vinyl phenol reductase from the wine spoilage yeast *Dekkera/Brettanomyces bruxellensis*. *Annals of Microbiology*, **65** (1), 321–329.
35. Conterno, L. and Henick-Kling, T. (2010) *Brettanomyces/Dekkera* off-flavours and other wine faults associated with microbial spoilage, in *Managing wine quality* (ed. Reynolds, A.G.), Woodhead Publishing, pp. 346–387.
36. Chatonnet, P., Dubourdie, D., Boidron, J.-N., Pons, M. (1992) The origin of ethylphenols in wines. *Journal of the Science of Food and Agriculture*, **60** (2), 165–178.
37. Scheffers, W.A. and Wiken, T.O. (1969) The Custers effect (negative Pasteur effect) as a diagnostic criterion for the genus *Brettanomyces*. *Antonie Van Leeuwenhoek*, **35** (Suppl. Yeast Symp.), A31–A32.
38. Suárez, R., Suárez-Lepe, J.A., Morata, A., Calderón, F. (2007) The production of ethylphenols in wine by yeasts of the genera *Brettanomyces* and *Dekkera*: a review. *Food Chemistry*, **102** (1), 10–21.
39. Conterno, L., Joseph, C.M.L., Arvik, T.J., *et al.* (2006) Genetic and physiological characterization of *Brettanomyces bruxellensis* strains isolated from wines. *American Journal of Enology and Viticulture*, **57** (2), 139–147.
40. Coulon, J., Perello, M.C., Lonvaud-Funel, A., *et al.* (2010) *Brettanomyces bruxellensis* evolution and volatile phenols production in red wines during storage in bottles. *Journal of Applied Microbiology*, **108** (4), 1450–1458.
41. Barata, A., Laureano, P., D'Antuono, I., *et al.* (2013) Enumeration and identification of 4-ethylphenol producing yeasts recovered from the wood of wine ageing barriques after different sanitation treatments. *Journal of Food Research*, **2** (1), 140–149.
42. Aguilar Solis, M. (2014) Evaluation of common and novel sanitizers against spoilage yeasts found in wine environments. PhD Thesis, Cornell Unversity, Ithaca, NY.
43. Dias, L., Pereira-da-Silva, S., Tavares, M., *et al.* (2003) Factors affecting the production of 4-ethylphenol by the yeast *Dekkera bruxellensis* in enological conditions. *Food Microbiology*, **20** (4), 377–384.

44. Juan, F.S., Cacho, J., Ferreira, V., Escudero, A. (2012) Aroma chemical composition of red wines from different price categories and its relationship to quality. *Journal of Agricultural and Food Chemistry*, **60** (20), 5045–5056.
45. Harris, V., Ford, C., Jiranek, V., Grbin, P. (2008) *Dekkera* and *Brettanomyces* growth and utilisation of hydroxycinnamic acids in synthetic media. *Applied Microbiology and Biotechnology*, **78** (6), 997–1006.
46. Marais, J. (1979) Effect of storage time and temperature on the formation of dimethyl sulfide and on white wine quality. *Vitis*, **18** (3), 254–260.
47. Segurel, M.A., Razungles, A.J., Riou, C., *et al.* (2005) Ability of possible DMS precursors to release DMS during wine aging and in the conditions of heat-alkaline treatment. *Journal of Agricultural and Food Chemistry*, **53** (7), 2637–2645.
48. Anness, B.J. and Bamforth, C.W. (1982) Dimethyl sulfide – a review. *Journal of the Institute of Brewing*, **88** (4), 244–252.
49. Cremer, D.R. and Eichner, K. (2000) Formation of volatile compounds during heating of spice paprika (*Capsicum annuum*) powder. *Journal of Agricultural and Food Chemistry*, **48** (6), 2454–2460.
50. Giovanelli, J., Mudd, S.H., Datko, A.H. (1980) Sulfur amino acids in plants, in *Amino acids and derivatives* (ed. Miflin, B.J.), Academic Press, pp. 453–505.
51. Scherb, J., Kreissl, J., Haupt, S., Schieberle, P. (2009) Quantitation of *S*-methylmethionine in raw vegetables and green malt by a stable isotope dilution assay using LC-MS/MS: comparison with dimethyl sulfide formation after heat treatment. *Journal of Agricultural and Food Chemistry*, **57** (19), 9091–9096.
52. Segurel, M.A., Razungles, A.J., Riou, C., *et al.* (2004) Contribution of dimethyl sulfide to the aroma of Syrah and Grenache noir wines and estimation of its potential in grapes of these varieties. *Journal of Agricultural and Food Chemistry*, **52** (23), 7084–7093.
53. Loscos, N., Segurel, M., Dagan, L., *et al.* (2008) Identification of *S*-methylmethionine in Petit Manseng grapes as dimethyl sulphide precursor in wine. *Analytica Chimica Acta*, **621** (1), 24–29.
54. Scheuren, H., Tippmann, J., Methner, F.J., Sommer, K. (2014) Decomposition kinetics of dimethyl sulphide. *Journal of the Institute of Brewing*, **120** (4), 474–476.

24

Wine Oxidation

24.1 Introduction

Oxidation is generally the limiting factor in the preservation of wine. In other words, most wines lose their palatability during long-term storage due to excessive oxidation rather than other causes like microbial spoilage. However, a small amount of oxygen can improve qualities of some wines, and large amounts of oxygen exposure are critical for the flavor of certain wines, such as Madeira, Sherry, Vino Santo, and Jura Vin Jaune. Several important historical wine authors have reflected on the significance of oxidation. In the earliest US text on wine production Rixford states, "[i]n a few instances, where the wines are strong enough to bear it, aging may be hastened by some exposure to the air, but great care must be taken that they are not left too long under its influence or disorganization may ensue"[1]. Almost 100 years later, in the early 1970s Amerine *et al.* wrote that "the principal changes in flavor and bouquet during aging in the wood are generally believed to be due to slow oxidation" [2]. Numerous other reviews address the importance of wine oxidation to wine stability, flavor and color, including those by Singleton [3], Cheynier [4], Danilewicz [5], du Toit *et al.* [6], Waterhouse and Laurie [7], Li *et al.* [8], Karbowiak [9], Oliveira *et al.* [10] Danilewicz [11], and Ugliano [12].

In summary, although there is now a greater awareness of other non-oxidative reactions that can occur during wine aging, for example, acid hydrolysis (Chapter 25), Louis Pasteur's nineteenth century observation, "*C'est l'oxygene qui fait le vin*" is still largely true [13].

24.2 Redox reactions

Redox refers to chemical reactions that involve the gain (reduction) and loss (oxidation) of electrons by the reactants. In the context of wine chemistry, oxidation is usually associated with an increase in the number of covalent bonds to oxygen and/or decrease in bonds made to hydrogen or other less electronegative atoms.

The tendency of compounds to be oxidized or reduced is described by their reduction potential, *E*. Tables of reduction potentials are compiled as *half-reactions* (see Table 24.1), which will include the oxidized form, reduced form, and electrons and/or protons gained during the reduction process. Reduction potentials for species are usually reported under standard conditions (e.g., E_0 values are reported for 1 M solute concentration or 101 325 Pa (1 atm)

Understanding Wine Chemistry, First Edition. Andrew L. Waterhouse, Gavin L. Sacks, and David W. Jeffery.

Table 24.1 Half-reactions for the reduction of some wine components

Oxidized form		Reduced form	E_0 (pH 0), 25 °C	$E_{3.5}$ (pH 3.5), 25 °C
SO_4^{2-}	$+2e^- + 2H^+ - H_2O$	HSO_3^-	−0.4V	−0.6V
Disulfides	$+2e^- + 2H^+$	Mercaptans	~0.1	~ −0.1
Acetaldehyde	$+2e^- + 2H^+$	Ethanol	0.2	0
Cu^{2+}	$+e^-$	Cu^+	0.15	0.15
Dehydro-ascorbic acid	$+2e^- + 2H^+$	Ascorbic acid	0.5	0.3
o-Quinone	$+2e^- + 2H^+$	*o*-Diphenol (catechol)	0.8	0.6
Fe^{3+}	$+e^-$	Fe^{2+}	0.77	0.77
H_2O_2	$+e^- + H^+$	$\cdot OH + H_2O$	0.38	0.28
½ O_2	$+2e^- + 2H^+$	H_2O	1.23	1.0

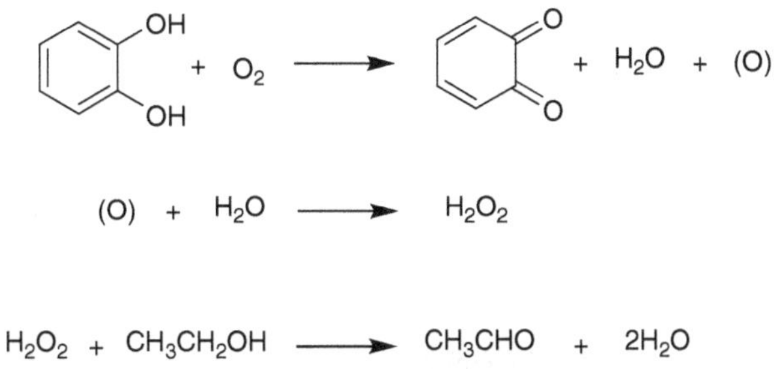

Figure 24.1 Formation of an o-quinone from an o-diphenol (top) and concurrent production of hydrogen peroxide (H_2O_2) through coupled oxidation (middle) in wine oxidation as proposed by Wildenradt and Singleton [17]. The resulting H_2O_2 can subsequently oxidize ethanol to form acetaldehyde (CH_3CHO, bottom)

gas concentration, 25 °C, pH 0). Half-reactions that include protons will have pH-dependent reduction potentials; typically, reduction potentials that involve protons will be about 0.2 V lower at wine pH than at pH 0. Higher values (more positive) for E indicate that the oxidized form is a stronger oxidizer, and the reduced form is a weaker reducer (weak antioxidant). Further information on this topic can be found in an introductory chemistry book.

The oxidizing or reducing nature of a solution is described by its redox potential, E, and under equilibrium conditions the ratio of oxidized to reduced ([Ox]/[Red]) forms of a species can be related to E by the Nernst equation, where F is Faraday's constant:

$$E = E_0 - \frac{RT}{F}\ln\left(\frac{[\mathrm{Ox}]}{[\mathrm{Red}]}\right) \tag{24.1}$$

In principle, redox potential measurements could be useful for evaluating whether wine is excessively reduced or oxidized. In practice, this is challenging because redox measurements are generally performed using platinum electrodes, and the redox potential is defined as the voltage applied across electrodes where there is no net current (reduction and oxidation reactions occur at the same rate). Danilewicz clarifies that this requires the presence of all participating substances at the electrode, but the key first step in wine oxidation, the reaction of oxygen and phenols, produces unstable *o*-quinone intermediates (Figure 24.1) [11]. Thus no equilibrium condition is present, so the Nernst equation cannot be utilized, and these measurements do not reveal a true redox potential. Instead, these platinum electrode measurements on wine primarily measure the direct reduction of oxygen, that is, the measured value is related only to the concentration of dissolved oxygen [14]. In addition, the redox potentials of metals are strongly influenced by the presence of metal chelators, one of which is tartrate, a strong Fe^{3+} chelator, so the use of redox potentials to assess thermodynamics must take such effects into account [15].

24.2.1 Thermodynamics of redox reactions

Full redox reactions involve combinations of half-reactions to balance out the exchange of electrons. For example, taking into account the reduction of oxygen and oxidation of ethanol, the overall reaction to form acetaldehyde under standard conditions would be:

$$\begin{array}{ll} \tfrac{1}{2}O_2 + 2H^+ + 2e^- \rightarrow H_2O & (E_0 = 1.23\,V)\ \text{Reduction} \\ \text{Ethanol} \rightarrow \text{Acetaldehyde} + 2e^- + 2H^+ & (E_0 = 0.2\,V)\ \text{Oxidation} \\ \hline \tfrac{1}{2}O_2 + \text{Ethanol} \rightarrow H_2O + \text{Acetaldehyde} & E_{cell} = 1.03\,V \end{array}$$

The electrochemical cell potential, E_{cell}, is related to the equilibrium constant for a reaction, K_{eq}, by the following equation, where n = number of electrons involved in the redox reaction:

$$\ln K_{eq} = E_{cell} \times \left(\frac{nF}{RT} \right) \tag{24.2}$$

Positive values of E_{cell}, as is the case for the aforementioned reaction of oxygen with ethanol, describe reactions that are thermodynamically favorable, and in the case of the full reaction shown above, $K_{eq} > 10^{34}$. The data shown in Table 24.1 can thus be used to determine what redox reactions could feasibly occur, which provides a *very* crude understanding of the events that will occur during wine oxidation, that is, sulfited wines exposed to oxygen should accumulate sulfate and/or acetaldehyde and lose oxygen over time. However, a strictly thermodynamic interpretation of wine oxidation is by itself inadequate. Thermodynamics alone would *incorrectly predict* that reducing compounds should be lost in reverse order of their reduction potentials; that is, nearly all sulfites must be converted to sulfate before diphenols could be oxidized to quinones or alcohols are oxidized to aldehydes, etc. Furthermore, thermodynamics alone can also not explain why vodka can sit on a shelf for years without displaying any signs of oxidation. Thus, in addition to a thermodynamic analysis, a kinetic, mechanistic understanding is necessary to rationalize the rate and products of real oxidations.

24.3 The central tenets of wine oxidation

The last decade has seen a growing consensus on the primary mechanisms involved in wine oxidation, which we will refer to as "The Central Tenets of Wine Oxidation."

1. Molecular oxygen (O_2) does not directly react with most wine components. The initial steps of oxygen consumption requires the presence of transition metal catalysts and good hydrogen donor molecules, particularly Fe(II) and *o*-diphenols.
2. Metal-catalyzed oxidation of *o*-diphenol groups results in the formation of quinones and H_2O_2. However, the thermodynamics of this reaction are unfavorable, and product consumption is essential to enable the reaction.
3. Quinones are strongly electrophilic and will react with wine nucleophiles including bisulfite, thiols, and flavan-3-ols. Antioxidants act in part by consuming the quinones, thus preserving flavors.

4. H_2O_2 reacts with bisulfite and oxidation reactions end. Alternatively, if there is insufficient bisulfite, H_2O_2 participates in the Fe(II)-catalyzed Fenton reaction to produce aldehydes and other oxidized species.
5. Aldehydes and related species react with phenolics, bisulfite, and other nucleophilic species. When the nucleophiles are exhausted, the aldehydes and related compounds accumulate, making the wine "oxidized."

This model has been particularly successful in describing why the rate of oxidation and product distribution changes with wine chemistry. Oxidation rates may be defined in several ways, including:

- The rate of loss of oxygen once introduced into wine.
- The rate of loss of reduced compounds, particularly SO_2.
- The rate of formation of compounds associated with oxidation, for example, brown pigments or acetaldehyde.

Early studies of wine chemistry observed that O_2 consumption roughly obeyed first-order kinetics. The half-life of O_2 in wine is usually on the order of several hours or days at room temperatures, or 100–1000-fold slower than oxygen consumption by polyphenol oxidase (PPO) activity in juice. Oxygen consumption was observed to be faster in red than in white wines, but there was little understanding as to how non-enzymatic oxidation occurred [16].

In 1974, Wildenradt and Singleton published one of the first papers discussing a reaction mechanism for oxidation. Using model wines, the authors observed that aldehyde formation and browning were only observed when polyphenolic compounds containing the *o*-diphenol moiety were added to the model system (e.g., caffeic acid, catechin); model systems containing only ethanol, water, and tartaric acid showed no aldehyde production [17].[1] The authors then hypothesized several important parts of the central tenets, namely that catechols react with oxygen to yield quinones and hydrogen peroxide, and that the peroxide subsequently reacts with ethanol to generate acetaldehyde [17] (Figure 24.1).

The discovery of quinone formation was an important milestone in rationalizing wine oxidation. The simultaneous formation of hydrogen peroxide was named *coupled oxidation* by Wildenradt and Singleton [17].

24.3.1 The role of metal catalysts in quinone formation

One problem with the scheme in Figure 24.1 is that the favored *ground-state triplet* form of O_2 cannot directly react with other ground-state organic molecules, as it is spin forbidden by Pauli's exclusion principle. In food systems, ground-state oxygen is typically activated to a reactive *singlet* state through one of two pathways:

- UV light exposure
- Transition metal catalysts.

Subsequent research has shown that transition metal catalysts, and particularly iron, appear to be critical to wine oxidation [5] (Chapter 4). Iron primarily exists as the reduced Fe(II) species in wine due to the presence of stronger reducing agents (Table 24.1). However, the precise mechanism by which the Fe(III)/Fe(II) couple facilitates oxidation is still difficult to discern, with the possibility of two one-electron steps, or one two-electron step, or some hybrid. A two-step scheme is shown in Figure 24.2.

Fe(II) can be rapidly oxidized to Fe(III), yielding a hydroperoxy radical capable of oxidizing *o*-diphenol to a semiquinone. The semiquinone can then be oxidized by Fe(III) to an *o*-quinone. Cu(I) may have a synergistic role in increasing oxidation rates in wine by regenerating Fe(II) (Table 24.1). This two-step process would involve the formation of the hydroperoxyl radical (protonated superoxide O_2^{-}), but an attempt to detect this

---

[1] The central role of *o*-diphenols in wine oxidation helps explain why an opened bottle of vodka can sit on a shelf for years with minimal evidence of aldehyde formation.

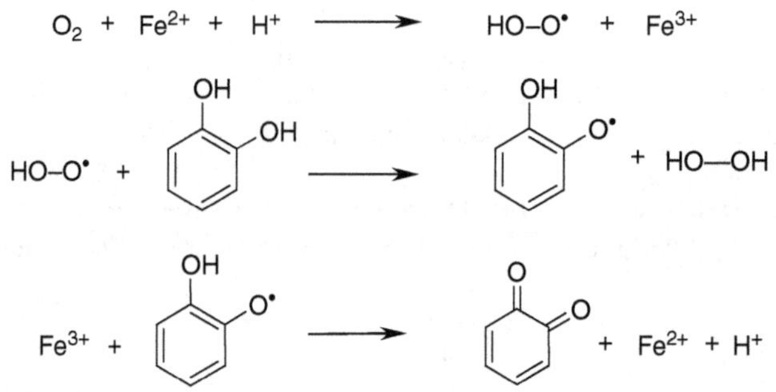

Figure 24.2 Hypothetical mechanism forming quinone and peroxide via the hydroperoxyl radical step, adapted from [4]

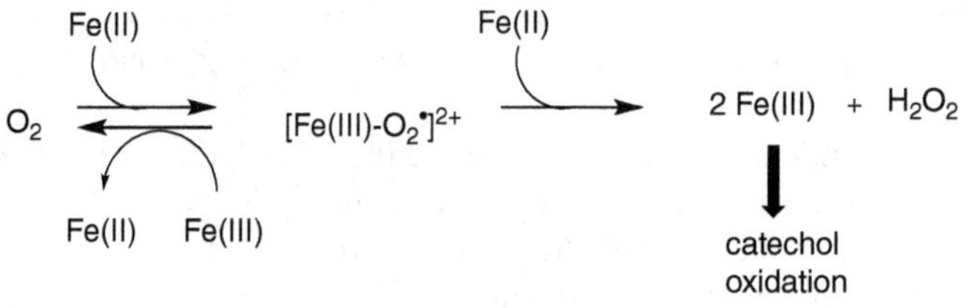

Figure 24.3 A two-step process for formation of o-quinone from an o-diphenol, showing potential roles of metal catalysts. Source: Danilewicz 2013 [19]. Reproduced with permission of AJEV

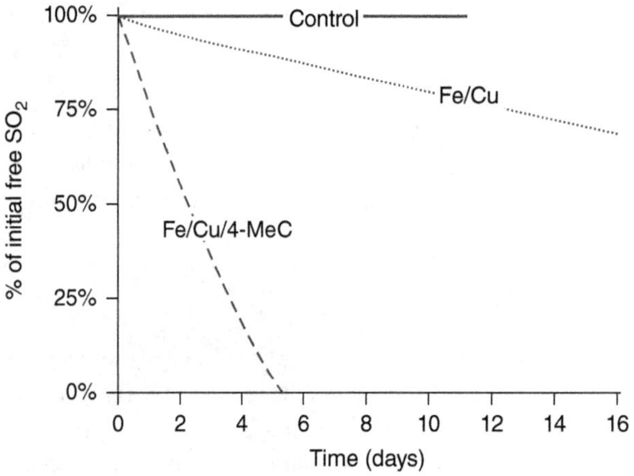

Figure 24.4 Effect of metals and phenolics on SO_2 consumption in model wine. 4-MeC is 4-methylcatechol. Source: Danilewicz 2007 [20]. Reproduced with permission of AJEV

species was not successful [18]. An alternative mechanism has been proposed that reduces oxygen with two Fe(II) species to generate hydrogen peroxide (Figure 24.3) and quinone [19]. Regardless of the mechanism, phenolic compounds and metal catalysts have a synergistic effect on the rate of wine oxidation [20] (Figure 24.4). Simple phenols have a much higher oxidation potential than catechols (Chapter 11), rendering them inaccessible to oxidation via this mechanism.

The reactivity of Fe(II) towards oxygen is determined by the redox potential of the Fe(III)/Fe(II) couple, which in an ideal solution will be $E_0=0.77$ V. However, this reduction potential can be altered by complexation of one or both of these two Fe forms.

- The major wine acid, tartaric, complexes Fe(III) and decreases the reduction potential of the Fe(III)/Fe(II) couple to 0.34 V, below that of the O_2/H_2O_2 couple. This favors Fe(III) and H_2O_2 formation and increases the rate of oxidation [15]. Similar pro-oxidative effects can be seen following addition of the Fe(III) chelating agent, ethylenediaminetetraacetic acid (EDTA) [21].
- Conversely, the decrease in Fe(III)/Fe(II) reduction potential brought about by tartrate chelation of Fe(III) makes it less able to directly oxidize *o*-diphenols to quinones, slowing oxidation [19]. This effect can be negated by the presence of strong Fe(II) chelators, which brings back the redox potential closer to the literature value for an Fe(III)/Fe(II) couple [22]. In addition, if a nucleophile is available to react with the quinone, the reaction can proceed even in systems with stabilized Fe(III), that is, tartrate buffered wine, based on Le Châtelier's principle.
- Strong Fe(II) complexing reagents, for example, ferrozine, slow oxidation reactions [21].[2]

In summary, if a wine contained no iron or related transition metals, oxygen would accumulate in barrels and bottles, and the wine would never oxidize!

24.4 The central tenets of quinone reactions

Subsequent investigations have shown that *o*-diphenols and metals are essential but not sufficient for wine oxidation. The driving force for the initial oxidation reactions of oxygen and catechol to quinone is not, surprisingly, the oxidation of catechol. The redox potential of the H_2O_2/O_2 couple at wine pH ($E_{3.5}$ ~0.45 V) is lower than a typical quinone/diphenol couple (0.6 V, Table 24.1) and equilibrium will strongly favor the 1,2-diphenol. Thus, little oxidation occurs unless a suitable nucleophile is available to consume the quinone (Figure 24.5) [19]. It is the consumption of the quinone by nucleophiles and the application of Le Chatelier's principle that leads to oxygen consumption [23].

In most cases, wine contains sufficient nucleophiles or other reducing agents and oxidation will proceed (Figure 24.5). Potential quinone reactions include:

- Condensation reaction with nucleophilic thiols
- Condensation reaction with other phenolics, particularly flavan-3-ol nucleophiles
- Condensation reaction with other nucleophiles, particularly bisulfite
- Reaction with reducing agents (ascorbic acid, bisulfite) to regenerate the original diphenol.

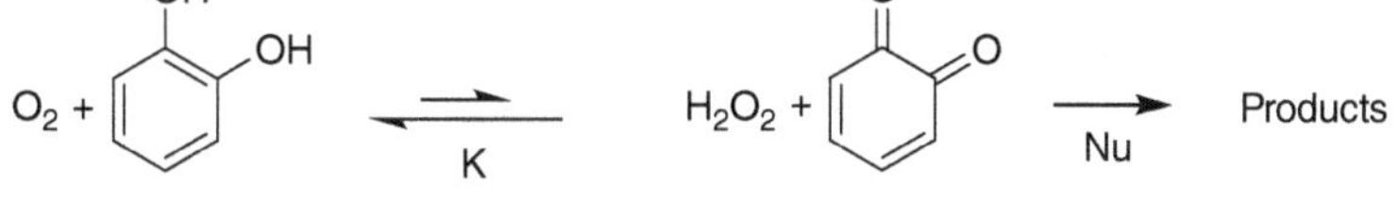

Figure 24.5 Equilibrium reaction of oxygen and catechol and the subsequent reaction with various nucleophiles. K of the equilibrium reaction is very small

[2] Although Fe(II) chelators represent a potential strategy for stabilizing wine against oxidation, the best studied versions (e.g., ferrozine) are somewhat toxic and are not approved for use in wine.

24.4.1 Quinone-thiol adducts

Thiols are very good nucleophiles and will react rapidly with quinones (Chapter 10). The first quinone-thiol adduct discovered in enology, and in fact the first specific wine oxidation product that was well characterized, resulted from an initial observation that caftaric acid is lost during must oxidation [24]. This was subsequently shown to be a result of the reaction between caftaric acid quinone and glutathione (GSH) [25]. The oxidation was proposed to be enzymatic, with grape polyphenol oxidase (PPO) catalyzing the oxidation of caftaric acid and its quinone reacting chemically with the thiol. This adduct, called the grape reaction product (GRP), is found in most commercial wines as a result of the oxygen exposure encountered during grape crushing and pressing. This outcome is not overly surprising given that GSH is the major thiol in grape juice, and considering that thiols are one of the most reactive nucleophiles present in wine. Others have reported that GSH can react with oxidized catechin to form a bond to the B-ring [26], in a manner similar to that shown for the thiol in Figure 24.6.

Although GSH is the most quantitatively important thiol in wine, there are several low-concentration odorous thiols that can also react with quinones [27] in an analogous manner.

24.4.2 Quinone-phenol adducts and browning in juice

The appearance of brown color in plant-based products is of broad interest to many different areas of food science [28]. In grape juice, browning is initiated largely by enzymatic oxidation, not chemical oxidation as observed in bottle aging [29], and particularly through PPO and related enzymes acting on hydroxycinnamates (Chapter 13). It can also be facilitated by a process called hyperoxidation, in which oxygen is intentionally added to the must to induce browning reactions [30]. The brown insoluble material is then separated from the must (e.g. by racking) prior to fermentation the finished wine is less susceptible to browning in the bottle. The first step of enzymatic browning is formation of the caftaric acid quinone via enzymatic oxidation by PPO. Browning will be decreased in musts containing high relative amounts of GSH as compared to hydroxycinnamates, since GSH can form GRP and arrest oxidation [31]. However, when solutions containing only hydroxycinnamates are oxidized, no browning appears, and, instead, some colorless dimeric products are formed (Figure 24.7) [32].

Browning in juice is, however, well correlated with flavan-3-ol concentration [33–35]. The pathway to enzymatic browning appears to follow coupled oxidation reactions where hydroxycinnamate quinones oxidize catechol groups on the flavan-3-ols to quinones, regenerating the hydroxycinnamates [36]. The electrophilic flavan-3-ol quinones are reactive and can couple with nucleophilic phloroglucinol A-rings on another flavanol (Chapter 10, boxed section). In the simplest example, one catechin quinone reacts with a catechin to form a dimer as reported by Guyot *et al.* [37] and shown in Figure 24.8. These authors also report related reactions with oxygen on the A-ring acting as a nucleophile. This type of reaction has since been confirmed by others [38].

The products shown in Figure 24.8 have no pigmentation, and formation of yellow or brown pigmented substances involves a second oxidation of a B-ring catechol on an initial dimeric product to form a new quinone.

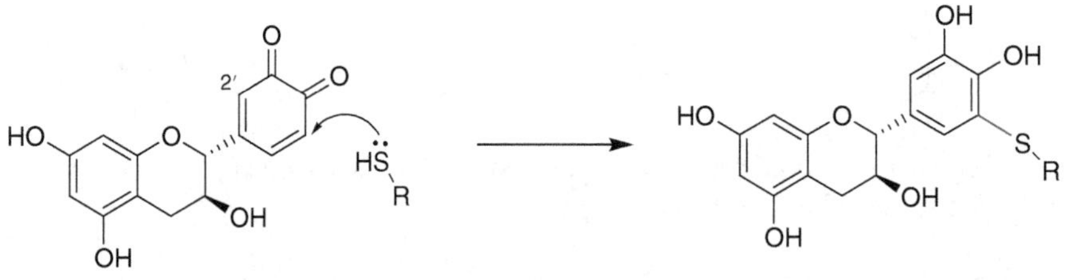

Figure 24.6 Thiol reaction with catechin quinone. All sites (2′, 5′, and 6′) are reactive and product ratios depend on the specific nucleophile and quinone

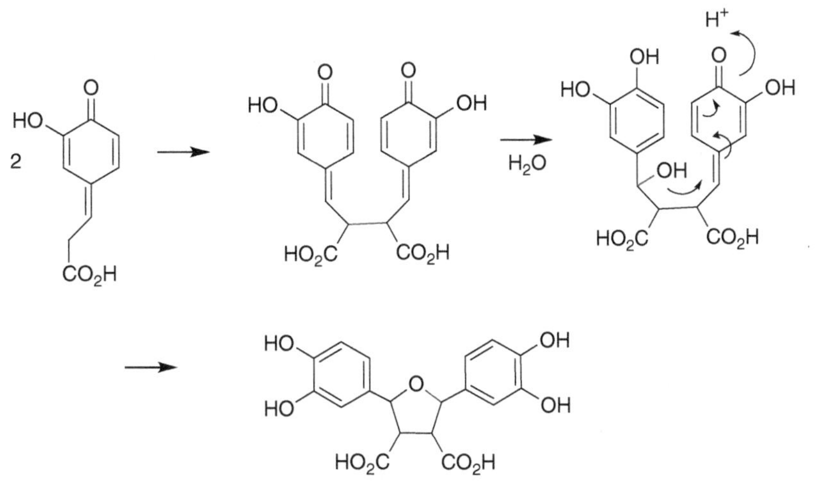

Figure 24.7 Hydroxycinnamate quinone dimerization producing a colorless product as reported by Fulcrand et al. [32]

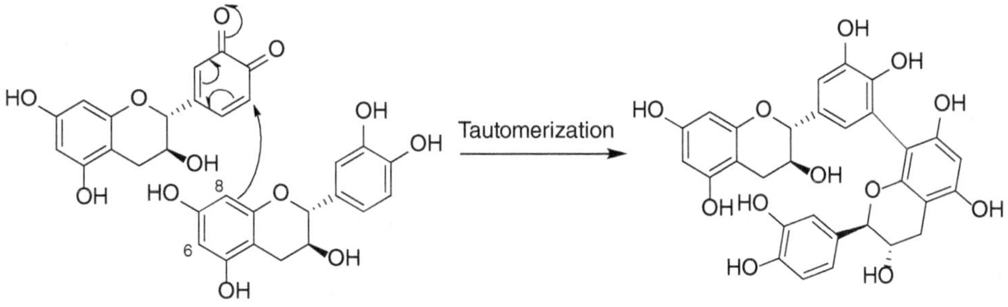

Figure 24.8 Flavan-3-ol condensation product arising from a nucleophilic attack of C8 of one catechin monomer on to the o-quinone of another. Reaction at C6 is also possible. Adapted from Reference [37]

This is followed by in intramolecular reaction at the 1′ position on the quinone with a nucleophilic A-ring oxygen from the other flavanol unit. This product has no hydrogen left at the addition site, so the product cannot rearrange via H isomerization back to an aromatic benzene structure, locking in a conjugated keto form as shown in Figure 24.9. These products can have a large number of conjugated double bonds, shifting the absorbance of light into the visible region[3] and thus yellow color is observed.

[3] Both the red flavylium form of anthocyanins and the yellow pigments formed from oxidation have extended networks (conjugation) of π-electrons (i.e., alternating single and double bonds). Conjugation lowers the energy required for the electronic transitions relevant to UV-vis spectroscopy, and shifts absorbance maxima to higher wavelengths (i.e., from UV to visible regions). Greater detail than this is beyond the scope of this work but can be found in organic chemistry or spectroscopy textbooks.

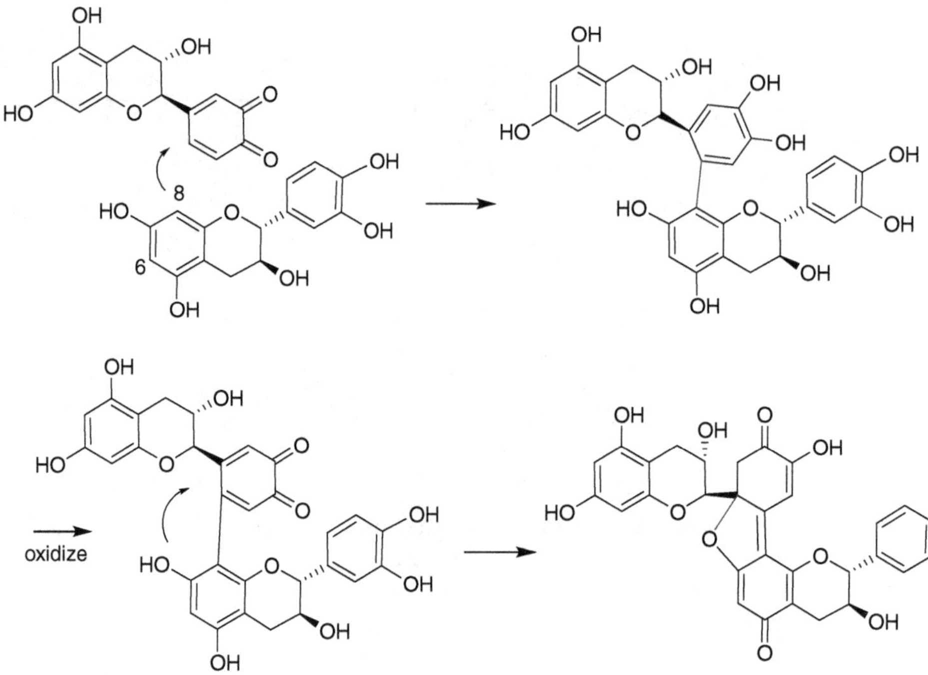

Figure 24.9 Formation of pigmented products of flavan-3-ols as described by Guyot et al. [37]. Products arising from the initial reaction at C6 are also observed

ortho-Quinone groups may also isomerize to the pseudo-*para*-quinone form (i.e., quinone methide), providing an electrophilic reaction center at the C2 on the C-ring, creating linkages at that position [39]. This can involve hydroxyl groups on a phloroglucinol ring, as these are also reactive as nucleophiles. Thus, in sterically constrained situations an intramolecular bond at the 2-position can be formed. Such products can also have extended π-bond conjugation, and thus be pigmented.

While such products are expected as a result of enzymatic oxidation in juice, in finished wine PPO is no longer active and oxidative browning occurs by other routes. In model wine solutions, the main yellow-brown compounds do not arise from a quinone electrophile, but appear to be xanthylium reaction products of aldehydes and flavan-3-ols, described in more detail below [40].

24.4.3 Reactions of quinones and other antioxidants: sulfur dioxide and ascorbic acid

Sulfur dioxide is well known to suppress enzymatic browning in a range of food systems, and early reports established that bisulfite reduced quinones and consequently lowered the level of oxidation products [41]. The general chemistry of bisulfite reactions with quinones was demonstrated some time ago with *p*-quinone, revealing that at low pH there are two reaction pathways: nucleophilic addition to the quinone, yielding about 25% of the sulfonate, and reduction to regenerate hydroquinone (*p*-dihydroxybenzene) as the major product. However, at higher pH values (neutral or slightly basic) the amount of hydroquinone decreased until the only product was the sulfonate adduct [42]. More recently, analogous results were observed with an *o*-quinone at wine pH values; that is, most of the quinone was reduced back to an *o*-diphenol, but the reaction yielded a small fraction of the sulfonate (Figure 24.10) [23].

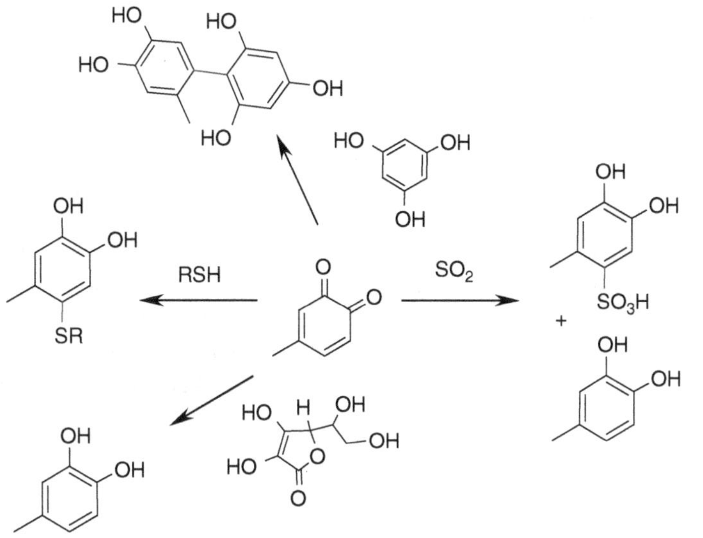

Figure 24.10 Alternate pathways for reactions of quinones with known nucleophiles and reducing agents

Another common antioxidant used in winemaking, ascorbic acid, is not as strong an antioxidant as SO_2, but can reduce *o*-quinones back to their catechol form just as quickly [43] (Figure 24.10). There is a small amount in grapes that is quickly lost during fermentation so that new wine has a negligible ascorbic acid content, but it can be added as a preservative, typically prior to bottling. Because of its ability to scavenge quinones, ascorbic acid can be used as an antioxidant adjuvant to SO_2, and its proper use can allow for lower usage of SO_2 in finished wine. However, ascorbic acid does not readily consume the other major product of wine oxidation (H_2O_2, see below) and its use without SO_2 can lead to enhanced browning [44].

24.4.4 Comparison of quinone scavenging reactions

When comparing the benefits of any particular antioxidant, a chemical comparison could entail studying the relative reaction rates of the antioxidant versus a desirable flavor substance, to show whether the antioxidant could prevent the loss of varietal flavor. Put another way, oxidation contributes reactive quinone electrophiles, so what important nucleophiles react with the quinones? A key category is the volatile thiols that contribute varietal aroma, such as 3-mercaptohexanol (3-MH) (Chapter 10). Knowledge of the relative reaction rates of the varietal thiols compared to potential protective antioxidants or sacrificial nucleophiles, which would consume quinones before the desirable thiols could react, has obvious importance. Some typical reactions of a model quinone with antioxidants and nucleophiles are summarized in Figure 24.10.

While this model ignores other oxidation-related reactions, the antioxidants SO_2, ascorbate, and GSH all react very quickly, and at a rate similar to hydrogen sulfide [43]. This was approximately six times faster than 3-MH at the same concentration. Based on these rate differentials, antioxidants should be fairly effective in protecting wines from loss of 3-MH. Furthermore, the concentration of GSH in juices is reported to be 50–320 μM [45], higher than cysteine, which is reported to be 8–60 μM [46]. Both these substances are found at orders of magnitude higher concentrations than the varietal thiols in wine, with those reported at about 100 nM (Chapter 10). Because GSH and SO_2 are more concentrated compared to the volatile thiols, and react more quickly, they should be effective in preventing the reaction of the varietal thiols with quinones until they are exhausted through oxidation. Alternatively, in the absence of SO_2 and other antioxidants, an oxidized flavan-3-ol should react with thiols, including 3-MH, H_2S, methanethiol, and so on.

24.5 The central tenets of the Fenton reaction and byproducts

24.5.1 Hydrogen peroxide reactions

The hydrogen peroxide formed via coupled oxidation of diphenols to quinones has two potential fates. The first, termed the Fenton reaction, involves Fe(II) catalysts again; hydrogen peroxide is decomposed to a hydroxyl radical (Figure 24.11), which can subsequently oxidize a large number of substrates, particularly alcohols to carbonyls.

The second pathway involves direct reaction of hydrogen peroxide with sulfur dioxide (in the form of bisulfite) [47]. This second pathway will avoid the Fenton step, but requires sufficient concentrations of bisulfite (Chapter 17). Other wine antioxidants, such as ascorbic acid [48] and GSH [49], can react with hydrogen peroxide at acid pH, but current information suggests their reaction rates with hydrogen peroxide are not competitive with Fe(II), and thus are unlikely to prevent the Fenton reaction.

$$\begin{array}{lrl} \text{Iron oxidation} & Fe^{2+} & \rightarrow Fe^{3+} + e^- \\ \text{Peroxide reduction} & H_2O_2 + H^+ + e^- & \rightarrow {}^{\cdot}OH + H_2O \\ \hline \text{Net reaction} & Fe^{2+} + H_2O_2 + H^+ & \rightarrow Fe^{3+} + {}^{\cdot}OH + H_2O \end{array}$$

The hydroxyl radical formed by the Fenton reaction is highly reactive and once formed it is not feasible to use antioxidants to slow or prevent its reaction with the majority of wine components. Instead, this radical will readily abstract a hydrogen or add to a substrate in a very non-selective manner [50]. Reaction of ${}^{\cdot}OH$ with water simply regenerates ${}^{\cdot}OH$, but reaction with the next highest concentration component of wine, ethanol, generates the major detectable product of the Fenton reaction, the 1-hydroxyethyl radical [18]. This radical goes on to produce acetaldehyde unless it can be quenched by other processes, such as reaction with hydroxycinnamates [51].

The indiscriminant reactivity of the hydroxyl radical towards wine compounds results in the Fenton reaction generating a large range of aldehydes and ketones, with acetaldehyde being the major product [17]. Other major oxidation products roughly parallel the concentrations of potential substrates in wine, include glyoxylic acid from tartaric acid [52], pyruvic acid from the oxidation of malic or lactic acid [53], and glyceraldehyde from the oxidation of glycerol [54]. In fact, as the hydroxyl radical is not selective, virtually all wine constituents will be oxidized in rough proportion to their concentration (see the Introduction chapter) so many additional oxidation products are formed [50].

Other aldehydes and ketones that arise from oxidation can have a major impact on wine aroma. In particular, these include methional and phenylacetaldehyde [55], sotolon [56], and 3-methyl-2,4-nonanedione [57] (Chapter 9). However, the origins of these aldehydes are not always clear and they do not necessarily arise by oxidation of the related alcohol. For instance, a route via the Strecker reaction is addressed elsewhere (Chapter 9). In addition, during grape crushing, lipid oxidase enzymes release (*E*)-2-alkenals [58] via lipid

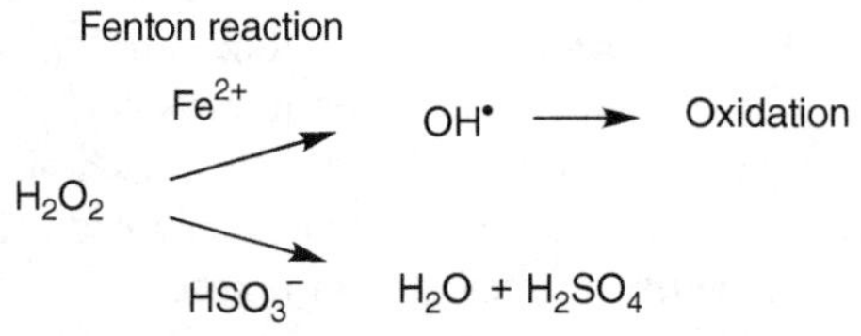

Figure 24.11 Key branch point for hydrogen peroxide

oxidation, as discussed earlier (Chapter 23.3). Finally, while not strictly a wine reaction, if a wine high in iron is consumed with foods high in omega-3 fatty acids, that is, fish, the ensuing oxidation reactions in the oral cavity (which include the Fenton reaction) can yield odorous mid-chain aldehydes that cause an off-flavor.[4]

24.5.2 Reactions of carbonyls

After the Fenton reaction creates an inventory of aldehydes and ketones, these reactive electrophiles go on to combine with nucleophilic substances in the wine. The well-known nucleophiles include thiols, alcohols, SO_2, and the phloroglucinol moiety (A ring) on flavonoids (Figure 24.12 and Chapter 10).

The reaction of aldehydes and ketones with SO_2 is well documented (see Chapters 9 and 17) and, notably, acetaldehyde forms a very stable hydroxysulfonate, while other carbonyls form weaker adducts. Acetaldehyde also reacts with alcohols under acidic conditions to form acetals (Chapter 9). These compounds have been particularly well documented in fortified wines as a result of extensive aging [60], and they appear to be important contributors to Port and Madeira aroma.

All flavonoids can react with electrophilic carbonyls and numerous products have been characterized. Desirable aging reactions dependent on oxidation involve the conversion of anthocyanins to derived wine pigments that arise in part from reactions with oxidation products. For instance, vitisin B is a product formed by reaction of malvidin-3-glucoside with acetaldehyde, and it is notably stable against bleaching by SO_2 or pH changes (Figure 24.12). These compounds and reactions are described in more detail in Chapters 16 and 25.

Reactions of oxidation products with other flavonoids are well known, and some of the products are pigmented due to unique cyclization and aromatization pathways. Condensation of flavan-3-ols with acetaldehyde results in colorless structures, with two flavan-3-ols bridged by an ethylene group (i.e., ethyl-linked). However, when tartaric acid is oxidized to glyoxal (also a yeast metabolite), the resulting bridged product

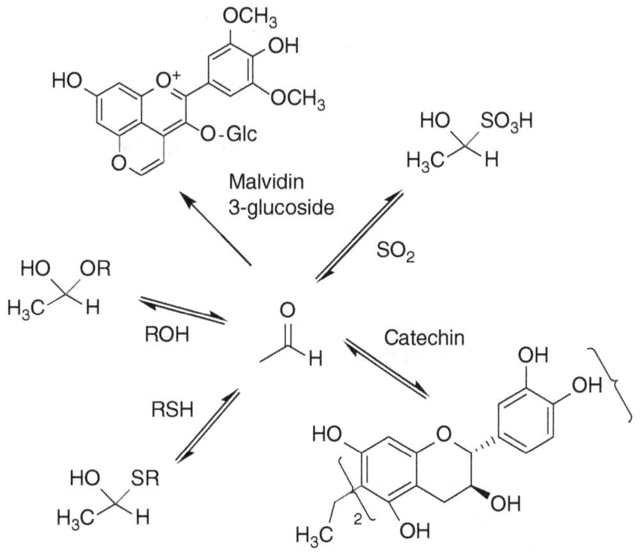

Figure 24.12 Alternative reaction pathways for acetaldehyde

[4] Tamura *et al.* concluded: "... ferrous ion is a key compound of the formation of fishy aftertaste in wine and seafood pairing" [59].

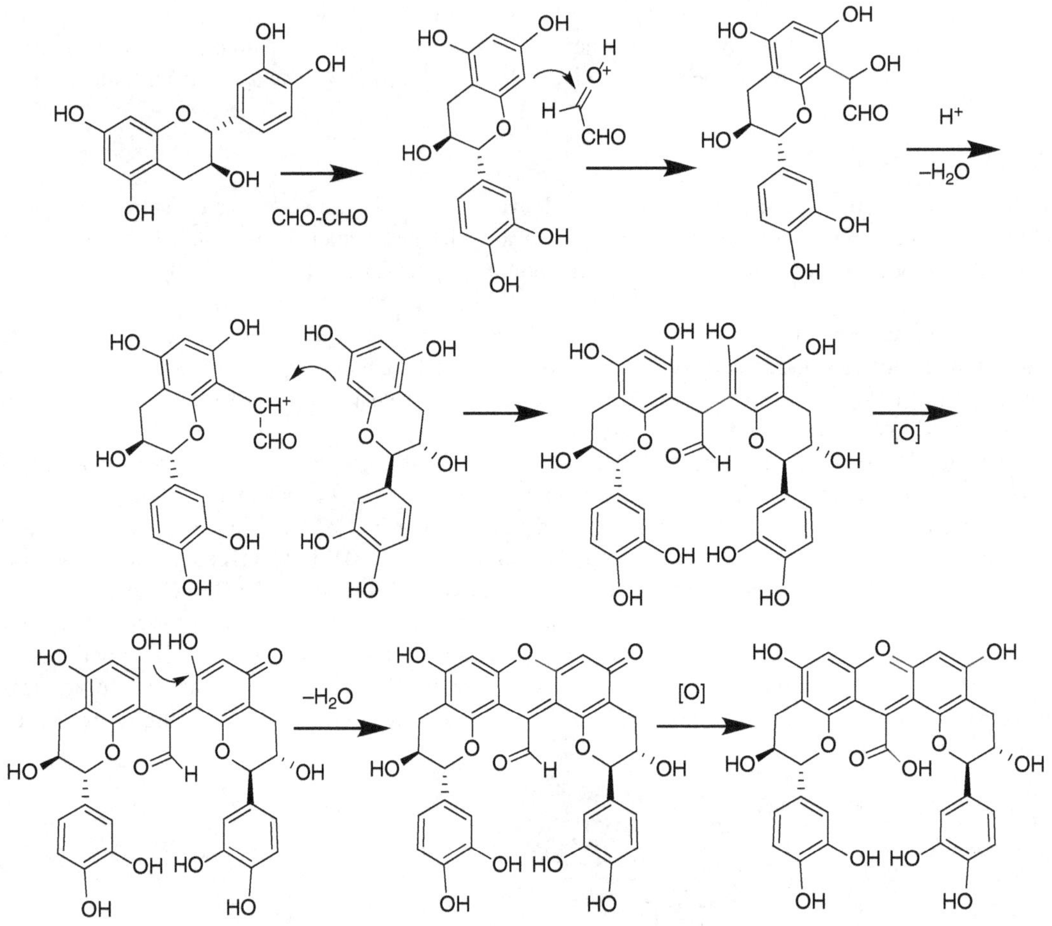

Figure 24.13 Reaction of catechin and glyoxal to colored xanthylium product as suggested in Reference [61]

continues to react, creating a xanthylium product (Figure 24.13) that absorbs in the visible region, and may contribute to the yellow hue of oxidized wines [61].

The quantities of flavonoid-acetaldehyde condensation products could potentially serve as a marker for the amount of oxygen to which a wine has been exposed. One approach to quantify these acetaldehyde-addition products is to hydrolyze them to release acetaldehyde in the presence of phloroglucinol, which forms the ethylene-bridged phloroglucinol dimer as a single substance that can be quantified [62]. The presence of condensation products increases as a wine ages, as might be expected from the generation of acetaldehyde by oxidation during aging [63]. However, the fraction of all bridging that could be detected with this method was small, at around 4–5%, suggesting that other substances such as glyoxal or pyruvate may be involved as well. Similar such reactions are expected to lead to higher molecular weight tannins in the wine and a study by Poncet-Legrand using X-ray scattering shows some confirmatory data [64]. The formation of these bridged species can be interrupted by the capture of the intermediate carbocation species by other nucleophiles, and this has been documented for GSH using NMR spectroscopy [26].

Studies describing the reaction of carbonyls with thiols are limited. An early study of acetaldehyde and a number of primary thiols observed equilibrium constants ranging from 20 to $36\,M^{-1}$. The pH had little effect on the equilibria but did affect product formation rate, which were generally not very fast [65]. One study of the effect of GSH on acetaldehyde initiated reactions in wine observed little GSH inhibition of reactions of acetaldehyde with other wine components, and reported a relatively weak equilibrium binding constant of $20\,M^{-1}$ [26]. These results suggest that thiols would not be capable of binding acetaldehyde adequately to diminish aroma or reactivity of either substrate, but further work could clarify the situation.

References

1. Rixford, E. (2008) *The wine press and the cellar*, Robert Mondavi Institute, Davis, CA.
2. Amerine, M.A., Berg, H.W., Cruess, W. (1972) *The technology of wine making*, 3rd edn, AVI, Westport, CT.
3. Singleton, V.L. (2001) A survey of wine aging reactions, especially with oxygen, in *Proceedings of the ASEV 50th Anniversary Annual Meeting* (ed. Rantz, J.M.), American Society for Enology and Viticulture, Seattle, WA, pp. 323–336.
4. Cheynier, V. (2002) Oxygen in wine and its role in phenolic reactions during aging, in *Uses of gases in winemaking* (eds Allen, M., Bell, S., Rowe, N., Wall, G.), Australian Society of Viticulture and Enology, Adelaide, SA, Australia, pp. 23–27.
5. Danilewicz, J.C. (2003) Review of reaction mechanisms of oxygen and proposed intermediate reduction products in wine: central role of iron and copper. *American Journal of Enology and Viticulture*, **54** (2), 73–85.
6. du Toit, W.J., Marais, J., Pretorius, I.S., du Toit, M. (2006) Oxygen in must and wine. *South African Journal of Enology and Viticulture*, **27** (1), 76–94.
7. Waterhouse, A.L. and Laurie, V.F. (2006) Oxidation of wine phenolics: a critical evaluation and hypotheses. *American Journal of Enology and Viticulture*, **57** (3), 306–313.
8. Li, H., Guo, A., Wang, H. (2008) Mechanisms of oxidative browning of wine. *Food Chemistry*, **108** (1), 1–13.
9. Karbowiak, T., Gougeon, R.D., Alinc, J.B., *et al.* (2010) Wine oxidation and the role of cork. *Critical Reviews in Food Science and Nutrition*, **50** (1), 20–52.
10. Oliveira, C.M., Ferreira, A.C.S., De Freitas, V., Silva, A.M.S. (2011) Oxidation mechanisms occurring in wines. *Food Research International*, **44** (5), 1115–1126.
11. Danilewicz, J.C. (2012) Review of oxidative processes in wine and value of reduction potentials in enology. *American Journal of Enology and Viticulture*, **63** (1), 1–10.
12. Ugliano, M. (2013) Oxygen contribution to wine aroma evolution during bottle aging. *Journal of Agricultural and Food Chemistry*, **61** (26), 6125–6136.
13. Pasteur, M.L. (1875) *Etudes sur le vin*, 2nd edn, Librairie F. Savy, Paris.
14. Kilmartin, P.A. (2010) Understanding and controlling non-enzymatic wine oxidation, in *Managing wine quality oenology and wine quality* (ed. Reynolds, A.G.), Woodhead Publishing, Oxford, pp. 432–458.
15. Danilewicz, J.C. (2014) Role of tartaric and malic acids in wine oxidation. *Journal of Agricultural and Food Chemistry*, **62** (22), 5149–5155.
16. Rossi, J.A.J. and Singleton, V.L. (1966) Contributions of grape phenols to oxygen absorption and browning of wines. *American Journal of Enology and Viticulture*, **17** (4), 231–239.
17. Wildenradt, H.L. and Singleton, V.L. (1974) The production of aldehydes as a result of oxidation of polyphenolic compounds and its relation to wine aging. *American Journal of Enology and Viticulture*, **25** (2), 119–126.
18. Elias, R.J., Andersen, M.L., Skibsted, L.H., Waterhouse, A.L. (2009) Identification of free radical intermediates in oxidized wine using electron paramagnetic resonance spin trapping. *Journal of Agricultural and Food Chemistry*, **57** (10), 4359–4365.
19. Danilewicz, J.C. (2013) Reactions involving iron in mediating catechol oxidation in model wine. *American Journal of Enology and Viticulture*, **64** (3), 316–324.
20. Danilewicz, J.C. (2007) Interaction of sulfur dioxide, polyphenols, and oxygen in a wine-model system: central role of iron and copper. *American Journal of Enology and Viticulture*, **58** (1), 53–60.
21. Kreitman, G.Y., Cantu, A., Waterhouse, A.L., Elias, R.J. (2013) Effect of metal chelators on the oxidative stability of model wine. *Journal of Agricultural and Food Chemistry*, **61** (39), 9480–9487.

22. Elias, R.J. and Waterhouse, A.L. (2010) Controlling the Fenton reaction in wine. *Journal of Agricultural and Food Chemistry*, **58** (3), 1699–1707.
23. Danilewicz, J.C., Seccombe, J.T., Whelan, J. (2008) Mechanism of interaction of polyphenols, oxygen, and sulfur dioxide in model wine and wine. *American Journal of Enology and Viticulture*, **59** (2), 128–136.
24. Singleton, V.L., Zaya, J., Trousdale, E., Salgues, M. (1984) Caftaric acid in grapes and conversion to a reaction product during processing. *Vitis*, **23**, 113–120.
25. Singleton, V.L., Salgues, M., Zaya, J., Trousdale, E. (1985) Caftaric acid disappearance and conversion to products of enzymic oxidation in grape must and wine. *American Journal of Enology and Viticulture*, **36** (1), 50–56.
26. Sonni, F., Moore, E.G., Clark, A.C., *et al.* (2011) Impact of glutathione on the formation of methylmethine- and carboxymethine-bridged (+)-catechin dimers in a model wine system. *Journal of Agricultural and Food Chemistry*, **59** (13), 7410–7418.
27. Nikolantonaki, M., Chichuc, I., Teissedre, P.L., Darriet, P. (2010) Reactivity of volatile thiols with polyphenols in a wine-model medium: impact of oxygen, iron, and sulfur dioxide. *Analytica Chimica Acta*, **660** (1–2), 102–109.
28. Lee, C.Y. and Whitaker, J.R. (1995) *Enzymatic browning and its prevention, ACS Symposium Series*, Vol. 600, American Chemical Society, Washington, DC.
29. Cheynier, V., Basire, N., Rigaud, J. (1989) Mechanism of *trans*-caffeoyltartaric acid and catechin oxidation in model solutions containing grape polyphenoloxidase. *Journal of Agricultural and Food Chemistry*, **37** (4), 1069–1071.
30. Schneider, V. (1998) Must hyperoxidation: a review. *American Journal of Enology and Viticulture*, **49** (1), 65–73.
31. Cheynier, V., Rigaud, J., Souquet, J.M., *et al.* (1990) Must browning in relation to the behavior of phenolic compounds during oxidation. *American Journal of Enology and Viticulture*, **41** (4), 346–349.
32. Fulcrand, H., Cheminat, A., Brouillard, R., Cheynier, V. (1994) Characterization of compounds obtained by chemical oxidation of caffeic acid in acidic conditions. *Phytochemistry*, **35** (2), 499–505.
33. Simpson, R.F. (1982) Factors affecting oxidative browning of white wine. *Vitis*, **21** (3), 233–239.
34. Cheynier, V., Rigaud, J., Souquet, J.M., *et al.* (1989) Effect of pomace contact and hyperoxidation on the phenolic composition and quality of Grenache and Chardonnay wines. *American Journal of Enology and Viticulture*, **40** (1), 36–42.
35. Fernandez-Zurbano, P., Ferreira, V., Escudero, A., Cacho, J. (1998) Role of hydroxycinnamic acids and flavanols in the oxidation and browning of white wines. *Journal of Agricultural and Food Chemistry*, **46** (12), 4937–4944.
36. Rigaud, J., Cheynier, V., Souquet, J.M., Moutounet, M. (1991) Influence of must composition on phenolic oxidation-kinetics. *Journal of the Science of Food and Agriculture*, **57** (1), 55–63.
37. Guyot, S., Vercauteren, J., Cheynier, V. (1996) Structural determination of colourless and yellow dimers resulting from (+)-catechin coupling catalysed by grape polyphenoloxidase. *Phytochemistry*, **42** (5), 1279–1288.
38. Jimenez-Atienzar, M., Cabanes, J., Gandia-Herrero, F., Garcia-Carmona, F. (2004) Kinetic analysis of catechin oxidation by polyphenol oxidase at neutral pH. *Biochemical and Biophysical Research Communications*, **319** (3), 902–910.
39. Mouls, L. and Fulcrand, H. (2012) UPLC-ESI-MS study of the oxidation markers released from tannin depolymerization: toward a better characterization of the tannin evolution over food and beverage processing. *Journal of Mass Spectrometry*, **47** (11), 1450–1457.
40. Es-Safi, N.E., Le Guerneve, C., Cheynier, V., Moutounet, M. (2000) New phenolic compounds formed by evolution of (+)-catechin and glyoxylic acid in hydroalcoholic solution and their implication in color changes of grape-derived foods. *Journal of Agricultural and Food Chemistry*, **48** (9), 4233–4240.
41. Embs, R.J. and Markakis, P. (1965) Mechanism of sulfite inhibition of browning caused by polyphenol oxidase. *Journal of Food Science*, **30** (5), 753.
42. Luvalle, J.E. (1952) The reaction of quinone and sulfite. 1. Intermediates. *Journal of the American Chemical Society*, **74** (12), 2970–2977.
43. Nikolantonaki, M. and Waterhouse, A.L. (2012) A method to quantify quinone reaction rates with wine relevant nucleophiles: a key to the understanding of oxidative loss of varietal thiols. *Journal of Agricultural and Food Chemistry*, **60** (34), 8484–8491.
44. Barril, C., Clark, A.C., Scollary, G.R. (2012) Chemistry of ascorbic acid and sulfur dioxide as an antioxidant system relevant to white wine. *Analytica Chimica Acta*, **732**, 186–193.

45. Cheynier, V., Souquet, J.M., Moutounet, M. (1989) Glutathione content and glutathione to hydroxycinnamic acid ratio in *Vitis vinifera* grapes and musts. *American Journal of Enology and Viticulture*, **40** (4), 320–324.
46. Bell, S.J. and Henschke, P.A. (2005) Implications of nitrogen nutrition for grapes, fermentation and wine. *Australian Journal of Grape and Wine Research*, **11** (3), 242–295.
47. McArdle, J.V. and Hoffmann, M.R. (1983) Kinetics and mechanism of the oxidation of aquated sulfur-dioxide by hydrogen-peroxide at low pH. *Journal of Physical Chemistry*, **87** (26), 5425–5429.
48. Deutsch, J.C. (1998) Ascorbic acid oxidation by hydrogen peroxide. *Analytical Biochemistry*, **255** (1), 1–7.
49. Pirie, N.W. (1931) The oxidation of sulphydryl compounds by hydrogen peroxide. I. Catalysis of oxidation of cysteine and glutathione by iron and copper. *Biochemical Journal*, **25** (5), 1565–1579.
50. Kaur, H., Halliwell, B., Lester, P. (1994) Detection of hydroxyl radicals by aromatic hydroxylation, in *Methods in enzymology*, Academic Press, pp. 67–82.
51. Gislason, N.E., Currie, B.L., Waterhouse, A.L. (2011) Novel antioxidant reactions of cinnamates in wine. *Journal of Agricultural and Food Chemistry*, **59** (11), 6221–6226.
52. Fulcrand, H., Cheynier, V., Oszmianski, J., Moutounet, M. (1997) An oxidized tartaric acid residue as a new bridge potentially competing with acetaldehyde in flavan-3-ol condensation. *Phytochemistry*, **46** (2), 223–227.
53. Fulcrand, H., Benabdeljalil, C., Rigaud, J., *et al.* (1998) A new class of wine pigments generated by reaction between pyruvic acid and grape anthocyanins. *Phytochemistry*, **47** (7), 1401–1407.
54. Laurie, V.F. and Waterhouse, A.L. (2006) Glycerol oxidation in wine and reactions with flavonoids. *American Journal of Enology and Viticulture*, **57** (3), 394A–395A.
55. Ferreira, A.C.S., Hogg, T., de Pinho, P.G. (2003) Identification of key odorants related to the typical aroma of oxidation-spoiled white wines. *Journal of Agricultural and Food Chemistry*, **51** (5), 1377–1381.
56. Cutzach, I., Chatonnet, P., Henry, R., *et al.* (1998) Study in aroma of sweet natural non Muscat wines: 2nd Part. Quantitative analysis of volatile compounds taking part in aroma of sweet natural wines during ageing. *Journal International des Sciences de la Vigne et du Vin*, **32** (4), 211–221.
57. Pons, A., Lavigne, V., Darriet, P., Dubourdieu, D. (2013) Role of 3-methyl-2,4-nonanedione in the flavor of aged red wines. *Journal of Agricultural and Food Chemistry*, **61** (30), 7373–7380.
58. Cullere, L., Cacho, J., Ferreira, V. (2007) An assessment of the role played by some oxidation-related aldehydes in wine aroma. *Journal of Agricultural and Food Chemistry*, **55** (3), 876–881.
59. Tamura, T., Taniguchi, K., Suzuki, Y., *et al.* (2009) Iron is an essential cause of fishy aftertaste formation in wine and seafood pairing. *Journal of Agricultural and Food Chemistry*, **57** (18), 8550–8556.
60. Ferreira, A.C.D., Barbe, J.C., Bertrand, A. (2002) Heterocyclic acetals from glycerol and acetaldehyde in port wines: evolution with aging. *Journal of Agricultural and Food Chemistry*, **50** (9), 2560–2564.
61. Es-Safi, N.E., Cheynier, V., Moutounet, M. (2003) Effect of copper on oxidation of (+)-catechin in a model solution system. *International Journal of Food Science and Technology*, **38** (2), 153–163.
62. Drinkine, J., Lopes, P., Kennedy, J.A., *et al.* (2007) Analysis of ethylidene-bridged flavan-3-ols in wine. *Journal of Agricultural and Food Chemistry*, **55** (4), 1109–1116.
63. Drinkine, J., Lopes, P., Kennedy, J.A., *et al.* (2007) Ethylidene-bridged flavan-3-ols in red wine and correlation with wine age. *Journal of Agricultural and Food Chemistry*, **55** (15), 6292–6299.
64. Poncet-Legrand, C., Cabane, B., Bautista-Ortin, A.B., *et al.* (2010) Tannin oxidation: intra- versus intermolecular reactions. *Biomacromolecules*, **11** (9), 2376–2386.
65. Lienhard, G.E. and Jencks, W.P. (1966) Thiol addition to carbonyl group. Equilibria and kinetics. *Journal of the American Chemical Society*, **88** (17), 3982–3995.

25

Topics Related to Aging

25.1 Introduction

Wine is a chemically dynamic system, and even after fermentation is complete wine composition contiunes to evolve during storage [1–9]. These post-fermentation changes are associated with the general term *aging*, but a useful distinction can be made between those changes that transpire during the maturation phase (e.g., bulk storage of wine in a tank or barrel), during which winemaker intervention can still readily occur, and an aging phase (post-packaging), where wine is sealed in its container and intervention is essentially limited to selection of storage conditions. Changes can be further distinguished by whether they will happen under completely anaerobic conditions or whether they require trace levels of oxygen. This chapter will examine some compositional and sensory effects of maturation and aging as a result of oxidation, the presence of nucleophiles and electrophiles, and acid-catalyzed reactions.

25.2 Reactions involving red wine pigments

One of the most dramatic chemical changes in red wine involves grape-derived monomeric anthocyanins. These can be a major component of red wine (approaching 1 g/L), but disappear as new pigments are produced (Figure 25.1a), resulting in a change in wine color from red-purple to brick- or orange-red [10–12]. This process begins during fermentation and continues during storage, such that, within two years, the majority of wine color derives from so-called "polymeric pigments" that are resistant to bleaching by SO_2. Mechanisms for the formation of derived pigments from monomeric anthocyanins, tannins, and other wine components were covered in Chapter 16, and these important reactions continue during maturation and aging [13]. As a common thread, the reactions that lead to the most stable forms of red wine color require trace amounts of oxygen and the presence of tannins. To summarize, the key types of compounds formed from reactions of anthocyanins are:

- *Pigmented species resistant to SO_2 bleaching (O_2 usually required).* The flavylium form of anthocyanins can act as an electrophile to react with nucleophiles (Chapter 16), including the enol form of aldehydes

Understanding Wine Chemistry, First Edition. Andrew L. Waterhouse, Gavin L. Sacks, and David W. Jeffery.

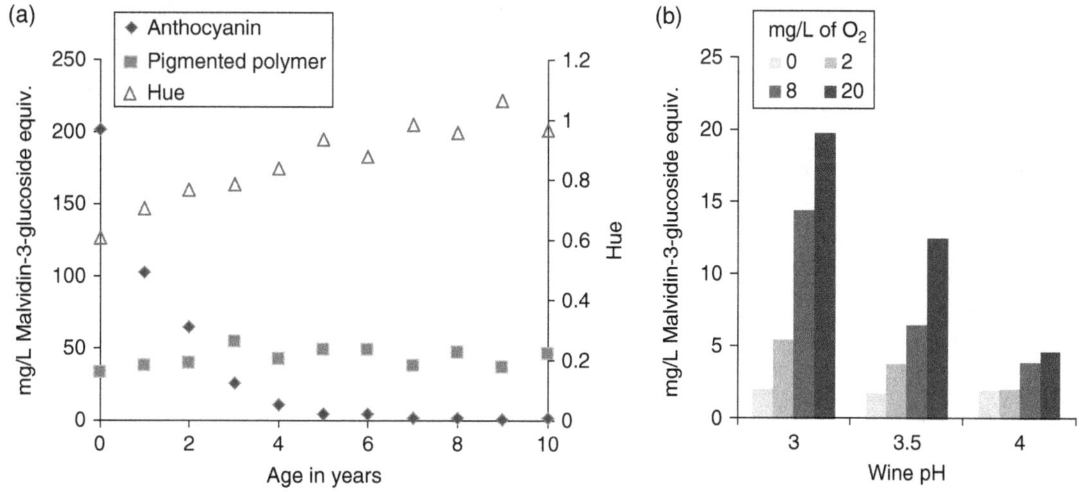

Figure 25.1 *Changes occurring in red wine due to (a) age for a vertical series of Cabernet Sauvignon wines from Coonawarra, South Australia, showing a decline in anthocyanins, and increases in pigmented polymer (high molecular weight pigmented compounds measured by HPLC, expressed as malvidin-3-glucoside equivalents) and hue (A420/A520; a higher value indicates more brick red versus red-purple color), and (b) pH adjustment and addition of oxygen at different rates (held constant over a 4 month period) for a 2011 Merlot wine from Bordeaux, France, showing increases in a pyranoanthocyanin pigment upon greater exposure to oxygen and as a result of a lower pH. Data obtained from References [22] and [23]*

(to yield pyranoanthocyanins), or tannins (to yield A-T type pigmented polymers). The former reactions will occur more rapidly at low pH (greater proportion of flavylium form, Figure 25.1b), and the latter will be favored at higher pH [14, 15]. However, all of these reactions will be favored in the presence of oxygen. In the case of A-T pigments, it is proposed that the flavene intermediate requires oxidation to regenerate the flavylium (colored) form of the anthocyanin [16, 17], which may cyclize to form a xanthylium cation (Figure 25.2) [15], while formation of aldehydes requires oxidation of other wine components (Chapter 24). Additionally, the presence of acetaldehyde (from fermentation or oxidation) can lead to the generation of ethyl-linked pigments (also favored at lower pH), which are not stable in the long term but are more resistant to bleaching than monomeric grape anthocyanins [18, 19].

- *Pigmented species that are not resistant, or are less resistant, to SO_2 bleaching (O_2 usually not required).* The major pathway for formation of pigmented polymers involves direct condensation of electrophilic flavan-3-ol cations (from tannin hydrolyis) with anthocyanins to form T-A adducts. This pathway does not require O_2, but will occur faster at lower wine pH due to faster hydrolysis of tannin interflavan bonds [14] (Chapters 14 and 16).
- *Non-pigmented or insoluble species (no O_2 or excess O_2 required).* Anthocyanins that do not form stable adducts can undergo acid-catalyzed transformation resulting in a loss of color, through opening of the heterocyclic C ring; hydrolysis of the glucoside also destablizes anthocyanins [18]. A-T adducts (flavene form, Chapter 16) can also be a source of color loss through intramolecular cyclization (to an A-type dimer) under anaerobic conditions or at low pH (Figure 25.2) [15, 20]. Alternatively, excessive oxidation and production of aldehydes and quinones (Chapter 24) can lead to formation of insoluble complexes, as well as high levels of browning [21].

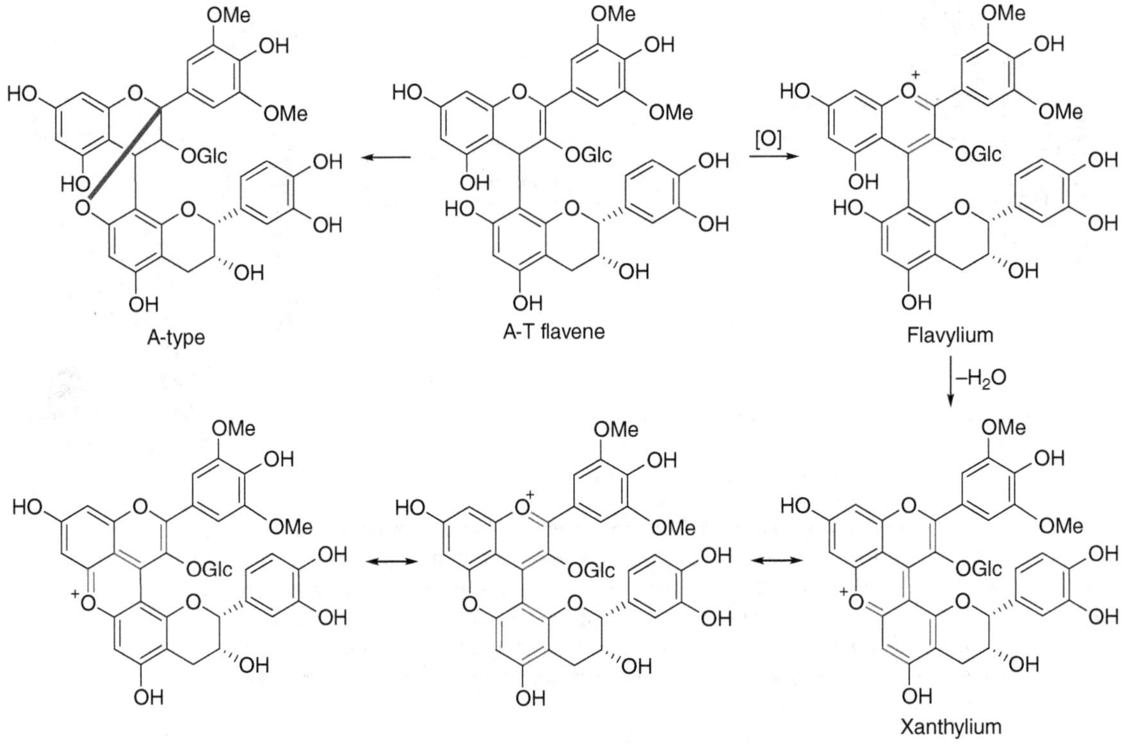

Figure 25.2 A colorless A-T flavene formed from malvidin-3-glucoside and (–)-epicatechin can react to form an A-type ether linkage (also colorless), or may oxidize to regenerate the flavylium cation (red color), which can undergo cyclization to produce a xanthylium cation (yellow/brown color)

The effects of oxygen on anthocyanins have also been studied through work on microoxygenation (MOX; for reviews on this technique see References [24] and [25])[1] and closures with differing oxygen transmission rates (OTRs) for wines at different pH values, as seen with the following examples. These studies highlight not only the continued evolution of phenolic composition and color, but also the impact of different closures and the prospect of choosing a closure OTR based on wine composition, expected storage conditions and time. For example, at 3 months post-bottling, Aglianico red wines (pH 3.46 and 3.64) that underwent a single MOX treatment at 2 mL of O_2/L for 8 weeks, or an additional MOX treatment of 1.5 mL O_2/L for another 8 weeks, had higher total anthocyanins (measured at pH 1) and color intensity compared to the control wines (Figure 25.3a and b) [26]. After 42 months of bottle storage the differences between MOX treatments and controls were no longer significant for color intensity (and hue), and total anthocyanins in the case of the pH 3.64 wine (Figure 25.3a and b). This moderating effect with time has been observed in other studies (e.g., References [27] and [28]), likely due to alternative degradation pathways of pigments during storage obscuring the early effects of the oxygen-dependent reactions. It is worth noting that while all treatments likely lost pigmented anthocyanins as compared to the initial wine (not reported in the study), the MOX-treated wines had proportionally higher concentrations of stable pigmented forms [26]. Such a decrease in total anthocyanins

[1] MOX is used to supply controlled amounts of oxygen over a period of time to red wine in a stainless steel tank to facilitate color-stabilizing reactions that would normally occur during extended maturation in porous oak barrels. Similarly, MOX can assist with reactions that induce tannin modifications, thereby positively improving mouthfeel by moderating astringency.

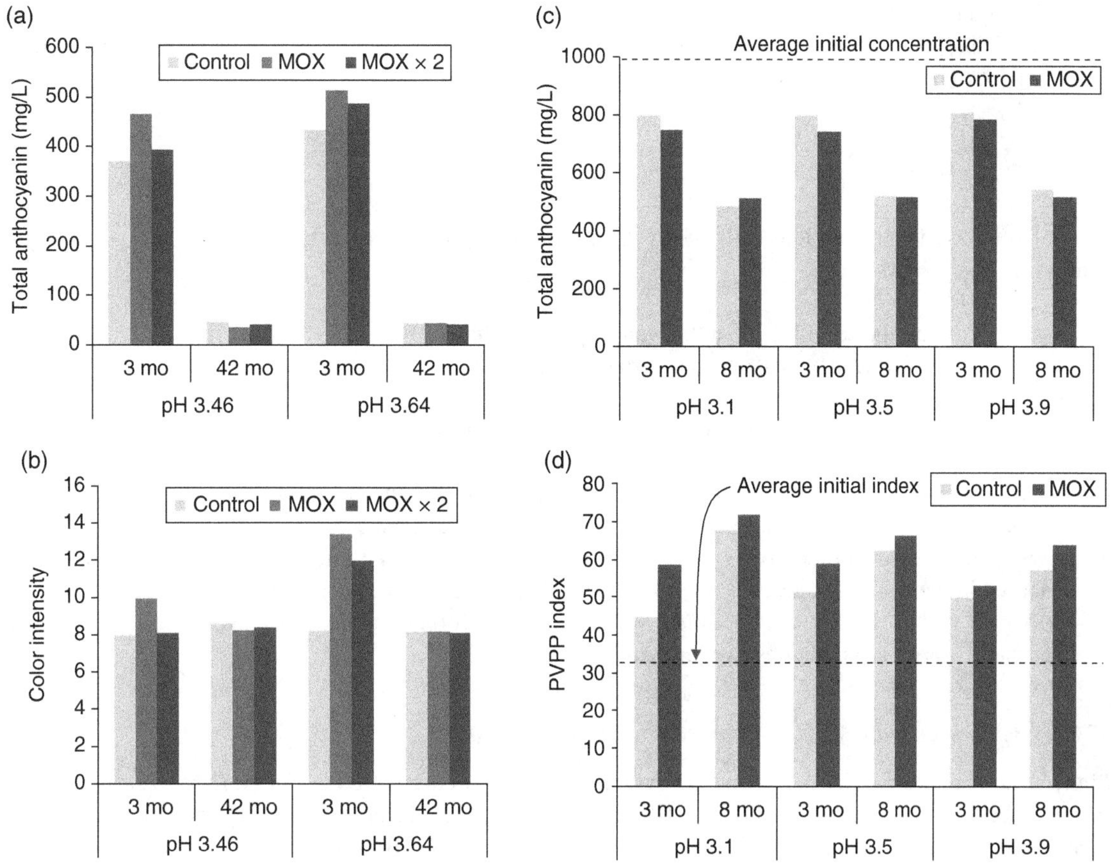

Figure 25.3 Impact of pH and oxygen combined with storage time (months) on (a) total anthocyanins and (b) color intensity for Aglianico red wines differening in pH treated with two different MOX regimes (left panel), and (c) total anthocyanins and (d) PVPP index (a measure of polymeric pigments) for Cabernet Sauvignon wine (pH 3.5) adjusted to pH 3.1 and 3.9 (right panel). Data obtained from References [26] and [28]

and increase in stable pigments can readily be seen with data from a study of Cabernet Sauvignon wine at different pH values that underwent MOX (15 mg/L per month for 3 months) followed by a period of storage (Figure 25.3c and d) [28].

The effect of synthetic closure OTR[2] was also studied for the Aglianico red wines mentioned above sealed with three different closures (low, medium, and high OTR), and having 6.5 or 9.8 mg/L of total package oxygen (TPO) at bottling [26]. After 10 months, the wine with 9.8 mg/L TPO revealed a decrease in total monomeric anthocyanins of about 40% for the closures with medium and high OTR, likely due to their incorporation into stable pigments, but the wine sealed with 6.5 mg/L TPO showed no signficant effect due to closure, likely because the differences were small compared to variations among treatments. Similar impacts of closure OTR and wine pH were also seen for pH-adjusted Cabernet Sauvignon stored under SaranTin (impermeable to oxygen) and Saranex (slightly permeable) screw cap closures [29], with greater formation

[2] Closure OTR can also have an impact (usually detrimental) on the color of white wine, as oxygen ingress leads to loss of free SO_2 and white wine browning, due to oxidation and reactions of phenolic compounds.

of stable pigments attributed to oxygen exposure and increased incorporation of anthocyanins into derived (non-bleachable) pigments at a lower pH. As with the effect of MOX over time, the impact of screw cap OTR appeared not to be a factor when evaluated after 24 months.

25.3 Hydrolytic and pH-dependent reactions

Several pH-dependent reactions have been discussed throughout the book; their relevance to aging is summarized as follows:

- Lower pH affects the equilibria of anthocyanin species (Chapter 16) and increases the hydrolytic cleavage rate of tannin interflavan bonds, which in turn impacts formation of different derived pigments as described above.
- Lower pH increases the rate of ester hydrolysis and esterification reactions (Chapters 7, 8, and 23.1), as discussed below.
- Through protonation of the carbonyl oxygen, lower pH increases the electrophilicity of carbonyl carbons and the overall reactivity of carbonyl species with nucleophiles such as phenolics (Chapter 9).
- Lower pH increases the rate of glycoside hydrolysis,[3] as well as the rate of isoprenoid rearrangements observed in the resulting aglycone intermediates (Chapter 23.1), as discussed below.
- Higher pH will increase the rate of dimethyl sulfide (DMS) formation, although this increase is negligible over the pH range typically observed in wine (Chapter 23.3).

25.3.1 Glycoside hydrolysis and aglycone rearrangements

Many wine odorants exist as bound precursors that will liberate free, volatile forms during storage (Chapter 23.1). Notably, 1,1,6-trimethyl-1,2-dihydronapthalene (TDN) in Riesling and monoterpenoid-related compounds in aromatic varieties such as Muscat both arise from hydrolysis and rearrangement of glycosidic precursors (Chapters 8 and 23.1). Proposed acid-catalyzed reaction mechanisms for the formation of TDN from Riesling acetal (itself an intermediate of C_{13}-norisoprenoid glycoconjugates from grapes) and wine lactone from monoterpenoid (linalool-8-carboxylate) precursors are shown in Figure 25.4. As is expected based on the mechanism, the rate of formation of these compounds will increase with decreasing pH (Figure 25.5) [30, 31].

25.3.2 Ester hydrolysis and esterification

Esters have a critical role in wine aroma (Chapter 7), and their formation and degradation by acid-catalyzed esterification and ester hydrolysis show strong pH dependence (Figure 25.6). As mentioned earlier, the ester/acid ratio will approach equilibrium during aging, which may result in either an increase or decrease of individual esters (Chapter 7). Typically, initial ester/acid ratios are such that ethyl esters of branched-chain fatty acids and fixed (non-volatile) acids increase, acetate esters decrease precipitously, and straight-chain fatty acids decrease slightly [34–36] (Figure 25.7). Notably, the hydrolysis of 3-mercaptohexyl acetate can be especially

[3] This includes hydrolysis of flavonoid glycosides, which liberates less-soluble aglycones such as flavonols. In the case of quercetin in particular, glycoside hydrolysis can lead to instabilities when the level of quercetin exceeds its solubility in wine. For some red grape varieties that respond to sunlight by producing high levels of flavonols, or for red wines that have not matured for long enough, there can be precipitation of a muddy looking material during aging, which is nearly pure quercetin despite its poor appearance. Natural precipitation of quercetin often goes unnoticed and is removed with wine lees, but can pose a problem if it occurs in bottled wine.

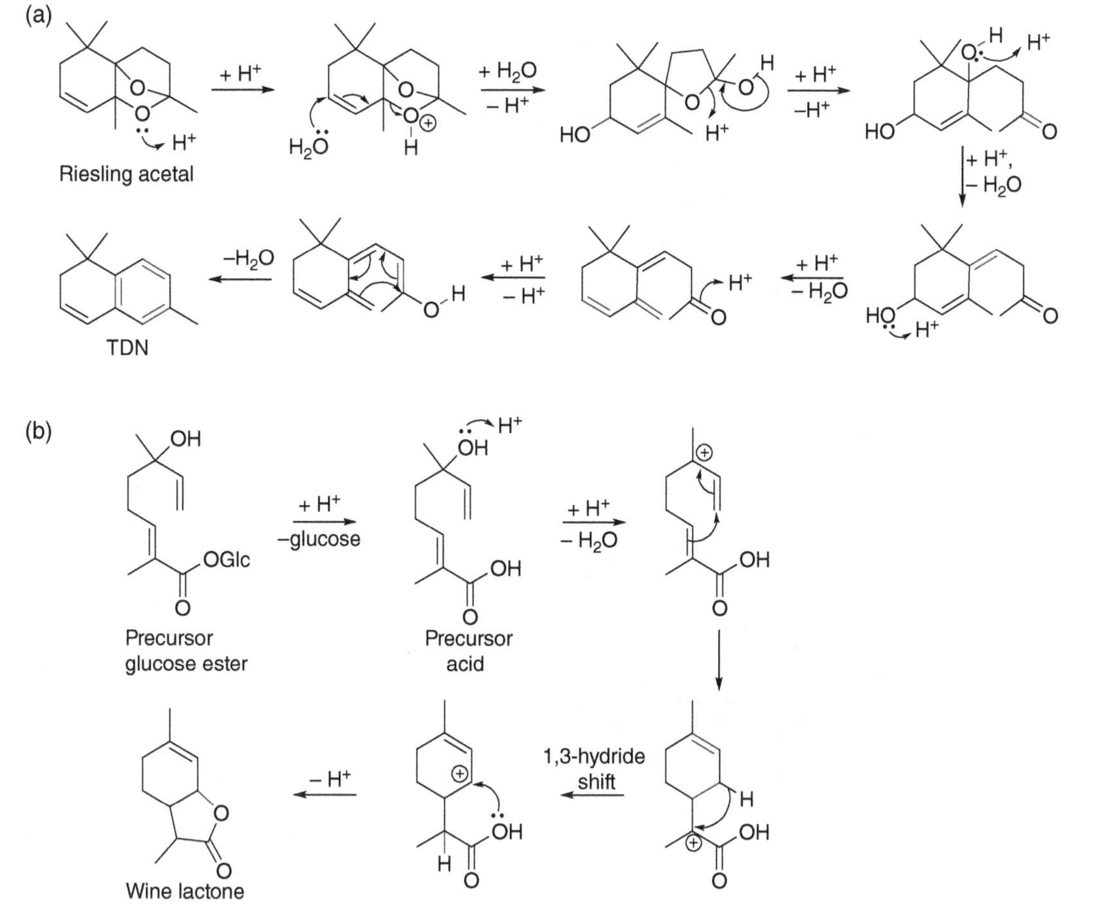

Figure 25.4 Proposed acid-catalyzed reaction mechanisms for conversion of (a) Riesling acetal into TDN via other known intermediates [32] and (b) precursor glucose ester and acid derivatives of linalool into wine lactone, invoking an interesting 1,3-hydride shift [33]

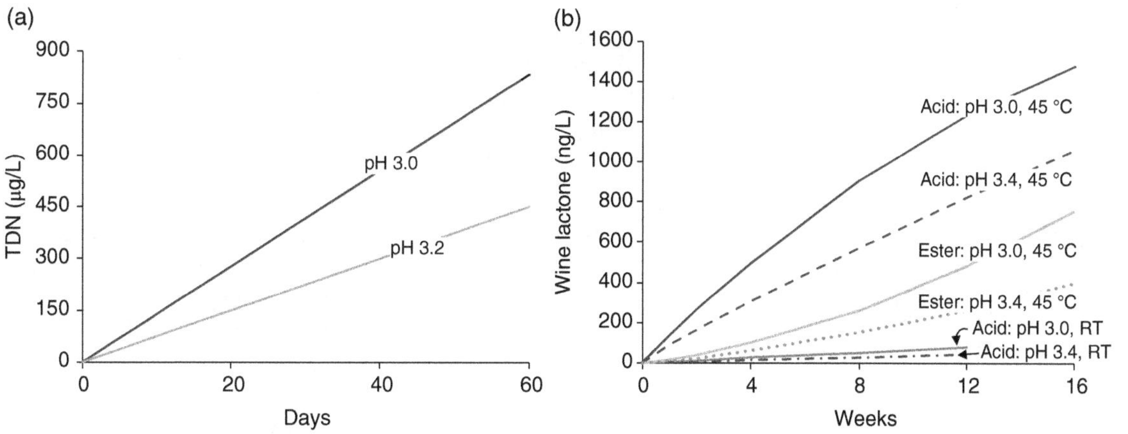

Figure 25.5 Acid-catalyzed formation of wine aroma compounds over time showing (a) concentration of TDN arising from Riesling acetal over 60 days in model wine at pH 3.0 and 3.2 stored at 45 °C and (b) concentration of wine lactone formed from precursor glucose ester or acid derivatives of linalool over 16 weeks in model wine at pH 3.0 and 3.4, and at room temperature (RT) and 45 °C. Note the precursor ester did not form wine lactone at room temperature over 12 weeks. Data derived from [30] and [31]

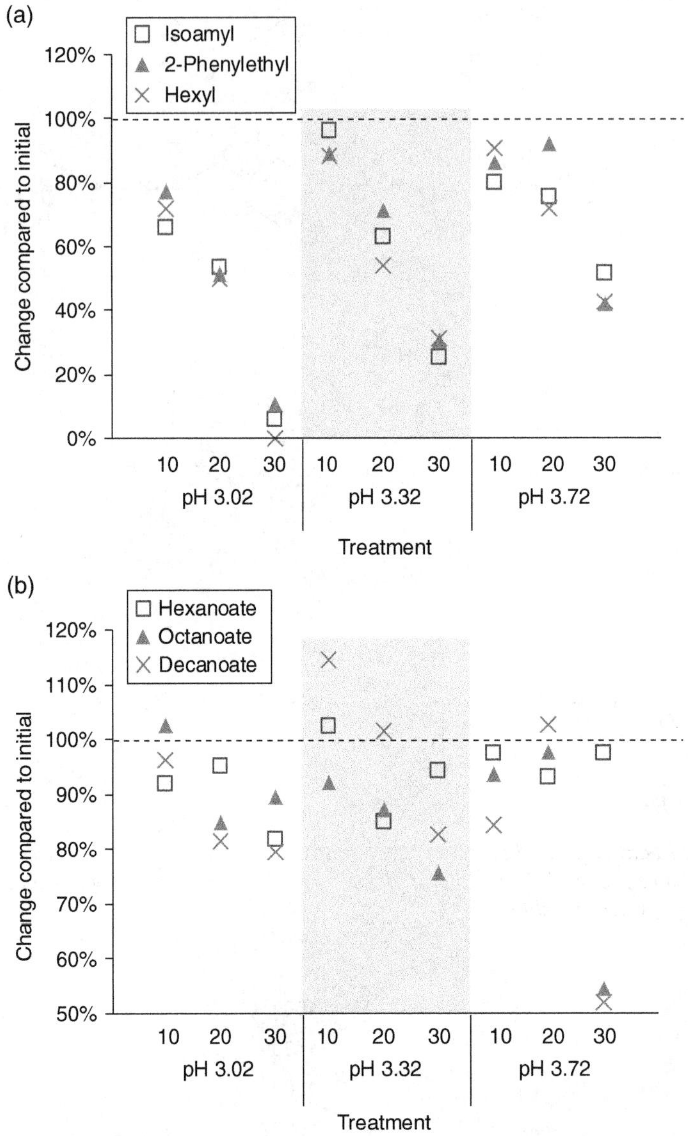

Figure 25.6 *Relative changes in concentration of volatiles 16 weeks after bottling for Colombard wine with altered pH values kept at different storage temperatures (10, 20, 30 °C), showing (a) sizeable decreases in acetate esters with increased temperature (as initial concentrations were far away from equilibrium concentrations), as well as an effect of lower pH, and (b) minimal changes in straight-chain ethyl esters (ignoring the aberrant results at 30 °C and pH 3.72) but a general decrease with increased temperature is relatively evident. Note the different y-axis scales. Data from Reference [38]*

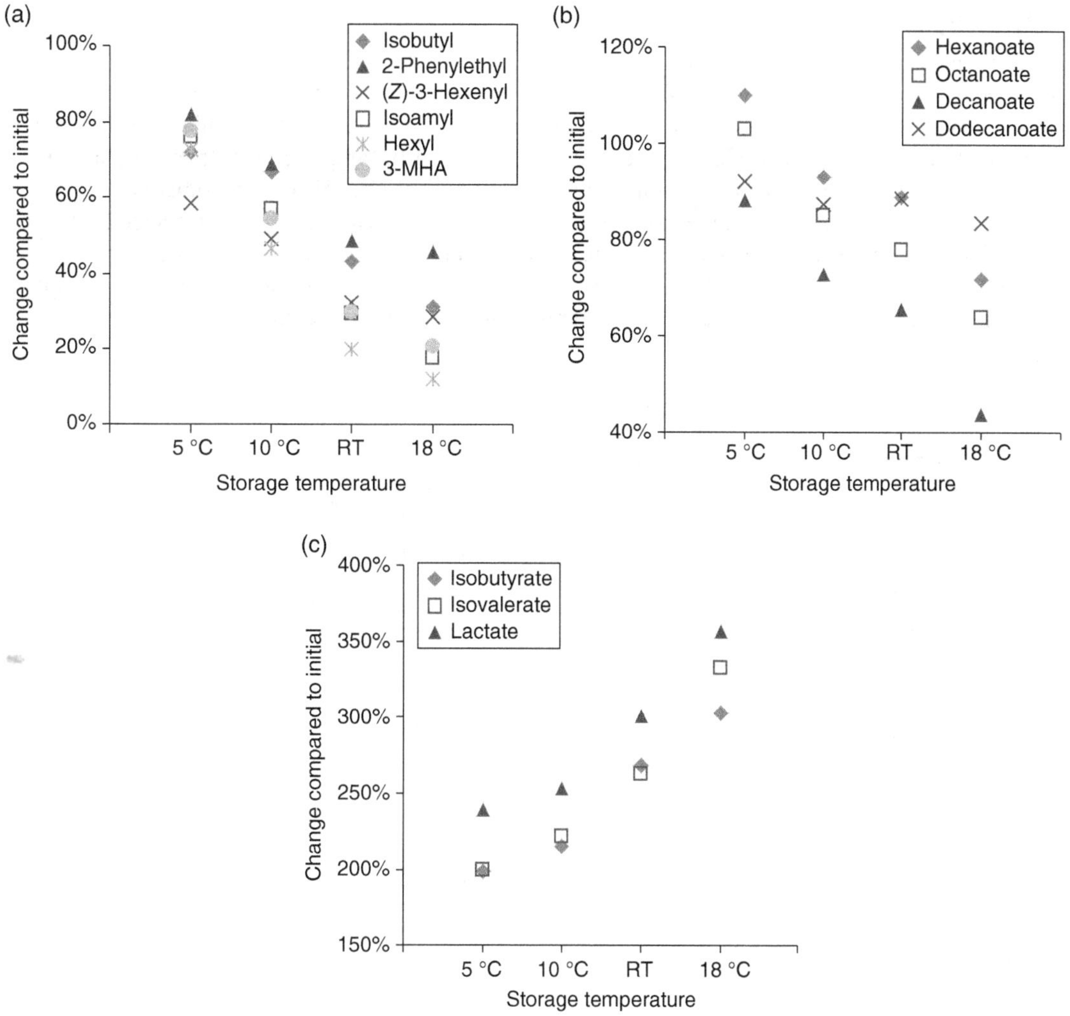

Figure 25.7 *Relative changes in concentration of volatiles one year after bottling for Sauvignon Blanc wine kept at different storage temperatures, showing (a) decrease in acetate esters, (b) decrease in straight-chain ethyl esters, and (c) increase in branched-chain ethyl esters. Room temperature (RT) fluctuated between 13 and 26 °C, with a mean of 19.5 °C. Despite the mean for RT being higher than the other treatments, the x axes are ordered according to the trends in the data. Note the different y-axis scales. Data from References [35] and [39]*

influential on the aroma of Sauvignon Blanc wines (Chapter 10),[4] and acetate esters in general hydrolyze faster than fatty acid ethyl esters of similar molecular weight (by around 2–4 times [37]). Overall, the storage temperature in conjunction with wine pH can have a strong influence on ester composition and wine sensory properties during aging, especially for young white wines reliant on the aroma contributions of volatile acetate esters, and ethyl esters to a lesser extent, derived from fermentation.

[4]This statement applies to other acid-labile potent aroma compounds that contribute important characters to certain wine styles (e.g., sulfanylalkyl acetates in Sauternes).

25.3.3 Other pH-dependent reactions

Apart from releasing grape-derived aroma volatiles, acid-catalyzed reactions can result in a loss of odor, such as in the conversion of linalool to the less potent monoterpenoid, α-terpineol (Chapter 8). Glycosidic precursors of oak lactones, present in wine from maturation in oak wood, can undergo similar transformations to their grape-derived counterparts (Chapter 7 and Chapter 23.1) [40]. Finally, sugars (pentoses and hexoses) present in wine may be converted to furfural, 5-(hydroxymethyl)furfural (HMF), and other volatiles (as well as brown colored polymers, i.e., caramelization) through acid-catalyzed degradation (Chapter 2), a pathway of particular importance to sweet wines or wines stored at higher temperatures [41]. This occurs through sugar enolization and elimination of water (dehydration), producing unsaturated dicarbonyl compounds, which can cyclize and further dehydrate to yield furans (and pyrans) (Figure 25.8) [42].

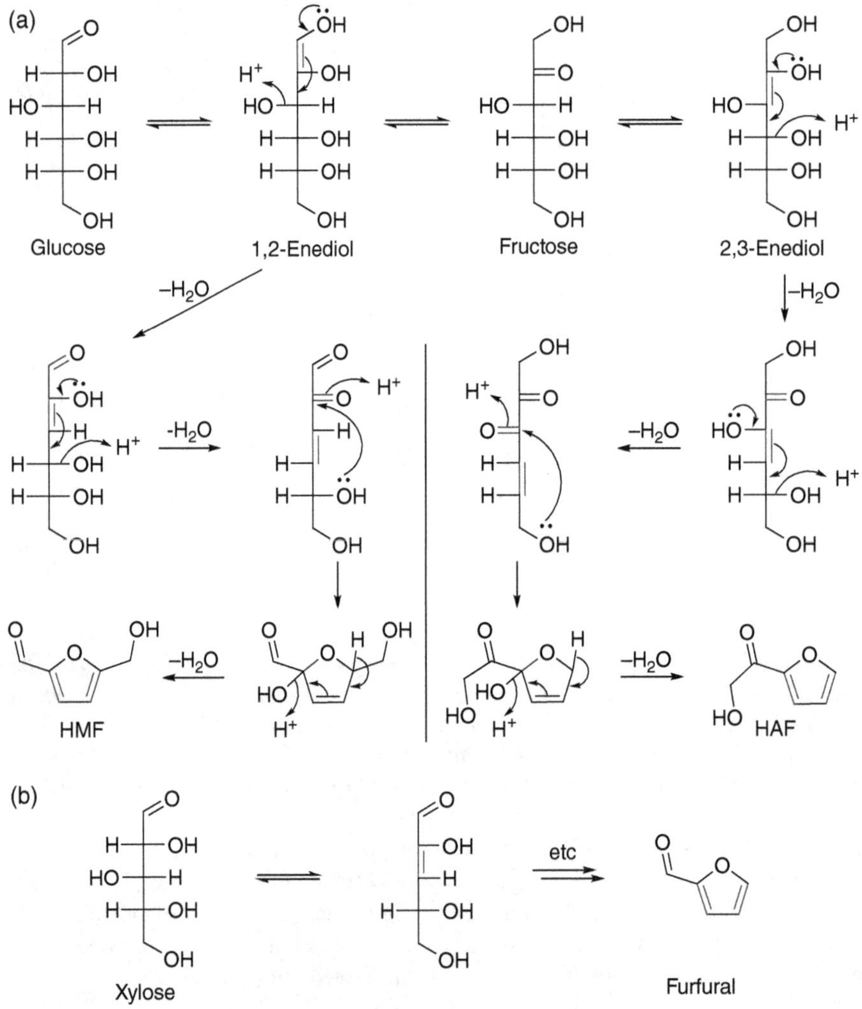

Figure 25.8 Acid-catalyzed enolization, dehydration, and cyclization reaction pathways of sugars leading to aroma compounds such as (a) HMF and 2-(hydroxyacetyl)furan (HAF) from hexoses including glucose and fructose and (b) furfural from pentoses such as xylose

25.4 Activation energy and temperature effects on aging

The rates of chemical reactions, including those observed in wine during storage, increase with increasing temperature. This arises due to the dependence of the rate constant (k) on temperature, as expressed by the Arrhenius equation:

$$k = Ae^{-E_a/RT} \tag{25.1}$$

where A is the Arrhenius constant (the proportion of molecules that will collide and react), E_a is the activation energy (the energy barrier to overcome for a reaction to occur), R is the gas constant (8.314 J/mol K), and T is the temperature in Kelvin. The logarithmic form of the Arrhenius equation can be used to calculate E_a and A when rate constants at different temperatures are known (Figure 25.9). Alternatively, E_a can be calculated from the logarithmic form of the Arrhenius equation, after subtraction and rearrangement, using rate constants determined at two temperatures as follows:

$$\ln\left(\frac{k_1}{k_2}\right) = \left(\frac{1}{T_2} - \left(\frac{1}{T_1}\right)\right)\frac{E_a}{R} \tag{25.2}$$

where k_1 and k_2 are the rate constants at temperatures T_1 and T_2, respectively.

For E_a values in the range of 40–70 kJ/mol (e.g., hydrolysis of acetate esters), the fold difference in reaction rates at two different temperatures can be approximated as $2^{\Delta T/10}$, which relates to the often-quoted rule of thumb that each 10 °C increase in temperature leads to roughly a doubling of the reaction rate. However, because of the exponential nature of the Arrhenius equation, this heuristic is less appropriate for reaction rates with higher E_a values, which will usually proceed slowly at room temperature but demonstrate a larger increase in rate upon heating. As shown in Table 25.1, increasing the storage temperature increases all reaction rates. However, the increase in acid hydrolysis of esters and tannins upon increasing storage temperature from 12 °C to 50 °C is about a factor of 10, while the same temperature increase results in almost a 20000-fold increase in HMF. Notably, the reactions that increase most dramatically at high temperatures (high E_a) are undesirable

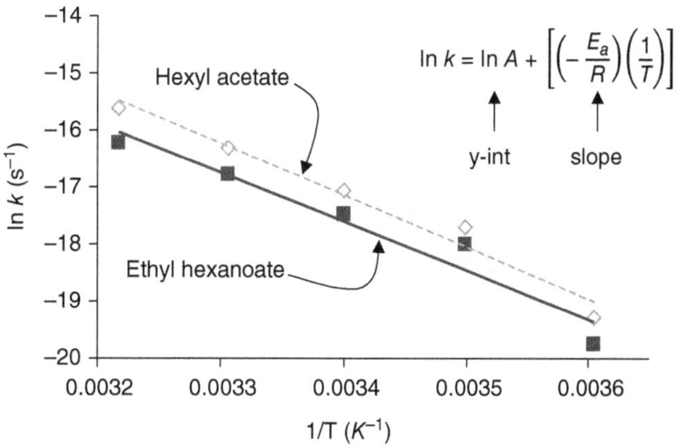

Figure 25.9 Determination of E_a and A from a plot of ln k versus $1/T$ for hydrolysis of two esters (hexyl acetate and ethyl hexanoate) based on the logarithmic form of Equation (25.1). Values of E_a are 62 and 53 kJ/mol for hexyl acetate and ethyl hexanoate, respectively. Data from Reference [43]

Table 25.1 Variation in activation energies and increases in reaction rate at elevated temperatures for a representative sample of reactions relevant to wine

Reaction	Conditions	Reference	Activation energy, E_a (kJ/mol)	Fold-increase in reaction rate as compared to 12 °C[a]	
				At 30 °C	At 50 °C
Acid hydrolysis of hexyl acetate	Model wine, pH 2.95–4.10	[43]	62	5	22
Hydrolysis of *S*-methylmethionine to dimethyl sulfide	Model beer, pH 5.2	[44]	186	106	10250
Acid degradation of fructose to HMF	Model orange juice, pH 2.5–4.5	[45]	199	147	19546
Formation of ethyl carbamate from urea and ethanol	White wines, pH 3.1–3.5	[46]	118	19	350
Acid hydrolysis of proanthocyanidins (tannins)	Model wine, pH 3.2	[47]	45	3	9
Acid hydrolysis of malvidin-3-glucoside	Model wine, pH 3.5	[48]	118	19	350

[a] Calculated from E_a and the Arrhenius equation.

in most table wines, including anthocyanin hydrolysis, and formation of dimethyl sulfide (DMS), HMF, and ethyl carbamate. This observation helps explain why expediting wine maturation by high-temperature storage will lead to different (and often inferior) outcomes than longer-term, lower-temperature storage.[5]

25.5 Effects of oak storage

The storage and transport of wine in wooden barrels has been practiced since antiquity. Historically, wooden barrels provided a robust (compared to an amphora), watertight, lightweight, and relatively convenient means to carry liquids, even across land – with the added benefit that the barrels could be broken down into staves for easy storage when not in use [49]. While different wood sources can be used for barrel production, oak heartwood is favored because of its large concentration of "tyloses" – plugs within the xylem that render it watertight.

25.5.1 Production of oak barrels and oak alternatives

Excellent technical discussions of the barrel production process have been published [50, 51], and the major steps can be summarized as follows:

- Mature white oak trees (genus *Quercus*) are harvested. The most commonly utilized species are native to Europe (*Q. robur* and *Q. petraea*) and North America (*Q. alba*).

[5] It is not clear if this statement continues to hold for extremely low storage temperatures and long storage times, for example, around the freezing point of water for centuries. Such conditions are experienced by wines recovered from shipwrecks in high latitudes, but (understandably) there is a lack of sensory data in the literature on these wines.

- Planks are cut from the oak heartwood and dried. Classically, this is done by stacking planks outside and air-drying ("seasoning") for long periods, typically 1–2 years. Alternate drying strategies that involve partial or complete usage of elevated temperatures also exist.
- Planks are cut into barrel staves.
- A *cooper* (barrel-maker) positions staves in metal hoops to form the barrel. The staves are heated during barrel production to increase their pliability. Traditionally, heating was done over a fire, although in modern production alternate heat sources (e.g., infrared lamps, steam) can be used.
- Once the barrel is formed, the cooper may continue to heat ("toast") the inside of the barrel to further alter flavor chemistry, as described below.
- The barrel is finished by addition of circular top and bottom pieces ("heads"), and cutting of a bung hole.

Alternative forms of oak – such as chips, shavings, cubes, and spiral rods – can be produced from waste material generated during stave production. Because of their smaller size, toasting of alternatives is often done in ovens as opposed to a point heat source, and is thus more uniform.

The most notable effect of maturing wine in contact with oak is the extraction of aroma compounds (Chapters 7, 10 and 12). The key compound classes and their precursors in oak (where relevant) are listed in Table 25.2.

Like other woods, oak is primarily composed of lignocellulose consisting of two classes of structural polymers (lignin and polysaccharides). These polymer classes have poor solubility in wine, but can serve as precursors for hundreds of volatiles under pyrolytic toasting conditions [52, 53]; these include several key aroma compounds (Table 25.2):

- Lignin, a highly crosslinked polymer of phenols, can pyrolyze to form volatile phenols and aldehydes with smoky, spicy, and vanilla aromas (Figure 25.10).
- Polysaccharides, particularly hemicellulose, can pyrolyze to form heterocyclic compounds with caramel and toasty aromas. The other major polysaccharide in wood, cellulose, appears to be of less importance to odorant formation due to its greater thermal stability [54].

Several factors will affect the concentrations of oak-derived volatiles for wine stored in contact with oak, but the three most important are wood source, barrel toasting protocols, and extraction conditions.

Table 25.2 Major components of oak heartwood and their role as precursors to key aroma compounds in oak barrels

Compound class	Concentration (dry weight) [50, 51]	Derived compounds	Aromas	Notes on derived compounds
Lignin	25–30%	Volatile phenols and phenolic aldehydes (e.g., guaiacol, vanillin)	Smoky, spicy, vanilla	Via toasting
Polysaccharides (cellulose, hemicellulose)	60–70%	Carbohydrate degradation products (e.g., furaneol)	Toast, caramel	Via toasting
Lipids	1–2%	Oak lactones	Coconut, sweet	Native, or via glycoside or hydroxyacid precursors
		Unsaturated aldehydes (e.g., (*E*)-2-nonenal)	Rancid oil, cardboard	Via unsaturated fatty acid oxidation
Hydrolyzable tannin	5–10%	–	–	–

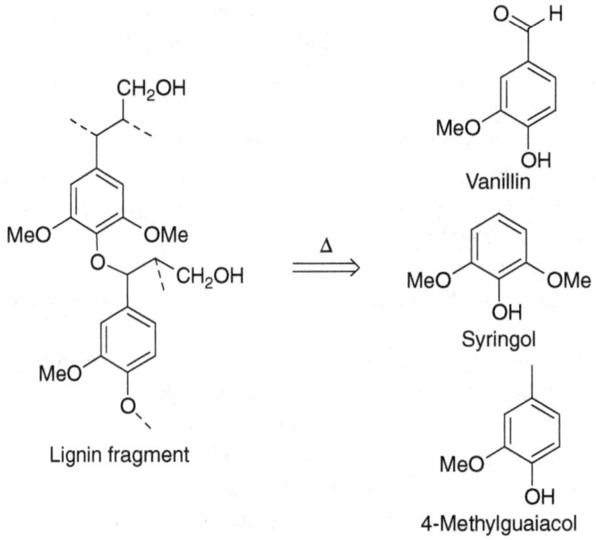

Figure 25.10 Thermal degradation of lignin yielding representative volatile phenols

25.5.2 Variation due to tree source

Several volatiles – particularly the oak lactones and their precursors – can vary considerably from tree to tree. Variation may also exist for the toast-derived compounds, but any source variation is usually overwhelmed by differences caused by the toasting process (see below). For the oak lactones:

- Species has a profound effect on concentrations in oak and resulting wines. In a survey of individual trees, *Q. petraea* averaged over 15-fold higher in total oak lactones than *Q. robur* (10.78 versus 0.61 μg/g) [55]. Similarly high concentrations of oak lactones have been observed for *Q. alba* [56], which likely contributes to American oak's anecdotal reputation as possessing more noticeable "coconut" and "woody" aromas.
- The localized growing environment will also have an effect, with standard deviations >100% and evidence of "high" and "low" regions within the same forest. The reason for higher oak lactone production in certain zones is unknown, although insect or disease pressure may contribute [55].

Grain tightness – that is, the spacing between tree growth rings – was historically believed to affect oak extractables, with a finer grain wood putatively having lower tannins and higher volatile content, although comparisons were often confounded by locational or species differences. More recent large-scale studies of hundreds of oak trees within and across sites have shown negligible correlations between volatiles and ring width [55, 57]. Similarly, negliglible changes were observed in sensory profiles of wines aged in oak barrels produced from different grain sizes [58].

25.5.3 Variation due to toasting

Toasting protocols vary in their length of time and in the temperatures achieved by the wood surface. Typically, the lexicon of "light," "medium," and "heavy" is used to describe increasing toast levels, although defining these conditions precisely is a challenge. One text has suggested a criteria of 120–180 °C surface temperature and a 5 min toast time for light, 200 °C and 10 min for medium, and 230 °C and 15 min for heavy [59], but other authors have suggested lower or higher maximum temperatures for medium and heavy toasts [58]. Furthermore, these reports do not generally account for the time the barrel is held below the maximum, during which toasting may still occur. Finally, the degree of toasting is typically heterogenous across a barrel, with the top and the bottom experiencing higher temperatures and more toasting than the middle of the barrel [58]. Detailed information about temperature profiles during toasting can be achieved with embedded thermocouples, but this is not a widespread practice. In sum, comparing absolute numbers on toast-derived volatiles across literature reports is challenging due to variations in cooperage methods.

As reported earlier in this book, both lignin-derived volatile phenols (Chapter 12) and carbohydrate degradation products (Chapter 2) are sensorially subthreshold in untoasted wood and only accumulate during toasting. The former (volatile phenols) are proportionally favored at higher toast temperatures or times (Figure 25.11). Carbohydrate degradation products appear to reach a maximum at lower toast temperatures, and in some cases they decline at high toast temperatures. The reason for this may be due to volatilization at high temperatures, but also because the majority of hemicellulose pyrolysis occurs at temperatures that are ~50 °C lower than lignin pyrolysis [54].

The large effects of the toast temperature on volatiles can help rationalize the high variability in toast-derived volatiles among barrels. High variability in toast profiles is seen not only among coopers but by the same cooper. A recent study of barrel toast patterns within a cooperage showed standard deviations of ±15 °C at various time points – in other words, the difference between medium and heavy toast levels [60].

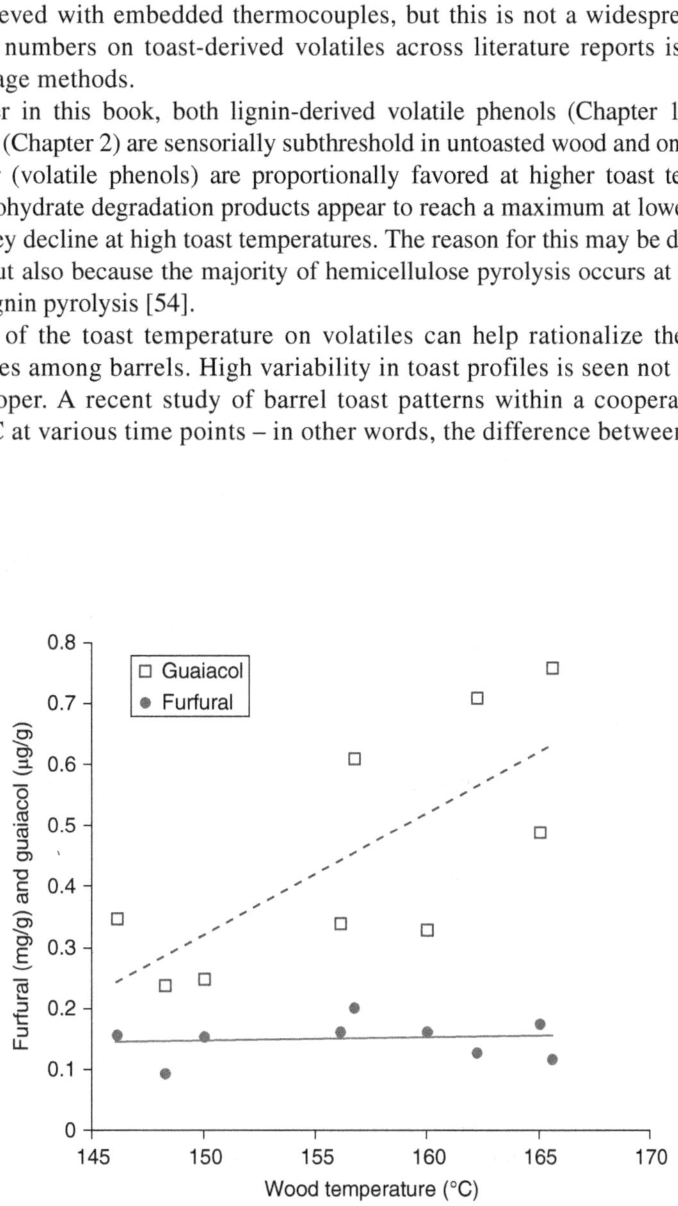

Figure 25.11 *The correlation of final wood temperature with concentrations of a lignin degradation product (guaiacol) and a carbohydrate degradation product (furfural) during barrel toasting. Temperatures were measured by thermocouples positioned within the barrel staves. Concentrations in untoasted portions of the wood averaged 0.03 µg/g for guaiacol and 0.003 mg/g for furfural. Data from Reference [60]*

In combination, the biological variability among trees and production variability during the toasting process explains the high variability in volatiles observed among barrels – for 10 barrel lots, standard deviations ranging from 15 to 40% were observed for several representative compounds (oak lactones, volatile phenols, furfural derivatives) [61]. Based on this, standard deviations of 50–125% are to be expected for key aroma compounds across individual barrels.[6]

25.5.4 Variation due to extraction

Volatiles are not extracted instantaneously from the barrel surface, and several factors vary the rate and amount of extraction:

- Many (but not all) phenols and carbohydrate degradation products obey roughly first-order kinetics, with concentrations eventually reaching an apparent equilibrium. In one study using traditional 225 L barrels, concentrations of guaiacol and syringol reached 50% of maximum within 3 weeks and furfural/5-methylfurfural within 2 months.
- Precursor-derived compounds, particularly oak lactone (Chapter 7), will continue to increase even once oak barrel contact ceases. For example, oak lactone increased by 25% during 3 months of bottle storage after an initial 9 month barrel storage period [62]. Glycosylated precursors of vanillin and other phenols have also been detected in oak wood [63], which may explain why vanillin increases linearly even after 2 years of barrel storage of a model wine [64].
- The amount of volatiles extracted will, of course, be lower in reused barrels – in one report, about 2-fold lower for guaiacol and 10-fold lower for furfuryl compounds after 180 days of storage [65]. The extraction rate of volatiles from reused barrels may also appear linear rather than asymptotic, presumably because volatiles must diffuse to the wood surface from the barrel interior.
- Oak alternatives with a small particle size, such as oak shavings, are reported to reach equilibrium faster than larger oak formats like chips [66]. In very general terms, this is credited to "a greater surface area-to-volume ratio." More specifically, this could be because small particles decrease the size of the boundary layer around the oak particle.

Finally, volatiles extracted from oak can undergo further (bio)chemical reactions in the wine. Aldehydes like vanillin and furfural are particularly prone to reactions, for example, reduction to alcohols by yeast (Chapter 22.1), reaction with H_2S to yield thiols during barrel-fermentation (Chapter 10), or reaction with anthocyanins or other phenolics during storage [67].

25.5.5 Other effects of oak

Beyond volatiles, oak storage can result in several other changes to wine, although the last two are only relevant to barrel storage as opposed to oak alternatives:

- Extraction of hydrolyzable tannins (Chapter 13) and other non-volatile components from the wood [50]. As mentioned earlier, tannins are usually extracted at concentrations far below the sensory threshold. However, recent reports have identified bitter tasting lignans [68] (a class of polyphenols) and sweet tasting triterpenoids [69] at concentrations around threshold.

[6] This variation helps explain the winemaking practice of tasting individual barrels, to identify components that could be used for different wines, such as a higher-priced reserve wine. In effect, the variability among barrels provides a natural experiment for winemakers to find ideal combinations of wine and oak flavor compounds.

- Adsorption of wine components by the wood [3, 70].
- Oxygen ingress through diffusion, either due to the porosity of the wood, the gaps between staves, or ullage when the bung is removed. In combination, these typically lead to oxygen pickup on the order of a few mg/L/month in a standard 225 L barrel, but the relative importance of these effects is contentious [50, 71, 72].
- A concentrating effect over time due to diffusion through staves and evaporation (around –5% per annum), along with slight changes in ethanol concentration (either an increase or decrease) as a function of relative humidity and factors (e.g., temperature, air speed) related to the storage environment [50, 73] – the *angels' share*.

25.6 Sensory effects of different aging conditions

As a general rule, bottle or tank aging of a wine results in a loss of fruity, floral, and fresh vegetable aromas and an increase in dried fruit, earthy, canned vegetable, woody, and either sulfurous (e.g., due to hydrolysis of *S*-methylmethionine, Chapter 23.3) or oxidation aromas (Figure 25.12). Oak-derived flavors may also increase in the case of barrel storage, among other changes as specified above. Greater oxygen exposure will increase concentrations of odorants with "oxidized" aromas (e.g., carbonyl compounds, Chapters 9 and 24), but also result in loss of thiols with "reduced" and "fruity" aromas (Chapter 10).

Higher storage temperatures and shorter times will partially simulate the effects of longer storage at cooler temperatures, particularly for the loss of primary and secondary characters observed for white wines, but also for some developed characters in red wines [74]. This information is relevant in a global market place where wines are shipped internationally, and by some estimates wine may appear to be aged an additional year or more during transit compared to isothermal storage at ordinary cellar temperatures [75]. However, as described in section 25.4, the relative rates of reactions important to wine flavor will vary with changes in temperature.

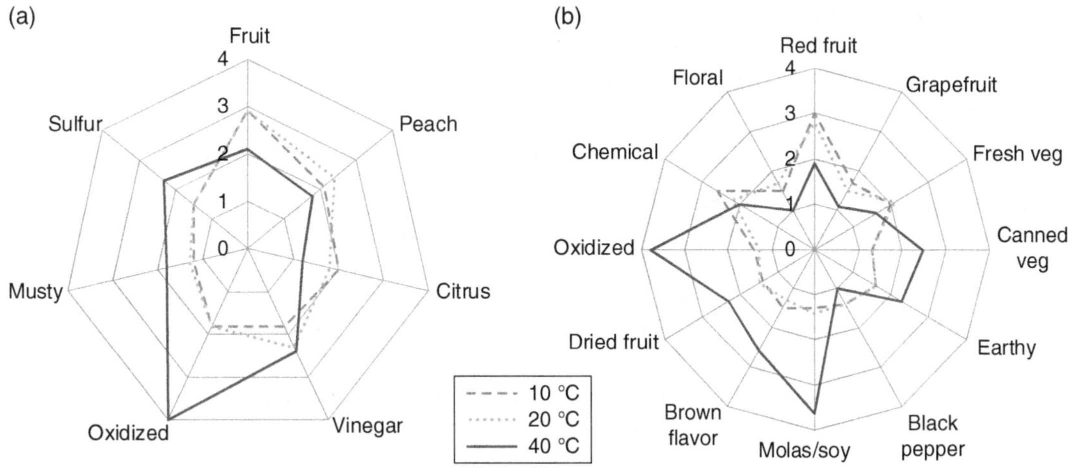

Figure 25.12 Effect of storage temperature (10, 20, 40 °C) on aroma attribute ratings (mean values for significant attributes) for (a) a Chardonnay wine stored in bottle for 3 months and (b) a Cabernet Sauvignon wine stored in bottle for 6 months (Molas/soy = molasses/soy sauce). Data from References [76] and [77]

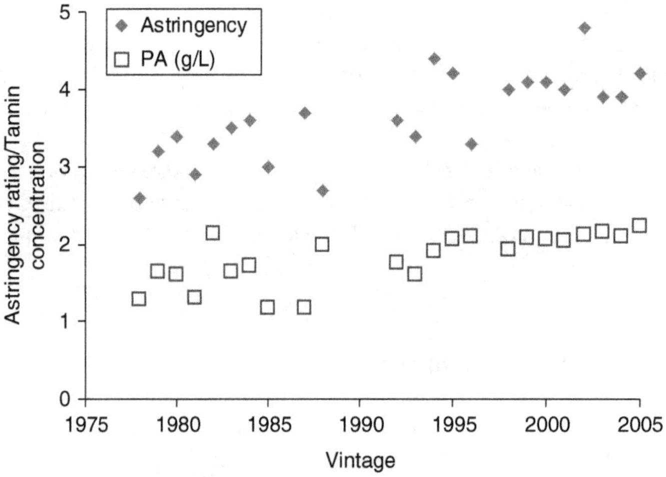

Figure 25.13 Links between wine age (vintage), mean astringency rating, and proanthocyanidin (PA) concentration (g/L), determined by depolymerization with acid and measuring absorbance at 550 nm, for a series of Cabernet Sauvignon wines from a single French chateau that were prepared and cellared under similar conditions. Data from Reference [82]

Tannin maturation in tank or barrel and aging in bottle leads not only to stabilization of color (as outlined above) but also results in decreased intensity of astringency (Chapters 14 and 24; also see References [78] and [79], and citations therein). Red wine quality is well correlated with having appropriate levels of astringency [80, 81]. Bottle aging is a classic approach to decrease the astringency of tannic young red wines, and some high tannin (and high price point wines) will be allowed to cellar-age for several years to allow the mouthfeel to "soften." For example, Cabernet Sauvignon wines produced by a single chateau in France and stored under similar conditions show a linear correlation ($R^2=0.598$) between vintage year and astringency scores [82] (Figure 25.13).[7] These changes arise from transformations of proanthocyanidins during storage that decrease their ability to react with proteins and, thus, decrease their astringency [22, 29, 83]. Many reactions involving proanthocyanidins will happen more readily in the presence of O_2 and at lower pH (as outlined above), including:

- Reactions with anthocyanins (to form T-A or A-T adducts).
- Reactions with aldehydes, vinylphenols, or other electrophiles.
- Acid hydrolysis, to yield smaller tannins (lower degree of polymerization) with decreased astringency.

Interestingly, the correlation of astringency with total proanthocyanidins was much weaker ($R^2=0.36$, Figure 25.13), which is evidence that most chemical measurements of tannins are imperfectly correlated with protein-binding capacity (Chapter 33).

Decisions about wine packaging are some of the most crucial to be made, in part because the winemaker typically has little recourse once a wine is packaged. The effects of material including glass and PET bottles,

[7] Note that the extended aging of red wines is the exception, not the norm: the vast majority of red wines are consumed within 2 years of the harvest. Winemakers with excessively tannic wines who do not wish to wait 5+ years to see a revenue stream can use alternate approaches to reduce astringency through the use of fining agents.

and bag-in-box pouches, along with natural cork, synthetic, and screw cap closures, have been investigated in several studies [7, 8, 29, 76, 77, 84–94]:

- Permeability of the packaging has a strong influence on sensory properties over time, as demonstrated with closure OTR. While there are reasonable differences within a closure type, they tend to obey the following pattern in increasing OTR:

$$\text{Screw caps} < \text{Technical} < \text{Natural} < \text{Synthetic}$$

  Increased closure OTR can facilitate beneficial reactions, as highlighted above with red wine color and mouthfeel, but it also leads to losses of SO_2. This can have particularly stark effects on white wine once the protective effects of SO_2 are diminished, with an increase in browning (A_{420}), loss of fruit characters, and an increase in "oxidized" and "developed" aromas (Figure 25.14). Conversely, "reduced" aroma attributes are more often ascribed to screw cap closures with a low OTR (Chapter 30), yet these closures preserve SO_2 and fruit characteristics. Similar considerations regarding permeability apply to the choice of container material; that is, whether it is made of glass or plastic (as in PET bottles or bag-in-box pouches).
- Packaging can cause *taints* due to migration of odorants from the packaging into the wine, or *scalping* due to absorption (or adsorption) of odorants into the packaging [95]. The best known of these conditions is "cork taint" (Chapter 18), but other packaging materials (particularly plastic-based) can also cause this. For example, after 3–6 months, sensory changes in white wine can be readily noted for certain plastics due to scalping or tainting [89, 91].

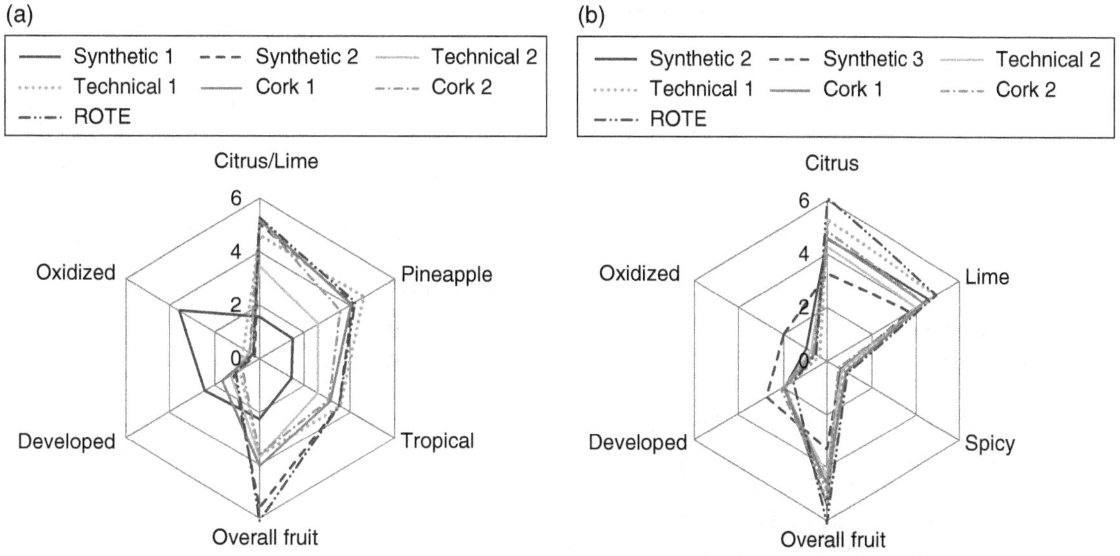

Figure 25.14 Mean aroma ratings for a Clare Valley Semillon wine bottled under a variety of closures, cellared at an average temperature of 17 °C and assessed by a sensory panel after (a) 6 months and (b) 12 months. ROTE = roll on tamper evident (i.e., screw cap). Selected data from Reference [84]

Finally, the role of lees aging (*sur lie*) deserves mention for its ability to influence wine sensory properties [3, 5, 96, 97]. Lees in this instance describes the autolysis of yeast (primarily) and bacteria cells, which releases a range of components into the wine. Among other things, these include:

- Enzymes that lead to glycoside hydrolysis (Chapter 23.1).
- Nitrogenous compounds such as amino acids, proteins and peptides, and mannoproteins, which can affect perceived sweetness and body (Chapter 6), stabilize foam in sparkling wines, as well as increase tartrate stability (Chapter 26.1).
- Aroma compounds (e.g., fatty acids, alcohols, and esters), which can lead to perception of the "yeasty" and "bread-like" aroma of sur lie aged wine.
- Polysaccharides that can impact mouthfeel (Chapter 2).

Additionally, storage in contact with lees under reductive conditions is implicated in the formation of sulfurous off aromas (Chapters 10 and 22.4) but lees can also be used to bind these compounds and remove them from wine. The differences between these roles of lees presumably relates to their relative degrees of oxidation. Similarly, lees can also adsorb other compounds, particularly non-polar volatiles and phenolic components, and, overall, aging on lees is aimed at improving wine quality through compositional changes that modify aroma, flavor, body and mouthfeel, and stabilize foam in sparkling wines. Commercial yeast autolysates (or other inactive dry yeast preparations) are available to complement or replace the lees arising from winery fermentations, potentially allowing a more rapid incorporation of beneficial components in a shorter time than would occur with ordinary storage on lees.

Closure Performance

Stoppers made from the bark of the cork oak tree date to the seventeenth century, and were the first closures that sealed bottles well enough to allow for long-term aging. Excessive problems with trichloroanisole (TCA) taint (Chapter 18) in the 1990s prompted a search for alternatives such that winemakers now have screw caps, synthetic stoppers, and a number of lesser known options to choose from. Producers also offer "technical cork" stoppers produced by extraction or chemical treatment of ground cork particles, which are then mixed with a resin before moulding. Natural cork remains the most popular closure option, and improvements in cork production have decreased the taint issue, but 1–3% of bottles are still contaminated. Several methods exist to extract TCA from ground cork, with supercritical carbon dioxide a notable success.

The principle chemical performance criterion for closures is the oxygen transfer rate (OTR), as the amount of oxygen entering the bottle is a key factor in aging chemistry. Wet corks transfer approximately 1 mg/O_2 per year, with the most common screw cap (Saranex liner) having similar properties. Synthetic corks are close, but generally higher and vary depending on their specifications. Tin-lined screw caps have the lowest reported OTR, at less than 0.05 mg/O_2 per year. Both manufactured types of closures can be obtained with specific OTR values that vary depending on their construction.

However, these average values obscure a more important issue – variability in OTR. Natural cork shows large variations, while the manufactured closures are very consistent. Data on screw cap performance also shows a large variation between reports, suggesting that variation in bottling operations is equally important. Compared to stoppers, the application of screw caps demands much more care in adjustment and management of the bottling hardware for successful sealing of the bottles.

References

1. Singleton, V.L. (2000) A survey of wine aging reactions, especially with oxygen. *Proceedings of the 50th Anniversary Annual Meeting*, Sęattle, Washington, pp. 323–336.
2. Ebeler, S.E. (2001) Analytical chemistry: unlocking the secrets of wine flavor. *Food Reviews International*, **17** (1), 45–64.
3. Garde-Cerdán, T. and Ancín-Azpilicueta, C. (2006) Review of quality factors on wine ageing in oak barrels. *Trends in Food Science and Technology*, **17** (8), 438–447.
4. Alexandre, H. and Guilloux-Benatier, M. (2006) Yeast autolysis in sparkling wine – a review. *Australian Journal of Grape and Wine Research*, **12** (2), 119–127.
5. Pérez-Serradilla, J.A. and de Castro, M.D.L. (2008) Role of lees in wine production: a review. *Food Chemistry*, **111** (2), 447–456.
6. Jackson, R.S. (2009) Nature and origins of wine quality, in *Wine tasting: a professional handbook*, 2nd edn, Academic Press, San Diego, CA, pp. 387–426.
7. Silva, M., Julien, M., Jourdes, M., Teissedre, P.-L. (2011) Impact of closures on wine post-bottling development: a review. *European Food Research and Technology*, **233** (6), 905–914.
8. Ugliano, M. (2013) Oxygen contribution to wine aroma evolution during bottle aging. *Journal of Agricultural and Food Chemistry*, **61** (26), 6125–6136.
9. Tao, Y., García, J.F., Sun, D.-W. (2014) Advances in wine aging technologies for enhancing wine quality and accelerating wine aging process. *Critical Reviews in Food Science and Nutrition*, **54** (6), 817–835.
10. Cheynier, V., Remy, S., Fulcrand, H. (2000) Mechanisms of anthocyanin and tannin changes during winemaking and aging, in *Proceedings of the ASEV 50th Anniversary Annual Meeting* (ed. Rantz, J.M.), June 19–23, 2000, Seattle, WA, American Society of Enology and Viticulture, Davis, CA, pp. 337–344.
11. Cheynier, V., Duenas-Paton, M., Salas, E., *et al.* (2006) Structure and properties of wine pigments and tannins. *American Journal of Enology and Viticulture*, **57** (3), 298–305.
12. He, F., Liang, N.-N., Mu, L., *et al.* (2012) Anthocyanins and their variation in red wines. II. Anthocyanin derived pigments and their color evolution. *Molecules*, **17** (2), 1483–1519.
13. de Freitas, V.A.P. and Mateus, N. (2010) Updating wine pigments, in *Recent advances in polyphenol research*, Wiley-Blackwell, pp. 59–80.
14. Salas, E., Fulcrand, H., Meudec, E., Cheynier, V. (2003) Reactions of anthocyanins and tannins in model solutions. *Journal of Agricultural and Food Chemistry*, **51** (27), 7951–7961.
15. Duenas M., Fulcrand H., Cheynier V. (2006) Formation of anthocyanin-flavanol adducts in model solutions. *Analytica Chimica Acta*, **563** (1–2), 15–25.
16. Somers, T.C. (1971) The polymeric nature of wine pigments. *Phytochemistry*, **10**, 2175–2189.
17. Malien-Aubert, C., Dangles, O., Amiot, M.J. (2002) Influence of procyanidins on the color stability of oenin solutions. *Journal of Agricultural and Food Chemistry*, **50** (11), 3299–3305.
18. Escribano-Bailón, T., Álvarez-García, M., Rivas-Gonzalo, J.C., *et al.* (2001) Color and stability of pigments derived from the acetaldehyde-mediated condensation between malvidin 3-*O*-glucoside and (+)-catechin. *Journal of Agricultural and Food Chemistry*, **49** (3), 1213–1217.
19. Atanasova, V., Fulcrand, H., Cheynier, W., Moutounet, M. (2002) Effect of oxygenation on polyphenol changes occurring in the course of wine-making. *Analytica Chimica Acta*, **458** (1), 15–27.
20. Remy-Tanneau, S., Le Guerneve, C., Meudec, E., Cheynier, V. (2003) Characterization of a colorless anthocyanin-flavan-3-ol dimer containing both carbon–carbon and ether interflavanoid linkages by NMR and mass spectrometry. *Journal of Agricultural and Food Chemistry*, **51** (12), 3592–3597.
21. Romero, C. and Bakker, J. (2000) Effect of acetaldehyde and several acids on the formation of vitisin A in model wine anthocyanin and colour evolution. *International Journal of Food Science and Technology*, **35** (1), 129–140.
22. McRae, J.M., Dambergs, R.G., Kassara, S., *et al.* (2012) Phenolic compositions of 50 and 30 year sequences of Australian red wines: the impact of wine age. *Journal of Agricultural and Food Chemistry*, **60** (40), 10093–10102.
23. Pechamat, L., Zeng, L., Jourdes, M., *et al.* (2014) Occurrence and formation kinetics of pyranomalvidin–procyanidin dimer pigment in Merlot red wine: impact of acidity and oxygen concentrations. *Journal of Agricultural and Food Chemistry*, **62** (7), 1701–1705.

24. Gómez-Plaza, E. and Cano-López, M. (2011) A review on micro-oxygenation of red wines: claims, benefits and the underlying chemistry. *Food Chemistry*, **125** (4), 1131–1140.
25. Anli, R.E. and Cavuldak, Ö.A. (2012) A review of microoxygenation application in wine. *Journal of the Institute of Brewing*, **118** (4), 368–385.
26. Gambuti, A., Rinaldi, A., Ugliano, M., Moio, L. (2013) Evolution of phenolic compounds and astringency during aging of red wine: effect of oxygen exposure before and after bottling. *Journal of Agricultural and Food Chemistry*, **61** (8), 1618–1627.
27. Gonzalez-del Pozo, A., Arozarena, I., Noriega, M.J., *et al.* (2010) Short- and long-term effects of micro-oxygenation treatments on the colour and phenolic composition of a Cabernet Sauvignon wine aged in barrels and/or bottles. *European Food Research and Technology*, **231** (4), 589–601.
28. Kontoudakis, N., González, E., Gil, M., *et al.* (2011) Influence of wine pH on changes in color and polyphenol composition induced by micro-oxygenation. *Journal of Agricultural and Food Chemistry*, **59** (5), 1974–1984.
29. McRae, J.M., Kassara, S., Kennedy, J.A., *et al.* (2013) Effect of wine pH and bottle closure on tannins. *Journal of Agricultural and Food Chemistry*, **61** (47), 11618–11627.
30. Daniel, M.A., Capone, D.L., Sefton, M.A., Elsey, G.M. (2009) Riesling acetal is a precursor to 1,1,6-trimethyl-1,2-dihydronaphthalene (TDN) in wine. *Australian Journal of Grape and Wine Research*, **15** (1), 93–96.
31. Giaccio, J., Capone, D.L., Håkansson, A.E., *et al.* (2011) The formation of wine lactone from grape-derived secondary metabolites. *Journal of Agricultural and Food Chemistry*, **59** (2), 660–664.
32. Winterhalter, P. (1991) 1,1,6-Trimethyl-1,2-dihydronaphthalene (TDN) formation in wine. 1. Studies on the hydrolysis of 2,6,10,10-tetramethyl-1-oxaspiro[4.5]dec-6-ene-2,8-diol rationalizing the origin of TDN and related C_{13} norisoprenoids in Riesling wine. *Journal of Agricultural and Food Chemistry*, **39** (10), 1825–1829.
33. Luan, F., Degenhardt, A., Mosandl, A., Wust, M. (2006) Mechanism of wine lactone formation: demonstration of stereoselective cyclization and 1,3-hydride shift. *Journal of Agricultural and Food Chemistry*, **54** (26), 10245–10252.
34. Diaz-Maroto, M.C., Schneider, R., Baumes, R. (2005) Formation pathways of ethyl esters of branched short-chain fatty acids during wine aging. *Journal of Agricultural and Food Chemistry*, **53** (9), 3503–3509.
35. Makhotkina, O. and Kilmartin, P.A. (2012) Hydrolysis and formation of volatile esters in New Zealand Sauvignon Blanc wine. *Food Chemistry*, **135** (2), 486–493.
36. Antalick, G., Perello, M.-C., de Revel, G. (2014) Esters in wines: new insight through the establishment of a database of French wines. *American Journal of Enology and Viticulture*, **65** (3), 293–304.
37. Rayne, S. and Forest, K. (2011) Estimated carboxylic acid ester hydrolysis rate constants for food and beverage aroma compounds. *Nature Precedings*. doi:10.1038/npre.2011.6471.1. Available from *Nature Precedings*, http://dx.doi.org/10.1038/npre.2011.6471.1.
38. Marais, J. (1978) Effect of pH on esters and quality of colombar wine during maturation. *Vitis*, **17** (4), 396–403.
39. Makhotkina, O., Pineau, B., Kilmartin, P.A. (2012) Effect of storage temperature on the chemical composition and sensory profile of Sauvignon Blanc wines. *Australian Journal of Grape and Wine Research*, **18** (1), 91–99.
40. Wilkinson, K.L., Prida, A., Hayasaka, Y. (2013) Role of glycoconjugates of 3-methyl-4-hydroxyoctanoic acid in the evolution of oak lactone in wine during oak maturation. *Journal of Agricultural and Food Chemistry*, **61** (18), 4411–4416.
41. Pereira, V., Albuquerque, F.M., Ferreira, A.C., *et al.* (2011) Evolution of 5-hydroxymethylfurfural (HMF) and furfural (F) in fortified wines submitted to overheating conditions. *Food Research International*, **44** (1), 71–76.
42. Belitz, H.-D., Grosch, W., Schieberle, P. (2009) Carbohydrates, in *Food chemistry* (eds Belitz, H.-D., Grosch, W., Schieberle, P.), Springer, Berlin, pp. 248–339.
43. Ramey, D.D. and Ough, C.S. (1980) Volatile ester hydrolysis or formation during storage of model solutions and wines. *Journal of Agricultural and Food Chemistry*, **28** (5), 928–934.
44. Scheuren, H., Tippmann, J., Methner, F.J., Sommer, K. (2014) Decomposition kinetics of dimethyl sulphide. *Journal of the Institute of Brewing*, **120** (4), 474–476.
45. Arena, E., Fallico, B., Maccarone, E. (2001) Thermal damage in blood orange juice: kinetics of 5-hydroxymethyl-2-furancarboxaldehyde formation. *International Journal of Food Science and Technology*, **36** (2), 145–151.
46. Kodama, S., Suzuki, T., Fujinawa, S., *et al.* (1994) Urea contribution to ethyl carbamate formation in commercial wines during storage. *American Journal of Enology and Viticulture*, **45** (1), 17–24.

47. Dallas, C., Hipólito-Reis, P., Ricardo-da-Silva, J.M., Laureano, O. (2003) Influence of acetaldehyde, pH, and temperature on transformation of procyanidins in model wine solutions. *American Journal of Enology and Viticulture*, **54** (2), 119–124.
48. Baranowski, E.S. and Nagel, C.W. (1983) Kinetics of malvidin-3-glucoside condensation in wine model systems. *Journal of Food Science*, **48** (2), 419–421.
49. McGovern, P.E. (2003) *Ancient wine: the search for the origins of viniculture*, Princeton University Press, Princeton, NJ.
50. Singleton, V.L. (1995) Maturation of wines and spirits: comparisons, facts, and hypotheses. *American Journal of Enology and Viticulture*, **46** (1), 98–115.
51. Schahinger, G. (2005) *Cooperage for winemakers: a manual on the construction, maintenance, and use of oak barrels* (ed. Rankine, B.C.), Winetitles, Adelaide, SA, Australia.
52. Faix, O., Fortmann, I., Bremer, J., Meier, D. (1991) Thermal degradation products of wood. *Holz als Roh- und Werkstoff*, **49** (7–8), 299–304.
53. Faix, O., Meier, D., Fortmann, I. (1990) Thermal degradation products of wood. *Holz als Roh- und Werkstoff*, **48** (7–8), 281–285.
54. Brebu, M. and Vasile, C. (2010) Thermal degradation of lignin – a review. *Cellulose Chemistry and Technology*, **44** (9), 353.
55. Prida, A., Ducousso, A., Petit, R., *et al.* (2007) Variation in wood volatile compounds in a mixed oak stand: strong species and spatial differentiation in whisky-lactone content. *Annals of Forest Science*, **64** (3), 313–320.
56. Masson, G., Guichard, E., Fournier, N., Puech, J.-L. (1995) Stereoisomers of ß-methyl-γ-octalactone. II. Contents in the wood of French (*Quercus robur* and *Quercus petraea*) and American (*Quercus alba*) oaks. *American Journal of Enology and Viticulture*, **46** (4), 424–428.
57. Mosedale, J.R. and Savill, P.S. (1996) Variation of heartwood phenolics and oak lactones between the species and phenological types of *Quercus petraea* and *Q. robur*. *Forestry*, **69** (1), 47–55.
58. Collins, T.S. (2012) The impact of variability in the toasting process at a commercial cooperage on the volatile composition of oak barrels and barrel aged wines, PhD Thesis, University of California, Davis.
59. Ribereau-Gayon, P., Glories, Y., Maujean, A., Dubourdieu, D. (2006) *Handbook of enology*, Vol. 2, *The chemistry of wine stabilization and treatments*, 2nd edn, John Wiley & Sons, Chichester, UK.
60. Collins, T.S., Miles, J.L., Boulton, R.B., Ebeler, S.E. (2015) Targeted volatile composition of oak wood samples taken during toasting at a commercial cooperage. *Tetrahedron*, **71** (20), 2971–2982.
61. Towey, J.P. and Waterhouse, A.L. (1996) Barrel-to-barrel variation of volatile oak extractives in barrel-fermented Chardonnay. *American Journal of Enology and Viticulture*, **47** (1), 17–20.
62. Perez-Prieto, L.J., Lopez-Roca, J.M., Martinez-Cutillas, A., *et al.* (2003) Extraction and formation dynamic of oak-related volatile compounds from different volume barrels to wine and their behavior during bottle storage. *Journal of Agricultural and Food Chemistry*, **51** (18), 5444–5449.
63. Slaghenaufi, D., Marchand-Marion, S., Richard, T., *et al.* (2013) Centrifugal partition chromatography applied to the isolation of oak wood aroma precursors. *Food Chemistry*, **141** (3), 2238–2245.
64. Spillman, P.J., Iland, P.G., Sefton, M.A. (1998) Accumulation of volatile oak compounds in a model wine stored in American and Limousin oak barrels. *Australian Journal of Grape and Wine Research*, **4** (2), 67–73.
65. Gómez-Plaza, E., Pérez-Prieto, L.J., Fernández-Fernández, J.I., López-Roca, J.M. (2004) The effect of successive uses of oak barrels on the extraction of oak-related volatile compounds from wine. *International Journal of Food Science and Technology*, **39** (10), 1069–1078.
66. Campbell, J., Pollnitz, A., Sefton, M., *et al.* (2006) Factors affecting the influence of oak chips on wine flavour. *Australian and New Zealand Wine Industry Journal*, **21** (4), 38–42.
67. Nonier, M.-F., Vivas, N., De Freitas, V., *et al.* (2011) A kinetic study of the reaction of (+)-catechin and malvidin-3-glucoside with aldehydes derived from toasted oak. *The Natural Products Journal*, **1** (1), 47–56.
68. Marchal, A., Cretin, B.N., Sindt, L., *et al.* (2015) Contribution of oak lignans to wine taste: chemical identification, sensory characterization and quantification. *Tetrahedron*, **71** (20), 3148–3156.
69. Marchal, A., Waffo-Teguo, P., Genin, E., *et al.* (2011) Identification of new natural sweet compounds in wine using centrifugal partition chromatography–gustatometry and Fourier transform mass spectrometry. *Analytical Chemistry*, **83** (24), 9629–9637.

70. Jarauta, I., Cacho, J., Ferreira, V. (2005) Concurrent phenomena contributing to the formation of the aroma of wine during aging in oak wood: an analytical study. *Journal of Agricultural and Food Chemistry*, **53** (10), 4166–4177.
71. Nevares, I., Crespo, R., Gonzalez, C., del Alamo-Sanza, M. (2014) Imaging of oxygen transmission in the oak wood of wine barrels using optical sensors and a colour camera. *Australian Journal of Grape and Wine Research*, **20** (3), 353–360.
72. Nevares, I. and del Alamo-Sanza, M. (2015) Oak stave oxygen permeation: a new tool to make barrels with different wine oxygenation potentials. *Journal of Agricultural and Food Chemistry*, **63** (4), 1268–1275.
73. Blazer, R.M. (1991) Influence of environment – wine evaporation from barrels. *Practical Winery and Vineyard*, **11**, 20–22.
74. Robinson, A.L., Mueller, M., Heymann, H., *et al.* (2010) Effect of simulated shipping conditions on sensory attributes and volatile composition of commercial white and red wines. *American Journal of Enology and Viticulture*, **61** (3), 337–347.
75. Butzke, C.E., Vogt, E.E., Chacón-Rodríguez, L. (2012) Effects of heat exposure on wine quality during transport and storage. *Journal of Wine Research*, **23** (1), 15–25.
76. Hopfer, H., Ebeler, S.E., Heymann, H. (2012) The combined effects of storage temperature and packaging type on the sensory and chemical properties of Chardonnay. *Journal of Agricultural and Food Chemistry*, **60** (43), 10743–10754.
77. Hopfer, H., Buffon, P.A., Ebeler, S.E., Heymann, H. (2013) The combined effects of storage temperature and packaging on the sensory, chemical, and physical properties of a Cabernet Sauvignon wine. *Journal of Agricultural and Food Chemistry*, **61** (13), 3320–3334.
78. del Carmen Llaudy, M., Canals, R., Gonzalez-Manzano, S., *et al.* (2006) Influence of micro-oxygenation treatment before oak aging on phenolic compounds composition, astringency, and color of red wine. *Journal of Agricultural and Food Chemistry*, **54** (12), 4246–4252.
79. Oberholster, A., Elmendorf, B.L., Lerno, L.A., *et al.* (2015) Barrel maturation, oak alternatives and micro-oxygenation: influence on red wine aging and quality. *Food Chemistry*, **173**, 1250–1258.
80. Fanzone, M., Peña-Neira, A., Gil, M., *et al.* (2012) Impact of phenolic and polysaccharidic composition on commercial value of Argentinean Malbec and Cabernet Sauvignon wines. *Food Research International*, **45** (1), 402–414.
81. Saenz-Navajas, M.P., Martin-Lopez, C., Ferreira, V., Fernandez-Zurbano, P. (2011) Sensory properties of premium Spanish red wines and their implication in wine quality perception. *Australian Journal of Grape and Wine Research*, **17** (1), 9–19.
82. Chira, K., Jourdes, M., Teissedre, P.-L. (2012) Cabernet Sauvignon red wine astringency quality control by tannin characterization and polymerization during storage. *European Food Research and Technology*, **234** (2), 253–261.
83. Chira, K., Pacella, N., Jourdes, M., Teissedre, P.-L. (2011) Chemical and sensory evaluation of Bordeaux wines (Cabernet-Sauvignon and Merlot) and correlation with wine age. *Food Chemistry*, **126** (4), 1971–1977.
84. Godden, P., Francis, L., Field, J., *et al.* (2001) Wine bottle closures: physical characteristics and effect on composition and sensory properties of a Semillon wine. 1. Performance up to 20 months post-bottling. *Australian Journal of Grape and Wine Research*, **7** (2), 64–105.
85. Capone, D., Sefton, M., Pretorius, I., Høj, P. (2003) Flavour "scalping" by wine bottle closures – the "winemaking" continues post vineyard and winery. *Australian and New Zealand Wine Industry Journal*, **18** (5), 16, 18–20.
86. Skouroumounis, G.K., Kwiatkowski, M.J., Francis, I.L., *et al.* (2005) The impact of closure type and storage conditions on the composition, colour and flavour properties of a Riesling and a wooded Chardonnay wine during five years' storage. *Australian Journal of Grape and Wine Research*, **11** (3), 369–384.
87. Brajkovich, M., Tibbits, N., Peron, G., *et al.* (2005) Effect of screwcap and cork closures on SO_2 levels and aromas in a Sauvignon Blanc wine. *Journal of Agricultural and Food Chemistry*, **53** (26), 10006–10011.
88. Sefton, M.A. and Simpson, R.F. (2005) Compounds causing cork taint and the factors affecting their transfer from natural cork closures to wine – a review. *Australian Journal of Grape and Wine Research*, **11** (2), 226–240.
89. Ghidossi, R., Poupot, C., Thibon, C., *et al.* (2012) The influence of packaging on wine conservation. *Food Control*, **23** (2), 302–311.
90. Wirth, J., Caillé, S., Souquet, J.M., *et al.* (2012) Impact of post-bottling oxygen exposure on the sensory characteristics and phenolic composition of Grenache rosé wines. *Food Chemistry*, **132** (4), 1861–1871.

91. Revi, M., Badeka, A., Kontakos, S., Kontominas, M.G. (2014) Effect of packaging material on enological parameters and volatile compounds of dry white wine. *Food Chemistry*, **152**, 331–339.
92. Karbowiak, T., Gougeon, R.D., Alinc, J.B., *et al.* (2010) Wine oxidation and the role of cork. *Critical Reviews in Food Science and Nutrition*, **50** (1), 20–52.
93. Guaita, M., Petrozziello, M., Motta, S., *et al.* (2013) Effect of the closure type on the evolution of the physical-chemical and sensory characteristics of a Montepulciano d'Abruzzo rosé wine. *Journal of Food Science*, **78** (2), C160–C169.
94. Silva, M.A., Jourdes, M., Darriet, P., Teissedre, P.-L. (2012) Scalping of light volatile sulfur compounds by wine closures. *Journal of Agricultural and Food Chemistry*, **60** (44), 10952–10956.
95. Sajilata, M.G., Savitha, K., Singhal, R.S., Kanetkar, V.R. (2007) Scalping of flavors in packaged foods. *Comprehensive Reviews in Food Science and Food Safety*, **6** (1), 17–35.
96. Torresi, S., Frangipane, M.T., Anelli, G. (2011) Biotechnologies in sparkling wine production. Interesting approaches for quality improvement: a review. *Food Chemistry*, **129** (3), 1232–1241.
97. Gambetta, J.M., Bastian, S.E.P., Cozzolino, D., Jeffery, D.W. (2014) Factors influencing the aroma composition of Chardonnay wines. *Journal of Agricultural and Food Chemistry*, **62** (28), 6512–6534.

26

The Chemistry of Post-fermentation Processing

Wine production does not end with alcoholic fermentation – most wines undergo further processing prior to bottling. This collection of chapters discusses physicochemical aspects of common cellar practices, which include:

- Cold stabilization involving tartrate crystallization phenomena and treatments associated with preventing precipitation in bottled wine
- Fining of wine to promote stability and desirable organoleptic qualities of wine, including tannin and protein fining as well as other fining treatments
- Particle filtration as a means of clarifying wine and the use of reverse osmosis for selective removal of small molecules
- Distillation of wine, which addresses the theoretical and practical aspects of converting wine into grape spirit.

Understanding Wine Chemistry, First Edition. Andrew L. Waterhouse, Gavin L. Sacks, and David W. Jeffery.

26.1

Cold Stabilization

26.1.1 Introduction

In winemaking and other areas of food production, *stability* is the attainment of a state or condition in which the product will have an acceptably low risk of demonstrating undesirable physical or sensory changes over a specified storage time and under a defined set of storage conditions.[1] In wine, instabilities that lead to visible changes to the wine are of particular concern to winemakers since they are readily detectable by consumers and lead to unnecessary concerns about product safety. The two most important visible defects that can occur during storage are tartrate instabilities (discussed in this chapter) and protein instabilities (Chapter 26.2).

The chemistries of free tartaric acid (H_2T) and its conjugate bases (bitartrate (HT^-), tartrate (T^{2-})) were discussed in Chapter 3. The total concentration of tartrate species is usually in the range of 2–6 g/L in wine (0.013–0.040 M). The relative distribution of the tartrate species is pH dependent, but across the range of typical wine pH values, HT^- represents the major species (50–70%) and reaches a maximum around pH 3.65 (Chapter 3, see Figure 3.2). Beyond affecting pH and TA, HT^- can react with K^+, the major metal cation in wine (0.003–0.07 M, Chapter 4), to form the poorly soluble potassium bitartrate (KHT, "cream of tartar"). Because the solubility of KHT is lower in ethanolic solutions than in water, and grape juice is nearly saturated in KHT, it is common to see the formation of KHT crystals along tank and barrel walls during the fermentation process; these can be readily separated from the wine by racking. As discussed below, KHT precipitation will decrease TA but may lead to either an increase or decrease in pH due to the amphoteric nature of HT^-. The low solubility of KHT also indirectly limits the maximum total concentration of tartaric acid species.

The majority of wines exist as metastable, supersaturated KHT solutions due to the presence of crystallization-inhibiting compounds and other factors. Colder temperatures decrease KHT solubility and accelerate crystal growth, and the potential for visible crystal formation increases upon chilling. Thus, the metastable state could be lost (and KHT crystallization observed) at a consumers' refrigerator temperature. The resulting KHT

[1] Note that stability does not imply that a product will have zero risk of undesirable changes. Similar to safety, product stability should be thought of as a probability of failure, not the impossibility of failure.

Understanding Wine Chemistry, First Edition. Andrew L. Waterhouse, Gavin L. Sacks, and David W. Jeffery.

crystals are harmless,[2] but because of their resemblance to broken glass they place a product at risk of consumer rejection. Several routine tests have been developed to predict if a wine is *cold-stable*, and most wineries will attempt to *cold-stabilize* their wines through one or more methods prior to bottling and commercial sale, particularly for white wines where refrigeration is more common and the resulting crystallization more evident.

Although most studies of KHT metastability focus on the phenomenon during cold storage, KHT precipitation can occur at room temperature due to changes in wine chemistry that are either intentional (e.g., pH adjustments) or spontaneous (e.g., precipitation or degradation of inhibiting compounds). Beyond KHT, other salts of organic acids can precipitate in bottle, particularly calcium tartrate (CaT).

26.1.2 KHT crystal properties and solubility

KHT forms orthorhombic crystals when precipitated from water or model hydroalcoholic solutions. Orthorhombic crystals appear as cubic crystals stretched along two of three axes, and can have several characteristic (*idiomorphic*) shapes (Figure 26.1.1a), including rectangular prisms and diamonds (hence the term "wine diamonds").

The solubility of KHT is defined by its solubility product (K_{sp}), the product of ion concentrations, and is a constant for a particular salt in a given matrix and temperature. In other words, K_{sp} can be used to calculate the maximum concentration of a salt that can be dissolved in wine or another solvent before saturation is reached. For KHT, the calculation can be expressed as follows:

$$K_{sp,KHT} = \left[K^+\right] \times \left[HT^-\right] \tag{26.1.1}$$

In real solutions, like wine, it is more appropriate to discuss the "activity" or "effective concentration" of a species rather than its concentration, where activity ≤ concentration ([X]). The activity coefficients, γ, are

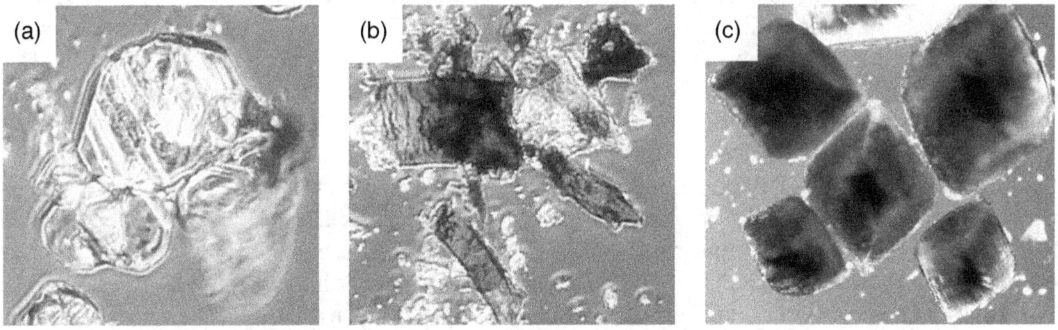

Figure 26.1.1 Light microscope images of (a) idiomorphic potassium bitartrate crystals from grape juice showing minimal co-precipitation and potassium bitartrate crystals from red grape juice and wine showing (b) irregular platelets and (c) rounding of crystal edges caused by phenolics and other wine components. (a) and (b) are from Alongi 2010 [1], reproduced with the permission of Journal of Agriculture and Food Chemistry. (c) is adapted from Anonymous 2015 [2], reproduced with the permission of Australian Wine Research Institute

[2] Although the dose eventually does make the poison. "This cream of tartar case is only another illustration of how necessary it is that some law should be passed for not allowing incompetent people to sell drugs…[it has been] only a few weeks since a grocer sold cream of tartar in mistake for arrowroot thus causing the death of a child," (*Chemist and Druggist*, February 15, 1890, Vol. 36, General and Provincial News: Incompetent drug sellers, p. 211).

calculated as the ratio of activity to concentration. A modified version of the solubility product can be expressed in terms of activities:

$$K'_{sp,KHT} = \gamma_{K^+}\left[K^+\right] \times \gamma_{HT^-}\left[HT^-\right] \tag{26.1.2}$$

For dilute concentrations of ions in pure water, $\gamma = 1$. In real solutions, γ is < 1 due to interactions of ions with oppositely charged ions, and lower values of γ are generally observed at higher ionic strength, a concept discussed in more detail earlier in the book (Chapter 3). For example, K^+ activity is approximately 20–30% lower (i.e., $\gamma = 0.7$–0.8) when measured by an ion selective electrode (which measures activity) as compared to $[K^+]$ measured by atomic absorption spectroscopy (which disrupts weak interactions and measures concentration) [3]. K^+ may form weak interactions with sulfate and anthocyanins [4], while HT^- or T^{2-} may weakly complex with cations in solution, for example, Ca^{2+} to form solubilized CaT [5].

In practice, K_{sp} is determined by adding excess KHT to a solvent to create a saturated solution, measuring $[K^+]$ using an appropriate technique (e.g., atomic absorption or emission spectroscopy), and taking the square root to calculate K_{sp} (since $[K^+] = [HT^-]$ upon dissolution of any KHT). Thus, most literature K_{sp} values of KHT incorporate any effects of the species under investigation (K^+, HT^-) on activity, although they will not account for effects of other ions.

The K_{sp} value for KHT in H_2O at room temperature is 1.07×10^{-3}, which equates to a solubility of 6.2 g/L. Tables of K_{sp} values and KHT solubility as a function of temperature and ethanol content can be found in the literature [6], and representative values are shown in Table 26.1.1. Key factors governing KHT solubility are:

- K_{sp} and solubility decrease with increasing ethanol concentration. The solubility of KHT is usually about 2–3-fold lower in model wine solutions than in water.
- Colder temperatures decrease K_{sp} and solubility. Typically, solubility decreases by a factor of 2–3 when a wine is chilled from 20 °C and 0 °C.

26.1.2.1 Concentration products and KHT metastability

For individual wines, it is possible to calculate the concentration product (CP) in a manner analogous to calculation of K_{sp}.[3]

Table 26.1.1 KHT solubility in representative matrices, expressed as the solubility product (K_{sp}) and as solubility in g/L (top), and typical concentration product (CP) values for cold-stable wines

Matrix	K_{sp} (mol²/L²)	Solubility (g/L)
Water, 20 °C	107×10^{-5}	6.2
12% (v/v) ethanol in water, 20 °C	16×10^{-5}	2.4
12% (v/v) ethanol in water, 0 °C	2.9×10^{-5}	1.0
Matrix	**Average CP (mol²/L²)**	
White wine, stable at 0 °C	4.1×10^{-5}	
Red wine, stable at 0 °C	7.7×10^{-5}	

[3] Concentration products (CPs) are an example of reaction quotients (Q), which can be compared against equilibrium (K) values to predict the direction a reaction will proceed in. General chemistry textbooks typically use Q rather than CP when discussing saturation and other equilibrium phenomena, but the CP terminology persists in usage in many wine texts.

$$CP_{wine} = \left[K^+\right]_{wine} \times \left[HT^-\right]_{wine} \quad (26.1.3)$$

These CP values can then be used to evaluate the susceptibility of a wine to precipitating KHT, such that:

- If $CP > K_{sp}$, the wine is supersaturated, and further precipitation of KHT is *possible*, although not guaranteed for reasons described later.
- If $CP = K_{sp}$, the wine is saturated, and no additional precipitation or dissolution is possible.
- If $CP < K_{sp}$, the wine is not saturated, and further dissolution of KHT would be possible (i.e., if it is added).

As mentioned above, $[K^+]$ is relatively straightforward to measure directly by spectroscopic methods. The concentration of $[HT^-]$ is usually not directly measured, but instead approximated by measuring total tartaric acid species, [Tartrate],[4] and then correcting for the fraction expected to exist as the HT^- species based on pH and ethanol determinations for the wine. Tables reporting $\%HT^-$ as a function of pH and ethanol concentration exist in the literature [6]. From these measurements, CP is then calculated:

$$CP = \left[K^+\right] \times \left[\text{Tartrate}\right] \times \%HT^- \quad (26.1.4)$$

Surprisingly, the CP values of stabilized commercial wines are comparable to or greater than K_{sp} values for KHT in analogous ethanol–water model solutions (Table 26.1.1), indicating that most wines are supersaturated in KHT and exist as metastable KHT solutions. The challenge to a winemaker is to ensure that this metastability persists until the point that the consumer opens the bottle – that is, several months at least, and potentially a number of years.

26.1.3 Critical factors for KHT precipitation

The process of formation of a visible KHT precipitate during storage, as summarized in Figure 26.1.2, involves three variables:

- Supersaturation based on *ion activities*
- Nucleation
- Crystal growth

26.1.3.1 Ion concentration versus ion activity

The solubility product is properly defined based on the product of ion activities, but, as mentioned earlier, ion activities are rarely measured during wine analyses even when feasible.[5] In the case of HT^-, appropriate techniques for measurement of activity in wine rather than concentration have not been described. The fact that concentration is greater than activity is one reason measured CP values can exceed K_{sp} values.

[4] Tartrate can be measured in several ways, including by HPLC, and by visible spectroscopy using a metavanadate reagent.

[5] The one routine exception is ion selective electrodes such as a pH probe, because pH is technically a measurement of proton activity rather than proton concentration.

26.1.3.2 Induction period and nucleation

KHT supersaturation creates the possibility of precipitation, but does not guarantee its occurrence. KHT crystals must first *nucleate* – that is, form crystals of sufficient size that they can grow further. Initially, a supersaturated solution will form small "embryos" of KHT crystals [7]. Formation of these embryos is usually energetically unfavorable – the small crystals will have a high surface-to-volume ratio and the surface KHT molecules will not be as tightly bound as KHT molecules in the interior.[6] As a result, these embryos will be more likely to dissolve than to grow (Figure 26.1.3). However, a small percentage of embryos may eventually

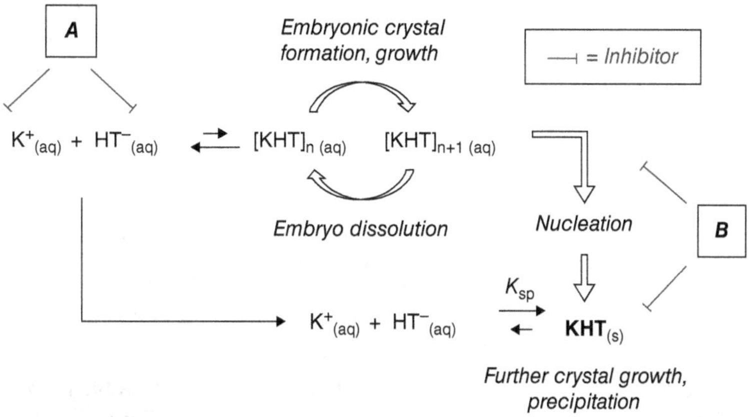

Figure 26.1.2 *Summary of steps necessary for KHT precipitation. Even when a wine is supersaturated with KHT, precipitation may be inhibited by other wine components (A) that decrease K^+ and HT^- activities or (B) that inhibit KHT nucleation and/or crystal growth*

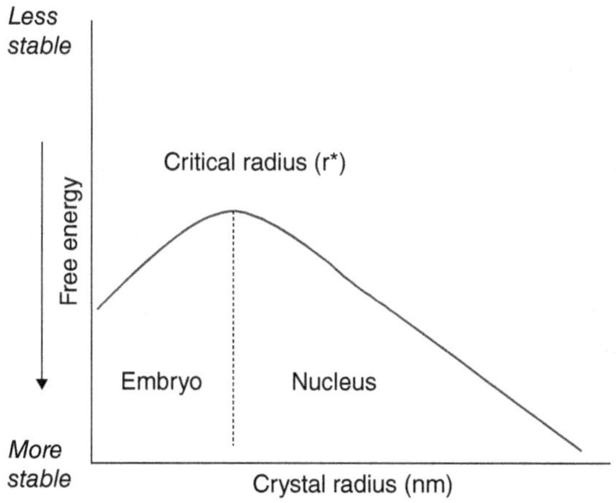

Figure 26.1.3 *A representation of free energy versus crystal size. Growth of crystals smaller than the critical radius is energetically unfavorable*

[6] The difference in free energy between molecules at the surface and interior is referred to as "interfacial free energy" or "surface free energy" in chemistry texts.

get large enough (i.e., have a sufficiently small surface-to-volume ratio) such that continued growth will result in a decrease in free energy. This minimum size is referred to as the *critical nucleus radius* and the transition is referred to as *nucleation*. The time prior to nucleation is referred to as the *induction period*, although commonly the induction time is determined by when crystals become visible and therefore includes both the induction period and subsequent crystal growth. In practical terms:

- The rate of nucleation increases as the exponent of $(-1/\sigma^2)$, where σ is the degree of supersaturation (calculated as $\ln(CP/K_{sp})$). Chilling wine decreases solubility of KHT and results in an increase in the nucleation rate and rate of KHT loss by increasing σ [8].
- In pure solutions, KHT nucleation can occur due to spontaneous growth of embryos in the bulk solution ("homogeneous nucleation"). However, it is much more common in wine and other systems for nucleation events to occur along surfaces ("heterogeneous nucleation") rather than in the bulk solution. This minimizes the crystal surface area and decreases the likelihood of redissolution of embryos. Thus, KHT crystals are frequently found in imperfections along the walls of tanks or barrels, or on components adhering to these walls, for example, yeast lees[7] [9].
- Addition of KHT seed crystals ("contact seeding", i.e., the contact process) is frequently used to expedite crystal growth. Assuming the same mass is used, smaller seed crystals, for example, 20 μm in place of 100 μm crystals, result in faster loss of KHT due to their greater surface area.

26.1.3.3 Crystal growth and surface fouling

Critical nucleus radii are typically on the order of 1–100 nm [7]. These are invisible to the human eye and are also several orders of magnitude smaller than typical crystal sizes recovered in wines. Following nucleation, KHT crystals are hypothesized to grow primarily by the separate addition of solvated K^+ and HT^- ions, rather than addition of neutral KHT units to the crystal [10]. The ions are first weakly adsorbed on to the crystal surface before they are eventually incorporated into the lattice, with incorporation of HT^- slower than K^+ due to its greater size and degree of solvation, and slower dehydration. Several factors will affect crystal growth rates:

- The rate of crystal growth in wines and juices is enhanced by agitation [11], either by facilitating transfer of ionized species to the surface or by facilitating dehydration (i.e., removal of solvent from the crystal surface).
- As with nucleation, higher degrees of supersaturation will yield faster crystal growth.
- Crystal growth can be slowed by crystal surface fouling or the adsorption of a wide range of molecules to KHT surfaces that inhibit further growth.

Crystal fouling is a major cause of the metastability and long induction periods of wines as compared to ethanolic solutions. Several species have been implicated for their ability to adsorb to crystal surfaces through ionic, hydrogen-bonding, and/or charge-transfer interactions [10, 12], as follows:

- Endogenous phenolics such as anthocyanins, which explains the greater KHT holding capacity of red wines [13].
- Endogenous macromolecules (e.g., proteins, polysaccharides), particularly mannoproteins, which explains the greater KHT holding capacity of sparkling and *sur lie* (lees) aged wines [14].

---

[7] A more rapid and enjoyable example of heterogeneous nucleation can be seen with the addition of a Mentos candy to a bottle of soda, the results of which can be found through an online search. The high surface area of the candy provides a large number of nucleation sites for the supersaturated CO_2 gas.

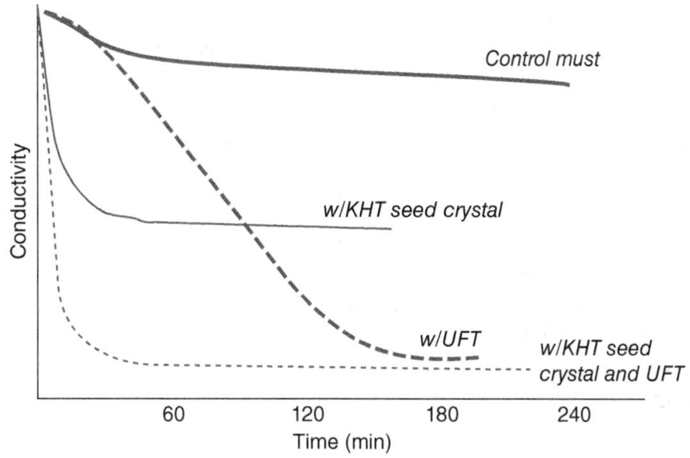

Figure 26.1.4 *The effect of KHT seeding and macromolecule removal (achieved through ultrafiltration, UFT) on KHT precipitation from grape must. A decrease in conductivity indicates that K^+ and HT^- ions have been lost from solution. Addition of 4 g/L of seed crystal results in a large increase in the rate of precipitation and decreases the time necessary to achieve equilibrium. UFT removes macromolecular colloidal compounds that can inhibit crystal growth past a certain size. Based on data from Reference [15]*

- Exogenous components added for the purpose of stabilizing the wine (mannoproteins again, as well as gum arabic, carboxymethylcellulose, and metatartaric acid), as mentioned later in this chapter.

Many of these stabilizing agents are macromolecules (e.g., polysaccharides), which will form colloidal dispersions both in isolation and following binding to KHT; as a result, it is common in the literature to see these stabilizers referred to as "protective colloids".[8]

In addition to slowing crystal growth, adsorption of fouling compounds will also change KHT crystal morphology due to the preferential binding to one or more of the crystal faces (Figure 26.1.1). For example, the presence of mannoproteins and other wine polysaccharides results in a rounding of crystal faces in white wines [10], while the presence of anthocyanins results in irregular platelets [9]. In fact, the amount of desirable wine components lost due to co-precipitation can be sizable. For example, KHT crystals from Carignan wines are reported to contain 0.2–0.3 % w/w anthocyanins and 1.9–2.5 % w/w tannin [9], and cold-stabilization of Concord grape juice can yield KHT crystals with >1% w/w anthocyanins and result in >30% loss of anthocyanins from the juice [1].

The effects of seed crystals and protective colloidal compounds on the rate and extent of KHT precipitation can be seen in Figure 26.1.4.

26.1.4 Testing for KHT stability

Prior to bottling, most commercial wines are assessed for their potential to form precipitates during storage, and particularly during refrigeration [16]. As described above, it is not appropriate to evaluate cold-stability by comparing CP values to K_{sp} due to the presence of crystal-inhibiting compounds in wine. However, the stability of a wine can be predicted by comparing its CP values to empirical observations generated from a large number of

[8] As a reminder, a colloidal solution is a homogeneous solution of large macromolecules or small particles that does not settle out to form two separate phases.

comparable wines – effectively an actuarial approach to predicting cold stability. These maximum CP values will vary with wine type, for example, red wines can have higher CP values due to the presence of phenolic species (Table 26.1.1). The maximum CP values recommended for stability vary with wine style, region, vintage, and variety, but as representative data, CP values for California wines were recommended to be $<16.5 \times 10^{-5}$ for table whites and $<30 \times 10^{-5}$ for table reds [17]. For lesser known wines or locations, specific CP limits may be unavailable.

An alternate approach to test for tartrate stability is to subject a wine to forcing conditions – that is, expose the wine to conditions that will accelerate the rate of KHT crystal growth. A widely used test involves agitating a wine sample at 0 °C in the presence of KHT seed crystals and measuring changes in conductivity at 5 min intervals for up to 30 min. A conductivity change of > 5% is reportedly indicative of likely KHT instability [17]. Several other variants exist, including visually inspecting for crystal formation rather than conductivity, omitting seeding, using longer or shorter hold times, or using cooler temperatures (Table 26.1.2).

As with other stability tests (e.g., protein stability, Chapter 26.2), the goal of these forcing tests is to predict long-term behavior of the wine (months to years) in an accelerated period (minutes to weeks). These run the risk of errors:

- False positives, especially if the testing conditions are more extreme than would be encountered post-bottling (e.g., seeding, freezing). This may lead to unnecessary time, energy, and material resources committed to stabilizing a wine that is at low risk for KHT precipitation.

Table 26.1.2 Approaches to testing for KHT cold stability

General approach	Examples or variations [16–20]	Notes
Compare wine CP value to recommended CP limits	Published tables for certain regions. Commercial labs may use proprietary values	Empirically derived predictions. Data not available for all regions or wine styles. Requires measuring [K^+], [Tartrate], pH, and ethanol
Chill wine, check for crystal formation ("cold-hold test")	Several variants exist: • Temperature: typically –4 to 0 °C • Time: days to weeks • Other variables: agitation	Simple, no specialized equipment required. Common in wineries. May be excessively stringent
Chill wine, monitor for change in conductivity	Several variants exist: • Temperature: typically, –4 to 0 °C • Time: minutes to hours • Other variables: often performed with seeding ("mini-contact") and agitation	Faster and more sensitive than visual inspection, but requires conductivity meter
Determine minimum temperature necessary for KHT dissolution ("saturation temperature," T_{sat})	Typically, a control sample and sample with added KHT are heated at a controlled rate, and T_{sat} is determined by when conductivity of the treated sample exceeds the control Variants: • Heating rate (°C/min) • T_{sat} limit: typically > 12.5 °C (white) and 22 °C (red) indicative of instability	Less dependent on crystal size and other parameters than mini-contact tests. Requires conductivity meter
Freeze wine, thaw, check for crystal formation	Conditions poorly defined. Overnight storage is common. Freezer temperatures can vary from –10 °C to –27 °C resulting in variation [20]	Common in wineries. Due to change of state, potential for false positives and negatives in predicting behavior under refrigeration conditions

- False negatives, either because of differences in physical state between testing and real conditions, or because other changes may occur to the wine during long-term room-temperature storage (e.g., loss of inhibitory compounds).

In summary, cold-stability is *not* an absolute concept; rather, cold-stability tests approximate the risk of a given wine to form visible KHT precipitates over a particular storage regimen, and the winemaker must balance this risk against the costs of preventative treatments.

26.1.5 Treatments for preventing KHT precipitation

If a wine is determined to be at sufficient risk for KHT instability before consumption, there are several approaches a winemaker can use to decrease the likelihood of this occurring, as summarized in Table 26.1.3, and reviewed in more detail elsewhere [16, 21–23].

Table 26.1.3 Strategies for achieving cold stability in wineries

General approach	Examples or variations	Notes
Decrease ion concentrations	Cold storage to induce KHT precipitation: • Often with seeding and agitation • Batch or continuous Nanofiltration to preconcentrate K^+ and HT^- and favor KHT precipitation	Batch process is simple, has the lowest wine losses of any technique, and uses readily available winery equipment. Energy intensive, although possibility for energy recovery. Ineffective for CaT instabilities. Equipment for continuous processes or nanofiltration can be expensive
	Ion exchange resins: • Polymeric resins capable of exchanging like-charged species • Most commonly as cation exchange (exchange H^+ or Na^+ for K^+, Ca^{2+}) • Less common: anion exchange (exchange citrate or OH^- for tartrate ions)	Resins must be regenerated with strong acid (or acid and NaCl) for re-use. Potential for non-selective losses of flavor and color compounds. May be other effects on sensory properties due to pH and TA changes
	Electrodialysis • High-pressure flow cell with membranes selectively permeable to small charged species	Expensive equipment. Minimal effect on red wine color compared to cold stabilization. Slight changes to [EtOH], TA and pH
Inhibit nucleation or crystal growth to maintain metastability	• Polysaccharides (carboxymethylcelluose, gum arabic, mannoproteins) • Peptides (polyaspartate) • Metatartaric acid	Treatments can be relatively expensive and dosage rate can be challenging to determine. Potential for changes to filterability, especially with polysaccharides Metatartaric will hydrolyze to tartaric acid during wine storage

26.1.5.1 Cold storage

The most straightforward means to decrease the chance of post-bottling cold instability is to subject the wine to forcing conditions – that is, store the wine at close to its freezing point to induce loss of KHT [18]. Subject to the prices of resources like energy and the price of the wine, cold storage may also be the most economical [22]. In addition to colder temperatures, the rate of precipitation can be enhanced by agitation and, in particular, the introduction of KHT seed crystals [24], as discussed above. Precipitated KHT can be pulverized and re-used to treat subsequent batches of wine.

Cold treatments generally yield wines that still have CP > K_{sp}, and thus continue to be metastable and at potential risk for KHT precipitate formation during subsequent storage. Further decreases in CP can be achieved by physiochemical treatments to remove K^+ and/or HT^-.

26.1.5.2 Ion-exchange resins

Polymeric resin beads (e.g., styrene–divinylbenzene copolymer) containing functional groups capable of exchanging like-charged ions can be used for decreasing ion species [21]. In winemaking operations, most resins are cation-exchange – that is, they contain negatively charged residues (sulfonic acid groups, $-SO_3H$, or carboxylic acid groups, –COOH, respectively giving strong and weak exchange capacities). A cation exchange resin can be prepared by washing it with strong acid prior to use, and wines will then exchange K^+ for H^+ in the cartridge. Alternatively, if no change in pH or TA is desired, the cartridge can be rinsed with a solution of a metal salt (e.g., NaCl or $MgCl_2$) prior to use to allow exchange of K^+ for Na^+ or Mg^{2+}, respectively [25].[9] The relative affinity of metals for resins varies between strong and weak exchanges.

- Weak exchangers generally show a preference for higher charge states ($Ca^{2+} > K^+$). Within a charge state, smaller ions (higher on the periodic table) are preferred, that is, greater charge density.
- Strong exchangers generally show a preference for lower charge states ($K^+ > Ca^{2+}$). Within a periodic table column, the higher number element will be preferred since they will have a smaller hydrated diameter.

An example specification sheet for a commercial cation-exchange resin is shown in Table 26.1.4. Frequently, these resins are sold to winemakers pre-packaged in cartridges. A critical parameter for evaluating the cost-effectiveness of a resin is its total exchange capacity expressed as charge equivalents per L of resin. The resin shown in Table 26.1.4 has a capacity of 1.7 eq/L; based on charge and atomic (or molecular)

Table 26.1.4 Example data sheet for a commercial cation exchange resin (Dow Amberlite®)

Matrix	Crosslinked polystyrene
Functional groups	Sulfonic acid
Physical form	Light grey beads
Ionic form as shipped	H^+
Total exchange capacity	=1.7 eq/L (H^+ form)
Harmonic mean size	0.600–0.800 mm

[9] Cation-exchange resins are typically found in homes as part of water-softening units. In this case the primary metals targeted are usually the divalent metals Ca^{2+} and Mg^{2+}, which can interfere with soap lathering and generate insoluble scales. These cations are exchanged for Na^+ on the resin.

weight this equates to either 66 g/L of monovalent K^+ cation (i.e., 1.7×39) or 34 g/L of the divalent Ca^{2+} cation (i.e., 1.7×40/2).

In practice, to avoid the need for precise calculations, it is common to treat only a portion of juice or wine with an excess of cation exchange resin. Under these conditions, >75% of metals (K^+, Ca^{2+}, etc.) will be removed, resulting in an increase in TA in the treated juice corresponding to the initial concentration of metal cation equivalents and a decrease in pH to near 2 [26]. This treated portion can then be back-blended. Because cation-exchange resins contain both hydrophobic and negatively charged regions, they can cause non-selective losses of other wine components, for example, the flavylium form of anthocyanin pigments or flavor compounds.

26.1.5.3 Electrodialysis

Another approach to decreasing ion concentrations is electrodialysis (ED), a technique widely used in food processing for removal or concentration of ionic compounds in foods or waste streams [27, 28].[10] An ED device consists of a central cell containing the product to be electrodialyzed (e.g., wine) separated from anode and cathode compartments, in which water flows, by anion- and cation-selective membranes. When an electric field is applied, ions in wine will migrate towards the appropriate electrode [26]:

- Bitartrate and tartrate will migrate through the anion-permeable membrane towards the anode. Sulfate (SO_4^{2-}) will be lost preferentially due to its small size and double charge. Malate/bimalate and lactate will also migrate, but to a lesser extent due to their higher pK_a values.
- Cations, primarily K^+ but also Mg^{2+}, Ca^{2+}, and others, will migrate towards the cathode through a cation-permeable membrane.

Similar to ion exchange, ED processing is typically done in a batch manner with a portion of wine being treated until the desired conductivity is reached, and then discharged and replaced with another batch. The net effect of electrodialysis is superficially similar to cold-stabilization – there will be a decrease in KHT. However, ED (like ion exchange) will remove Ca^{2+} and other metals as well [29], and thus is potentially more effective at preventing CaT instability. In addition, the removal of sulfate may affect the stability of other wine components, such as proteins (Chapter 26.2) and the elimination of some T^{2-} in addition to HT^- results in a slight pH decrease, ~0.2 pH units, in contrast to cold stabilization where either a decrease or increase can occur (Chapter 3).

Finally, a variation on conventional electrodialysis is bipolar electrodialysis (BPED), which incorporates bipolar membranes capable of allowing both anionic and cationic exchange [28]. BPED systems can achieve H^+ and OH^- exchange, which means that they can be designed to selectively remove either free acids (to prevent a pH decrease and minimize TA) or metal hydroxide salts (to increase TA and decrease pH, similar to cation exchange resins). The approach is widely used in the fruit juice industry and has also been applied to wine [30].

26.1.5.4 Additives for preventing nucleation and crystal growth

Certain additives can be used to prevent KHT crystal formation [14] (Table 26.1.3), although their legality varies among countries (Chapter 27). Generally, these work by inhibiting nucleation or fouling crystal faces to prevent formation of visible crystals, as described earlier. Many of these additives are large macromolecules, especially polysaccharides.

[10] Other examples of applications in the food industry include demineralization of whey and concentration of plant proteins.

26.1.6 CaT and related precipitates

While KHT precipitation is the most common (and best studied) cause of in-bottle crystalline deposits, others are occasionally observed. For example, calcium oxalate precipitation has been reported to result following addition of tartaric acid contaminated with oxalic acid, and calcium mucate has been reported in wines with botrytis infections. High concentrations of ferric ions can lead to ferric phosphate precipitation (*iron casse*) as described elsewhere in this book (Chapters 4 and 26.2).

A more common crystalline instability is CaT, most commonly observed when calcium salts are used for deacidification (Chapter 3) or when wines are stored in concrete tanks (Chapter 4). The solubility of CaT is lower than KHT, 0.1 versus 2.4 g/L at 20 °C in 12% v/v ethanol. Similar to KHT, CaT often exists in supersaturation in wines, with $CP > K_{sp}$ by a factor of 2–5 [31]. Several components in wine are capable of inhibiting CaT nucleation or crystal growth, including malic acid and uronic acids (i.e., pectin) [32]. However, the occurrence of CaT instability is usually lower because $[T^{2-}]$ (i.e., divalent tartrate required to form CaT) is generally lower than $[HT^-]$ at wine pH, and because $[K^+]$ is usually 10-fold higher than $[Ca^{2+}]$ in grapes (Chapter 4). Additionally, Ca^{2+} activity in wine is decreased in the presence of T^{2-}, and a portion of CaT will be molecularly dispersed in solution without precipitating [5]. In contrast to KHT, the rate and amount of CaT precipitation from wine is virtually unaffected by temperature [33], and cold-storage is thus not an appropriate strategy for CaT stabilization or for testing CaT stability of wine. Seeding and agitation greatly increase the rate of CaT precipitation in wine [34], indicating that the nucleation rate is likely a limiting step, but seeding is not often used in wineries due to the unavailability of pure seed crystals. Ion-exchange resins or electrodialysis are appropriate ways to lower $[Ca^{2+}]$ and $[T^{2-}]$, but in practice the best way to avoid CaT precipitation is to avoid the use of calcium salts during wine production.

References

1. Alongi, K.S., Padilla-Zakour, O.I., Sacks, G.L. (2010) Effects of concentration prior to cold-stabilization on anthocyanin stability in Concord grape juice. *Journal of Agricultural and Food Chemistry*, **58** (21), 11325–11332.
2. Anonymous (2015) Hazes and deposits picture gallery. Australian Wine Research Institute. Available from: https://www.awri.com.au/industry_support/winemaking_resources/wine_instabilities/hazes_and_deposits/picture_gallery/.
3. Gerbaud, V., Gabas, N., Blouin, J., Laguerie, C. (1996) Nucleation studies of potassium hydrogen tartrate in model solutions and wines. *Journal of Crystal Growth*, **166** (1–4), 172–178.
4. Bertrand, G.L., Carroll, W.R., Foltyn, E.M. (1978) Tartrate stability of wines. I. Potassium complexes with pigments, sulfate, and tartrate ions. *American Journal of Enology and Viticulture*, **29** (1), 25–29.
5. McKinnon, A.J., Scollary, G.R., Solomon, D.H., Williams, P.J. (1994) The mechanism of precipitation of calcium L(+)-tartrate in a model wine solution. *Colloids and Surfaces A – Physicochemical and Engineering Aspects*, **82** (3), 225–235.
6. Berg, H.W. and Keefer, R.M. (1958) Analytical determination of tartrate stability in wine I. Potassium bitartrate. *American Journal of Enology and Viticulture*, **9** (4), 180–193.
7. De Yoreo, J.J. and Vekilov, P.G. (2003) Principles of crystal nucleation and growth. Reviews in Mineralogy and Geochemistry, **54** (1), 57–93.
8. Dunsford, P. and Boulton, R. (1981) The kinetics of potassium bitartrate crystallization from table wines II. Effect of temperature and cultivar. *American Journal of Enology and Viticulture*, **32** (2), 106–110.
9. Vernhet, A., Dupre, K., Boulange-Petermann, L., *et al.* (1999) Composition of tartrate precipitates deposited on stainless steel tanks during the cold stabilization of wines. Part II. Red wines. *American Journal of Enology and Viticulture*, **50** (4), 398–403.
10. Rodriguez-Clemente, R. and Correa-Gorospe, I. (1988) Structural, morphological, and kinetic aspects of potassium hydrogen tartrate precipitation from wines and ethanolic solutions. *American Journal of Enology and Viticulture*, **39** (2), 169–179.

11. Dunsford, P. and Boulton, R. (1981) The kinetics of potassium bitartrate crystallization from table wines. I. Effect of particle size, particle surface area and agitation. *American Journal of Enology and Viticulture*, **32** (2), 100–105.
12. Celotti, E., Bornia, L., Zoccolan, E. (1999) Evaluation of the electrical properties of some products used in the tartaric stabilization of wines. *American Journal of Enology and Viticulture*, **50** (3), 343–350.
13. Balakian, S. and Berg, H. (1968) The role of polyphenols in the behavior of potassium bitartrate in red wines. *American Journal of Enology and Viticulture*, **19** (2), 91–100.
14. Marchal, R. and Jeandet, P. (2009) Use of enological additives for colloid and tartrate salt stabilization in white wines and for improvement of sparkling wine foaming properties, in *Wine chemistry and biochemistry* (eds Moreno-Arribas, M.V. and Polo, M.C.), Springer, pp. 127–158.
15. Bott, E. (1986) Centrifugal separation of tartrate from wines stabilized by the contact process. *Wine Industry Journa*l, (August), 35–38.
16. Howe, P. (2013) Potassium bitartrate/calcium tartrate: cold stability of wines, Part 2. *Practical Winery and Vineyard*, (Winter), 34–39.
17. Zoecklein, B.W., Fugelsang, K.C., Gump, B.H., Nury, F.S. (1999) *Wine analysis and production*, Kluwer Academic/Plenum Publishers, New York.
18. Iland, P., Bruer, N., Ewart, A., Markides, A., Sitters, J. (2004) Monitoring the winemaking process from grapes to wine techniques and concepts. Patrick Iland Wine Promotions Pty, Ltd. Adelaide, SA, Australia.
19. dos Santos, P.C., Gonçalves, F., De Pinho, M.N. (2002) Optimisation of the method for determination of the temperature of saturation in wines. *Analytica Chimica Acta*, **458** (1), 257–261.
20. Leske, P.A., Bruer, N.G.C., Coulter, A.D. (eds) (1995) Potassium tartrate – how stable is stable? *9th Australian Wine Industry Technical Conference,* 1995, Adelaide, SA, Australia.
21. Lasanta, C. and Gómez, J. (2012) Tartrate stabilization of wines. *Trends in Food Science and Technology*, **28** (1), 52–59.
22. Low, L.L., O'Neill, B., Ford, C., *et al.* (2008) Economic evaluation of alternative technologies for tartrate stabilisation of wines. *International Journal of Food Science and Technology*, **43** (7), 1202–1216.
23. Bosso, A., Panero, L., Petrozziello, M., *et al.* (2015) Use of polyaspartate as inhibitor of tartaric precipitations in wines. *Food Chemistry*, **185**, 1–6.
24. Boulton, R.B., Singleton, V.L., Bisson, L.F., Kunkee, R.E. (1999) *Principles and practices of winemaking*, Kluwer Academic/Plenum Publishers, New York.
25. Mira, H., Leite, P., Ricardo da Silva, M., Curvelo-Garcia, A. (2006) Use of ion exchange resins for tartrate wine stabilization. *Journal International Des Sciences de la Vigne et du Vin*, **40** (4), 223–246.
26. Escudier, J.L., Cauchy, B., Lutin, F., Moutounet, M. (2012) Acidification and tartaric stabilization. Technological comparison of ion exchange resins by extraction and ion membrane. *Progres Agricole et Viticole*, **129** (13/14), 324–332.
27. Hestekin, J., Ho, T., Potts, T. (2010) Electrodialysis in the food industry, in *Membrane technology*, Wiley-VCH Verlag GmbH & Co. KGaA, pp. 75–104.
28. Mondor, M., Ippersiel, D., Lamarche, F. (2012) Electrodialysis in food processing, in *Green technologies in food production and processing* (eds Boye, J.I. and Arcand, Y.), Springer US, pp. 295–326.
29. Gómez Benítez, J., Palacios Macías, V.M., Szekely Gorostiaga, P., *et al.* (2003) Comparison of electrodialysis and cold treatment on an industrial scale for tartrate stabilization of sherry wines. *Journal of Food Engineering*, **58** (4), 373–378.
30. Rozoy, E., Bazinet, L., Gagne, F., *et al.* (2013) Development of a deacidification of wine by a bipolar membrane electrodialysis: a feasibility study of the laboratory scale. *Bulletin de l'OIV*, **86** (986/987/988), 187–208.
31. Berg, H.W. and Keefer, R.M. (1959) Analytical determination of tartrate stability in wine. II. Calcium tartrate. American Journal of Enology and Viticulture, **10** (3), 105–109.
32. McKinnon, A.J., Williams, P.J., Scollary, G.R. (1996) Influence of uronic acids on the spontaneous precipitation of calcium L-(+)-tartrate in a model wine solution. *Journal of Agricultural and Food Chemistry*, **44** (6), 1382–1386.
33. Clark, J.P., Fugelsang, K.C., Gump, B.H. (1988) Factors affecting induced calcium tartrate precipitation from wine. *American Journal of Enology and Viticulture*, **39** (2), 155–161.
34. Abguéguen, O. and Boulton, R.B. (1993) The crystallization kinetics of calcium tartrate from model solutions and wines. *American Journal of Enology and Viticulture*, **44** (1), 65–75.

26.2
Fining

26.2.1 Introduction

In the production of wine and other beverages, *fining* refers to addition of material(s) to remove undesired substances from the beverage. During fining, chemical interactions between fining agent(s) and the targeted component(s) leads to an insoluble product that precipitates from the wine and is removed. Fining agents are usually classified as processing aids rather than additives (Chapter 27) since they are not expected to stay in the wine. In some cases fining is conducted to remove faults in flavor, color, or aroma of the wine, and in other cases the goal is primarily clarification (Table 26.2.1). Fining is rarely entirely selective, and desirable wine components may be partially removed, or "stripped," from the wine, particularly when excess fining agent is used. Excessive additions of fining agents may also lead to incomplete precipitation of the fining agent (*overfining*). Overfining can also cause a visible (and undesirable) haze, and can be a health concern for agents with allergenic potential, such that some jurisdictions require labeling when such agents have been used.

From a chemistry perspective, the key requirements for a useful fining agent are as follows:

- The fining agent must interact with the target compound to form an aggregate. Types of interactions include:
  - Covalent or ionic bonds, as is observed during copper fining of thiols[1]
  - Electrostatic bonds, as is observed between bentonite (negative charge at wine pH) and proteins (positive charge)
  - Hydrogen bonding, as can occur during fining of proteins (H-bond donor) and polyphenols (H-bond acceptor)
  - Hydrophobic interactions, as occurs between activated carbon and non-polar wine constituents.
- The fining agent should remove a reasonable portion of the targeted compound, but only in light of the following point.

[1] For simplicity, many practical wine texts imply that all interactions are electrostatic, that is, all fining agents and target compounds are either positively or negatively charged. While appropriate for an introductory level, this approximation is not useful in describing more complex behavior.

Understanding Wine Chemistry, First Edition. Andrew L. Waterhouse, Gavin L. Sacks, and David W. Jeffery.

Table 26.2.1 *Common fining treatments*

Winemaking problem	Agent	Issues
Excessive tannin and astringency in red wine	Proteins: gelatin, albumen, isinglass, casein, others	Residual protein, undesired effects on taste and color, allergic potential
Browning or bitterness in white wine	Proteins as above (especially casein and isinglass); polyvinlpolypyrrolidone (PVPP)	Residual protein as above, potentially poor clarification with PVPP
Red color removal for certain white wines (e.g., Pinot Gris, sparkling wines made with Pinot Noir)	Activated carbon	Non-selective, removes other components
Potential for hazy wine due to heat-unstable proteins	Bentonite	Metal contamination from bentonite, waste disposal
Off aroma from hydrogen sulfide or volatile thiols	Copper salts (especially sulfate)	Residual copper, reappearance of off aromas, wine oxidation
Off aromas – general	Activated carbon	Non-selective, loss of other flavors
Poor settling during fining, or visible hazes	Silica gel or kieselsol (*co-fining* agents)	Wine pH will affect surface charge

- The fining agent should display selectivity towards the targeted compound as compared to other organoleptically important compounds. Selectivity may be characterized by a separation factor, or the ratio of the distribution ratios for the targeted and untargeted compounds. In other words, if the fining agent simply absorbs all organic substances, its utility is limited by its poor selectivity.
- The resulting aggregate should have low solubility to ensure precipitation, which is usually accomplished by formation of larger complexes. In some cases, settling aids will be added to neutralize surface charge to allow small aggregates to coalesce, for example, addition of alginates (to neutralize positive surface charge).
- The precipitate should have a density considerably different from wine (usually larger) and a compact structure to favor rapid settling and minimize wine losses. This separation can be facilitated by use of settling aids, for example, SiO_2 (to increase density). Less commonly, flotation using an inert gas like CO_2 can be used to bring the precipitate to the surface to facilitate its removal.

26.2.2 Tannin fining with proteins

During red wine production, the maceration process can result in extraction of excessive tannin or phenolics that contribute to astringency (Chapter 21). The removal of tannin from red wine is generally carried out by addition of protein to yield an insoluble precipitate, decreasing both tannin and perceived astringency. Protein addition may also be used for removing phenolics from white wines, especially in press fractions where they can lead to bitterness or browning.

Several different proteins are employed for removal of tannins and other phenolics, most of animal origin. A common thread of all of these proteins is that they tend to be high in proline and hydroxyproline, which increases their tannin-binding efficiency as described later in the chapter [1].

- Gelatin is perhaps the most widely used proteinaceous fining agent. Gelatin is produced by partial hydrolysis of collagen, the main structural protein in skin, tendon, and bone. Because of its production method, gelatin does not have a well-defined chemical structure, but can be classified by its mean molecular weight and gelling strength (bloom number). Typical wine fining applications use gelatin with slightly higher molecular weight (40–50 kDa) and gelling strength ("bloom number" = 100–150) [2] than would typically be used for production of confectionaries.

- Egg whites or albumen purified from the egg whites.
- Casein (or potassium caseinate) sourced from milk. Milk (or milk solids) can also be used, and will possess lipophilicity in addition to tannin binding character; thus, it is often used to co-remove off-odorants or other hydrophobic compounds.
- Isinglass, derived from the collagen-like substance in the swim bladder of certain fish.
- Animal blood (bloodmeal) has been used historically [3], though its use was prohibited after the origin of Creutzfeld-Jacob ("mad cow") disease was identified.
- Plant proteins (e.g., from wheat, peas, etc.). Due to the potential for allergies to some of the animal proteins, there is an effort to introduce the use of plant proteins, although some under consideration may also have allergenic issues as well.
- Finally, although technically not a protein, PVPP has structural similarities to proteins and is often used to remove lower molecular weight phenolics.

26.2.2.1 Mechanism of protein–tannin interactions

The mechanism for protein binding to tannin has been proposed to arise from both hydrophobic and polar interactions, and especially hydrogen bonding in the case of proanthocyanidins (Figure 26.2.1). Based on the work of Hagerman and Butler, and McManus and colleagues, Haslam concluded that the binding of tannins to proteins is based on several factors [4]:

- Hydrogen-bonding appears to be of greater importance for interactions of proanthocyanidins (condensed tannins, Chapter 14) and proteins, with hydrophobic interactions of secondary importance. The reverse appears to be the case for interactions of hydrolyzable tannins (Chapter 13) with proteins [5].
- Binding is stronger when one or both components has a high degree of flexibility, so as to facilitate multi-dentate binding. For instance, proteins possessing random coil conformations were more effective at

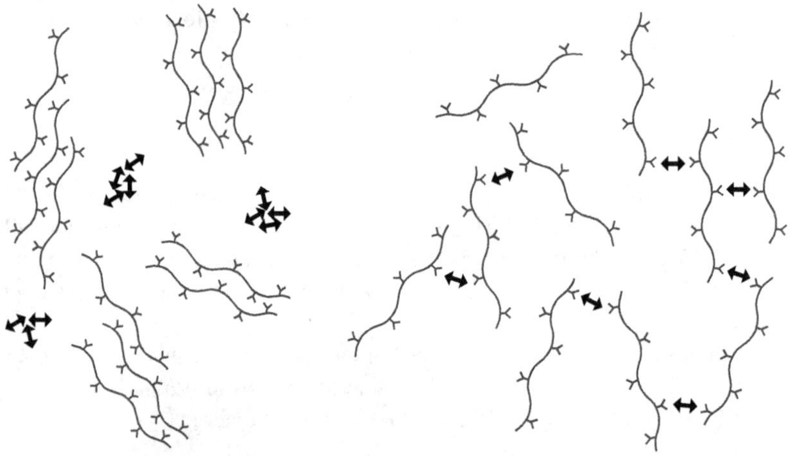

Figure 26.2.1 *Models showing the binding of epigallocatechin gallate to polyproline after 5 minutes (left) and 3 hours (right), as seen via cryo-TEM images. Reprinted from Food Hydrocolloids, Vol. 20, Poncet-Legrand* et al.*, Poly(L-proline) interactions with flavan-3-ols units, © 2006, p. 687, with permission from Elsevier [8]*

binding than tight globular proteins, and flexible polyphenolics, such as proanthocyanidins or pentagalloyl glucose, bind more tightly to proteins than rigid phenolics such as castalagin [6].

- The degree of polyphenol–protein interactions is highly correlated with the proline content of the protein. This appears to be because proline residues provide a site for hydrophobic interactions that initiate binding, which then facilitates hydrogen bonding between protein amide groups and phenolic hydroxyl groups [7].

Yokotsuka and Singleton treated different proanthocyanidin fractions with 15 g/hL gelatin, and observed that the highest MW fraction ("polymeric tannin") was best adsorbed (60% removed) [9], as compared to 48% of oligomeric tannins and only 8% of dimeric proanthocyanidins. The smallest oligomers are thus negligibly affected by fining with proteins [3]. Sarni-Manchado *et al.* noted that gelatin was fairly selective in removing larger, gallated proanthocyanidins [10]. Analogously, low molecular weight fractions (2–10 kDa) of gelatin precipitated only 55–70% of the amount of tannin precipitated by standard commercial gelatin (average MW = 70 kDa). Temperatures lower than 25 °C generally resulted in slightly better tannin removal (by ~20%). Typical treatments for gelatin are in the 2–15 g/hL range.

Increasing protein additions results in increased tannin precipitation. However, tannin fining with proteins does not follow Langmuir-type adsorption behavior, in which adding increasing amounts of protein would result in a limit to what the protein can adsorb. Instead, adsorption of tannins by proteins follows the Freundlich equation (Equation 26.2.1), where addition of tannin results in a decreasing fraction of tannin adsorbed, but no plateau in the amount of tannin bound to the protein (Figure 26.2.2).

$$\frac{x}{m} = K_F C^{1/n} \text{ or, in logarithmic form, } \log\frac{x}{m} = \log K_F + \frac{1}{n}\log C, \text{ as used in Figure 26.2.2} \qquad (26.2.1)$$

In the Freundlich equation, x = total solute concentration (tannin), m is the adsorbent concentration (protein), K is the Freundlich constant (a binding constant value), C is the equilibrium concentration of solute, and n is the Freundlich index for the adsorbent. Values of n less than 1 generally indicate cooperative binding, which is common for protein-polyphenol interactions.

Key parameters that affect adsorption include the pH of the wine and the pI or isoelectric point (Chapter 5) of the protein. Binding is at a maximum for most protein used in fining when it has a net neutral charge, when the pH is at or near its pI value. However, this is a general rule and the effective pH ranges for binding will depend on protein type [11]. For example, BSA binds only in a narrow pH range (±1.5 pH unit) near its pI, while pepsin, pI = 1.0, is effective from pH 2 to 7, and trypsin, pI = 10.1, has high tannin binding from pH 3 to 10.

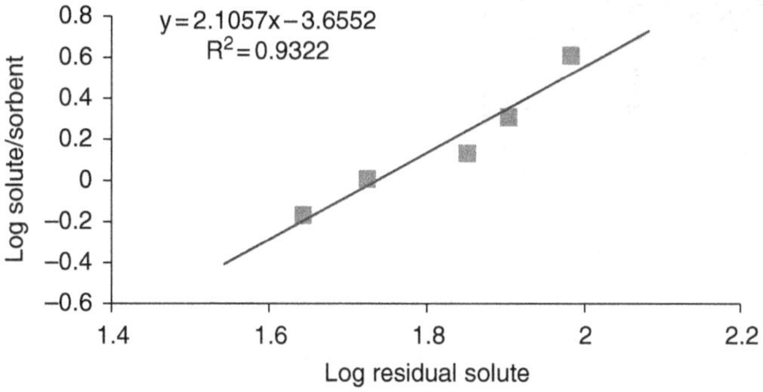

Figure 26.2.2 Freundlich plot of polymeric tannin adsorption by gelatin (data from Reference [9]). Under Freundlich conditions, a plot of log [solute]/[adsorbent] to log [residual solute] will show a linear correlation with the slope equal to the Freundlich index (1/n) and the intercept the log of the Freundlich constant. See Equation (26.2.1)

The production of white wine often includes fining treatments to remove phenolic components to reduce browning, astringency, or bitterness. One report showed that fining grape juice with protein (K caseinate) decreased polyphenol levels and browning potential when used in conjunction with bentonite and microcrystalline cellulose (as a commercial preparation), compared to bentonite alone [12], with few other effects aside from a slight decrease in protein content and lower concentrations of some volatiles. Casein appears to be much more effective in reducing turbidity, color, and browning potential than gelatin, isinglass, or albumen [13] – possibly because it is more hydrophobic than these other proteins and thus non-selectively removes other non-polar compounds. In white wine, none have significant effects on total phenolics or the hydroxycinnamates, the major phenolic fraction in whites (Figure 26.2.3).

26.2.2.2 Synthetic polymers for phenolic fining

Apart from naturally derived proteins, synthetic polymers with a polyamide structure similar to that of proteins have also been used to remove tannins and other phenolics. Due to its rigidity and porous nature, PVPP demonstrates selectivity towards smaller flavan-3-ols, such as monomers and small oligomers, through both hydrogen bonding and non-polar interactions (Figure 26.2.4). This behavior is in contrast to the preference shown by most proteins towards larger tannins, and PVPP can therefore be used to remove color or decrease browning potential in white wine [14, 15], although its practical application is challenging due to its poor settling characteristics [16]. A typical treatment level is 10 g/hL.

26.2.2.3 Co-fining agents

Protein and related fining agents possess positive surface charges that inhibit aggregation, and also have densities that are only slightly greater than wine (1.2–1.3 g/L versus 1 g/L), resulting in poor settling behavior. This problem can be remedied with co-fining agents, most commonly silica gel, kieselguhr (a different

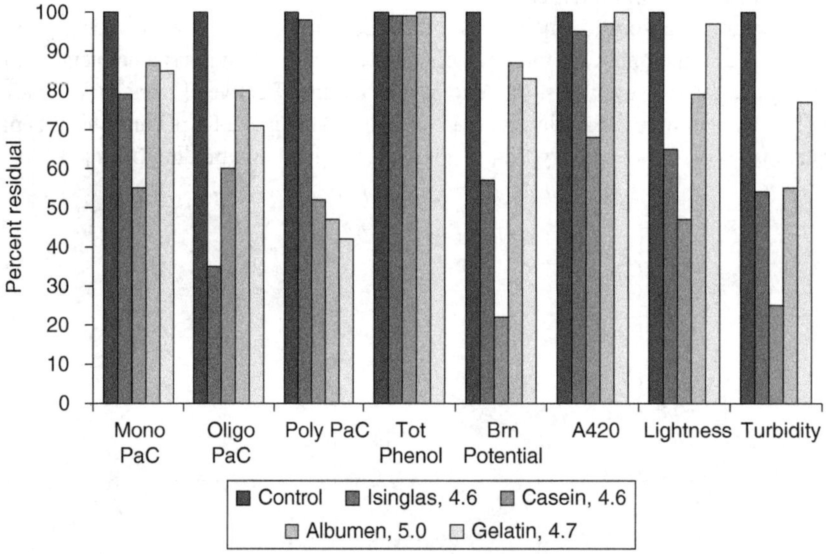

Figure 26.2.3 Effect of fining agents on white wine. Percent residual for monomeric, oligomeric, and polymeric proanthocyanidins, total phenol, browning potential, A_{420}, removed color in terms of lightness, and turbidity after protein fining treatment. Isoelectric point listed in legend for each protein. Data from Cosme et al., *2008 [13]*

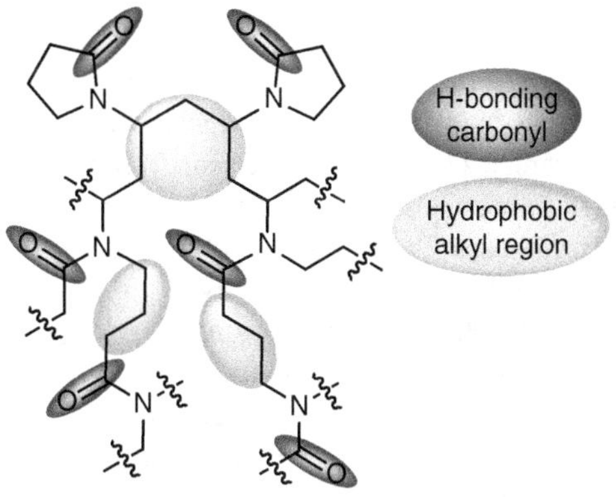

Figure 26.2.4 Hypothetical structure of PVPP, shown as a fragment (six subunits), with hydrophobic and hydrogen-bonding regions highlighted

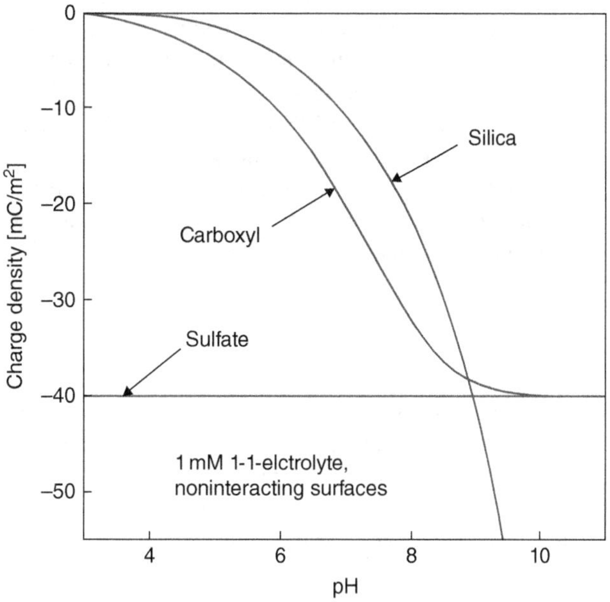

Figure 26.2.5 Surface charge of silica gel versus pH. Net charge is near zero at wine pH. Source: Behren 2001 [18]. Reproduced with permission from AIP Publishing LLC

preparation of silica gel), or polysaccharides such as alginates. Silica gel is often added to wine with or following proteinaceous fining agents, and the two are often combined in proprietary mixtures to improve settling of the fining agent [17]. While silica gel has a negative charge when prepared as kieselsol, the most common type, it appears that at wine pH, the charge is largely neutralized (Figure 26.2.5) [18]. Despite that, it improves flocculation and results in more effective precipitation, probably via hydrogen bonding. When the precipitation occurs, there is an additional clarifying effect due to entrainment of other small particles that are precipitated with the silica–protein complex. Typical treatments are based on commercial suspensions, at rates of 20–100 mL/hL of wine.

26.2.2.4 Selectivity

Proteins (and protein-tannin complexes) can bind to components other than tannin (Table 26.2.1). In general, color is reduced as a result of protein fining treatments, largely because the agents can adsorb pigmented tannin–anthocyanin conjugates (Chapter 24) [19] – greater adsorption is expected from more hydrophobic proteins. On the other hand, protein fining agents do not appear to affect the amount of protein in wine or alter the tendency for that protein to form hazes [20]. However, other reports show that fining with proteins can reduce levels of some aroma compounds by 10–20%, and even more of the glycosides of some important volatiles, potentially diminishing varietal character during aging [21]. Puig-Diu *et al.* observed large (25–49%) losses of esters, alcohols, and monoterpenoids when gelatin or bentonite (see below) was used as a fining agent [22]. Sanborn saw a few similar changes in aromatics, but a sensory panel was able to detect only a few differences between control and fined wines in a 26-attribute descriptive analysis [23].

26.2.2.5 Proteins as allergens

Proteins have allergenic potential, and allergies are known to the various common animal proteins used in fining wine. A trial where standard proteins were added to wines and provided to sensitive subjects elicited very weak responses, possibly because the wine's acidity denatures the proteins [24]. Regardless, there are now labeling mandates when using such fining agents in winemaking, for instance in Canada [25]. The basis for the labeling is the opportunity for residual proteins to persist when overfining occurs, first reported in 1981 [26]. A recent investigation using a sensitive ELISA assay was unable to detect albumen in a wine that had been normally treated with egg whites [27]. The authors rightly point out that labeling for allergenic content should be based on analytical testing for the presence of allergens. Some investigations have evaluated alternative plant proteins, and while some problems have been noted, such as off-taste, there are some promising results with pea protein [16] or palatin [28].

26.2.2.6 Inactivated yeast fractions (IVFs) as an alternative to proteins

The inherent association between tannins and other polyphenolics with cell wall mannoproteins [29] is the basis for investigations using IVF, or yeast hulls, as materials to remove phenolics or decrease their astringency in juice or wine. However, the effects observed to date have conflicting outcomes, with explanations that the mannoproteins are either stabilizing or precipitating tannins and pigments [30]. A review of the use of IVF materials demonstrated that the effect is yeast-strain dependent, and thus the outcome could potentially be difficult to predict, but could be modified by yeast selection [31].

26.2.3 Protein fining with bentonite

One issue with white (or rosé) wine stability is the cloudiness (haze) that can occur when wine is heated (Figure 26.2.6) [32].[2] This is a result of proteins (Chapter 5) in the wine denaturing and aggregating into larger particles that scatter visible light (the Tyndall effect). This is less of a concern for red wines, since extracted polyphenolic compounds will partially precipitate grape proteins during the winemaking process, and also because the haze is less evident in darkly colored wines. It was discovered early on that only

[2] This is usually inadvertent, such as from incorrect storage conditions or during transit. Formation of protein haze is primarily a visual nuisance as most consumers expect wines to be clear, irrespective of whether the wine may be abused once it leaves the retailer (e.g., placed in the trunk (boot) of a car on a hot day).

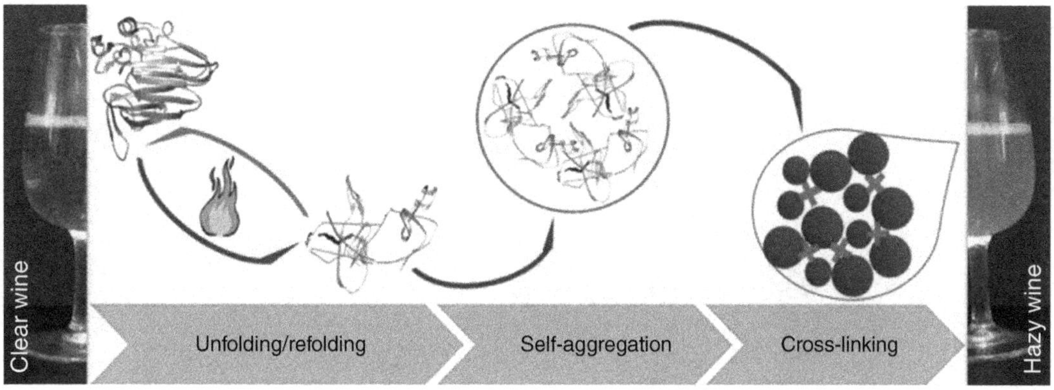

Figure 26.2.6 Mechanism for wine haze formation. Source: Van Sluyter 2015 [32]. Reproduced with permission from American Chemical Society

particular proteins contribute to this instability [33, 34]. More recent work points to proteins produced by the grape as a reaction to pathogens [35], particularly thaumatin-like proteins (TLP) and chitinases. Collectively, these are known as pathogenesis-related (PR) proteins and can persist through the winemaking process due to their resistance to acid hydrolysis and proteases [36].

Aside from the PR proteins, other factors can increase the extent of haze formation.

- The most critical factor appears to be sulfate concentration [37]. Sulfate is a *kosmotropic* ion (i.e. it interacts strongly with water, making the water less able to solubilize the protein), and high sulfate increases protein denaturation and aggregation.[3]
- Higher pH will facilitate aggregation, since PR proteins will be positively charged at wine pH (pI>4).
- High concentrations of phenolics (such as tannins, phenolic acids/esters, and small flavonoids) have also been shown to enhance haze formation [38].[4]

The widely applied remedy to eliminate PR proteins and prevent associated haze problems is to treat wines with bentonite clays, referred to as montmorillonites (the main active component).[5] Bentonites consist of hydrated aluminum silicate flakes bearing a net negative charge, with the incorporation of exchangeable cations within layers in the crystal structure. Those cations vary according to the source of the bentonite, with some regions predominantly having calcium (e.g., Germany), and others sodium (e.g., Wyoming). The cation content can affect final wine mineral content (Chapter 4), but also affect the performance of the treatment; sodium bentonites swell considerably more when made into a slurry due to the lower charge density of Na^+ compared to Ca^{2+} in the interlayers (Figure 26.2.7).

Bentonite can be viewed as a cationic (Na^+ or Ca^{2+}) exchange adsorbent that substitutes its cations for PR proteins, which will be positively charged at wine pH, as described previously. Bentonite contains a

---

[3] For similar reasons, ammonium sulfate gradients are widely used in biochemistry to induce precipitation ("salt out") different fractions of proteins.

[4] Complexes of polyphenols (particularly flavan-3-ol dimers) with proteins appears to be the major cause of beer haze [39].

[5] The use of related aluminosilicate clays (kaolin, Spanish clay) is frequently described in nineteenth century winemaking texts, although the application was to clarify hazy wines rather than in its current role as a prophylactic treatment. The use of bentonite for clarifying wine was first described in the literature by Saywell in 1934 [40], and it has largely displaced the use of other clays due to its higher adsorptive capacity.

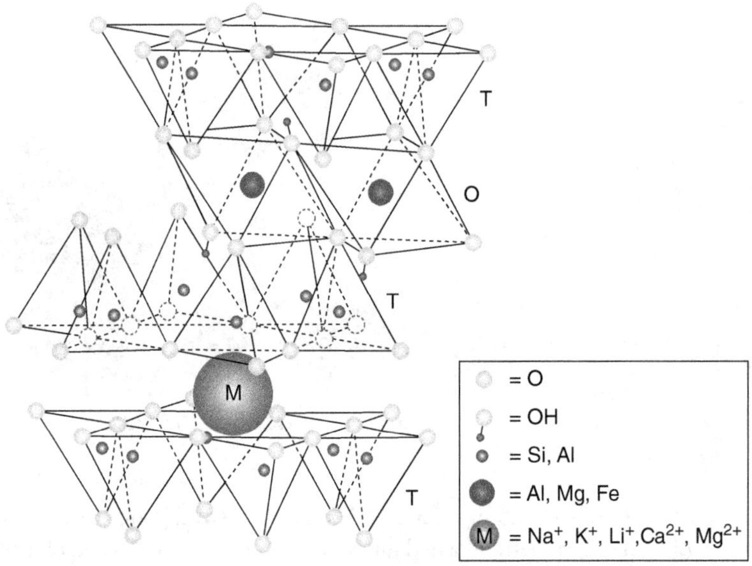

Figure 26.2.7 Schematic of the structure of montmorillonite, the mineral in bentonite that provides the adsorptive function. The exchangeable cation, Na^+, K^+, etc., is replaced by the protein. Source: Pusch 2012 [41]. Used under CC-BY http://creativecommons.org/licenses/by/4.0/

quantifiable number of binding sites with a particular binding affinity, and these values can be determined using models developed for enzyme kinetics [42]. Accordingly, protein adsorption most often follows the Langmuir equation with a clear saturation effect (Figure 26.2.8).

$$\frac{x}{m} = \left(\frac{x}{m}\right)_{\text{max}} \times \frac{C'}{K_L + C'} \tag{26.2.2}$$

The Langmuir equation for adsorption reaches a specific limit. K_L is Langmuir's constant, C' is the equilibrium concentration of the solute, x is the total amount of solute, m the concentration of adsorbent, and $(x/m)_{max}$ the maximum amount that can be absorbed. By plotting C' versus x/m, one can derive $(x/m)_{max}$ and K_L. Those values can be obtained from linear regression of a double reciprocal plot [42].

The different cationic forms of bentonite lead to differences in protein binding affinity. Beyond having a greater swelling ability, sodium bentonites can also adsorb a larger amount of protein for a given mass (e.g., about twice as much as bentonites containing calcium, Figure 26.2.8). On the other hand, calcium bentonites give more compact lees due to their lesser swelling. Typical treatment levels are 20–50 g/hL in white wines.

Sodium or calcium will be released during bentonite fining, as is expected based on the ion-exchange mechanism, which could potentially cause legal and flavor issues (sodium) or stability issues (calcium). Bentonites contain between 3 and 12% sodium and 3 and 17% calcium, depending on whether the sample is a calcium or a sodium type. The amount of sodium that appears in a wine as the result of treatment is reportedly 15–25 mg/L and 20–30 mg/L for calcium, when the wine is treated with high levels of 400 g/hL of sodium-containing bentonites [43], although others have used exceptionally high levels of treatment for experimental purposes, 2500 g/hL, and observed an increase of 300–400 mg/L of sodium [44]. Beyond that, these reports also show that other contaminant metals also appear (Chapter 4), such as lead, aluminum, or iron.

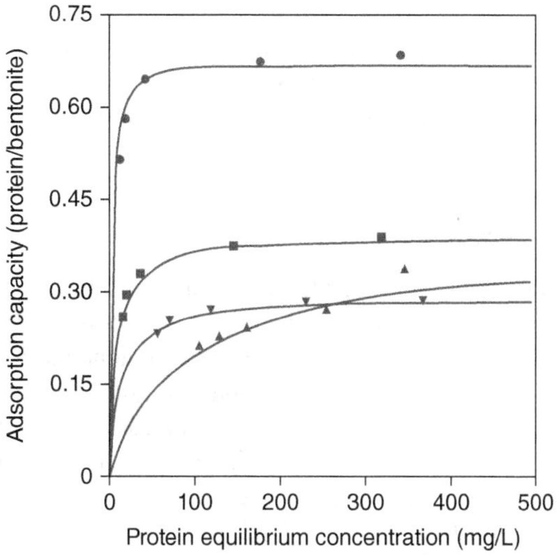

Figure 26.2.8 Normal Langmuir plots for the bentonite types: sodium (•), sodium/calcium (■), calcium/sodium (▼), and calcium (▲) forms. Source: Blade 1988 [42]. Reproduced with permission of AJEV

26.2.3.1 Selectivity

In addition to removing proteins, bentonite can strip other wine components in a non-selective fashion, but a review suggests only marginal effects that wine consumers might be able to discern when used at normal levels [35].

26.2.3.2 Alternatives to bentonite

Bentonite disposal is a problem for wineries and various alternatives for protein removal (including more effective uses of bentonite) have been proposed, but none have been adopted to any significant extent. A report reviewed potential substitutes for bentonite [32]. Proteases were considered promising when used with flash pasteurization, as there were few sensory changes and the resulting wines did not turn hazy. Carrageenan has some potential but it has some hazing potential itself, while chitin has several drawbacks. Zirconium oxide could function as substitute ion exchange material for bentonite, and it could be reused as well, but it must be tested in a commercial setting to establish viability.

26.2.4 Miscellaneous fining and related treatments

26.2.4.1 Copper fining and sulfurous malodors

Copper salts are used to treat some sulfur aromas in wine that have potent aroma effects. In particular, low molecular weight sulfur compounds including H_2S and thiols are particularly noxious (Chapters 10 and 30). H_2S will react with copper salts (such as $CuSO_4$) to form an exceptionally insoluble copper sulfide (CuS, $K_{sp} = 4 \times 10^{-36}$). In contrast to H_2S and thiols, copper will not form a complex with sulfides, disulfides, or thioacetates. Although CuS and related complexes are highly insoluble, they appear to remain dispersed in wine. Commercial white wines adjusted to ~1 mg/L Cu and equimolar H_2S showed essentially unchanged Cu content (>95% of original value) following filtering or 5 days storage and racking [45] (Chapter 24). Furthermore, copper complexes appear to be capable of regenerating H_2S and possibly thiols during storage [46]. Thus, copper additions are not necessarily a "fining" operation, as the copper remains, albeit in a different form.

26.2.4.2 Removal of transition metals

Historically, a number of methods have been used to remove iron and copper in order to avoid the formation of precipitates (*casses*) [47]. However, the universal usage of stainless steel and wood cooperage in wine production today has essentially eliminated the need for such treatments. During the middle parts of the twentieth century, the use of epoxy-coated steel tanks or concrete tanks with exposed steel reinforcement led to dissolved iron at levels of 5–20 mg/L [48]. At those levels iron casse (iron (III) phosphate [white casse] or tannate [blue casse]) could readily form in bottled wine, but with iron levels in modern times at 0.5–3 mg/L (Chapter 4), winemakers have largely forgotten about the problem or its treatment. The most effective treatment involves fining with potassium ferrocyanide, which forms an insoluble complex with metals such as iron and copper [47]. The treatment is complicated because the reagent has the potential to release cyanide if not used properly (i.e., if overfined), so its permitted use is tightly regulated. Ion exchange resins also have some utility in removing metals, which was discussed in Chapter 26.1.

26.2.4.3 Activated charcoal and non-polar sorbents

When a wine becomes contaminated with various undesirable components, such as aroma defects from rotten grapes (Chapter 18), a winemaker's last resort is often activated carbon or charcoal. Activated carbon is a non-specific adsorbent that has a very high surface area and can adsorb non-polar and polar substances. It will remove many components in addition to the problematic substance, so it is generally only used in small quantities or applied when a neutral tasting product is better than the untreated option [49]. Similar effects can be achieved through the use of non-polar resins or neutral oils (milk or cream), although these are not approved for use in winemaking in some regions (Chapter 27).

Although various taints can be removed by activated carbon or non-polar materials, the efficacy of these treatments is compromised by their poor selectivity. For example, 3-isopropyl-2-methoxypyrazine (IPMP, "MALB taint," Chapter 5) can be decreased by >30% with 0.2 g/L activated charcoal [49]. However, sensory testing revealed no change in the MALB taint character, likely because other desirable odorants (e.g., esters) with similar or greater hydrophobicity were also removed. For juice taints, these problems can be avoided by treating the juice with a non-polar sorbent rather than the wine, since most wine odorants are formed during fermentation (see the Introduction in this chapter) [50].

Hydrophobic materials are particularly well suited for removing highly hydrophobic compounds, including several undesirable components in wine. For example, TCA ("cork taint," Chapter 18) has a log P ~4.0, and is thus one of the most non-polar compounds in wine – by comparison, the log P of ethyl hexanoate is estimated to be 2.8. Both cork particles and polyethylene have been reported to be fairly effective at diminishing levels of TCA and several related compounds [51], and a number of companies supply filtration materials with hydrophobic resins designed to remove cork taint [52]. The highly non-polar TDN (log P ~ 4.8, "petrol," Chapter 9) is reported to decrease 50% due to scalping by synthetic closures, while several less polar aroma compounds were negligibly affected [53]. Analogously, activated carbon has been applied to the removal of ochratoxin A (Chapter 18) [54]. In all of these cases, relatively small rates of hydrophobic sorbent are necessary to remove sufficient quantities of the targeted highly hydrophobic compounds, which minimizes the effects on non-targeted compounds.

References

1. Siebert, K.J. (2009) Haze in beverages, Chapter 2, in *Advances in food and nutrition research* (ed. Henry, C.J.), Academic Press, pp. 53–86.
2. Marchal, R. and Jeandet, P. (2009) Use of enological additives for colloid and tartrate salt stabilization in white wines and for improvement of sparkling wine foaming properties, in *Wine chemistry and biochemistry* (eds Moreno-Arribas, M.V. and Polo, M.C.), Springer, New York, pp. 127–158.
3. Ricardo-da-Silva, J.M., Cheynier, V., Souquet, J.M., *et al.* (1991) Interaction of grape seed procyanidins with various proteins in relation to wine fining. *Journal of the Science of Food and Agriculture*, **57** (1), 111–125.
4. Haslam, E. (1998) *Practical polyphenolics*, Cambridge University Press, Cambridge.
5. Hagerman, A.E., Rice, M.E., Ritchard, N.T. (1998) Mechanisms of protein precipitation for two tannins, pentagalloyl glucose and epicatechin16 (4–>8) catechin (procyanidin). *Journal Agriculture Food Chemistry*, **46** (7), 2590–2595.
6. McManus, J.P., Davis, K.G., Beart, J.E., *et al.* (1985) Polyphenol interactions. Part 1. Introduction; some observations on the reversible complexation of polyphenols with proteins and polysaccharides. *Journal of the Chemical Society, Perkin Transactions II*, (9), 1429–1438.
7. Murray, N.J., Williamson, M.P., Lilley, T.H., Haslam, E. (1994) Study of the interaction between salivary proline-rich proteins and a polyphenol by 1H-NMR spectroscopy. *European Journal of Biochemistry*, **219** (3), 923–935.
8. Poncet-Legrand, C., Edelmann, A., Putaux, J.L., *et al.* (2006) Poly(L-proline) interactions with flavan-3-ols units: Influence of the molecular structure and the polyphenol/protein ratio. Food Hydrocolloids, **20** (5), 687–697.
9. Yokotsuka, K. and Singleton, V.L. (1995) Interactive precipitation between phenolic fractions and peptides in wine-like model solutions: turbidity, particle size, and residual content as influenced by pH,temperature and peptide concentration. *American Journal of Enology and Viticulture*, **46** (3), 329–338.
10. Sarni-Manchado, P., Deleris, A., Avallone, S., *et al.* (1999) Analysis and characterization of wine condensed tannins precipitated by proteins used as fining agent in enology. *American Journal of Enology and Viticulture*, **50** (1), 81–86.
11. Hagerman, A.E. and Butler, L.G. (1978) Protein precipitation method for the quantitative determination of tannins. *Journal of Agricultural and Food Chemistry*, **26** (4), 809–812.
12. Puig-Deu, M., Lopez-Tamames, E., Buxaderas, S., Torre-Boronat, M.C. (1999) Quality of base and sparkling wines as influenced by the type of fining agent added pre-fermentation. *Food Chemistry*, **66** (1), 35–42.
13. Cosme, F., Ricardo-da-Silva, J.M., Laureano, O. (2008) Interactions between protein fining agents and proantho-cyanidins in white wine. *Food Chemistry*, **106** (2), 536–544.
14. Sims, C.A., Eastridge, J.S., Bates, R.P. (1995) Changes in phenols, color, and sensory characteristics of muscadine wines by prefermentation and postfermentation additions of PVPP, casein, and gelatin. *American Journal of Enology and Viticulture*, **46** (2), 155–158.
15. Caceres-Mella, A., Pena-Neira, A., Parraguez, J., *et al.* (2013) Effect of inert gas and prefermentative treatment with polyvinylpolypyrrolidone on the phenolic composition of Chilean Sauvignon blanc wines. *Journal of the Science of Food and Agriculture*, **93** (8), 1928–1934.
16. Cosme, F., Capao, I., Filipe-Ribeiro, L., *et al.* (2012) Evaluating potential alternatives to potassium caseinate for white wine fining: effects on physicochemical and sensory characteristics. *LWT – Food Science and Technology*, **46** (2), 382–387.
17. Hahn, G.D. and Possmann, P. (1977) Colloidal silicon dioxide as a fining agent for wine. *American Journal of Enology and Viticulture*, **28** (2), 108–112.
18. Behrens, S.H. and Grier, D.G. (2001) The charge of glass and silica surfaces. *Journal of Chemical Physics*, **115** (14), 6716–6721.
19. Castillo-Sanchez, J.J., Mejuto, J.C., Garrido, J., Garcia-Falcon, S. (2006) Influence of wine-making protocol and fining agents on the evolution of the anthocyanin content, colour and general organoleptic quality of Vinhao wines. *Food Chemistry*, **97** (1), 130–136.
20. Chagas, R., Monteiro, S., Ferreira, R.B. (2012) Assessment of potential effects of common fining agents used for white wine protein stabilization. *American Journal of Enology and Viticulture*, **63** (4), 574–578.
21. Cabaroglu, T., Razungles, A., Baumes, R., Gunata, Z. (2003) Effect of fining treatments on the aromatic potential of white wines from Muscat Ottonel and Gewurztraminer cultivars. *Sciences des Aliments*, **23** (3), 411–423.

22. Puig Deu, M., Lopez Tamames, E., Buxaderas, S., Torre Boronat, M.C. (1996) Influence of must racking and fining procedures on the composition of white wine. *Vitis*, **35** (3), 141–145.
23. Sanborn, M., Edwards, C.G., Ross, C.F. (2010) Impact of fining on chemical and sensory properties of Washington State Chardonnay and Gewurztraminer wines. *American Journal of Enology and Viticulture*, **61** (1), 31–41.
24. Marinkovich, V.A. (1981) Allergic symptoms from fining agents used in winemaking, in *Wine, Health and Society Symposium*, 1981, San Francisco, GRT Book Publishing, pp. 119–123.
25. Vintage Wine and Application of Enhanced Allergen Regulations (2012) Bureau of Chemical Safety Food Directorate (ed.), Health Canada, Ottowa.
26. Watts, D.A., Ough, C.S., Brown, W.D. (1981) Residual amounts of proteinaceous additives in table wine. *Journal of Food Science*, **46** (3), 681–683, 687.
27. Uberti, F., Danzi, R., Stockley, C., *et al.* (2014) Immunochemical investigation of allergenic residues in experimental and commercially-available wines fined with egg white proteins. *Food Chemistry*, **159**, 343–352.
28. Gambuti, A., Rinaldi, A., Moio, L. (2012) Use of patatin, a protein extracted from potato, as alternative to animal proteins in fining of red wine. *European Food Research and Technology*, **235** (4), 753–765.
29. Mekoue Nguela, J., Sieczkowski, N., Roi, S., Vernhet, A. (2015) Sorption of grape proanthocyanidins and wine polyphenols by yeasts, inactivated yeasts, and yeast cell walls. *Journal of Agricultural and Food Chemistry*, **63** (2), 660–670.
30. Ángeles Pozo-Bayón, M., Andújar-Ortiz, I., Moreno-Arribas, M.V. (2009) Scientific evidences beyond the application of inactive dry yeast preparations in winemaking. *Food Research International*, **42** (7), 754–761.
31. Caridi, A. (2007) New perspectives in safety and quality enhancement of wine through selection of yeasts based on the parietal adsorption activity. *International Journal of Food Microbiology*, **120** (1–2), 167–172.
32. Van Sluyter, S.C., McRae, J.M., Falconer, R.J., *et al.* (2015) Wine protein haze: mechanisms of formation and advances in prevention. *Journal of Agricultural and Food Chemistry*, **63** (16), 4020–4030.
33. Bayly, F.C. and Berg, H.W. (1967) Grape and wine proteins of white wine varietals. *American Journal of Enology and Viticulture*, **18** (1), 18–32.
34. Hsu, J.-C. and Heatherbell, D.A. (1987) Heat-unstable proteins in wine. I. Characterization and removal by bentonite fining and heat treatment. *American Journal of Enology and Viticulture*, **38** (1), 11–16.
35. Waters, E.J. and Colby, C.B. (2009) Proteins, in *Wine chemistry and biochemistry* (eds Moreno-Arribas, M.V. and Polo, M.C.), Springer, New York, pp. 213–230.
36. Linthorst, H.J.M. (1991) Pathogenesis-related proteins of plants. *Critical Reviews in Plant Sciences*, **10** (2), 123–150.
37. Pocock, K.F., Alexander, G.M., Hayasaka, Y., *et al.* (2007) Sulfate – a candidate for the missing essential factor that is required for the formation of protein haze in white wine. *Journal of Agricultural and Food Chemistry*, **55** (5), 1799–1807.
38. Esteruelas, M., Kontoudakis, N., Gil, M., *et al.* (2011) Phenolic compounds present in natural haze protein of Sauvignon white wine. *Food Research International*, **44** (1), 77–83.
39. Siebert, K.J. (1999) Effects of protein–polyphenol interactions on beverage haze. Stabilization and analysis. *Journal of Agricultural and Food Chemistry*, **47** (2), 353–362.
40. Saywell, L.G. (1934) The clarification of wine. *Industrial and Engineering Chemistry*, **26** (9), 981–982.
41. Pusch, R., Knutsson, S., Al-Taie, L., Hatem, M. (2012) Optimal ways of disposal of highly radioactive waste. *Natural Science*, **4**, 906–918.
42. Blade, W.H. and Boulton, R. (1988) Adsorption of protein by bentonite in a model wine solution. *American Journal of Enology and Viticulture*, **39** (3), 193–199.
43. Postel, W., Meier, B., Markert, R. (1986) Influence of processing aids on the content of mineral compounds in wine. I. Bentonite. *Mitteilungen Klosterneuberg*, **36**, 20–27.
44. Catarino, S., Madeira, M., Monteiro, F., *et al.* (2008) Effect of bentonite characteristics on the elemental composition of wine. *Journal of Agricultural and Food Chemistry*, **56** (1), 158–165.
45. Clark, A.C., Grant-Preece, P., Cleghorn, N., Scollary, G.R. (2015) Copper (II) addition to white wines containing hydrogen sulfide: residual copper concentration and activity. *Australian Journal of Grape and Wine Research*, **21** (1), 30–39.
46. Franco-Luesma, E. and Ferreira, V. (2014) Quantitative analysis of free and bonded forms of volatile sulfur compounds in wine. Basic methodologies and evidences showing the existence of reversible cation-complexed forms. *Journal of Chromatography A*, **1359**, 8–15.

47. Ribereau-Gayon, P., Yves, G., Maujean, A., Dubourdieu, D. (2000) *Handbook of enology*, Vol. **2**, John Wiley & Sons Ltd, Chichester, UK.
48. Curvelo-Garcia, A.S. and Catarino, S. (1998) Os metais contaminantes dos vinhos: origens da sua presença, teores, influência dos factores tecnológicos e definição de limites (revisão bibliográfica crítica). *Ciencia E Tecnica Vitivinicola*, **13**, 49–70.
49. Pickering G., Lin J., Reynolds A., *et al.* (2006) The evaluation of remedial treatments for wine affected by *Harmonia axyridis*. *International Journal of Food Science and Technology*, **41** (1), 77–86.
50. Ryona I., Reinhardt J., Sacks G.L. (2012) Treatment of grape juice or must with silicone reduces 3-alkyl-2-methoxypyrazine concentrations in resulting wines without altering fermentation volatiles. *Food Research International*, **47** (1), 70–79.
51. Capone, D.L., Skouroumounis, G.K., Barker, D.A., *et al.* (1999) Absorption of chloroanisoles from wine by corks and by other materials. *Australian Journal of Grape and Wine Research*, **5** (3), 91–98.
52. Eder, R., Hutterer, E.-M., Weingartund, G., Brandes, W. (2008) Reduction of 2,4,6-trichloranisole and geosmin contents in wine by means of special filter layers. *Mitteilungen Klosterneuburg*, **58** (1), 12–16.
53. Capone, D.L., Simpson, R.F., Cox, A., *et al.* (eds) (2005) New insights into wine bottle closure performance – flavour "scalping" and cork taint, in *Proceedings of the Australian Wine Industry Technical Conference*, 2005, Urrbrae, Australian Wine Industry Technical Conference Inc.
54. Quintela, S., Villarán, M.C., López de Armentia, I., Elejalde, E. (2013) Ochratoxin A removal in wine: a review. *Food Control*, **30** (2), 439–445.

26.3

Particle Filtration and Reverse Osmosis

26.3.1 Introduction

The process of liquid filtration is used in a range of industries to remove suspended particles or microorganisms by passing a fluid through a porous filter medium.[1] In some cases, the suspended matter is the desired product, for example, precipitation or recrystallization of a reaction product. However, in production of beverages such as wine and beer (or municipal water) the primary goal is to clarify, stabilize, and/or sterilize the liquid phase [1, 2]. Other means to achieve a degree of clarification are available to the winemaker (e.g., settling and racking, centrifugation (Chapter 19), fining (Chapter 26.2)) but this chapter focuses on common approaches using filtration to achieve more rigorous removal of suspended particles – or exclusion of particular compounds in the case of reverse osmosis (RO) – at different stages of the winemaking process. There are many types of filtration systems and porous filter media [3], with the choice depending on the components being removed and the level of clarity required at the particular stage of winemaking. Typically, there are less stringent requirements early in production, but at bottling a high degree of wine clarity (extremely low turbidity) is almost always essential.[2] Filtration can remove particles either through physical exclusion (i.e., the particles are larger than the filter pores) or by adsorption. The latter is of particular interest to wine chemists because: (i) adsorbed compounds are often responsible for blocking (*fouling*) filters, decreasing the efficiency of filtration, and (ii) adsorption could potentially result in losses of flavor or color compounds.

[1] Filtration also commonly features in everyday life in the home and workplace. Depth filters are used in swimming pools and water purifiers, coffee percolators utilize cake filtration (the coffee grounds form the cake), and a pasta colander is like a membrane filter.

[2] A notable exception to this relates to "natural" wines, in which case minimal intervention is used so wines may be bottled with some cloudiness. In general, however, consumers expect to see clear wine in the bottle, and proper (i.e., sterile) filtration helps to ensure the wine does not suffer from microbial instabilities after bottling.

Understanding Wine Chemistry, First Edition. Andrew L. Waterhouse, Gavin L. Sacks, and David W. Jeffery.

26.3.2 Definitions, principles, and characteristics of winery filtration

A range of definitions and equations are used to describe filtration in engineering and fluid dynamics texts (e.g., [4] and [5]). The *feed* is the liquid to be filtered and the *filtrate* is the liquid that exits the filtration system. The clarity of liquids is described by their *turbidity* – the ability of the liquid to scatter light – and is typically measured by a nephelometer (also called a turbidity meter, Figure 26.3.1a), and expressed as nephelometric turbidity units (NTUs).[3] Turbidity is monitored to assess the effectiveness of winery filtration (and other clarification) operations. Filtration may be described as *coarse* (removal of the largest particles,

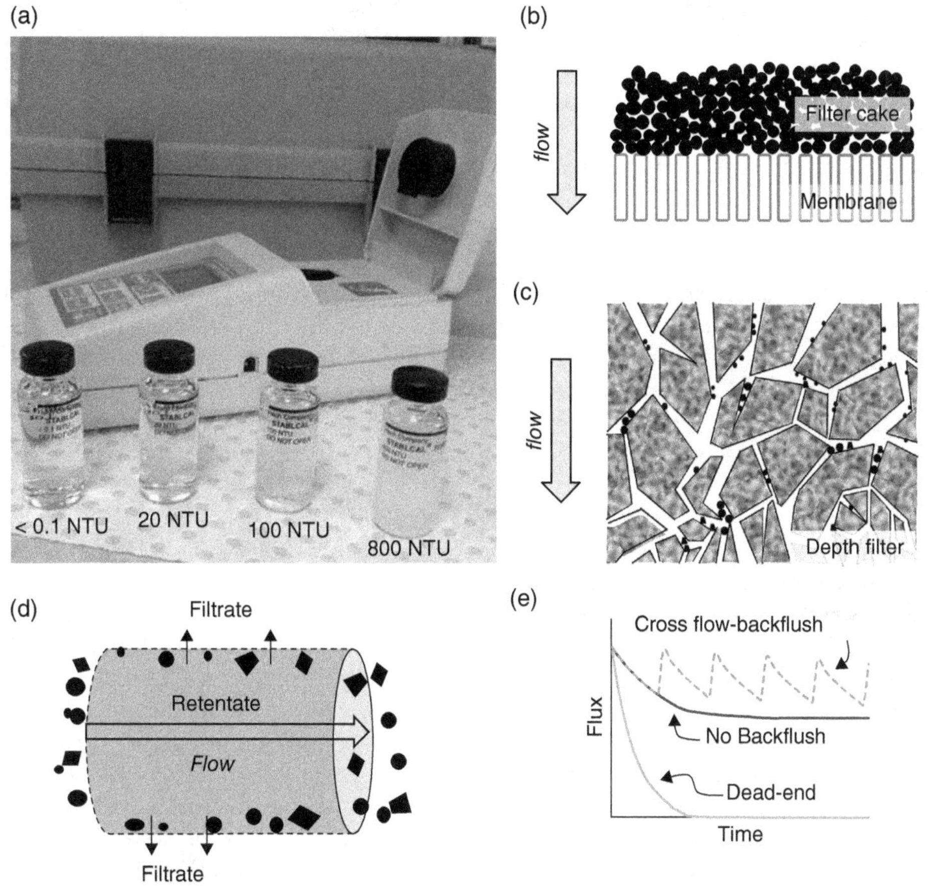

Figure 26.3.1 *Image of (a) a portable nephelometer and formazin standards of increasing NTU as indicated, and representative diagrams of filtration models showing (b) dead-end surface filtration with screening of particles at the membrane filter surface as well as formation of a filter cake, (c) dead-end depth filtration, in which particles are mechanically trapped within the tortuous paths of the filter matrix or electrostatically adsorbed, (d) tangential crossflow filtration, where particles are swept along the filter surface, and (e) decrease in flux during filtration of a wine by dead-end and crossflow filtration (with and without periodic backflushing), assuming constant pressure and constant concentration of particulates*

[3] A typical nephelometer design is similar to a spectrophotometer, but with a detector off-axis from the light source (often, at a 90 degree angle) so that only scattered light is measured. Note that a red wine could have high absorbance but still have low turbidity. In addition to NTU, turbidity may also be expressed as formazin turbidity units (FTU) and European brewing convention (EBC) turbidity units.

leaving a hazy filtrate, around 100 NTU), *tight* (or polishing/fine, removal of small suspended solids, leaving a "cellar bright" filtrate, at around 10 NTU or less), and *sterilizing* (removal of microorganisms, giving a biologically stable, bright wine, typically < 1 NTU, which is suitable for bottling).

Several approaches are used to classify filtration. A common approach is to distinguish filtration based on whether particles are filtered within the filter media (*depth*) or above the surface of the filter media (*surface, cake*) (Figure 26.3.1) [1, 2].

- In *depth filtration*, particles (solids) in the feed are mechanically trapped or adsorbed *within the filter medium* along *tortuous paths* making up the depth of the filter, for example a cellulose fiber pad (in a winery) or piece of filter paper (in a lab). Resins can be included in some depth filter media to introduce a surface charge and increase adsorption through electrostatic effects due to the *zeta potential* of the filter matrix (*electrokinetic capture* sites) and the charges on the particles (e.g., colloids and microorganisms). Depth filters do not have well-defined pore sizes.
- In *surface filtration*, filtration takes place upstream, *outside of the filter medium*. Particles too large to pass through pores on the filter surface are retained. Typically, surface filtration is achieved through use of a *membrane*, or a semi-permeable layer of material with a defined maximum pore size, although larger particles can be filtered at the surface of a depth filter.

During operation, both surface and depth filtration will lead to build up of a *filter cake* composed of inert solid material (e.g., cells, grape solids) deposited on the surface of the filter media, resulting in *cake filtration*.[4] Similar to depth filtration, cake filtration will result in particles being removed by adsorption or trapped within channels. In winemaking and other areas of the beverage industry, filter cakes may be intentionally generated with a *pre-coat* of a rigid solid (termed a *filter aid*, Chapter 27) like diatomaceous earth (DE).[5]

A second approach to classifying filtration is based on whether the feed has to pass through the filter medium or can travel parallel to it.

- In *dead-end filtration*, the filtrate exits a filter medium positioned perpendicular to the feed stream (Figure 26.3.1b and c). Dead-end filtration systems are simpler and less expensive to implement, and thus far more common in wineries. However, the flow rate through a dead-end filter will eventually drop to near zero due to fouling unless the filtering region is cleaned; for example, the filter cake can be removed or the filter media replaced. Dead-end filtration can be conducted with depth or membrane filters (Figures 26.3.2 and 26.3.3).
- In *tangential (crossflow) filtration*, the filtrate exits through a filter medium that is parallel to the feed stream (Figure 26.3.1d). In addition to the filtrate, crossflow filtration generates a *retentate* stream that does not pass through the filter pores and is recirculated through the crossflow filter multiple times. While more costly to establish, crossflow filtration provides several advantages, particularly that the parallel flow of feed decreases the extent of cake build-up on the filter surface, so the decrease in flow rate over the course of filtration is less severe in crossflow than in dead-end filtration (Figure 26.3.1e). This means that liquids of higher turbidity can be filtered, including lees, therefore eliminating the need for multiple filtration steps. Crossflow filtration is performed with membranes (Figure 26.3.4).

Finally, filtration may be classified by the size of particles designed to be retained, referred to as the *cut-off* (usually given in micrometers, Table 26.3.1). An *absolute cut-off* indicates that 100% of particles greater than the rated particle size will be filtered out. Absolute cut-off values are only valid if the medium has entirely

---

[4] Some texts classify cake filtration as a variation of depth filtration, or else as an entirely separate type of filtration, rather than a type of surface filtration.

[5] Diatomaceous earth (DE) is the fossilized remains of algae (diatoms). Because of its classification as a probable carcinogen and hazardous waste, DE use as a filter aid has been replaced in some wineries with perlite (thermally processed and milled volcanic rock).

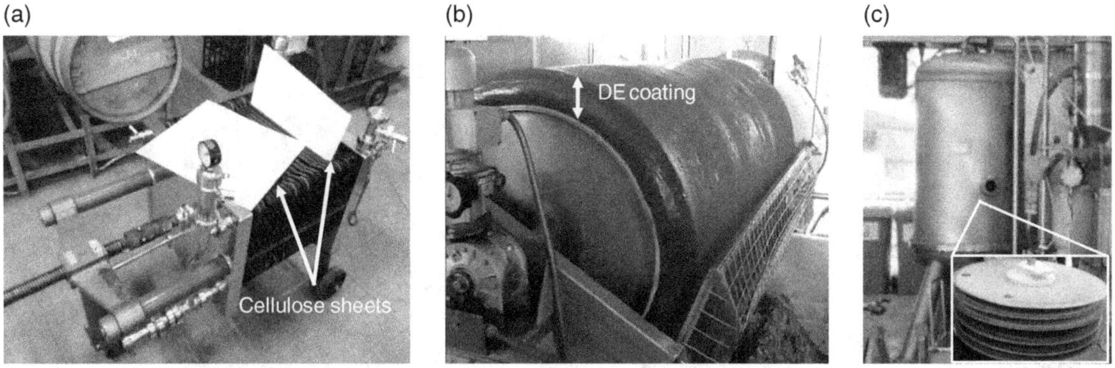

Figure 26.3.2 Winery depth and cake filtration systems include (a) flat sheet filter press comprising 40 cm × 40 cm cellulose sheets positioned within the black frames on the press, which offers variable capacity by using more or fewer frames, (b) rotary drum vacuum filter (RDV) coated with DE or perlite (approximately 100 mm thick) suitable for large volumes of turbid juice or wine (e.g., lees from the bottom of tanks), and (c) pressure leaf filter housing, with the inset at bottom right showing horizontal leaves (screens) used for supporting DE filter cakes

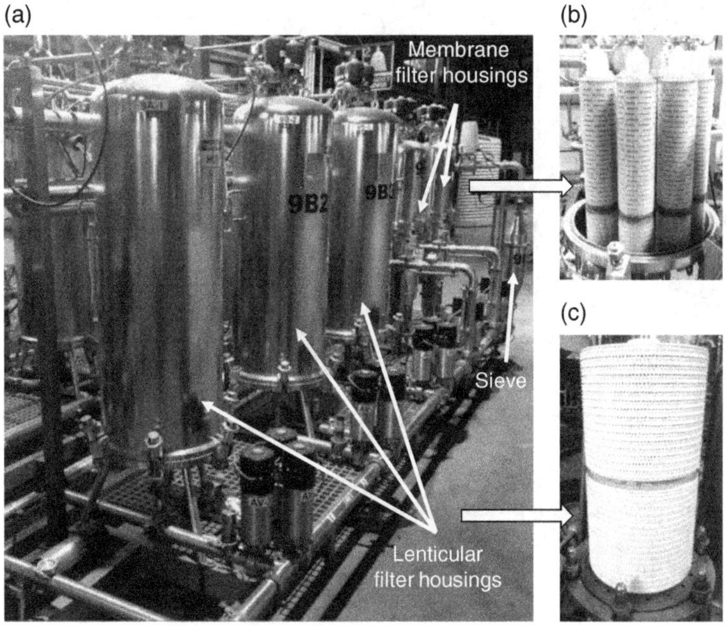

Figure 26.3.3 Winery automated filtration skid consisting of (a) a coarse sieve as well as depth and surface filter housings containing (b) pleated polymer membrane filter cartridges and (c) cellulose lenticular filter modules, respectively (not to scale). Source: Reproduced with permission of Luke Wilson

consistent pore sizes and the particles are spherical. Many filter media and filtrates do not strictly satisfy this description, and *nominal cut-off* is used to define a minimum percentage of particles larger than the filter rating that will be retained. Because depth filtration does not employ defined pore sizes, it is invariably described by nominal cut-off values, and will not be used to achieve sterile filtration (no microbes in the filtrate). A range of characteristics related to winery filtration is summarized in Table 26.3.1 and typical

Figure 26.3.4 Winery crossflow filters include (a) front view and (b) side view of a polymer membrane system containing hollow fibers in either (c) normal high flow and (d) high solids lees (note wider fiber bore) housings, and (e) front view and (f) side view of a ceramic membrane system containing (g) numerous filter candles in a housing. Source: (c), (d), (e), and (g) reproduced with permission of Luke Wilson

filtration sequences for different wine types are shown in Table 26.3.2. A more thorough discussion of filtration topics can be found in practical wine texts [6, 7].

26.3.3 Filtration and fouling

Most filtration models utilize Darcy's Law or one of its variations to explain how flow will vary during filtration or with different filtration media. This can be expressed in several ways, including:

$$J = \frac{\Delta P}{\mu R_t} \tag{26.3.1}$$

Table 26.3.1 *Characteristics of separations using different filtration processes*

Process	Nominal cut-off		Typical use	Typical filter type/media
	μm	g/mol		
Particle filtration (macro)	100–1000	–	Juice or wine fractions with high solids content, e.g., yeast lees, grape solids. Often used soon after fermentation	Cake filtration with diatomaceous earth (DE) or perlite
Particle filtration (micro)	1–100	–	Removal of dirt, crystalline matter, wine lees, insoluble fining agents such as bentonite. During wine cellaring or prior to microfiltration	Depth filter (DE, perlite, or cellulose sheet)
Microfiltration	0.1–10	>100 000	Yeast, bacteria, and particle removal. Haze removal or wine stabilization prior to bottling	Depth filter (cellulose sheet or lenticular module) or membrane filter (crossflow or pleated polymer cartridge [sterile])
Ultrafiltration/ nanofiltration	>0.001–0.1	200–100 000	Colloids, large molecules, viruses. Typically of little use in a winery	Membrane filter (crossflow)
Reverse osmosis	0.0001–0.001	1–200	Removal of small molecules. Alcohol adjustment or fault removal throughout winemaking process	Membrane filter (crossflow)

Table 26.3.2 *Typical filtration sequences used on different white and red wines prior to bottling. Crossflow filtration can replace the lenticular filtration steps if it has been conducted within several weeks of the final filtration sequence*

Wine	Sequence of filtration (from left to right)			
	Medium lenticular	Fine lenticular	Membranes[a]	Sieve[b]
White				
Typical	×	✓	✓	✓
Sparkling	×	✓	✓	×
Red				
Typical	✓	×	×	✓
Sweet/low SO_2	✓	×	✓	✓
Premium/super premium	×	×	×	✓

[a]Consists of pre-filter (0.65 μm) and final filter (0.45 μm).

[b]Stainless steel gauze with a large cut-off (e.g., 45 μm) that serves as a final catch-point for foreign objects.

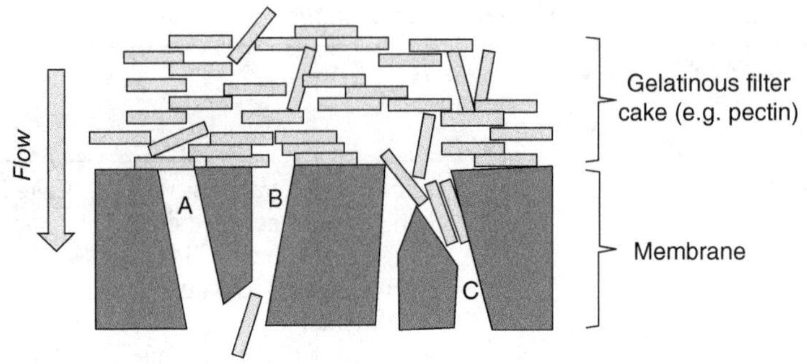

Figure 26.3.5 Representation of mechanisms responsible for filter fouling during wine filtrations showing a gelatinous filter cake and direct blockage of pores by particles [8], either through complete (A) or bridging (B) obstruction at the surface, or internally ("standard" blocking model, C)

where J is the flux, or the filtrate volume per unit surface area of the filter, and μ is the viscosity of the feed. The flow is created by a pressure drop (ΔP) across the filter, that is, the difference in pressure between the feed and filtrate sides of the filter.[6] R_t is the total resistance of the system, and comprises both the resistance created by the filter medium as well as additional resistance from the filter cake or blocking. The inverse of the resistance is the *permeability*, expressed as Darcy units,[7] and is a measure of how easily a filter medium allows liquid to pass through and is widely used to evaluate filter performance. Generally speaking, lower resistance (and higher flux) is more desirable in production settings.

As filtration is performed on a wine, R_t will increase due to deposition of solids on the filter media or filter cake. Unless pressure is increased over a filtration run, the flux will decrease (see Figure 26.3.1). Several mechanisms can explain this increase in resistance (Figure 26.3.5):

- Deposition of filter cake. This results in an increase in resistance as the liquid must traverse a longer distance through pores within the cake. Cakes may be *incompressible*, as would be formed from rigid particles like DE, or *compressible*, as would be formed by organic material. In particular, flexible macromolecules like pectins can generate a gelatinous, compressible filter cake at the membrane surface. The resistance of compressible cakes increases with increasing pressure drop, and often results in a greater loss of flux than incompressible cakes.
- Particles can block pores directly. This blockage may occur within the pores of the filter medium (*standard blocking*) or as a result of complete or partial (*bridging*) obstruction of pores at the filter medium surface.

Several empirical and theoretical models have been developed to characterize the change in resistance (and thus flux) over the course of a dead-end filtration.[8] A review of all models is well outside the scope of this book, but many wine filtrations can be well-modeled through the power model [6], where α and b are empirically derived constants, and V/A is the total volume filtered per unit area of the filter:

$$R_t = R_m + \alpha \left(V/A \right)^b \tag{26.3.2}$$

[6] In a winery, this pressure drop can be created by gravity, but is more commonly created through the use of pumps.

[7] A permeability of one Darcy (roughly that of sand) permits a flow of 1 mL/s of a fluid with the viscosity of water (1 cP) under a pressure gradient of 1 atm/cm acting across an area of 1 cm^2.

[8] Similar statements can be made about crossflow filtration.

Incompressible cakes, as occur when filter aids like DE are added, can be well modeled by $b = 1$ (i.e., the resistance increases linearly with the volume of juice filtered). The various blocking models use $b = 2$–3 and filtration of most real wines by pad filtration falls in the range of $b = 1.5$–4 [9].

The change in resistance over time has a profound effect on the *filterability* of a wine. Wines with poor filterability will show a strong (and undesirable) exponential increase in resistance (high b values) over the course of filtration. Filterability is often characterized empirically by a filterability index (FI) for a specific wine and a specific filter (usually a membrane) [10]. Several methods for calculating FI exist and generally involve comparing the time necessary to filter two successive volumes of wine through a microfiltration membrane of defined characteristics [10–12]. For example, one approach uses FI < 20 as a criteria for acceptable filterability:

$$\text{FI} = \left(\text{Time (s) to filter 400 mL}\right) - \left(1.66 \times \text{Time (s) to filter 200 mL}\right) \qquad (26.3.3)$$

Filterability for cellared wines prior to bottling is often poorly correlated with the amount of solids in the wine (Figure 26.3.6).

Several studies have investigated the cause of fouling by studying the composition of foulants and through simulations using synthetic wines, primarily in membrane filtrations (reviewed in Reference [13]). This work has implicated colloidal material formed from macromolecules,[9] particularly:

- Grape-derived polysaccharides (primary cause)
- Polyphenols, for example, proanthocyanidins
- Mannoproteins.

Interestingly, the particles most commonly implicated in fouling are smaller (1–100 nm) than the typical pore sizes encountered in microfiltration (0.2 or 0.45 μm). As shown in Figure 26.3.7, fouling (decrease in flux)

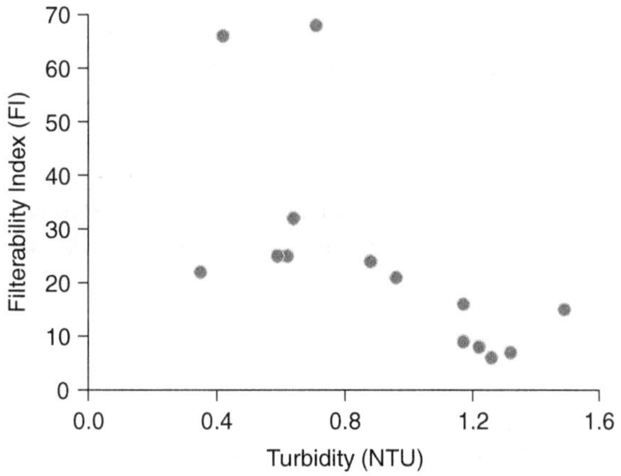

Figure 26.3.6 *Comparison of turbidity (NTU, proxy for suspended solids) and filterability index (FI) for 14 red and white wines prior to membrane filtration and bottling. Data for three wines with intermediate NTU and FI > 500 are not shown. Data from Reference [12]*

[9] Macromolecules are also implicated in fouling of other beverages. β-Glucans represent a major challenge in brewing filtrations, and proteins are a primary cause of fouling in milk filtrations.

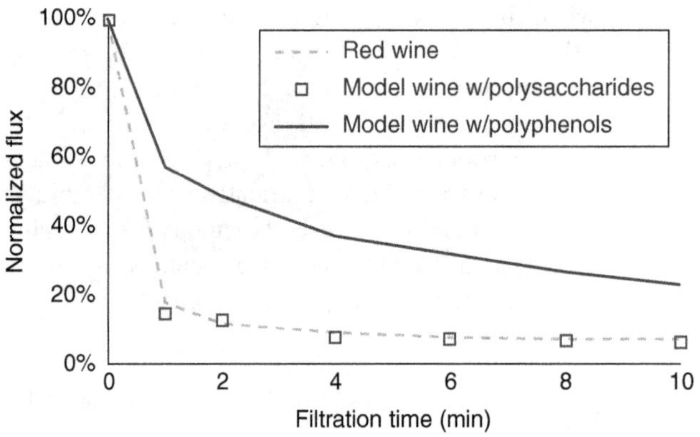

Figure 26.3.7 Change in flux (a proxy for fouling) versus time for crossflow microfiltration (0.2 μm) of a red wine, and two synthetic wines containing either wine polysaccharide or polyphenol fractions. Data from Reference [15]

can occur even in a synthetic wine containing polysaccharides or polyphenols and with no particles >0.2 μm [8]. Studies of juice indicate that fouling begins with blocking of pores, with cake formation occurring later [14]. Unless incompressible filter aids like DE are used, the cakes formed during wine filtrations tend to be gelatinous and highly compressible. Having a more turbid wine (e.g., more microbial lees) can potentially improve filterability by generating a cake more rapidly and preventing pore fouling (see Figure 26.3.5) [14].

Both the physical and chemical properties of membranes will affect their adsorption of macromolecules:

- Membrane surface roughness increases the rate of fouling during microfiltration, presumably by creating areas that are not well swept by the filtration flux [16].
- Alumina (Al_2O_3) membranes and many other ceramics will have a positive surface charge at wine pH. This can lead to electrostatic interactions with wine components, particularly acidic macromolecules like pectic polysaccharide fragments (Chapter 2) [17]. Because other ions in the wine can shield the positive surface charge of alumina through formation of a *double layer*, these interactions will be stronger at higher flow rates, which disrupt the shielding. This type of interaction is a form of electrokinetic capture (or zeta-potential filtering).
- Adsorption of macromolecules on organic membranes will involve non-covalent interactions. For example, polyethersulfone (PES) membranes are reported to adsorb polysaccharides more strongly (and foul more rapidly) than polypropylene (PP) membranes because of their H-bonding capability [13].

26.3.4 Reverse osmosis

RO is used to selectively remove small dissolved molecules (i.e., around 200 g/mol or less[10]) to concentrate juice or must (i.e., remove water) or make corrections to wine composition (e.g., decrease volatile acidity

[10] Depending on the membrane cut-off there can be some overlap with nanofiltration, which may be thought of as "loose RO" in a winery context. However, some applications, such as the removal of taints, may be classified as nanofiltration in different resources (e.g., government regulations). A useful distinction is the ability to pass ions – nanofiltration can pass monovalent and some divalent ions whereas RO generally cannot (hence RO is used for desalination of water for drinking purposes).

(VA), acids, taints, and alcohol) [18–20]. In comparison to membrane microfiltration, RO membranes have much smaller nominal cut-offs in the nm range (Table 26.3.1). The lower permeability means RO membranes must operate at much higher pressures, up to about 80 bar, compared to around 1 bar for most microfiltration membranes. Selectivity is governed by molecular size of the solute, but is also strongly affected by the chemistry of the membrane and its interactions with solutes [21,22]. For compounds of similar molecular weight, factors tending to increase *rejection* (i.e., retention by the membrane) are:

- Molecules with higher hydrophobicity/lower polarity through binding to the membrane
- Larger apparent "size" of solutes, either due to hydration or branching of molecules
- Like charges on membrane and solute leading to electrostatic repulsion
- Ionization and extent of charge on solutes causing greater hydration.

During normal osmosis water will travel through a membrane from low (permeate) to high (retentate) solution concentrations, and the drive experienced by water is referred to as the *osmotic pressure*. As the name implies, reverse osmosis requires that the process of osmosis happens in reverse, where water or other small solutes must flow from the more concentrated solution (retained wine) to a less concentrated solution. This can only occur when the *transmembrane pressure* (i.e., the pressure difference between retentate and permeate) is greater than the osmotic pressure. Due to practical limitations on pressures achievable with commercial pumps and membranes, the osmotic pressure effectively limits the degree of concentration achievable, such as when increasing sugar concentration along with other components in a juice by RO (to a maximum of around 27 °Brix) [23,24].

The permeate from RO filtration of wine will contain water as well as other low molecular weight compounds such as ethanol and acetic acid. Unlike treating juice, the typical goal during RO of wine is not to concentrate the retentate, but instead to prepare the permeate for a subsequent step that removes an unwanted component. Following that treatment, such as recovery of ethanol by distillation in the case of lowering wine alcohol content [25] (Chapter 26.4), the permeate can be recombined with the concentrated retentate, thereby reconstituting the wine. RO is thus a useful pre-treatment process for decreasing unwanted low molecular weight components because it affords a level of selectivity. In addition to decreasing alcohol in wine, treatments include removing VA by anion exchange resin [26] and eliminating taints caused by volatile phenols (i.e., wines affected by *Brettanomyces* or bushfire smoke) through their adsorption on resins [27, 28].

26.3.5 Sensory effects of filtration

The typical cut-offs for haze removal and microbial stabilization (>0.2 μm) are larger than all compounds known to have a flavor impact on wine. However, as described above, filter fouling typically arises from adsorption and bridging by smaller macromolecules like polysaccharides and polyphenols that can contribute to the mouthfeel of wine. Adsorptive losses of smaller molecules (e.g., odorants, bitter compounds) is also well documented in the food industry.[11] Potentially, large flavorless solutes that could serve as flavor precursors could also be removed. As a result, filtration – and particularly membrane microfiltration – is viewed with suspicion by some wine producers. However, anecdotal accounts (and, to some extent, the peer-reviewed literature) regarding the chemosensory effects of filtration are confounded by several factors:

[11] "Debittering" with cellulose acetate sheets to remove low molecular weight polyphenols like naringin is a common process in the orange juice industry.

- Lack of filtration leaves the wine susceptible to growth of spoilage organisms (e.g., *Brettanomyces*, lactic acid bacteria), which can change wines' sensory properties.
- Contaminated or poorly cleaned filter media can taint wines.
- Transfer operations during filtration can introduce oxygen.

With these caveats, existing literature does indicate that membrane microfiltration can result in small but significant decreases in flavor or color compounds (Table 26.3.3), particularly anthocyanins and other polyphenols [13, 29–32]. Significant differences in some aroma properties are occasionally noted, but studies on the effects of microfiltration on wine odorants are scarce, and may be of lesser importance – one report has shown significant (and minor) decreases for only three odorants out of over 100 quantified following microfiltration on hydrophobic membranes [29].

Table 26.3.3 *A selection of data from studies assessing changes to wine sensory and chemical properties arising from different filtration processes. Data from References [29] to [32]*

Sample	Description of Filter Media	Parameters Affected
Cabernet Sauvignon wine *Compared before/after filtration*	Membrane filter with polypropylene pre-filter (1.2 μm) and PVDF final filter (0.65 μm)	Lower in color intensity (2%), total polyphenols (9–13%), anthocyanins (2–3%) and tannins (2–6%) Lower concentrations for a small number of volatiles Significant sensorial differences related to body and aroma
Synthetic red wine containing marc extract and model polyphenol and polysaccharide *Compared binding by polymer types*	PES or PP hollow fiber membranes (0.2 μm) as used in a crossflow filter	Greater binding of polysaccharide compared to polyphenol on PES (2-fold) and PP (4-fold) Greater binding of polyphenol (10-fold) and polysaccharide (17-fold) by PES membrane
Red wine blend *Compared to unfiltered wine*	Crossflow with hollow fiber PES membrane (0.2 μm)	Sensory profile up to two months was relatively constant for all samples; in control wines after this point berry and stone fruit aromas decreased and oak, grassy, earthy and smoke aromas increased Lower in A420, A520 and color density (1–10%) compared to control Lower in the concentrations of tannin (8–26%) and anthocyanin (5–10%) compared to control
Malbec and rosé wines *Compared to unfiltered wine*	Sequential filtration through coarse and tight pads (cellulose or cellulose+DE), then two membrane filters (0.45 μm, PES or Nylon)	Lower in color intensity in all cases, as follows: *Malbec* Cellulose pads then PES membranes, 11.0% Cellulose+DE pads then Nylon membranes, 17.5% *Rosé* Cellulose pads then PES membranes, 21.8% Cellulose+DE pads then Nylon membranes, 46.5%

References

1. Holdich, R.G. (2002) Filtration of liquids, in *Fundamentals of particle technology*, Midland Information Technology and Publishing, Shepshed, UK, pp. 29–44.
2. Ripperger, S., Gösele, W., Alt, C., Loewe, T. (2013) Filtration, 1. Fundamentals, in *Ullmann's Encyclopedia of industrial chemistry*, Wiley-VCH Verlag GmbH & Co. KGaA, Weinheim, Germany, pp. 1–38.
3. Ripperger, S., Gösele, W., Alt, C., Loewe, T. (2013) Filtration, 2. Equipment, in *Ullmann's Encyclopedia of industrial chemistry*, Wiley-VCH Verlag GmbH & Co. KGaA, Weinheim, Germany, pp. 1–40.
4. Tilton, J.N. (2008) Fluid and particle dynamics, in *Perry's chemical engineers' handbook*, 8th edn (eds Green, D.W. and Perry, R.H.), McGraw-Hill, New York.
5. Genck, W.J., Dickey, D.S., Baczek, F.A., *et al.* (2008) Liquid–solid operations and equipment, in *Perry's chemical engineers' handbook*, 8 edn (eds Green, D.W. and Perry, R.H.), McGraw-Hill, New York.
6. Boulton, R.B., Singleton, V.L., Bisson, L.F., Kunkee, R.E. (1999) *Principles and practices of winemaking*, Kluwer Academic/Plenum Publishers, New York.
7. Zoecklein, B.W., Fugelsang, K.C., Gump, B.H., Nury, F.S. (1999) *Wine analysis and production*, Kluwer Academic/ Plenum Publishers, New York.
8. El Rayess, Y., Albasi, C., Bacchin, P., *et al.* (2011) Cross-flow microfiltration of wine: effect of colloids on critical fouling conditions. *Journal of Membrane Science*, **385–386**, 177–186.
9. de la Garza, F. and Boulton, R. (1984) The modeling of wine filtrations. *American Journal of Enology and Viticulture*, **35** (4), 189–195.
10. Peleg, Y., Brown, R.C., Starcevich, P.W., Asher, R. (1979) Method for evaluating the filterability of wine and similar fluids. *American Journal of Enology and Viticulture*, **30** (3), 174–178.
11. Alarcon-Mendez, A. and Boulton, R. (2001) Automated measurement and interpretation of wine filterability. *American Journal of Enology and Viticulture*, **52** (3), 191–197.
12. Bowyer, P., Edwards, G., Eyre, A. (2012) NTU vs wine filterability index – what does it mean for you? *The Australian and New Zealand Grapegrower and Winemaker*, **585**, 76–80.
13. El Rayess, Y., Albasi, C., Bacchin, P., *et al.* (2011) Cross-flow microfiltration applied to oenology: a review. *Journal of Membrane Science*, **382** (1–2), 1–19.
14. Salgado, C., Palacio, L., Carmona, F.J., *et al.* (2013) Influence of low and high molecular weight compounds on the permeate flux decline in nanofiltration of red grape must. *Desalination*, **315**, 124–134.
15. Vernhet, A. and Moutounet, M. (2002) Fouling of organic microfiltration membranes by wine constituents: importance, relative impact of wine polysccharides and polyphenols and incidence of membrane properties. *Journal of Membrane Science*, **201** (1), 103–122.
16. Lee, N., Amy, G., Croué, J.-P., Buisson, H. (2004) Identification and understanding of fouling in low-pressure membrane (MF/UF) filtration by natural organic matter (NOM). *Water Research*, **38** (20), 4511–4523.
17. Belleville, M.P., Brillouet, J.M., De La Fuente, B.T., Moutounet, M. (1992) Fouling colloids during microporous alumina membrane filtration of wine. *Journal of Food Science*, **57** (2), 396–400.
18. Massot, A., Mietton-Peuchot, M., Peuchot, C., Milisic, V. (2008) Nanofiltration and reverse osmosis in winemaking. *Desalination*, **231** (1–3), 283–289.
19. Wollan, D. (2010) Membrane and other techniques for the management of wine composition, in *Managing wine quality* (ed, Reynolds, A.G.), Woodhead Publishing, Cambridge, UK, pp. 133–163.
20. El Rayess, Y. and Mietton-Peuchot, M. (2015) Membrane technologies in wine industry: an overview of applications and recent developments. *Critical Reviews in Food Science and Nutrition*. doi: 10.1080/ 10408398.2013.809566.
21. Bellona, C., Drewes, J.E., Xu, P., Amy, G. (2004) Factors affecting the rejection of organic solutes during NF/RO treatment – a literature review. *Water Research*, **38** (12), 2795–2809.
22. Kucera, J. (2010) Basic terms and definitions, in *Reverse osmosis: design, processes, and applications for engineers*, John Wiley & Sons, Inc., Hoboken, NJ, pp. 21–40.
23. Mietton-Peuchot, M., Milisic, V., Noilet, P. (2002) Grape must concentration by using reverse osmosis. Comparison with chaptalization. *Desalination*, **148** (1–3), 125–129.

24. Kiss, I., Vatai, G., Bekassy-Molnar, E. (2004) Must concentrate using membrane technology. *Desalination*, **162**, 295–300.
25. Schmidtke, L.M., Blackman, J.W., Agboola, S.O. (2012) Production technologies for reduced alcoholic wines. *Journal of Food Science*, **77** (1), R25–R41.
26. Smith, C.R. (inventor) (1996) Apparatus and method for removing compounds from a solution. US Patent US5480665 A, 1996.
27. Ugarte, P., Agosin, E., Bordeu, E., Villalobos, J.I. (2005) Reduction of 4-ethylphenol and 4-ethylguaiacol concentration in red wines using reverse osmosis and adsorption. *American Journal of Enology and Viticulture*, **56** (1), 30–36.
28. Fudge, A.L., Ristic, R., Wollan, D., Wilkinson, K.L. (2011) Amelioration of smoke taint in wine by reverse osmosis and solid phase adsorption. *Australian Journal of Grape and Wine Research*, **17** (2), S41–S48.
29. Arriagada-Carrazana, J.P., Sáez-Navarrete, C., Bordeu, E. (2005) Membrane filtration effects on aromatic and phenolic quality of Cabernet Sauvignon wines. *Journal of Food Engineering*, **68** (3), 363–368.
30. Ulbricht, M., Ansorge, W., Danielzik, I., *et al.* (2009) Fouling in microfiltration of wine: the influence of the membrane polymer on adsorption of polyphenols and polysaccharides. *Separation and Purification Technology*, **68** (3), 335–342.
31. Bowyer, P., Edwards, G., Eyre, A. (2013) Wine filtration and filterability – a review of what's new. *The Australian and New Zealand Grapegrower and Winemaker*, **599**, 76–80.
32. Buffon, P., Heymann, H., Block, D.E. (2014) Sensory and chemical effects of cross-flow filtration on white and red wines. *American Journal of Enology and Viticulture*, **65** (3), 305–314.

26.4
Distillation

26.4.1 Introduction

Distillation is an industrially important process for partially or completely separating (fractionating) volatile compounds in a liquid from each other or from non-volatile components. At a basic level, distillation involves boiling a liquid and condensing the vapors to produce *distillate* fractions, with separation of components arising from differences in vapor composition relative to that of the boiling liquid. Crude forms of distillation have been around for thousands of years, and the use of distillation in Asia may date to as early 800 BC. While there is conjecture that alcoholic beverages were used to produce distillates in these very earlier stills, others date the first distillation of alcohol to China in the first century [1]. Distillation was also refined by the Islamic alchemists of the eighth to fifteenth centuries, and Arabic has given us several distillation-related terms, including alcohol (al kohl/kohol) and alembic (the historical term for a still).

In Europe, the concept of distillation (and associated equipment) was first documented by the ancient Greeks around the first century AD, initially in relation to distilling sea water to produce potable water [2, 3]. Alcohol distillation in Europe was first reported in a recipe for obtaining "aqua ardens" (burning water) from wine in 1200 AD [3]. Contemporaneous accounts of "aqua vitae" (water of life) were also reported, and although the alcohol was used primarily for medicinal purposes, without doubt such distillates were also consumed as a beverage. These early distillations involved some degree of sophistication – fractions were collected and redistillation was undertaken, and at times herbs and flowers were steeped in the alcohol prior to a final distillation. Consequently, the technology of distillation progressed and spread across Europe and other parts of the world, and fermented beverages produced from sources other than grapes, such as those from grains and other fruits, were also utilized as *feedstocks*. Thus began the evolution of the spectrum of contemporary distilled beverages such as whisky, gin, rum, vodka, and of course grape-derived spirits such as brandy and variations of it, as well as high strength (i.e., highly *rectified*) grape alcohol used for fortifying wines [4]. This chapter describes some of the principles of distillation [e.g., 5–7] as they relate to water–ethanol solutions [e.g., 8–10] (i.e., the main components of wine), and focuses on the production of alcohol from grapes and grape byproducts [e.g., 11–14].

Understanding Wine Chemistry, First Edition. Andrew L. Waterhouse, Gavin L. Sacks, and David W. Jeffery.

26.4.2 Vapor–liquid equilibria

Separation by distillation is an interphase (liquid–vapor) mass transfer process that exploits differences in volatility of components within a liquid mixture. Volatility is the ratio of mole fractions of a single substance between vapor and liquid phases (i.e., equilibrium ratio K, Table 26.4.1)[1] and signifies the ease by which a compound is *vaporized* (evaporated). This relates to vapor–liquid equilibrium (VLE) – the coexistence of liquid and vapor phases – and the equilibrium vapor pressure of a pure compound, which is the pressure caused by molecules in the saturated vapor phase in equilibrium with the saturated liquid phase.[2] The relative volatilities of different components in a mixture determine the ability to separate those components by distillation. The concept of volatility was presented in Chapter 1 in terms of Henry's Law (Table 26.4.1), which is appropriate for low-concentration solutes where volatility is primarily dependent on solvent–solute interactions. For components at high concentrations – like ethanol and water – volatility will depend more on self-interactions and is better described by Raoult's Law (Table 26.4.1).

26.4.2.1 Raoult's and Dalton's Laws

Vapor pressures differ based on the type of liquid and strength of attractive forces holding the molecules together. The vapor pressure of pure liquids is a direct function of temperature, and as the temperature increases so too does vapor pressure.[3] For liquid mixtures, each component will exert its own vapor pressure (i.e., partial vapor pressure) as a function of its concentration in a liquid of given composition. This is the basis of Raoult's Law, which states that the ratio of partial vapor pressure of a component to its vapor pressure as a pure liquid is equal to the mole fraction of that component in the liquid mixture. Put another way – a typical wine is around 95% water on a molar basis, so we expect the vapor pressure of water in a wine head-space to be about 95% that of the vapor pressure of pure water, assuming ideal Raoult's Law behavior. Summing partial vapor pressures gives the total vapor pressure exerted by the liquid (i.e., Dalton's Law, the pressure exerted by a mixture of gases is the sum of their partial pressures, Table 26.4.1). Raoult's, Dalton's and Henry's Laws are represented diagrammatically in Figure 26.4.1 (a) and (d).

26.4.2.2 Ideal and non-ideal behavior

For a binary solution containing similar chemical components (e.g., ethanol and methanol or benzene and toluene), the intensity of the molecular interactions between these different entities is comparable to the interactions of either pure liquid alone; this system behaves as an ideal solution and obeys Raoult's Law throughout the composition range (Figure 26.4.1a–c). The situation is not the same in the case of water and ethanol mixtures that feature in wine distillation, as these two components are chemically quite different

---

[1] Mole fractions for liquid and vapour are denoted by X and Y, respectively. The expression for volatility is explained in Table 26.4.1.

[2] The kinetic energy of molecules in the liquid phase results in some molecules leaving the surface of the liquid to form a vapor. Molecules in the vapor phase also condense and return to the liquid phase. At equilibrium, the rate of molecules evaporating is balanced by the rate of molecules condensing, giving a certain vapor pressure. Input of energy (heat) increases both the temperature of the liquid and the kinetic energy of the molecules, meaning they can evaporate more easily, which increases the vapor pressure above the liquid and translates into higher volatility. At a given temperature a compound with a lower boiling point has a higher vapor pressure and is more volatile than one with a higher boiling point. Exploiting these differences enables separation by distillation.

[3] If enough heat is added a liquid will reach boiling point, which occurs when the vapor pressure has reached atmospheric pressure. That is why water will boil at a lower temperature atop a mountain, for example, compared to sea level where the atmospheric pressure is higher, and why pressure cookers can achieve higher temperatures and faster cooking times than conventional boiling. Conversely, boiling points are lowered by applying a vacuum because the "system" pressure is lower than atmospheric pressure.

Table 26.4.1 Useful equations and explanations of their relationships as they apply to distillation

Name	Equation	Explanation
Henry's Law	$P_a = X_a H_a$	P_a is the partial vapor pressure of component *a* in the mixture, X_a is the mole fraction of component *a*, and H_a is the Henry's law constant of component *a* for a given temperature and solvent.
Raoult's Law	$P_a = X_a P_a^o$	P_a is the partial vapor pressure of component *a* in the mixture, X_a is the mole fraction of component *a*, and P_a^o is the vapor pressure of pure component *a* at a given temperature. For a non-ideal solution the activity coefficient is included as a multiplier (modified Raoult's Law).
Activity coefficient (γ)	$\gamma_a = P_a(\text{actual})/P_a(\text{ideal})$	Used for non-ideal solutions to account for molecular interactions (and partial vapor pressures) that vary with the composition of the liquid. The activity coefficient for each component in a mixture may be greater than 1 (positive deviation from ideality) or less than 1 (negative deviation from ideality).
Dalton's Law	$P = P_a + P_b + \ldots$	P is the total vapor pressure, P_a is the partial vapor pressure of component *a*, P_b is the partial vapor pressure of component *b*, etc, at a given temperature.
	$P = X_a P_a^o + X_b P_b^o + \ldots$	This expresses Dalton's Law in terms of Raoult's Law by substituting in $X_a P_a^o$ in place of P_a, etc.
	$P_a = Y_a P$	The Dalton's Law relationship between P_a and P can also be expressed in terms of vapor mole fraction, Y_a.
	$Y_a = X_a P_a^o / P$	This equation can be derived by substituting in $X_a P_a^o$ in place of P_a in the equation above, which allows liquid and vapor compositions to be related with pressure.
Volatility (K)	$K_a = Y_a / X_a$	K_a is the volatility of component *a*, and Y_a and X_a are the mole fractions of component *a* in the vapor and liquid, respectively. From this relationship it can be seen that vapor enrichment rises as K increases above 1.
	$K_a = P_a^o / P$	K_a can also be expressed in terms of Raoult's and Dalton's Law expressions given above. This form of the equation shows that the volatility of component *a* is the ratio between the vapor pressure of pure *a* and the total pressure. For non-ideal solutions the activity coefficient of the component is included as a multiplier for the vapor pressure of the pure component.
Relative volatility (α)	$\alpha_{ab} = K_a / K_b$	K_a and K_b are the volatilities of components *a* and *b*, respectively. It is necessary to specify which components are being compared when assessing relative volatility.
	$\alpha_{ab} = P_a^o / P_b^o$	α_{ab} can also be expressed using K_a and K_b written in terms of Raoult's and Dalton's Laws as shown above. This reveals that the relative volatility of an ideal solution is given by the ratio of vapor pressures of the pure components (note that temperature dependence is ignored). For non-ideal solutions the activity coefficients of each component are included as multipliers for the respective vapor pressures of the pure components.

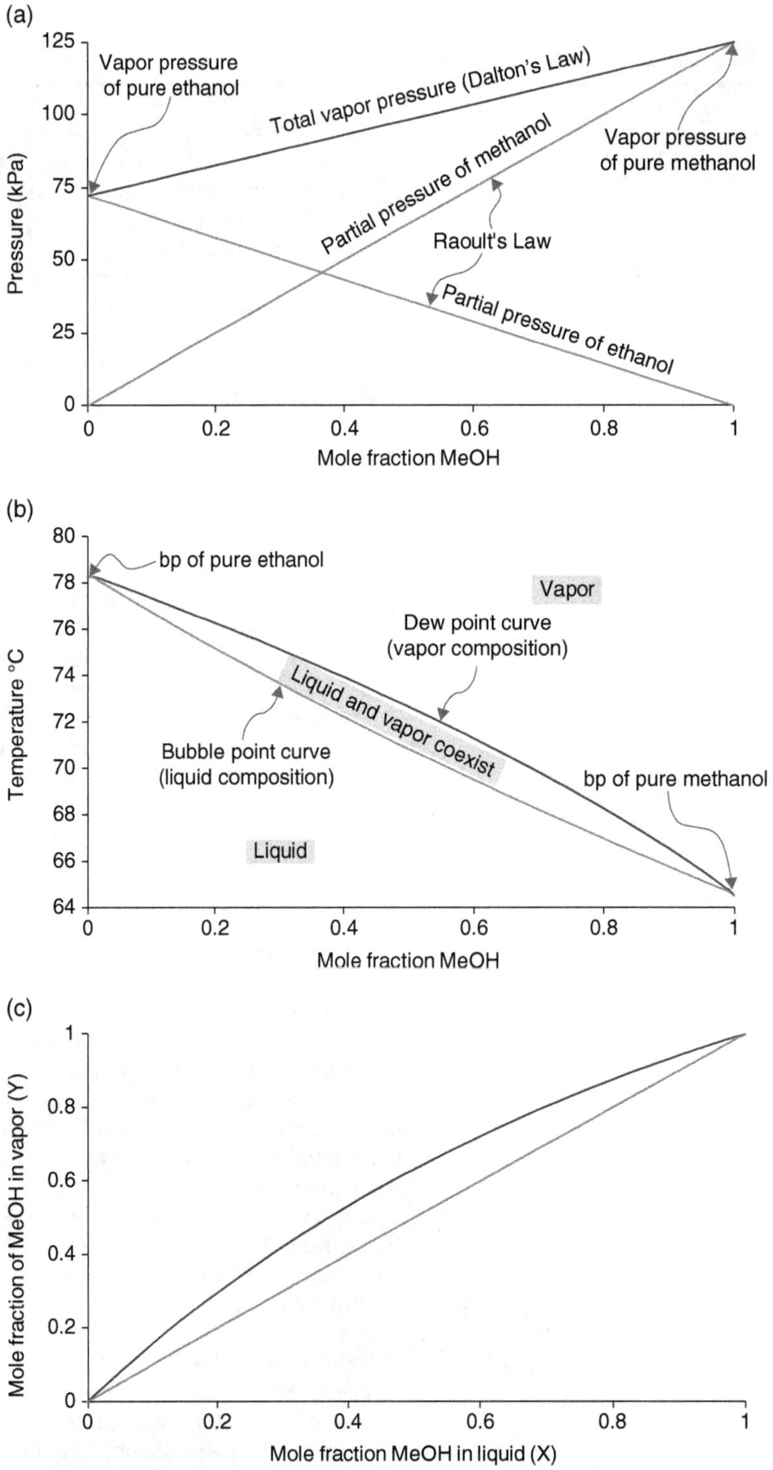

Figure 26.4.1 *VLE diagrams showing (a) isothermal (arbitrary temperature of 70 °C) pressure–composition (P–X) plot (showing partial and total vapor pressures, and Dalton's and Raoult's Laws), (b) isobaric (101.325 kPa) temperature–composition (T–X) plot (highlighting regions for liquid, vapor, and vapor + liquid), (c) X–Y diagram (X = Y diagonal shown for reference) for an ideal mixture of methanol–ethanol,*

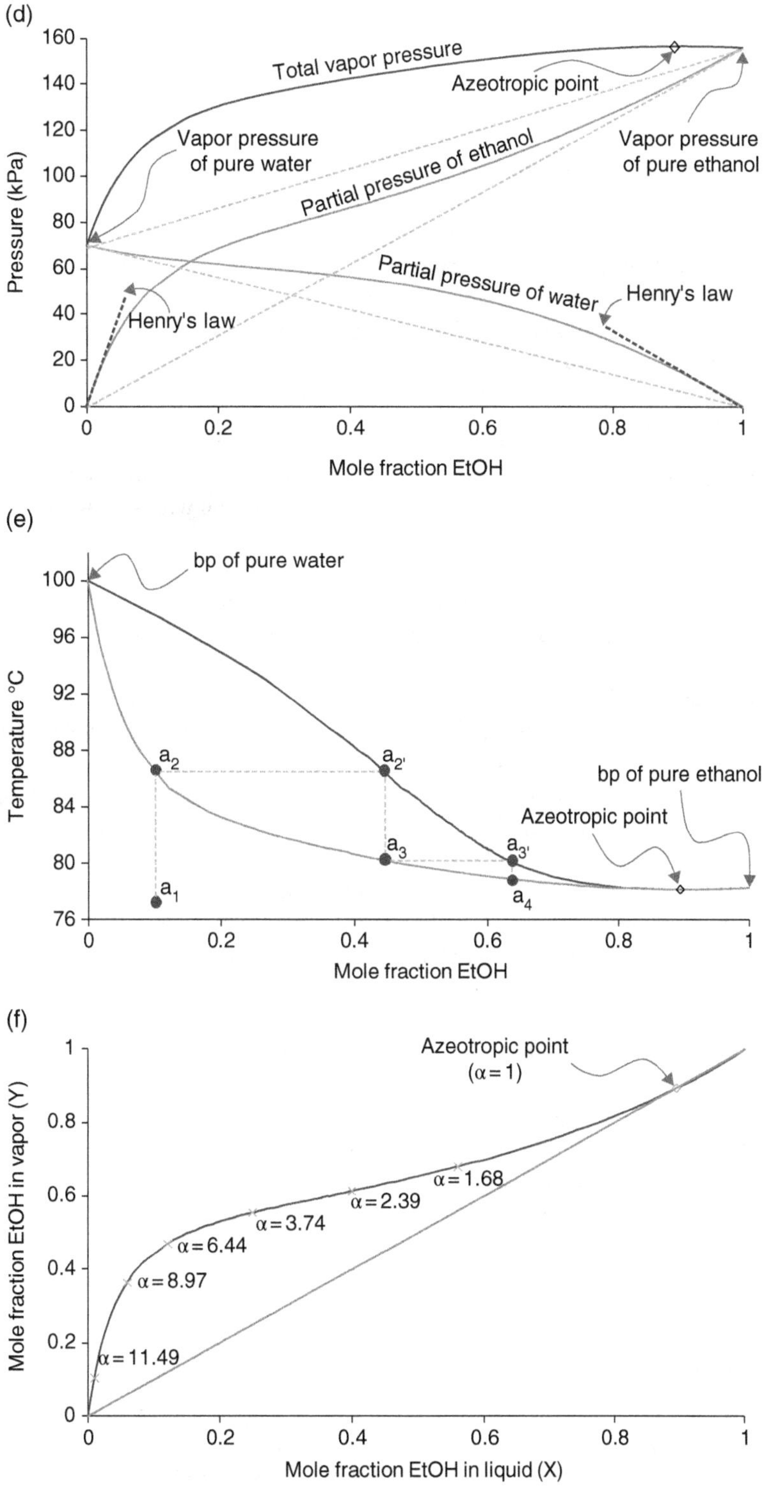

Figure 26.4.1 *(continued) (d) P–X plot (dashed lines show the ideal behavior scenario, dashed red lines show where Henry's Law would apply), (e) T–X plot (dashed lines indicate how distillation separates components based on an arbitrary starting temperature and liquid composition, a_1), and (f) X–Y diagram (with some relative volatilities indicated), for a non-ideal mixture of ethanol–water. Data obtained from www.vle-calc.com/phase_diagram.html or calculated[4] in some instances*

[4] The calculations are beyond the scope of this chapter but involved the use of readily available constants, Antoine and van Laar equations, and Raoult's and Dalton's Laws.

(Chapter 1),[5] This leads to non-ideal behavior and a positive deviation from Raoult's Law (Figure 26.4.1d–f) because the attractive forces between water and ethanol are weaker than those of either pure component.[6] Compared to an ideal solution, ethanol–water mixtures demonstrate higher partial vapor pressures for each component of the mixture, meaning a higher total pressure (Figure 26.4.1d). In a certain ethanol–water composition, these mixtures will also form a solution that has a lower boiling point than either pure component (Figure 26.4.1e).[7] At the azeotropic point, liquid and vapor compositions are identical and further enrichment is not possible with ordinary fractional distillation. Hence, under normal conditions ethanol can only be distilled to 96% alcohol by volume (abv).

26.4.2.3 Separation and enrichment

A plot of temperature versus composition (T-X plot) for ethanol–water (Figure 26.4.1e) has some features added to show how separation by distillation is possible, using a solution with an ethanol mole fraction of 0.1 (roughly 20% abv) as an example:

1. Beginning at point a_1 (or any temperature below the bubble point for this particular composition) the mixture is heated towards boiling until the temperature rises to its bubble point (when the first bubble of vapor forms) of about 87 °C (point a_2).
2. Drawing a horizontal "tie line" to meet the dew point curve (when the first drop of vapor condenses) shows the vapor in equilibrium with this liquid (an equilibrium "stage") is richer in the more volatile component with composition a_2' (ethanol mole fraction of about 0.44).
3. The condensed liquid shown at point a_3, now at a lower bubble point of about 80 °C (approaching that of pure ethanol), also has the same composition as the vapor at a_2', showing that ethanol is enriched compared to the initial liquid composition of 0.1 mole fraction.
4. Repeating this cycle of vaporizing and condensing produces a new VLE stage[8] and vapor composition with an ethanol mole fraction of around 0.64 (point a_3'), which condenses to a liquid with a lower bubble point again (around 79 °C, point a_4); ultimately the azeotrope for ethanol–water can be reached in this way using enough stages.

The above example assumes that each distillation stage, or "plate," is 100% efficient (i.e., equilibrium is reached at each theoretical stage). In practice, most distillation stages are not so efficient, and production of highly purified alcohol from wine requires stills with a large number of physical (actual) stages (i.e., columns containing plates or packing material, see Section 26.4.3 below) that maximize mass transfer rates and more closely approach equilibrium through enhanced contact time and mixing of phases. Nonetheless, the example serves to highlight how the process of distillation separates the components in a liquid mixture because of differences in composition between liquid and vapor phases in equilibrium.

---

[5] Note the relative symmetry of the methanol-ethanol VLE diagram as compared to the ethanol–water diagram (Figure 26.4.1).

[6] In this situation a modified version of Raoult's Law can be used which introduces an activity coefficient, the ratio between actual and ideal partial vapor pressures, for each component to account for the altered behavior.

[7] Conversely, non-ideal behavior of liquid mixtures showing a negative deviation from Raoult's Law can lead to a maximum boiling azeotrope, where the vapor pressures are lower than expected if the solution was ideal, and the boiling point of the mixture at a specific composition is higher than that of either pure component.

[8] This can be thought of as if the vapors were condensed once the still had reached equilibrium and the resulting liquid was distilled again, this time boiling at a lower temperature and producing more enriched vapor. In reality, this occurs within the distillation system with separate stages (mass transfer zones) where rising enriched vapors interact with cooler liquid higher within the still, enabling more ethanol to transfer to the vapor phase while more water condenses into the liquid phase at each stage. A greater number of stages results in more enriched vapor exiting the still.

26.4.2.4 Relative volatility and ease of separations

X–Y diagrams are particularly useful when considering the separation of components by distillation. The distance of the curve from the $X=Y$ diagonal (drawn for reference in Figure 26.4.1c and f) indicates the possible enrichment of the vapor phase with the more volatile component as compared to the liquid phase. Greater enrichment leads to better separation of the components in solution. More specifically, the relative volatilities for the components in a mixture (i.e., the ratio of volatility of one component with that of the other, designated α, Table 26.4.1)[9] can be used to determine the point in distillation where the greatest enrichment occurs. For ethanol–water mixtures, the greatest enrichment occurs for dilute ethanol solutions (high α values, Figure 26.4.1f), whereas it is marginal at higher ethanol concentrations (the curve approaches the diagonal as the composition nears the azeotropic value where $\alpha=1$).

26.4.2.5 VLE of a multicomponent mixture

While considering the distillation of wine as a binary system of ethanol and water is a useful exercise, many other volatile compounds exist in wines and distillates. Termed *congeners*, these are the aroma and flavor components of wine that can be carried over to provide the characteristic sensory attributes of the distillate. They are minor components, so it is assumed that they do not affect the volatility of the ethanol–water matrix or that of each other. Because of their low concentrations, their volatility is often better described by Henry's Law than Raoult's Law, and is usually a function of the ethanol strength of the liquid in addition to temperature (Chapter 1). When ethanol strength is high, many congeners have lower volatility than ethanol itself, and, conversely, in dilute ethanol solutions some congeners have higher volatility than ethanol (Figure 26.4.2). For example:

- Alcohols such as 2-methyl-1-propanol, 2-methyl-1-butanol, and 3-methyl-1-butanol (i.e., higher alcohols, Chapter 6), commonly called fusel alcohols/oils by distillers, are more volatile than ethanol below about 20% abv.
- Methanol is more volatile than ethanol only above around 80% abv (Figure 26.4.2a).
- For ethyl esters (Chapter 7), ethyl acetate is more volatile than ethanol across the composition range (as is diethyl acetal, Chapter 9), ethyl hexanoate is more volatile than ethanol below about 65% abv, and ethyl lactate is always less volatile than ethanol at any ethanol concentration (Figure 26.4.2b).

This knowledge is crucial to understanding the separation of components by distillation of wine and determines the point at which certain congeners are best separated from ethanol depending on the distillation system, as elaborated below.

26.4.3 Batch and continuous distillation

Production of grape spirit involves the separation and enrichment of alcohol from the predominantly aqueous wine matrix. Distillation of wine (approx. 9–12% abv) is used to produce a product with up to 83% abv in the case of brandy and variants (i.e., flavored spirit containing congeners), or up to 96% abv (the azeotropic point) in the case of neutral grape spirit (spiritus vini rectificatus, SVR) used for fortifying wines such as port and sherry. The two methods used to produce these spirits are batch distillation using a pot still and continuous distillation using one or more fractionating columns [16–19].

---

[9] Note the relative volatility of a non-ideal ethanol–water solution strongly depends on the liquid composition because the activity coefficients change with composition. As such, the relative volatility of ethanol with respect to water varies markedly across the composition range.

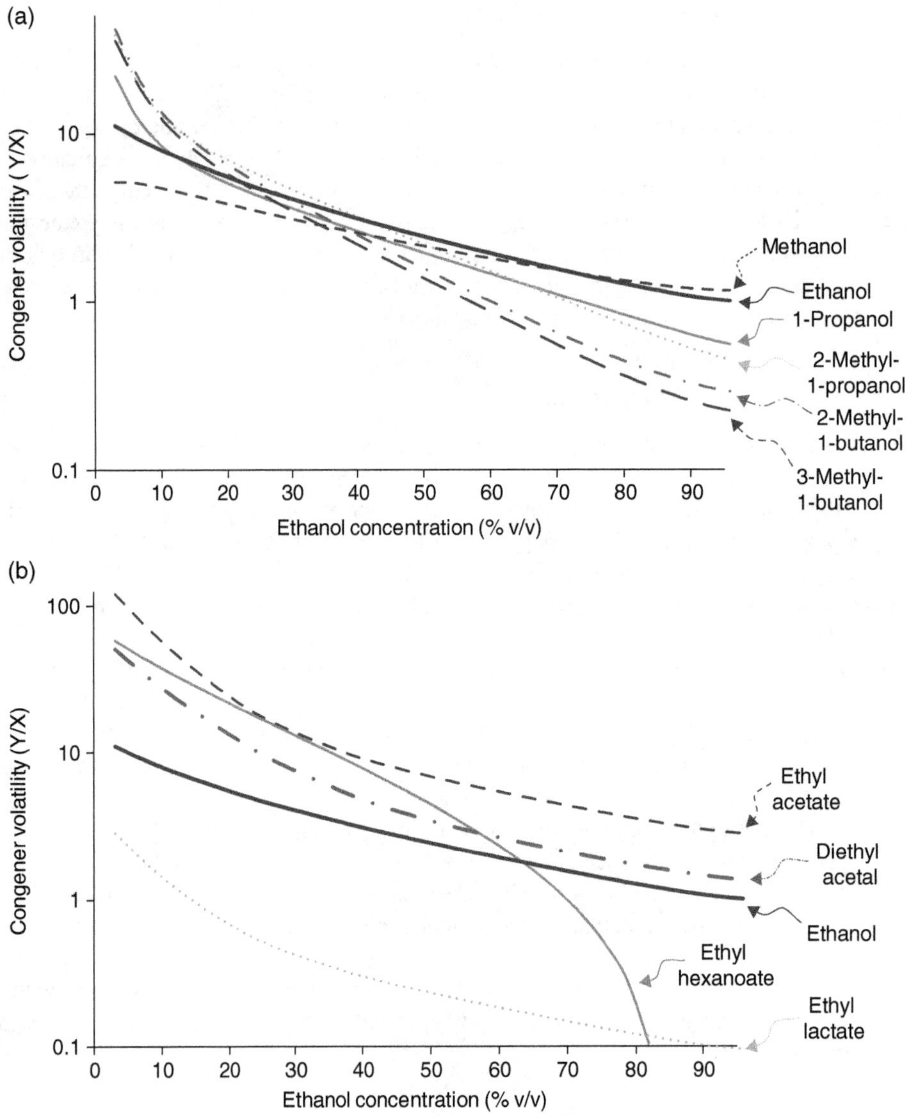

Figure 26.4.2 Volatility data as a function of ethanol concentration in wine distillates for (a) representative alcohols and (b) representative esters and acetal. Calculated from data in Reference [15]

26.4.3.1 Reflux and reflux ratio

Condensing and returning of vapors to the system – known as reflux – can improve the efficiency of a separation and reduce the number of stages required to achieve a certain ethanol concentration. The two types of reflux are:

- Internal reflux, where vapors have cooled and condensed before leaving the column.
- External reflux, where vapors that have exited the column are condensed by a partial condenser (dephlegmator) and returned to the still.

Reflux improves the effectiveness of the separation by allowing greater vapor–liquid contact time, which allows the system to more closely approach equilibrium. For external reflux, the volume of distillate returned to the column divided by the volume being taken off as product gives the reflux ratio, which is an important operational factor used to modify the separation of components. A high reflux ratio, which returns a greater amount of externally condensed distillate, gives greater separation of components (especially important for those with similar boiling points) but also increases the energy requirements and distillation time to produce a given amount of product.

26.4.3.2 Batch distillation with a pot still

Batch distillation for producing grape spirit predates the invention of continuous stills. As the name implies, distillation is performed batch-wise, in a pot still almost always constructed from copper (Figure 26.4.3).[10] A pot still consists of:

- Container (i.e., pot) for the distillation substrate (e.g., wine)
- Heat source
- Column and still head
- Vapor pipe
- Condenser.

Double distillation is performed with a pot still to obtain the required alcohol strength and separation of components – that is, a portion of the initial distillate is returned to the pot and redistilled [11, 12, 20, 21].

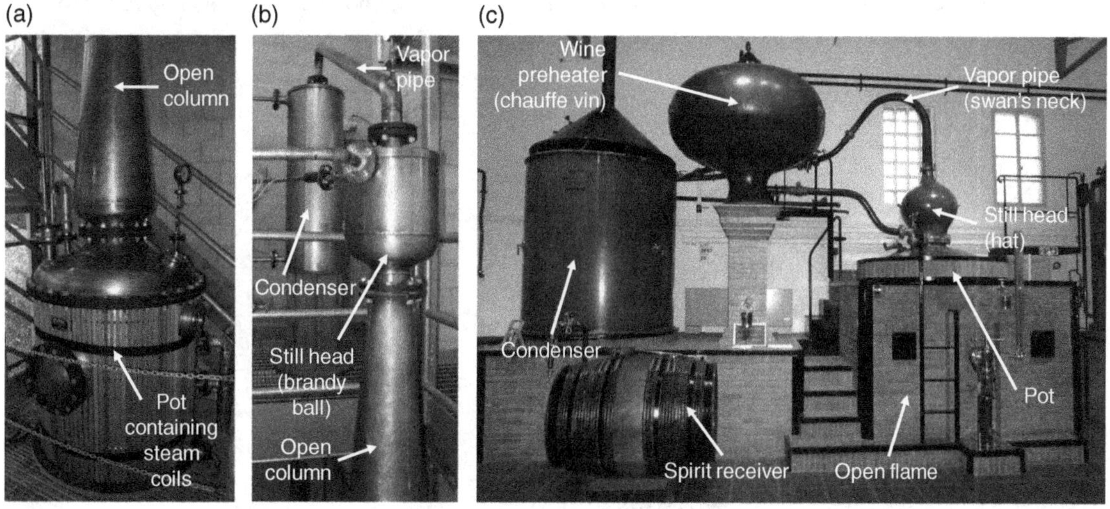

Figure 26.4.3 Images of pot stills used for distillation showing (a) a copper pot still with open column, (b) the top portion of the open column, still head, vapor pipe, and condenser (on the next floor above in the still house), and (c) a Cognac charentais pot still consisting of open flame (gas or wood fire, hidden), copper pot (only the top is visible), still head, vapor pipe which passes through a wine pre-heater (to pre-heat the following batch of wine), and condenser. Source: Reproduced with permission of Michael Hage

[10] Copper provides several advantages for use in stills: it is highly malleable during manufacturing, it has excellent thermal conductivity, and it can strip out malodorous thiols from the vapors.

This method is used for commercial production of wine spirits (e.g., eau de vie, brandy, Cognac, Armagnac, pisco) and grape marc spirits (e.g., grappa and rakia). During both distillation steps, fractions ("cuts") are taken by switching the flow of distillate from one receiving container to another. Typically, these cuts are classified as "heads" (beginning), "hearts" (middle), and "tails" (end). The hearts cut of the first distillation produces "low wine" (*brouillis*) – approximately 30% of initial pot volume – which contains 20–50% abv; heads (60% abv) from the first distillation run are less than 1% of pot volume whereas tails (3% abv) are approximately 6% of pot volume (Figure 26.4.4). These cuts contain various congeners relative to their volatilities (see Figure 26.4.2). The heads cut contains a relatively high concentration of ethanol, but is enriched in highly volatile, malodorous compounds like acetaldehyde and ethyl acetate.[11] Conversely, the tails cut is low in alcohol and contains low-volatility congeners such as acetic acid, ethyl lactate, and 2-phenylethanol. The alcohol in the heads and tails cuts can be recovered by recycling these cuts into the next batch of wine destined for distillation (or simply by diluting with water and redistilling). The first distillation is also referred to as a "stripping run" since it will strip out the majority of the ethanol from water and non-volatiles; this generally requires recovery of about 35% of the initial volume in the pot.

The batch process requires four first distillations (or a larger pot still, as is often the case in Cognac) to produce enough low wine to fill the pot for the second distillation, where again heads, hearts (sometimes several cuts), and tails are collected. The hearts cut contains the spirit used for brandy – up to about 40% of initial pot volume and about 70–80% abv – whereas the heads (around 1% of pot volume and 75% abv) and tails (about 6% of pot volume and 3% abv) can be recycled, as mentioned above. Late hearts cuts may be redistilled with another batch of low wine. The behavior of congeners during the second distillation is quite different from that of the first, due to the higher ethanol strength of both the low wine and subsequent distillate, and provides greater ability to separate esters and aldehydes (heads cut) and fusel alcohols (tails cut) from the spirit (see Figure 26.4.2). Depending on the pot still design, cool water can also be run through a

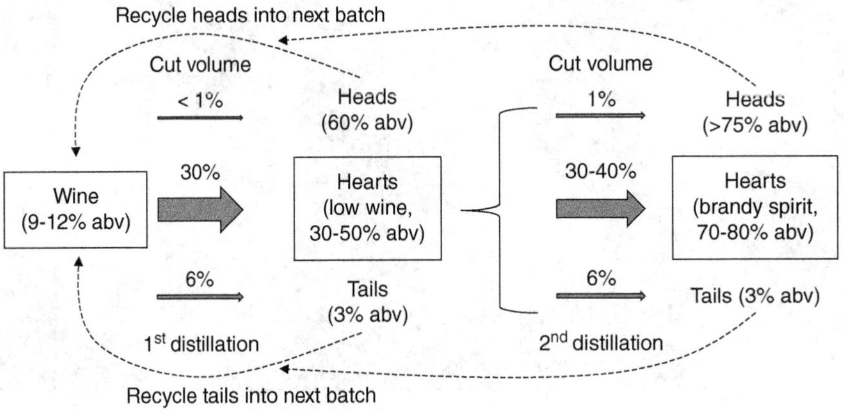

Figure 26.4.4 Schematic of cuts made during the double distillation process for transforming wine into low wine and then brandy spirit, showing alcohol strength and cut volume (as a % of initial pot content), and how the heads and tails can be recycled (or diluted with water and redistilled). Late hearts cuts made during the second distillation may be redistilled with another batch of low wine if necessary

[11] More acetaldehyde in a base wine would thus result in a larger heads fraction and a less efficient distillation process. Brandy producers will typically not use SO_2 additions before fermentation to avoid high levels of bound acetaldehyde accumulation (Chapter 22.1).

brandy ball (Figure 26.4.2b) during the second distillation to provide external reflux, further improving the separation of components.

Instead of an open column and brandy ball, a pot still may have a rectifier column containing 20–30 distillation trays (actual stages) that improve internal reflux, as well as a reflux condenser to enhance separation. This configuration is a batch process but overcomes the need for double distillation, and can provide higher strength spirit (around 92% abv) with reduced volume of the cuts taken. Brandy strength spirit can be produced by lowering the reflux ratio with less cooling to the reflux condenser. An Armagnac still[12] is similar to the pot rectifier, having a tray-filled rectifying column but operating as a continuous still, as described in the next section.

26.4.3.3 Continuous distillation with fractionating columns

In contrast to batch distillation in a pot still, continuous distillation involves a constant feed of wine (or low wine) and input of energy balanced with a continual production of distillates and stillage (waste), that is, a dynamic equilibrium. Continuous distillation can have lower operating costs than batch distillation due to higher rates of throughput and ability to automate, but is also more expensive to install and complex to operate. Continuous stills are usually made of stainless steel components, and a system may utilize one column, as is often the case for brandy production (Figure 26.4.5), or can consist of multiple columns (Figure 26.4.6) to produce highly rectified, neutral SVR [22]. There are many column designs and operational details that are beyond the scope of this chapter and can be found elsewhere [8, 23]. Unlike the open column typical of a pot still, a continuous still column can contain horizontal trays of different designs (e.g., sieve, bubble cap, fixed, or floating valves) or packings (random such as Raschig/Pall rings and saddles, or structured such as gauze and mesh bundles) [5, 24]. These devices within a fractionating column provide very large surface areas for liquid and vapor to interact, increasing internal reflux and improving the vapor–liquid mass transfer efficiency. The distillate may be removed at multiple locations along a column, based on the volatility of congeners as a function of ethanol strength. Highly volatile congeners are taken off overhead, much like with a pot still. Less-volatile congeners (with higher boiling points) such as fusel alcohols can concentrate on trays lower in the column, below the take-off point of the desired spirit.

A continuous single-column analyzer with bubble cap trays (Figure 26.4.5) can be used for brandy production. This may be part of a larger multicolumn still arrangement or can stand alone with some variations in design (e.g., use of sieve or valve trays, and ability to produce high-strength spirit or draw off fusel oils due to tray configuration) [11]. Wine is pre-heated and enters a few trays below the top of the column, where it is heated by live steam. The bottom section of the column (stripping section) contains more bubble cap trays than the top, where rectification takes place (concentrating section). The more volatile heads (e.g., acetaldehyde) are cooled and separated, and some vapors are returned as external reflux. The remaining vapors containing ethanol and congeners are cooled and sent to a smaller column to further remove heads components, with brandy spirit exiting from the bottom of the column. Depending on the operating conditions (e.g., external reflux, feed rate, heat input) the still can be run flexibly enough to produce different distillate strengths, so if the quality of the feed is high enough[13] then brandy spirit with an alcohol content >80% abv

---

[12] Strict regulations govern the production of Armagnac and Cognac. Batch distillation in a pot still is mandatory for Cognac whereas Armagnac can be produced using a pot still, much like those used for Cognac, or more commonly with a continuous Armagnac still (armagnacais).

[13] Production of quality brandy typically requires wine that is made especially for distillation, as is the case for Cognac and Armagnac. Wines of neutral grape varieties from a clean fermentation (one where minimal fusel oils, aldehydes, and sulfur compounds are produced), having moderate alcohol levels, high acid (microbial stability, better aroma quality), and no added SO_2, are distilled soon after fermentation. Some lees may be included during distillation for additional flavor (e.g., from fatty acids and their ethyl esters).

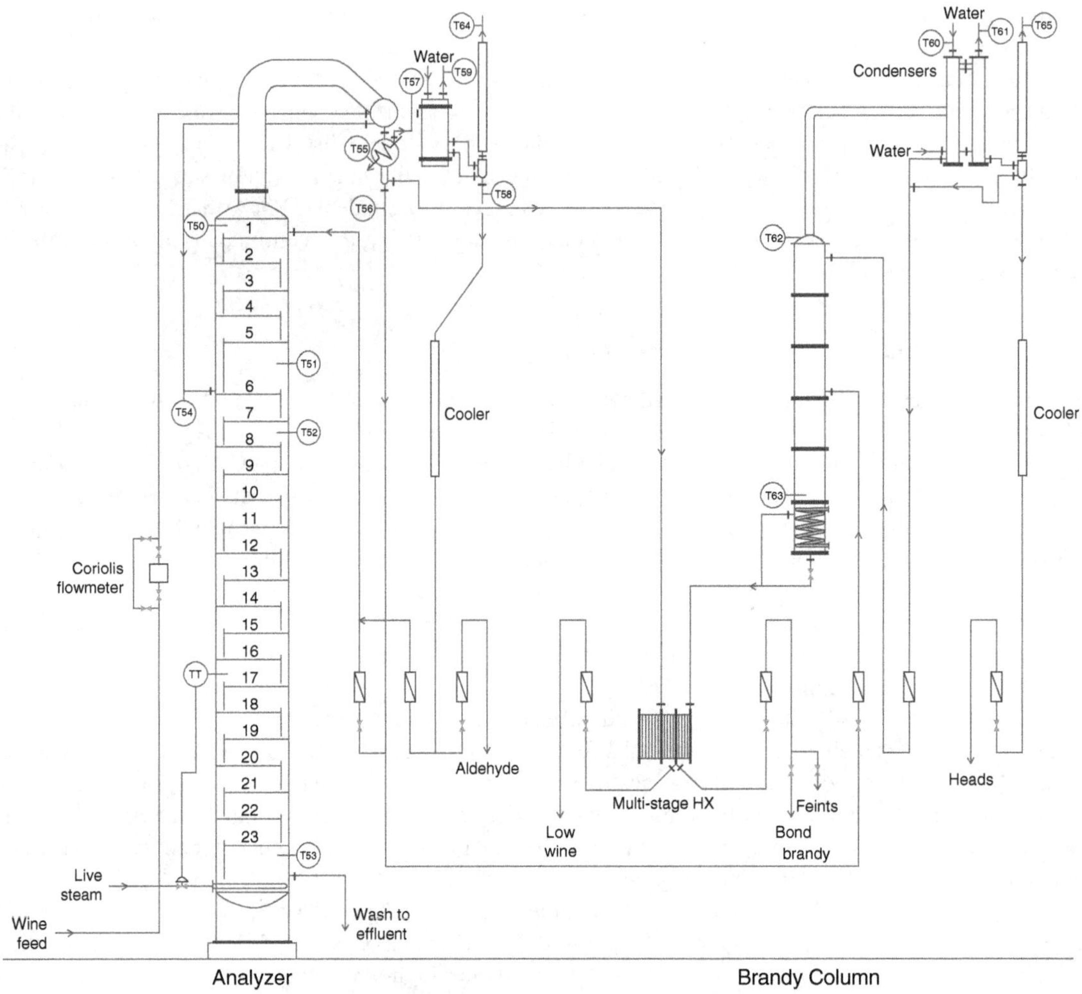

Figure 26.4.5 *Single-column continuous still (analyzer) containing bubble cap trays and smaller column containing supports and packing (brandy column, concentrates heads) used to produce brandy spirit at up to 83% abv. When lower quality wine or lees are used the still can be operated to produce low wine at around 40% abv destined for multicolumn distillation. Source: Reproduced with permission of Tarac Technologies*

can be targeted (maximum % abv depends on local regulations). On the other hand, if lees or low-quality wine are distilled, then the still is run so as to produce a low wine at about 40% abv, which is then further distilled in a multicolumn system such as the one described below.

The three-column still example in Figure 26.4.6 consists of a purifier column to remove heads, a rectifier column to concentrate the alcohol, and a methanol column to remove methanol. The feed enters near the top few bubble cap trays of a purifier column heated by live steam, and hot hydroselection water is added to the trays above to enable extractive distillation to occur. Thus, the purpose of the hydroselection water (and live steam) is to raise the boiling point of the mixture by diluting the ethanol to about 20% abv or less. This increases the relative volatility of the heads components (refer to Figure 26.4.2) and reduces their miscibility in the diluted solution, allowing them a greater chance to boil off. Heads vapors pass through a small packed

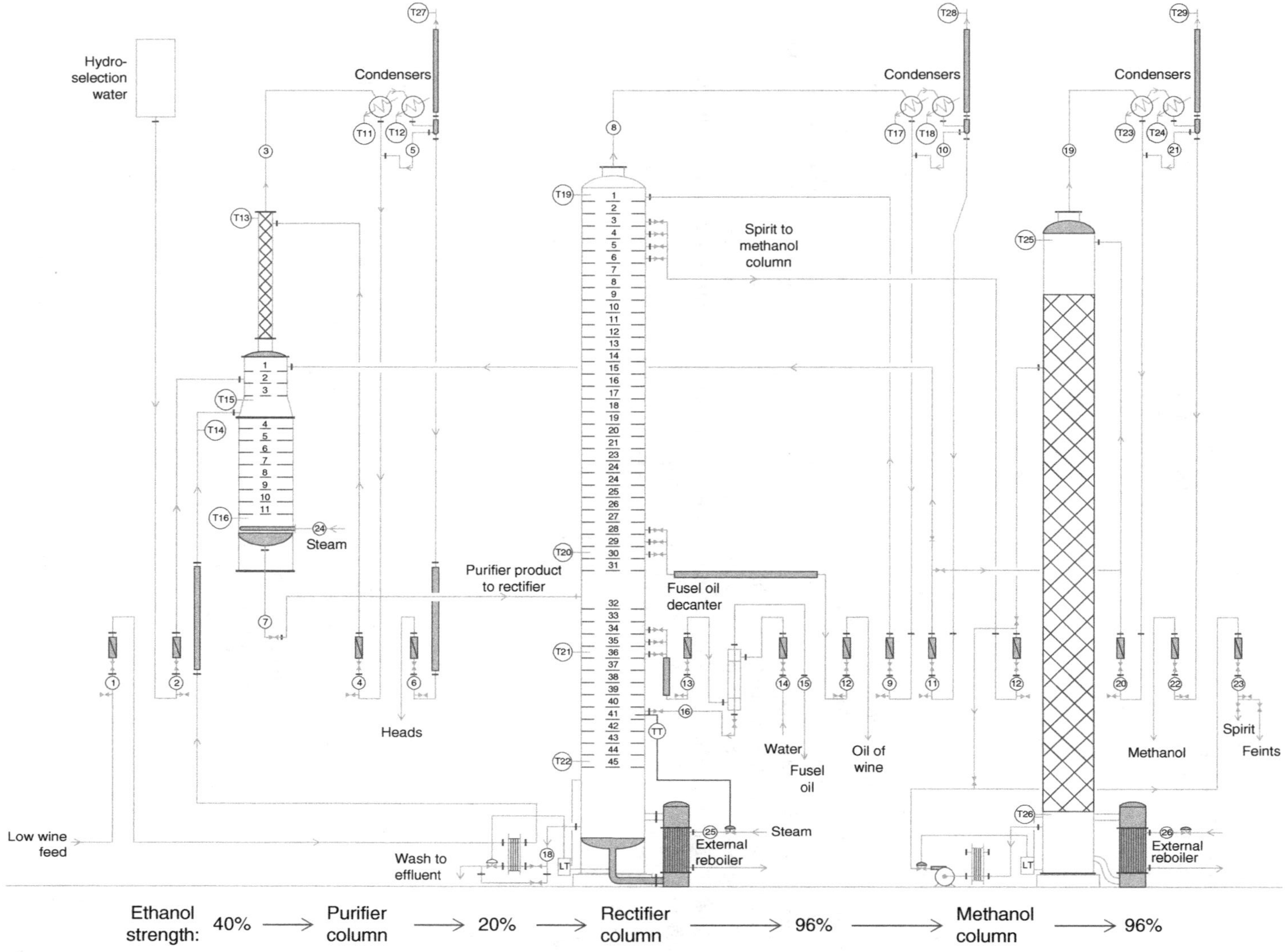

Figure 26.4.6 Three-column continuous still consisting of purifier, rectifier, and methanol columns used to produce 96% abv neutral spirit from approximately 40% abv low wine. Note the alcohol strengths entering and leaving the columns and that live steam is used to heat the purifier (where hydroselection water is also added) but not in the rectifier or methanol column, which feature external reboilers for heating. Source: Reproduced with permission of Tarac Technologies

column and are returned as external reflux or cooled and separated. The weaker ethanol solution leaves from the bottom of the purifier and enters the rectifier column at a bubble cap tray with about the same ethanol content (feed tray), where it is heated by a reboiler to rectify the ethanol to 96% abv. The trays below the feed point strip ethanol from the dilute solution (stripping section, fewer trays) whereas the trays above concentrate the ethanol and remove other congeners (concentrating section, contains twice the number of trays to enable rectification to 96% abv; compare this with the analyzer column in Figure 26.4.5).

The concentrating section enables separation of so-called "oils of wine" (i.e., C_3 and C_4 alcohols concentrating at around 75% abv) and "fusel oils" (i.e., C_5 alcohols concentrating at around 40% abv) due to their lower volatility than ethanol at 96% abv (refer to Figure 26.4.2). These congeners collect at a higher temperature (i.e., closer to the reboiler, at around 80–90 °C) on trays lower down in the column and can be drawn off (oils of wine above the feed plate and fusel oils below).[14] Heads are condensed overhead and returned to the rectifier as external reflux or sent to the purifier to be redistilled, and the spent waste exits from the bottom, pre-heating the low wine entering the purifier.

The ethanol fraction that emerges from the spirit trays near the top of the rectifier can contain a relatively high concentration of methanol (see Chapter 6 for more on methanol).[15] Methanol is more volatile than ethanol at >70% abv, but the difference in volatilities is not that great (refer to Figure 26.4.2), and a very efficient packed column[16] called a demethylizer (or methanol column) is used to obtain the separation. The highly rectified ethanol is gently reboiled as it flows down to exit from the bottom of the methanol column, whereas methanol and other remaining congeners are condensed overhead. Again, heads from the column can be returned as external reflux or recycled back to the purifier.

Variations are possible in the column designs of multicolumn stills, with different numbers of trays and combinations of columns, such as a packed column purifier or a methanol column with trays in the lower region and packing in higher regions. In any case, the principles are the same, and differences in volatility relative to the ethanol strength of the liquid in the column are exploited to remove congeners and concentrate the ethanol. Wine, lees, and other lower quality feedstocks with higher concentrations of unwanted congeners, for example grape marc with higher amounts of aldehydes and methanol, require greater effort to separate the ethanol and produce neutral SVR. In these cases additional columns need to be used, such as a five-column still with packed and bubble cap tray columns (wash or analyzer columns similar to Figure 26.4.5) to produce low wine and remove volatiles at low ethanol strength, followed by a rectifier with bubble cap trays and two packed methanol columns to produce SVR with a very low congener content.

26.4.3.4 Alcohol reduction in wine

Beyond its role in spirits production, different forms of distillation can feature in the production of reduced alcohol wines [25, 26]. Specialized distillation equipment in the form of an evaporator or spinning cone column use lower temperature distillation under vacuum for alcohol removal, producing an ethanol-rich distillate that may be fractionally distilled in the conventional way to rectify and recover the ethanol. Permeate obtained from wine that has undergone alcohol reduction by reverse osmosis or other membrane techniques (Chapter 26.3) can also be distilled in the conventional way to remove the ethanol, where the water-rich

---

[14] Components such as heads and fusel oils that are separated and drawn off during distillation also contain ethanol. Collectively termed feints, these components are sent to a feints still to recover the ethanol prior to disposal of the remaining waste.

[15] Although higher trays also contain high-strength alcohol, the spirit is not drawn from those trays because it contains a small amount of the more volatile heads components not removed by the purifier, and these are quite soluble at high ethanol strength.

[16] Instead of actual trays, the number of which determines the ability to separate components, packed columns are described in terms of their height, which determines the number of equilibrium stages. The height equivalent of a theoretical plate (HETP) is used to express efficiency of a packed column; the smaller the HETP, the higher the theoretical plate count and the better the separation for a given column length.

remainder is returned to the original wine. This approach has the advantage that only the low molecular weight fraction of the wine will be subjected to distillation, and should result in fewer changes to the wine flavor. Whatever the distillation technique, the principles remain the same and involve mass transfer as a result of liquid and vapor interacting.

26.4.4 Spirit composition and cask maturation

Whereas processing decisions and equipment selection can have many effects on distillate composition, the degree of rectification (ethanol content) is of particular importance because higher rectification indicates lower congener concentrations. Brandy production involves lower rectification, since congeners contribute to the desirable sensory properties of the product, while SVR is used when only the alcohol is required and not any of the flavors, such as when fortifying some ports and sherries. Table 26.4.2 provides an idea of the concentrations of congeners found in brandies (including Cognac and Armagnac) and SVR arising from pot or continuous stills. In particular, there are lower levels of heads components (i.e., acetaldehyde, diethyl acetal, ethyl acetate) and furfural in continuous still spirits, and substantially elevated levels of methanol and heads components in pot still marc spirit. Furthermore, some difference in the concentration of a number of congeners (especially fusel oils and ethyl lactate) between brandies and Armagnacs obtained from continuous stills is evident due to variations in still design. Note that the maximum limits of congeners permitted in the European Union for SVR (96% abv) are markedly lower than typical values observed in other spirits as a result of the multicolumn stills used to produce highly rectified neutral alcohol.

Brandy spirit is usually aged in oak casks where some oak components, including aroma compounds are extracted [14, 17]. Reactions that occur during wine aging (Chapter 25) can also potentially contribute to aging of spirits, with some important differences:

- Brandy has a higher pH than wines (4.2–4.4), and thus acid-catalyzed reactions like ester hydrolysis and formation will occur proportionally more slowly. Interestingly, these reactions appear to occur more rapidly in barrel-aged brandies than those aged in glass [32], possibly because organic acids that lower the pH can be extracted from oak wood.
- Brandy will have a lower transition metal concentration, a lower phenolic content, and negligible SO_2 content as compared to most wines, and thus will consume oxygen much more gradually. While aldehyde concentrations are typically higher in brandy than in wine, this is because brandy is often stored semi-aerobically. When both wine and brandy are exposed to large amounts of air, aldehydes will accumulate in wine more rapidly.

The combination of aging in new and old oak of different origins for various periods of time (decades in some cases) and blending of different spirits produces the characteristics of the commercial spirit. Additionally, most spirits will undergo *reduction* with distilled water to dilute it from cask to bottle strength (often stepwise over a period of time) [20, 21]. Very old brandies cellared in oak for many years will naturally lose alcohol over time due to evaporation, and may already be at the appropriate strength at the time of blending and bottling so no reduction is required.

Beyond ethanol, the aroma of Cognac is reported to originate largely from fermentation derived congeners – particularly branched-chain ethyl esters, and fusel alcohols and their corresponding aldehydes – and wood-derived odorants. The only high-odor activity odorant of likely grape origin reported in Cognac was β-damascenone [33], and concentrations over 100 μg/L have been reported in some brandies [34], which is an order of magnitude higher than most wines (Chapter 8). Such elevated concentrations of β-damascenone likely arise due to its liberation from glycosylated precursors as a result of heating of the wine (Chapter 23.1).

Table 26.4.2 *Comparison between generic wine data and the composition of major volatiles in a selection of grape spirits produced by batch or continuous distillation. Data from Section A and references [12] and [27] to [31]*

Congener (mg/L of absolute ethanol)[a]		Pot Still				Continuous Still		
	Wine	Brandy	Cognac	Armagnac	Marc	Brandy	Armagnac	SVR[b]
Acetaldehyde	110	103–202	140–276	191–247	517–767	57–101	59–69	50
Diethyl acetal	445	58–102	58–101	58–81	277–432	21–100	33–54	–
Methanol	370	372–542	102–510	389–408	5769–7917	138–490	392–433	300
1-Propanol	280	408–495	294–372	277–285	439–697	285–490	269–287	50
2-Methyl-1-propanol	600	385–640	1092–1335	907–1030	626–878	248–352	884–920	
2-Methyl-1-butanol	465	1742–2022[c]	2797–2910[c]	642–729	427–579	985–1342[c]	659–669	
3-Methyl-1-butanol	2020			2660–2860	1185–2007		2420–2530	
Ethyl acetate	370	390–535	440–512	254–331	738–1669	148–225	228–246	13
Ethyl hexanoate	4	6–10	6–9	3	5–14	2–7	3	
Ethyl decanoate	3	31–50	15–44	4–6	16–69	12–24	4	
Ethyl lactate	3030	55–69	184	340–347	171–257	5–72	390–505	
Furfural	90	33–34	24–29	10–14	–	2–4	2	ND[d]

[a] Congeners are often expressed in terms of liters of pure alcohol, so for a spirit that is 40% abv, the listed mg/L value would be multiplied by 0.40 to get the actual mg/L of congener in that spirit. The data for wine are based on a value of 11% abv.

[b] Maximum values according to European Union regulations. Total aldehydes expressed as acetaldehyde, total higher alcohols as sum of individuals, and total esters expressed as ethyl acetate.

[c] Isomers of 2-methyl-1-butanol and 3-methyl-1-butanol reported together.

[d] Not detectable.

References

1. Huang, H.T. (2000) Evolution of distilled wines in China, in *Part V: Fermentations and food science*, Cambridge University Press, Cambridge, UK, pp. 203–231.
2. Sherwood, T.F. (1945) The evolution of the still. *Annals of Science*, **5** (3), 185–202.
3. Liebmann, A.J. (1956) History of distillation. *Journal of Chemical Education*, **33** (4), 166.
4. Lea, A.G.H. and Piggott, J.R. (eds) (2003) *Fermented beverage production*, 2nd edn, Kluwer Academic/Plenum Publishers, New York.
5. Bennett, B.L. and Kovak, K.W. (2000) Optimize distillation columns. *Chemical Engineering Progress*, **96**, 19–34.
6. Luyben, W.L. (2006) Fundamentals of vapor–liquid phase equilibrium (VLE), in *Distillation design and control ssing aspen simulation*, 2nd edn (ed. Luyben, W.L.), John Wiley & Sons, Inc., Hoboken, New Jersey, pp. 1–26.
7. Kiss, A.A. (2013) Basic concepts in distillation, in *Advanced distillation technologies*, John Wiley & Sons, Ltd, Chichester, UK, pp. 1–35.
8. Jacques, K.A., Lyons, T.P., Kelsall, D.R. (eds) (2003) *The alcohol textbook*, 4th edn, Nottingham University Press, Nottingham.
9. Meirelles, A.J.A., Batista, A.C., Scanavini, H.F.A., *et al.* (2009) Distillation applied to the processing of spirits and aromas, in *Extracting bioactive compounds for food products* (ed. Meireles, M.A.A.), CRC Press, Boca Raton, FL, pp. 75–135.
10. Buglass, A.J. (2011) *Handbook of alcoholic beverages: technical, analytical and nutritional aspects*, John Wiley & Sons, Ltd, Chichester, West Sussex, UK.
11. Guymon, J.F. (1974) Chemical aspects of distilling wines into brandy, in *Chemistry of winemaking*, American Chemical Society, Washington, DC, pp. 232–253.
12. Léauté, R. (1990) Distillation in alambic. *American Journal of Enology and Viticulture*, **41** (1), 90–103.
13. Silva, M.L., Macedo, A.C., Malcata, F.X. (2000) Review: steam distilled spirits from fermented grape pomace. Revision: Bebidas destiladas obtenidas de la fermentación del orujo de uva. *Food Science and Technology International*, **6** (4), 285–300.
14. Tsakiris, A., Kallithraka, S., Kourkoutas, Y. (2014) Grape brandy production, composition and sensory evaluation. *Journal of the Science of Food and Agriculture*, **94** (3), 404–414.
15. Martin, A., Carrillo, F., Trillo, L.M., Rosello, A. (2009) A quick method for obtaining partition factor of congeners in spirits. *European Food Research and Technology*, **229** (4), 697–703.
16. Piggot, R. (2003) From pot stills to continuous stills: flavor modification by distillation, in *The alcohol textbook*, 4th edn (eds Jacques, K.A., Lyons, T.P., Kelsall, D.R.), Nottingham University Press, Nottingham, pp. 255–266.
17. Buglass, A.J., McKay, M., Lee, C.G. (2011) Brandy, in *Handbook of alcoholic beverages: technical, analytical and nutritional aspects* (ed. Buglass, A.J.), John Wiley & Sons, Ltd, Chichester, West Sussex, UK, pp. 574–594.
18. Buglass, A.J., McKay, M., Lee, C.G. (2011) Grape and other pomace spirits, in *Handbook of alcoholic beverages: technical, analytical and nutritional aspects* (ed. Buglass, A.J.), John Wiley & Sons, Ltd, Chichester, West Sussex, UK, pp. 595–601.
19. Bertrand, A. (2003) Armagnac, brandy, and Cognac and their manufacture, in *Encyclopedia of food sciences and nutrition*, 2nd edn (ed. Caballero, B.), Academic Press, Oxford, pp. 584–601.
20. Cantagrel, R. and Galy, B. (2003) From vine to Cognac, in *Fermented beverage production*, 2nd edn (eds Lea, A.G.H. and Piggott, J.R.), Kluwer Academic/Plenum Publishers, New York, pp. 195–212.
21. Bertrand, A. (2003) Armagnac and wine-spirits, in *Fermented beverage production*, 2nd edn (eds Lea, A.G.H. and Piggott, J.R.), Kluwer Academic/Plenum Publishers, New York, pp. 213–238.
22. *Technical Symposium on Australian grape spirit and brandy production* (1978) Waite Agricultural Research Station, Urrbrae SA, Adelaide, SA, 12–13 June 1978, The Australian Wine Resarch Institute.
23. Kiss, A.A. (2013) Design, control and economics of distillation, in *Advanced distillation technologies*, John Wiley & Sons, Ltd, Chichester, UK, pp. 37–65.
24. Pilling, M. and Holden, B.S. (2009) Choosing trays and packings for distillation. *Chemical Engineering Progress*, **105**, 44–50.
25. Pickering, G.J. (2000) Low- and reduced-alcohol wine: a review. *Journal of Wine Research*, **11** (2), 129–144.

26. Schmidtke, L.M., Blackman, J.W., Agboola, S.O. (2012) Production technologies for reduced alcoholic wines. *Journal of Food Science*, **77** (1), R25–R41.
27. Bertrand, A. (1989) Role of continuous distillation process on the quality of Armagnac, in *Distilled beverage flavour: recent develoments* (eds Piggott, J.R. and Paterson, A.), Ellis Horwood Ltd., Chichester, UK, pp. 97–115.
28. Panosyan, A.G., Mamikonyan, G.V., Torosyan, M., *et al.* (2001) Determination of the composition of volatiles in cognac (brandy) by headspace gas chromatography–mass spectrometry. *Journal of Analytical Chemistry*, **56** (10), 945–952.
29. López-Vázquez, C., Herminia Bollaín, M., Berstsch, K., Orriols, I. (2010) Fast determination of principal volatile compounds in distilled spirits. *Food Control*, **21** (11), 1436–1441.
30. EC Regulation No. 110/2008 (2008).
31. Etiévant, P.X. (1991) Wine, in *Volatile compounds in foods and beverages*, 1st edn (ed. Maarse, H.), Marcel Dekker, New York, pp. 483–546.
32. Onishi, M., Guymon, J.F., Crowell, E.A. (1977) Changes in some volatile constituents of brandy during aging. *American Journal of Enology and Viticulture*, **28** (3), 152–158.
33. Uselmann, V. and Schieberle, P. (2015) Decoding the combinatorial aroma code of a commercial Cognac by application of the sensomics concept and first insights into differences from a German brandy. *Journal of Agricultural and Food Chemistry*, **63** (7), 1948–1956.
34. Sefton, M.A., Skouroumounis, G.K., Elsey, G.M., Taylor, D.K. (2011) Occurrence, sensory impact, formation, and fate of damascenone in grapes, wines, and other foods and beverages. *Journal of Agricultural and Food Chemistry*, **59** (18), 9717–9746.

27

Additives and Processing Aids

27.1 Introduction

The standard of identity for wine in most wine-producing countries requires that wine be solely the product of complete or partial alcoholic fermentation of grapes, but also allows for additions of other materials to juice or wine. Materials used in winemaking must be *explicitly allowed*, that is, if a material is not expressly permitted for use in winemaking, then it is assumed to be forbidden.[1] Some of these additions have been practiced for centuries or even millennia, as in the case of "plastering" a wine with gypsum ($CaSO_4$) to lower wine pH,[2] but others date to the last decade, and new additions are approved periodically.[3]

27.2 Regulations and terminology

As a general rule, approved materials are expected to leave the wine recognizable as wine, or as stated in the United States Code of Federal Regulations, 27 CFR 24.246 [1]:

> *Materials used in the process of filtering, clarifying, or purifying wine may remove cloudiness, precipitation, and undesirable odors and flavors, but the addition of any substance foreign to wine which changes the character of the wine, or the abstraction of ingredients which will change its character, to the extent inconsistent with good commercial practice, is not permitted....*

[1] Most countries will allow for wineries to petition for the use of unapproved additions, particularly in cases where an addition is approved for use in other foods and also used in winemaking in other countries. Furthermore, countries will often have separate categories for wines that are produced with unapproved (but food-safe additives) so long as the wine is labeled appropriately, for example, "wine with natural flavors."

[2] As one example, *The Rational Manufacture of American Wines*, published by Osterreicher & Co. in 1872, describes the use of many modern wine additives, including tartaric acid, egg whites, isinglass (for fining), sulfur dioxide (through burning of sulfur wicks), and cellulose paper (as a filtration aid). Several other questionable – and currently illegal – additions are described, including glycerol, orrisroot, butyric ether (aka ethyl butyrate!), and sulfuric acid. A chapter on "compounding" to simulate prestigious European wines like red Bordeaux is also included.

[3] Similar statements can be made about approval of new winemaking processes, such as electrodialysis or reverse osmosis.

Understanding Wine Chemistry, First Edition. Andrew L. Waterhouse, Gavin L. Sacks, and David W. Jeffery.

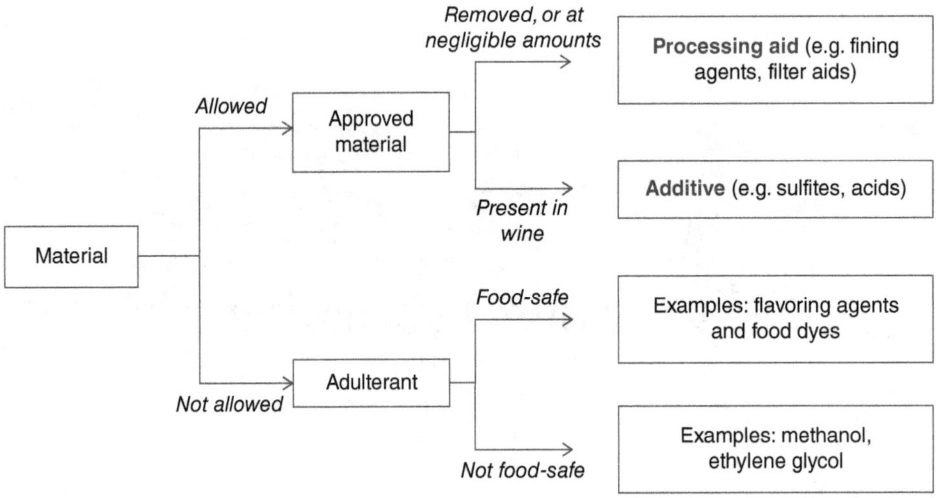

Figure 27.1 *Flow chart showing the distinctions among additives, processing aids, and adulterants*

Wine additions are distinguished from wine *adulteration* (Figure 27.1), in which unallowable compounds are added to improve the flavor or color of a wine or to decrease production costs, for example, because the adulterant is less expensive than the wine. Adulterants may be safe for use in food but disallowed because they are not a natural component of a wine (e.g., food dyes to increase red color), or they may be unsafe for use in food (e.g., ethylene glycol to increase perceived body, Chapter 28). An additional distinction can be made between true *additives* versus *processing aids* [2]:

- Additives describe compounds that are expected to stay in the finished product to exert their effect (e.g., bisulfite, tartaric acid).
- Processing aids are substances that are added during processing but remain at undetectable or insignificant concentrations in the finished product, either because they are removed intentionally, are converted into other naturally occurring substituents, or else were present at minimal concentrations throughout processing. Some processing aids leave traces that are problematic for a segment of the population, such as allergenic protein fining agents (Chapter 26.2).

Regulations that allow for additions will usually specify additional restrictions regarding usage, which may include:

- The maximum allowable addition and/or the maximum residual amount in the finished wine.
- A description of when the addition may take place (e.g., before fermentation versus in the finished wine), or the types of conditions that the material may be used to remedy.
- Limits on impurities, origins, or methods of preparation of the materials.

Such restrictions can be seen in the following specific examples from different regions. Acacia (gum arabic) is approved for use in the United States (27 CFR 24.246) [1] at concentrations up to 1.92 g/L to stabilize wine by preventing potassium bitartrate precipitation (Chapter 26.1). Other uses, such as improving mouthfeel, are not stated and thus not allowed. Ferrocyanide fining (Chapter 26.2) is legal according to Australian and New Zealand regulations (Standard 4.5.1) [3] but winemakers are required to leave residual iron in the wine after treatment (as an assurance that the wine is not overfined, Chapter 26.2). Furthermore, among a multitude of other limits, OIV regulations (International Oenological Codex) [4] stipulate a maximum level of free cyanide upon testing a solution of potassium ferrocyanide.

Finally, most wine regions require that winemakers record the identity, amount, and date of any material used during winemaking (with the exception of filtration aids), and that this information be available for audit. The reason for this is not just to ensure that only approved materials are used, but also to provide for traceability of the use of specific additive lots and to account for any losses of what amounts to a highly taxed and regulated product.

27.3 Additives and processing aids: functions and comparison across regions

The function of most additives and processing aids fits into one of the following four categories, although a few processing aids fit into more than one:

- *Antimicrobial.* The major antimicrobial agent used in wine production is SO_2 (usually added as potassium metabisulfite, Chapter 17), which displays efficacy against a broad spectrum of microorganisms. Other antimicrobials are often more selective: sorbic acid is effective against yeast but not lactic acid bacteria (LAB), and the converse is true for lysozyme.

  The antimicrobials are additives except for the processing aid dimethyl dicarbonate (DMDC), which is a broad-spectrum sterilizing agent; where legal it may be added to wine (or other beverages), typically at concentrations up to 200 mg/L. The activity of DMDC is believed to result from its electrophilicity and ability to react with a broad range of nucleophiles, including thiols and amine groups in the active sites of enzymes [5]. The half-life of DMDC in model wine is ~30 min at room temperature, as it will react with other wine components to yield less harmful products in insignificant amounts, for example, with water to produce methanol and CO_2 (Figure 27.2). As a result, DMDC is typically only used just prior to bottling to ensure a sterile product (particularly for sweeter wines that are more prone to spoilage).
- *Stabilizing, clarifying, and purifying.* Most of the materials in this category are fining agents that serve as processing aids (Chapter 26.2), although an important exception is inhibitors of potassium bitartrate precipitation like gum arabic that function as additives (Chapter 26.1). Materials to prevent oxidation are also in this category, which includes both inert gases as well as antioxidant additives (such as SO_2 mentioned above, Chapter 24).
- *Correcting natural deficiencies of the must or wine.* Regions may allow for the addition of sugar, acid, tannin, and water, although the timing and extent of these adjustments can vary across regions and within a region. For example, in the United States, federal regulations allow for amelioration of must or wine with sugar, acid, and water up to 35% by volume, so long as the titratable acidity does not fall below 5 g/L (27 CFR 24.178). However, the state of California Administrative Code (Section 17010(a)) does not allow for sugar additions, and allows water addition to musts to "facilitate winemaking," including to decrease the potential alcohol content of must and prevent stuck fermentations.
- *Yeast nutrients and microorganisms.* Regions generally allow for inoculation with yeast and LAB, and for addition of some nutrients, particularly those most often deficient in grape juice, like yeast assimilable nitrogen (often in the form of diammonium phosphate, Chapters 5 and 22.3).

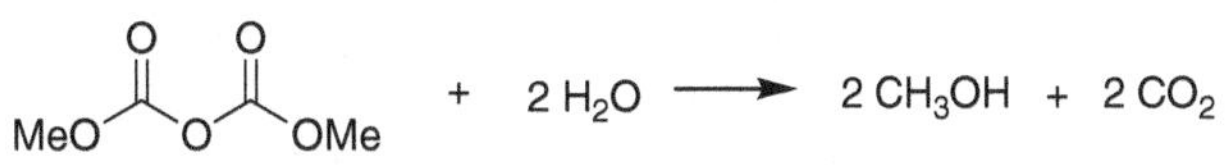

Figure 27.2 Reaction of dimethyldicarbonate (DMDC, left) with water to yield methanol and carbon dioxide

Table 27.1 Comparison of permitted materials for winemaking according to OIV, Australia, and US regulations. Additions of water, juice, sugar, distilled grape spirits, and active microorganism cultures are also permitted, but are not included in this table. Category: M = preventing microbial spoilage; S = stabilizing, purifying, or clarifying wine; C = correction of natural deficiencies; N = fermentation nutrient; P = processing aid, does not remain in wine. Regulations: A = Australia; U = United States; O = OIV; Blank = permissible under all three jurisdictions at the time of printing

Reference	Additive/Aid	Category					Major uses	Regulations
		M	S	C	N	P		
26.1	Acacia (gum arabic)		x				Inhibit tartrate precipitation	
9, 24	Acetaldehyde		x			x	Color stabilization (polymeric pigment formation), but only in juice prior to concentration	U
26.2	Activated carbon		x			x	General fining agent	
26.2	Albumen (or whole egg whites)		x			x	Polyphenol removal	
	Argon	x	x				Protection of wine from oxidation	A, O
22.2	Ammonium salts, e.g., $(NH_4)_2PO_4$ or $(NH_4)_2SO_4$				x	x	Nitrogen source. Specific salts allowed can vary among countries	
24	Ascorbic acid, erythorbic acid		x				Antioxidant	
26.2	Bentonite		x			x	Protein removal, settling aid (helps lees compaction)	
3, 26.1	$CaCO_3$		x	x		x	Deacidification, tartrate stabilization	
22.2	Ca pantothenate				x		Vitamin B_5 supplement	A, U
26.1	Carboxymethyl cellulose salts		x				Tartrate stabilization	A
3	$CaSO_4$			x			Lowering pH in production of sherry-type wines	U
	CO_2	x	x				Protection of wine from oxidation; flavor	
26.1	Ca tartrate		x				Seeding for Ca tartrate stabilization	A
26.2	Casein	x					Polyphenol removal	
	Chitin-glucan		x				Fining of metals and other components	O
	Chitosan	x	x				Fining of metals and other components; antimicrobial (*Brett.*)	A, O
3, 22.1	Citric acid	x		x			Acidulant	
10, 26.2	Cu (II) salts (sulfate, citrate)		x				Removal of odorous H_2S and mercaptans, cupric citrate allowed in AUS, OIV	
	Defoaming agents (e.g., PDMS)					x	Control of foaming during fermentation	
	Dimethyldicarbonate (DMDC)	x				x	Broad-spectrum antimicrobial, forms methanol and CO_2 upon hydrolysis	
	Enzyme – amylases					x	Conversion of starch to fermentable sugars	
2, 19	Enzyme – carbohydrases (pectinases, hemicellulases, cellulases)					x	Increase juice extraction, clarify and stabilize wine	
	Enzyme – catalase and glucose oxidase (GOX)	x				x	Glucose removal by enzymatic conversion to gluconic acid; process forms H_2O_2 which can be removed by catalase	U
	Enzyme – lysozyme	x					Active against LAB	

5	Enzyme – protease		x			x	Protein stabilization	A, U
5	Enzyme – urease		x			x	Removal of urea to prevent ethyl carbamate formation	
	Ethyl maltol, maltol		x				"To stabilize wine" as a flavor enhancer; may not be used in *vinifera* wines	U
26.2	Ferrocyanides and $FeSO_4$		x			x	Removal of copper or other metals, mercpatans	
26.3	Filtration aids (diatomaceous earth, cellulose, etc.)			x		x	Assist in clarification	
3	Fumaric acid	x		x			Acidulant, general antimicrobial	U
26.2	Gelatin/collagen		x			x	Polyphenol removal	
	Cork granules		x			x	Flavor[a]	U
17	Hydrogen peroxide		x			x	Removal of excess SO_2	A
	Ion-exchange resins			x			Acid adjustment, considered a "process" in US	
26.2	Isinglass/fish glue		x			x	Polyphenol removal	
3	Lactic acid			x			Acidulant	
3	Malic acid[b]			x			Acidulant	
26.1	Metatartaric acid		x				Inhibit tartrate precipitation	A, O
26.2	Milk products (e.g., skim milk powder)		x			x	Polyphenol removal	
24	N_2	x	x			x	Prevent oxidation	
25	Oak/oak chips					x	Flavor[a]	
24	O_2					x	Controlled oxidation	
26.2	Phytates		x			x	Removal of transition metals	A, O
26.2	Plant proteins		x			x	Polyphenol removal	A, O
26.2	Polyvinylpolypyrrolidone (PVPP)		x			x	Polyphenol removal	
26.1	Potassium bitartrate			x		x	Seed crystal for tartrate stabilization	
3, 26.1	K_2CO_3/$KHCO_3$				x	x	Deacidification, tartrate removal	
17	Potassium metabisulfite ($K_2S_2O_7$), SO_2	x	x				Antimicrobial and antioxidant	
26.2	Silica gel (SiO_2)		x			x	Cofining with protein fining agents, assists with settling prior to filtration	
18	Sorbic acid or K sorbate	x					Prevent yeast growth	
22.2	Soy flour (defatted)				x	x	Nitrogen source	U
14	Tannin		x				Mouthfeel, cofining with protein fining agents	
3, 26.4	Tartaric acid[b]			x			Acidulant	
22.2	Thiamine				x	x	Vitamin B_1 supplement	
22.2	Yeast, autolyzed				x	x	Nitrogen, other nutrients; can bind inhibiting compounds	
26.1	Yeast mannoproteins		x				Tartrate stabilization	A, O

[a]The US Code of Federal Regulations states that oak chips and cork granules may be used to "smooth wine," a fantastically ambiguous definition.

[b]The US Code of Federal Regulations requires that tartaric acid be a "byproduct of wine manufacture," but no such restriction applies to malic acid, which can be produced by chemical synthesis and be used as a racemic mixture.

The materials approved for use in each country are found in regulations associated with wine production. Differences among regulations and the resulting impacts on trade have been reviewed [6]. With a few exceptions, the lists tend to be similar, although the names or specific forms or amounts of active compounds may vary; for example, both the citrate and sulfate salts of copper are permitted in Australia, but only copper sulfate is permitted in the United States. Several databases are available for accessing or comparing up-to-date regulations in a more convenient fashion, including FIVS-Abridge and the Australian Wine Research Institute (AWRI) – Permitted Additives and Processing Aids databases.

A comparison of allowed additives and processing aids is shown in Table 27.1 for three sets of regulations: 27 CFR 24, subparts F and L, of the United States of America [1]; Standard 4.5.1 of the Commonwealth of Australia [3]; and the Oenological Codex of the International Organization of Vine and Wine (OIV) [4], which had 46 member countries as of 2015. As a caveat, many subregions or appellations within these larger jurisdictions will often have more stringent regulations that would disallow specific additions, as was described in the case of water and sugar additions in California.

References

1. Title 27-Alcohol, Tobacco Products and Firearms, Chapter I, Subchapter A, Part 24. Wine. United States Code of Federal Regulations, Washington, DC, 2015.
2. Robin, A.-L. and Sankhla, D. (2013) European legislative framework controlling the use of food additives, in *Essential guide to food additives*, 4th edn (ed. Saltmarsh, M.), The Royal Society of Chemistry, pp. 44–64.
3. Australia and New Zealand Food Standards Code – Standard 4.5.1 – Wine Production Requirements. Australian Government, Federal Register of Legislative Instruments, 2014.
4. International Oenological Codex. Organisation Internationale de la Vigne et du Vin, Paris, France, 2014.
5. Golden, D.A., Worobo, R.W., Ough, C.S. (2005) Dimethyl dicarbonate and diethyl dicarbonate, in *Antimicrobials in food*, 3rd edn (eds Davidson, P.M., Sofos, J.N., Branen, A.L.), CRC Press, pp. 305–326.
6. Juban, Y. (2000) Oenological practices: the new global situation. *Bulletin de l'OIV (France)*, **73**, 20–56.

Part C

Case Studies: Recent Advances in Wine Chemistry

28
Authentication

28.1 Introduction

The contemporary media is filled with spectacular examples of fraudulent wines. In China, the widespread practice of refilling the empty bottles of prestigious Bordeaux labels with cheaper wine – or other concoctions – has led auction houses to destroy bottles following tastings [1]. A billionaire wine collector sued an antique wine dealer after questions were raised about the authenticity of bottles, including those putatively owned by Thomas Jefferson [2] or from extraordinarily rare Bordeaux and Burgundian vintages [3]. Prior to export, forgers replaced Tuscan wines costing over $100 per bottle with wine costing about a $1 per liter [4]. Wine is among the more common foodstuffs subject to economically motivated fraud [5][1] – not only because of its cost, but also because it may go through an opaque supply and distribution chain to reach a consumer. The attractiveness of wine to counterfeiters is not a new phenomenon – a noted wine consumer, Jefferson commented on the prevalence of fraud in his era [6], and well before this The Satyricon depicts the character of Trimalchio serving an apparently fraudulent bottle of expensive Falernian wine during a Roman banquet [7]. Fraud is often discovered through auditing or whistle-blowers, but may also be detected (or confirmed) chemically. This chapter will discuss routine and emerging strategies for wine authentication. Also see Chapter 4 for a discussion of metals and authenticity.

28.2 Fraud – categories and detection approaches

While product and brand fraud is headline-grabbing, it is only one of several classes of fraud that occur in wine and other foods (Table 28.1). Beyond counterfeiting, producers may substitute grapes from a less prestigious (and less expensive) grape-growing region or a less prestigious (and less expensive) grape variety. Currently, fraud involving brand, region, or grape variety is most often detected by regulatory bodies through prosaic means (audits, insider tips, suspicious buyers), since most wineries – and some growing regions – are

[1] Everstine *et al.* [5] report that the food product categories responsible for the most adulteration incidents are seafood and fish, primarily through substitution of less expensive species; dairy products, to adulterate apparent protein content; and fats and oils, particularly adulteration of olive oil with less expensive oils.

Understanding Wine Chemistry, First Edition. Andrew L. Waterhouse, Gavin L. Sacks, and David W. Jeffery.

Table 28.1 *Overview of types of fraud observed in the wine and elsewhere in the food industry*

Type of fraud	Representative examples in wine	Relative difficulty in detecting	Common literature approaches
Brand/product label (counterfeiting)	Refilling of authentically labeled bottles or forgery of labels with inexpensive wine and/or simulated products [5]	Easy to medium (if authentic products available, or if comparing wine to non-wine) Hard (if authentic products unavailable)	Wide range of appropriate techniques, including GC-MS or LC-MS, FT-IR, NMR
Adulteration			
Naturally occurring	Addition of sugar, acid, or water during winemaking where forbidden [5]	Medium	Isotope ratio mass spectrometry (IRMS)
Exogenous	Addition of diethylene glycol to Austrian wines to increase body [5]	Easy (if expected) Hard (if unexpected)	GC-MS or LC-MS
Process	Use of unauthorized processing aids or winemaking techniques, i.e., the use of microoxygenation in regions where forbidden	Medium to hard	Not commonly evaluated
Species/variety	Substitution of less expensive Merlot and other red grapes for Pinot Noir in Languedoc (France) [9]	Easy (in grapes) Medium to hard (wine)	Wide range of appropriate techniques, including GC-MS or LC-MS, FT-IR, NMR
Place of origin	Substitution of less expensive Central Valley grapes for Napa grapes in California [10]	Medium to hard	Isotopic and elemental analyses

required to keep extensive records regarding production volume and practices.[2] Potential frauds can also be confirmed or detected through the use of chemical analyses. Detection strategies based on a single chemical component are prone to manipulation, and *fingerprinting* approaches, in which patterns of chemical components are correlated with different product types, are far more powerful. Fingerprinting typically relies on similar multivariate approaches to those described for metabolomics application (Chapter 32), and involves the following steps:

- A large number of authentic and/or fraudulent samples of known type (grape variety, region, brand, etc.) are sourced.
- Samples are analyzed by one or more techniques, and key features are determined and quantified.
- A classification model is developed based on multivariate statistics ("pattern recognition").
- Finally, the classification model must be validated, preferably using other samples of known provenance that were not used in the original model development [8].

---

[2] In the case involving substitution of Merlot and other grapes for Pinot Noir in Languedoc (Table 28.1), the fraud was originally detected by a bureaucrat who became suspicious because the total amount of Pinot Noir wine sold by a single co-operative exceeded the total registered production for the region.

Once developed, the database and model can be improved by inclusion of new authentic samples. Most of the analytical tools described in Chapter 32 (e.g., NMR, GC-MS, HPLC-UV/VIS, NIR, etc.) are appropriate for fingerprinting of wines and other foodstuffs [11–13], with the common thread being that these tools measure a large number of sample attributes to increase difficulty for potential adulterators. Fingerprinting approaches have been used for authentication of alcoholic beverages and other foodstuffs since at least the 1970s, and have become increasingly widespread with the advent of inexpensive computing power [14]. Analytical power has also increased during this period – an early report on Scotch whisky authentication detected and quantified 17 peaks (termed "features," Chapter 32) by GC-FID to discriminate among samples [15], while a more recent report using UPLC-QTOF-MS required 50% less time to detect *7600* features in various whiskey samples, of which 43 were selected to discriminate samples [16].

Fraud can also occur when producers use illegal additives (*adulterants*, Chapter 27), such as food dyes or flavorants. Adulterants that are not ordinarily observed in wine, or are found in only trace amounts, can be quantified by targeted analyses using an appropriate tool (e.g., HPLC-UV/VIS for artificial food colors, GC-MS for aroma compounds). Adulterants that would naturally be present in wine – such as glycerol – are more challenging to detect, and detection of fraud in these cases must often rely on subtle differences in stable isotope ratios [17] (see below).

Finally, products may be fraudulent because they utilize unauthorized production practices or processing aids. Processing frauds are neither widely reported nor widely studied, possibly because such frauds are not as commonly perpetuated or are seen as less serious. Additionally, processing frauds are generally more challenging to detect than adulteration since the material or equipment used, for example, a disallowed resin, is not designed to remain in the wine. Targeted approaches for detecting residual concentrations of the processing material are appropriate in some cases, and fingerprinting analyses could possibly detect some illicit techniques.

28.3 Stable isotope ratio analysis to detect glycerol adulteration

Addition of compounds naturally present in wine or juice (sugars, acids, ethanol, CO_2, glycerol, etc.) is highly restricted in most wine-producing regions (Chapter 27). In some cases, it is possible to establish authenticity by detecting impurities from the industrially produced versions (e.g., denaturants present in industrial ethanol, or non-native enantiomers like D-malic acid). Adulteration may also result in unnaturally high concentrations of the adulterant.[3] However, if the final concentration is within normal values, adulteration using compounds normally found in wine cannot be readily detected using conventional analytical techniques. An alternate approach to authenticate wines or their components is based on very small, naturally occurring differences in stable isotope ratios – for example, the range of naturally occurring $^{13}C/^{12}C$ ratios is 0.0105 to 0.0115 – which are usually expressed in delta (δ) notation in units of per mille (‰) to emphasize differences [18]:

$$\delta^{13}C_x = \frac{{}^{13}R_x - {}^{13}R_{VPDB}}{{}^{13}R_{VPDB}} \times 1000\,‰ \qquad (28.1)$$

$^{13}R_x$ is the isotope ratio of a sample, which can represent a bulk sample, an individual compound, or a specific carbon position on a compound, and $^{13}R_{VDPB}$ is the consensus isotope ratio of an international standard

[3] In the 2000s, analysis of suspect South African Sauvignon Blanc wines revealed 3-isobutyl-2-methoxypyrazine (IBMP, "green pepper") concentrations over 200 ng/L – much higher than the original grapes, and 4-fold higher than any values previously reported in the literature.

($^{13}R_{VDPB}$ = 0.0112372). A typical $\delta^{13}C$ value for sugars from plants like grapes or other fruits is in the range of −25 to −30‰, as compared to −12 to −18‰ for cane or corn sugar [19]. These subtle differences arise from differences in photosynthesis pathways (C3 plants versus C4 plants, respectively), and can be exploited to detect the addition of sugar before (chaptalization) or after (back-sweetening) fermentation in regions where this practice is forbidden [20]. Stable isotope ratios of oxygen ($^{18}O/^{16}O$) and hydrogen (D/H) can also be used for determining illegal additions, for example, of water. However, because these ratios are highly dependent on latitude, temperature, and other climactic parameters, they are of particular interest for authenticating the place of origin [17]. In the European Union, databases for site-specific D/H and $^{13}C/^{12}C$ ratios of ethanol and $^{18}O/^{16}O$ ratios for water have been established since the 1980s [17].

Detecting naturally occurring variation in $\delta^{13}C$ and other isotope ratios requires high-precision measurements beyond what is achievable with standard instrumentation used for small molecule quantification, for example, a gas chromatograph (GC) with a quadrupole mass spectrometer detector (Chapter 32). Many high-precision isotope analyses are performed by a dedicated isotope ratio mass spectrometer (IRMS). Certain applications – particularly measurement of D/H ratios for the methyl and methylene groups of ethanol to detect sugar additions – are routinely done by NMR [17]. In IRMS analyses, samples must first be

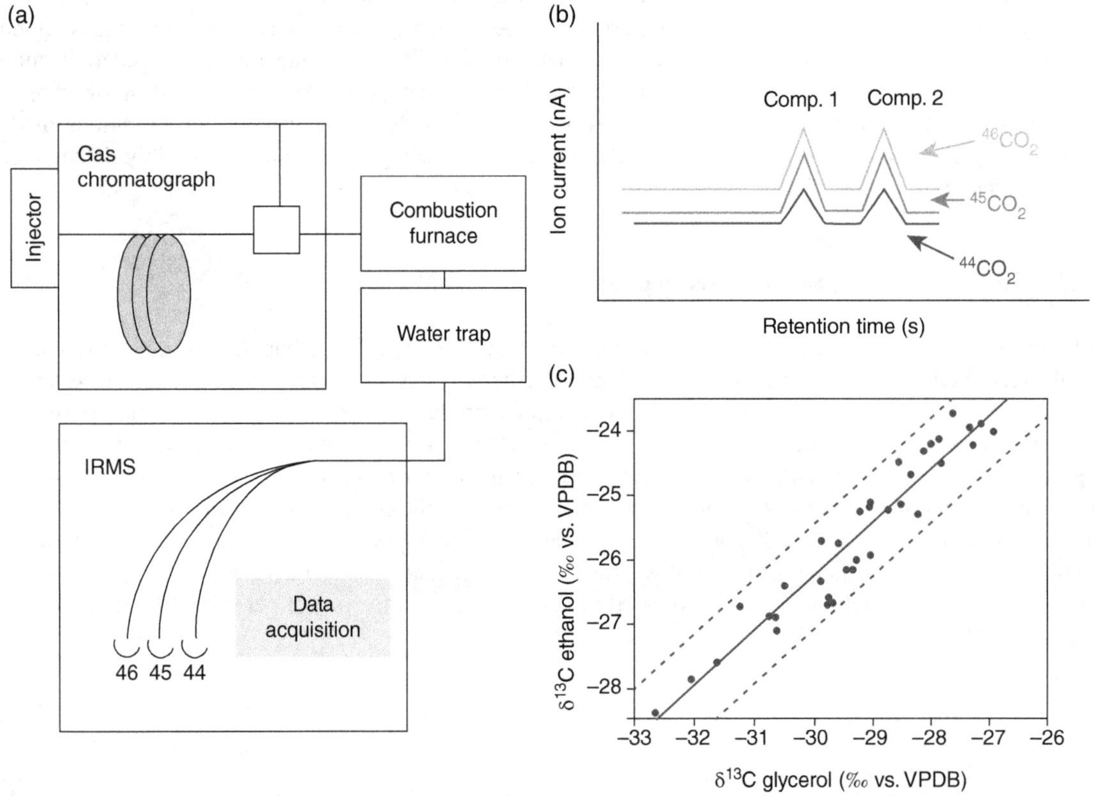

Figure 28.1 (a) Schematic of a gas chromatograph–combustion–isotope ratio mass spectrometer (GC-C-IRMS) from Reference [18], (b) Representative chromatographic data from a GC-C-IRMS system, showing CO_2 isotopologue traces for two compounds, 1 and 2, and (c) $\delta^{13}C$ values for ethanol (measured by GC-C-IRMS and LC-IRMS) and glycerol (measured by LC-IRMS) for 35 Italian wines. The solid line represents the best fit and the dashed lines represent the 95% confidence intervals. Data from Reference [21]

converted to a gas (e.g., combustion of organic C to CO_2 for $\delta^{13}C$ measurements) prior to analysis [19]. For stable isotope analysis of individual compounds, a GC can be coupled to the IRMS via a combustion furnace (Figure 28.1a). The IRMS is designed to measure only ions associated with the targeted isotopologues, for example, $^{44}CO_2$, $^{45}CO_2$, and $^{46}CO_2$ (Figure 28.1b).

A potential application of high-precision isotope ratio measurements involves detection of glycerol adulteration (Figure 28.1c). Glycerol increases perceived sweetness and body (Chapter 2), but its addition is forbidden in most countries. Because both glycerol and ethanol are formed through glycolysis (Chapter 22.1) their $\delta^{13}C$ values are tightly correlated [21]. Addition of industrial glycerol – typically produced by fermentation of corn – would perturb the expected relationship.[4] While a skilled counterfeiter could potentially also use IRMS to target and modify a $\delta^{13}C$ value of ethanol, such an exercise would be costly and diminish the economic incentive to adulterate the wine. In addition, the fraud could still be detected by using additional endogenous markers.

28.4 Future challenges in wine authentication

A major challenge in all areas of foodstuff authentication is the detection of unexpected adulterants [22].[5] While it is typically easy to devise tests for known adulterants, non-standard adulterants will often evade detection – as initially occurred with diethylene glycol additions in Austria in the mid-1980s [5]. Other challenges in authentication include:

- Determining the authenticity of rare wines can be difficult because of the challenge (or impossibility) in finding authentic standards for comparison. One recent approach for checking the age of wines purportedly vinted before World War II is to check the ^{14}C content, since this radioisotope is much higher in post-1945 wines due to nuclear weapons testing (the "bomb pulse") [23].
- Many fingerprinting techniques are expensive and require the destruction of the wine. Only a fraction of commercial wines are "chemically audited" once in the supply chain. Inexpensive and non-invasive spectroscopic approaches to evaluating wine composition, such as IR [24], may be useful in this respect.
- Quantitative determination of the grape varieties used to make a wine is still not routine. Several studies have used fingerprinting tools to discriminate among young monovarietal wines. For example, anthocyanin profiles have been used by several authors to profile varietal wines [25–27], and low proportions of coumaroylated anthocyanins are typically found in Pinot Noir as compared to most other varietal wines. However, authentication of blended and older wines are a challenge. Recent reports show that it may be possible to extract grape DNA from monovarietal wines for the purpose of identification [28]. This approach could potentially be extended for quantitative analysis of grape variety blend in the future, but the susceptibility of certain DNA bases to hydrolysis ("depurination") at low pH may limit the feasibility of the approach – half-lives are on the order of days at room temperature and wine pH [29].
- Similarly, fraud involving the place of origin (appellation) of grapes is still challenging to detect routinely. In addition to stable isotope analysis, elemental analyses by ICP-MS, ICP-AES and related techniques have shown promise for discriminating wine regions [30–32], but the close proximity of many wine regions makes their discrimination challenging.

---

[4] The strategy of comparing an authenticity marker to other compounds is widely recommended in forensic type studies to correct for naturally occurring variation [18]. For example, in the detection of testosterone doping in athletes, $\delta^{13}C$ values of testosterone metabolites are compared to $\delta^{13}C$ values of other endogenous metabolites.

[5] An example from the food industry is the adulteration of milk powder and pet food with melamine to increase apparent protein content. Common approaches to determining protein content were based on total nitrogen concentration, which did not distinguish between melamine (66% N by mass) and dietary protein. Once its identity was known, specific tests for melamine were relatively easy to develop.

- Although fingerprinting approaches are powerful, they are also harder to standardize across labs. Certification processes and databases for certain measurements have been developed – particularly isotope ratio measurements – but the uncertain robustness of other promising fingerprinting approaches in the literature has limited their widespread adoption [8].

As a final note, all types of fraud – particularly brand/product fraud in the distribution chain – may be best prevented not only through improved (bio)-chemical analysis, but also by increasing the traceability of the wine. For example, the nascent use of radio frequency ID (RFID) tags let wineries, distributors, sellers, and consumers account for and authenticate products in the distribution chain [33].

References

1. Pierson, D. Pricey counterfeit labels proliferate as China wine market booms. *Los Angeles Times,* January **14**, 2012.
2. Wallace, B. (2008) *The billionaire's vinegar: the mystery of the world's most expensive bottle of wine*, Crown Publishers, New York.
3. Secret, M. Jury convicts wine dealer in fraud case. *New York Times,* December **18**, 2013.
4. Squires, N. Andrea Bocelli embroiled in Italian wine scam. *The Daily Telegraph,* May **30**, 2014.
5. Everstine, K., Spink, J., Kennedy, S. (2013) Economically motivated adulteration (EMA) of food: common characteristics of EMA incidents. *Journal of Food Protection*, **76** (4), 723–735.
6. Hailman, J.R. (2006) *Thomas Jefferson on wine*, University Press of Mississippi, Jackson.
7. Arbiter, P. (1996) *The Satyricon* (ed. Walsh, P.G.), Clarendon Press, Oxford, UK.
8. Riedl, J., Esslinger, S., Fauhl-Hassek, C. (2015) Review of validation and reporting of non-targeted fingerprinting approaches for food authentication. *Analytica Chimica Acta*, **885**, 17–32.
9. Randall, D. and Walton, E. Sour grapes! Gallo victim of wine world's biggest con. *The Independent,* February **7**, 2010.
10. Goel, V. In vino veritas. In Napa, deceit. *New York Times,* January **25**, 2015.
11. Fotakis, C., Kokkotou, K., Zoumpoulakis, P., Zervou, M. (2013) NMR metabolite fingerprinting in grape derived products: an overview. *Food Research International*, **54** (1), 1184–1194.
12. Versari, A., Laurie, V.F., Ricci, A., *et al.* (2014) Progress in authentication, typification and traceability of grapes and wines by chemometric approaches. *Food Research International*, **60**, 2–18.
13. Ebeler, S.E. and Takeoka, G.R. (2011) Progress in authentication of wine and food, in *Progress in authentication of food and wine* (eds Ebeler, S.E., Takeoka, G.R., Winterhalter, P.), American Chemical Society, pp. 3–13.
14. Brereton, R.G. (2007) *Applied chemometrics for scientists*. John Wiley & Sons, Ltd, Chichester, UK, and Hoboken, NJ.
15. Saxberg, B.E., Duewer, D.L., Booker, J.L., Kowalski, B.R. (1978) Pattern recognition and blind assay techniques applied to forensic separation of whiskies. *Analytica Chimica Acta*, **103** (3), 201–212.
16. Collins, T.S., Zweigenbaum, J., Ebeler, S.E. (2014) Profiling of nonvolatiles in whiskeys using ultra high pressure liquid chromatography quadrupole time-of-flight mass spectrometry (UHPLC–QTOF MS). *Food Chemistry*, **163**, 186–196.
17. Christoph, N., Hermann, A., Wachter, H. (2015) 25 Years authentication of wine with stable isotope analysis in the European Union – review and outlook. *BIO Web of Conferences*, **5**, 02020.
18. Zhang, Y., Tobias, H.J., Sacks, G.L., Brenna, J.T. (2012) Calibration and data processing in gas chromatography combustion isotope ratio mass spectrometry. *Drug Testing and Analysis*, **4** (12), 912–922.
19. Brand, W.A. (1996) High precision isotope ratio monitoring techniques in mass spectrometry. *Journal of Mass Spectrometry*, **31** (3), 225–235.
20. Asche, S., Michaud, A.L., Brenna, J.T. (2003) Sourcing organic compounds based on natural isotopic variations measured by high precision isotope ratio mass spectrometry. *Current Organic Chemistry*, **7** (15), 1527–1543.
21. Cabañero, A.I., Recio, J.L., Rupérez, M. (2010) Simultaneous stable carbon isotopic analysis of wine glycerol and ethanol by liquid chromatography coupled to isotope ratio mass spectrometry. *Journal of Agricultural and Food Chemistry*, **58** (2), 722–728.
22. Ashurst, P.R. and Dennis, M.J. (1996) Introduction to food authentication, in *Food authentication* (eds Ashurst, P.R. and Dennis, M.J.), Springer, pp. 1–14.

23. Asenstorfer, R.E., Jones, G.P., Laurence, G., Zoppi, U. (2011) Authentication of red wine vintage using bomb-pulse ^{14}C, in *Progress in authentication of food and wine* (eds Ebeler, S.E., Takeoka, G.R., Winterhalter, P.), American Chemical Society, pp. 89–99.
24. Cozzolino, D., Cynkar, W., Kennedy, E., *et al.* (2011) R&D in action in Australia: non-destructive analysis of wine. *NIR News*, **22** (1), 10–11.
25. Berente, B., De la Calle García, D., Reichenbächer, M., Danzer, K. (2000) Method development for the determination of anthocyanins in red wines by high-performance liquid chromatography and classification of German red wines by means of multivariate statistical methods. *Journal of Chromatography A*, **871** (1–2), 95–103.
26. Eder, R., Wendelin, S., Barna, J. (1994) Classification of red wine cultivars by means of anthocyanin analysis. 1st report: Application of multivariate statistical methods for differentiation of grape samples. *Mitteilungen Klosterneuburg*, **44** (6), 201–212.
27. von Baer, D., Rentzsch, M., Hitschfeld, M.A., *et al.* (2008) Relevance of chromatographic efficiency in varietal authenticity verification of red wines based on their anthocyanin profiles: interference of pyranoanthocyanins formed during wine ageing. *Analytica Chimica Acta*, **621** (1), 52–56.
28. Bigliazzi, J., Scali, M., Paolucci, E., *et al.* (2012) DNA extracted with optimized protocols can be genotyped to reconstruct the varietal composition of monovarietal wines. *American Journal of Enology and Viticulture*, **63** (4), 568–573.
29. An, R., Jia, Y., Wan, B., *et al.* (2014) Non-enzymatic depurination of nucleic acids: factors and mechanisms. *PloS One*, **9** (12), e115950.
30. Hopfer, H., Nelson, J., Collins, T.S., *et al.* (2015) The combined impact of vineyard origin and processing winery on the elemental profile of red wines. *Food Chemistry*, **172**, 486–496.
31. Martin, A.E., Watling, R.J., Lee, G.S. (2012) The multi-element determination and regional discrimination of Australian wines. *Food Chemistry*, **133** (3), 1081–1089.
32. Baxter, M.J., Crews, H.M., Dennis, M.J., *et al.* (1997) The determination of the authenticity of wine from its trace element composition. *Food Chemistry*, **60** (3), 443–450.
33. Przyswa, E. (2014) Protecting your wine. *Wines and Vines*, August, 38–48.

29

Optimizing White Wine Aromas

29.1 Introduction

Wine producers often speak about making the "best wine," but such statements obscure the fact that consumers (and critics) do not have uniform opinions in their wine preferences. For example, Sauvignon Blanc wines can possess a diverse range of aromas (e.g., tropical/fruity versus green/herbaceous), and controlled studies have identified specific consumer segments that prefer each of these styles [1]. "Optimization" of aromas or other sensory attributes of a wine – whether by changing vineyard practices, the winemaking process, or even packaging decisions – first requires defining a desired outcome, which is often one that the winemaker believes will lead to better sales.

Because of legal restrictions regarding additives and production practices (Chapter 27) the chemistry of a finished wine is closely related to the initial grape composition, which will in turn be dependent on grape variety, growing region, and viticultural practices. Despite the marketing parlance that the best winemakers "let the fruit speak for itself," winemakers have a high level of influence over wine quality and style, and research has been conducted into all manner of aspects, from harvest decisions to pre-fermentation operations and adjustment of juice/must composition, and then to fermentation and storage conditions (Chapters 21 to 25).

Several techniques for optimizing wine aromas – particularly white wine aromas – have been described throughout the book, such as the practice of fermenting at low temperatures to increase "fruity" aromas due to esters (Chapter 22.1) and minimizing oxygen exposure post-fermentation to avoid oxidation of varietal thiols that impart tropical aromas (Chapter 24). Selection of yeast (and bacteria) will also critically affect many white wine sensory attributes (e.g., References [2] to [8]). This chapter focuses on one emerging research area within this broad topic – optimization of white wine aromas through yeast selection, with particular emphasis on varietal thiols.

29.2 Enhancement of varietal thiols

The highly potent odorants known as polyfunctional (varietal) thiols contribute to the tropical aromas of many wines, and particularly Sauvignon Blanc (Chapter 10). As discussed previously, these thiols appear to be formed by microbial metabolism of grape-derived, non-volatile precursors (*S*-conjugates) in the

Understanding Wine Chemistry, First Edition. Andrew L. Waterhouse, Gavin L. Sacks, and David W. Jeffery.

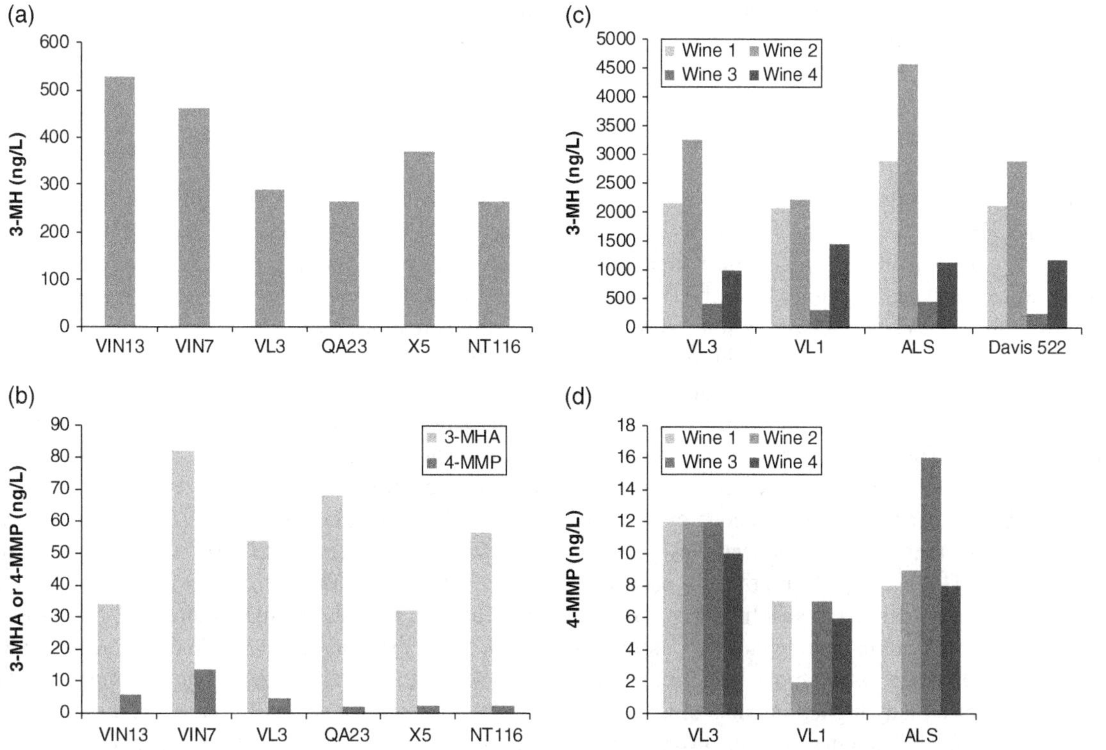

Figure 29.1 Influence of yeast strain on varietal thiol concentrations for (a and b) a single Sauvignon Blanc juice fermented on a 20 L scale (left panel), and (c and d) four different Sauvignon Blanc juices fermented on a 225 L scale (4-MMP was undetectable in each wine using Davis 522 strain, right panel). Data from References [10] and [12]

juice (Chapter 23.2). Certain consumer segments prefer Sauvignon Blanc with strong tropical aromas and less intense green/herbaceous aromas [1, 9]. As noted by Swiegers *et al.*, vineyard management is effective at controlling the primary odorants responsible for herbaceous aromas (particularly methoxypyrazines, Chapter 5), but the secondary aromas of "tropical"-smelling varietal thiols and other compounds are controlled through fermentation parameters [10]. Because only a small fraction of the *S*-conjugate pool is liberated during a typical fermentation (Chapter 23.2), there has been considerable interest in characterizing the ability of various yeast strains to release different varietal thiols during fermentation [10–12].

As shown in Figure 29.1, yeast strains can vary both in the total amount of key varietal thiols they produce (3-mercaptohexan-1-ol [3-MH], 3-mercaptohexyl acetate [3-MHA], and 4-mercapto-4-methylpentan-2-one [4-MMP]), but also in their relative concentrations. This information can allow winemakers to optimize production of wines like Sauvignon Blanc with certain characteristics, that is, more or less 3-MHA ("passionfruit") as compared to 4-MMP ("cat pee"), depending on the targeted style. Due to the relative stability of many of the thiols under reductive conditions, the sensory effects of yeast strain can still be evident after some years of bottle aging [13].

Key enzymes involved in the liberation of varietal thiols during fermentation have been identified through studies of genetically modified (GM) yeasts, and non-commercial GM yeasts overexpressing certain enzymes can produce greater amounts of varietal thiols (Chapter 23.2). This knowledge has helped drive research in molecular breeding[1] to produce yeasts with superior varietal thiol production in commercial wine strains [14, 15].

29.3 Cofermentation and spontaneous fermentation

The genetic diversity of commercial *S. cerevisiae* yeast strains provides tools to optimize wine aroma attributes. However, an even greater diversity may be accessible through the myriad indigenous yeasts (including non-*Saccharomyces*) present in and around wineries and vineyards. Winemakers may choose not to inoculate with a commercial yeast, but instead execute a *spontaneous* (or wild) fermentation (more properly, a spontaneous *cofermentation*, since multiple species/strains will be present).[2] Alternatively, commercial non-*Saccharomyces* yeasts – or hybrids of *Saccharomyces* and non-*Saccharomyces* yeasts – may be employed alone or in cofermentation with conventional yeasts [16–20].

The use of spontaneous fermentation or commercial non-*Saccharomyces* yeasts often produces sensorially distinct wines.[3] These wines are not necessarily appealing to a broad range of consumers – the use of "native" *Saccharomyces* isolated from oak trees yielded wines with strong sulfurous aromas and low fruitiness [21]. However, other work with spontaneous or non-*Saccharomyces* fermentations has shown characteristics that would be appealing to particular segments, such as greater wine complexity or increased intensity of certain fruity aromas [18, 20, 22]. Unsurprisingly, sensory differences often appear to be related to differences in common fermentation-derived metabolites (e.g., fatty acid ethyl esters, acetate esters, higher alcohols, fatty acids), but grape-derived aroma compounds can also be influenced. A comparison of three indigenous yeast strains (inoculated separately) and a spontaneous fermentation using Albariño juice showed that one strain in particular produced substantially higher amounts of C_{13}-norisoprenoids (β-damascenone and β-ionone) and monoterpenoids (geraniol and linalool), yielding a wine that was rated highest in quality by a sensory panel [23].

Cofermentations have also been demonstrated to result in unexpected outcomes due to interactions between different yeasts. In one study, Sauvignon Blanc wines produced by coinoculated commercial *S. cerevisiae* and non-*Saccharomyces* natural isolates were compared to the same wines produced from yeast monocultures (Figure 29.2). Concentrations of grape-derived compounds (C_{13}-norisoprenoids, monoterpenoids, and varietal thiols) produced from cofermentation were not necessarily intermediary to other values, for example, β-damascenone production was double in an Mp-Sc cofermentation as compared to monocultures of Mp or Sc. However, in some cases monocultures stood out as higher producers, especially Mp and Sc for 3-MH and 3-MHA, and Cz for linalool, geraniol, and β-damascenone (Figure 29.2) [24]. Significantly, differences in varietal thiol production (and sensory characters) have also been observed for Sauvignon Blanc fermented with combinations of two or three commercial *S. cerevisiae* strains [25]. Considering that many yeast suppliers offer over 30 strains, winemakers potentially have thousands of possible combinations to explore even if they limit themselves to no more than three strains per cofermentation. Future work is expected to lead to a better molecular understanding of why yeast strain interactions occur, and how they can be employed by winemakers to optimize for particular outcomes.

---

[1] Despite GM yeasts having been used for research over many years, a number of countries ban their use in commercial winemaking. Among other things, consumer acceptance is a major hurdle. Nonetheless, GM yeast research has allowed for certain genetic markers to be identified that relate to varietal thiol production. Desirable traits can then be selected for using non-GM approaches, such as classical breeding between different yeast strains and backcrossing of hybrids with the commercial parent strain.

[2] As a caveat, avoiding inoculation does not guarantee that commercial yeasts will not be present. Several genotyping studies on spontaneous fermentations indicate that the major yeasts at the end of fermentation are often identical to (or closely related to) commercial wine strains, since these yeasts are able to tolerate the higher ethanol environments.

[3] By their very nature, spontaneous fermentations with indigenous yeasts are relatively unpredictable, so obtaining a reproducible outcome could be difficult (especially when compared to inoculated fermentation).

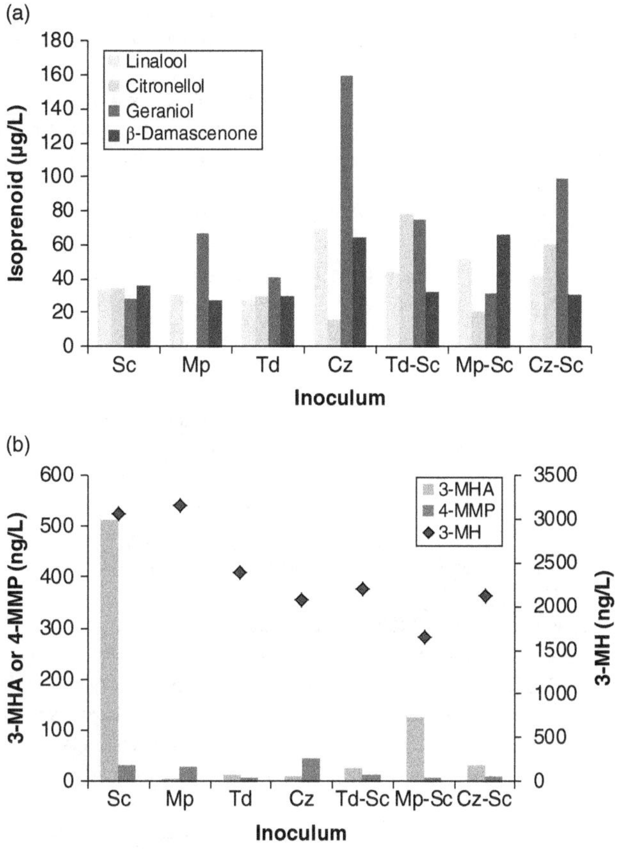

Figure 29.2 Influence of yeast species and coinoculation on concentrations of (a) monoterpenoids and β-damascenone, and (b) varietal thiols, in fermentations of Sauvignon Blanc juice. Sc, Saccharomyces cerevisiae; Mp, Metchnikowia pulcherrima; Td, Torulaspora delbrueckii; Cz, Candida zemplinina. Selected data from Reference [24]

References

1. King, E.S., Osidacz, P., Curtin, C., *et al.* (2011) Assessing desirable levels of sensory properties in Sauvignon Blanc wines – consumer preferences and contribution of key aroma compounds. *Australian Journal of Grape and Wine Research*, **17** (2), 169–180.
2. Lambrechts, M.G. and Pretorius, I.S. (2000) Yeast and its importance to wine aroma – a review. *South African Journal for Enology and Viticulture*, **21**, 97–129.
3. Swiegers, J.H., Bartowsky, E.J., Henschke, P.A., Pretorius, I.S. (2005) Yeast and bacterial modulation of wine aroma and flavour. *Australian Journal of Grape and Wine Research*, **11** (2), 139–173.
4. Swiegers, J.H. and Pretorius, I.S. (2007) Modulation of volatile sulfur compounds by wine yeast. *Applied Microbiology and Biotechnology*, **74** (5), 954–960.
5. Lerm, E., Engelbrecht, L., du Toit, M. (2010) Malolactic fermentation: the ABC's of MLF. *South African Journal of Enology and Viticulture*, **31** (2), 186–212.
6. Cordente, A., Curtin, C., Varela, C., Pretorius, I. (2012) Flavour-active wine yeasts. *Applied Microbiology and Biotechnology*, **96** (3), 601–618.
7. Steyer, D., Ambroset, C., Brion, C., *et al.* (2012) QTL mapping of the production of wine aroma compounds by yeast. *BMC Genomics*, **13** (1), 573.

8. Styger, G., Jacobson, D., Prior, B., Bauer, F. (2013) Genetic analysis of the metabolic pathways responsible for aroma metabolite production by *Saccharomyces cerevisiae*. *Applied Microbiology and Biotechnology*, **97** (10), 4429–4442.
9. Lund, C.M., Thompson, M.K., Benkwitz, F., *et al.* (2009) New Zealand Sauvignon Blanc distinct flavor characteristics: sensory, chemical, and consumer aspects. *American Journal of Enology and Viticulture*, **60** (1), 1–12.
10. Swiegers, J.H., Kievit, R.L., Siebert, T., *et al.* (2009) The influence of yeast on the aroma of Sauvignon Blanc wine. *Food Microbiology*, **26** (2), 204–211.
11. Dubourdieu, D., Tominaga, T., Masneuf, I., *et al.* (2006) The role of yeasts in grape flavor development during fermentation: the example of Sauvignon Blanc. *American Journal of Enology and Viticulture*, **57** (1), 81–88.
12. Murat, M.-L., Masneuf, I., Darriet, P., *et al.* (2001) Effect of *Saccharomyces cerevisiae* yeast strains on the liberation of volatile thiols in Sauvignon Blanc wine. *American Journal of Enology and Viticulture*, **52** (2), 136–139.
13. King, E.S., Francis, I.L., Swiegers, J.H., Curtin, C. (2011) Yeast strain-derived sensory differences retained in Sauvignon Blanc wines after extended bottle storage. *American Journal of Enology and Viticulture*, **62** (3), 366–370.
14. Dufour, M., Zimmer, A., Thibon, C., Marullo, P. (2013) Enhancement of volatile thiol release of *Saccharomyces cerevisiae* strains using molecular breeding. *Applied Microbiology and Biotechnology*, **97** (13), 5893–5905.
15. Pretorius, I.S., Curtin, C.D., Chambers, P.J. (2015) Designing wine yeast for the future, in *Advances in fermented foods and beverages* (ed. Holzapfel, W.), Woodhead Publishing, Cambridge, UK, pp. 197–226.
16. Varela, C., Siebert, T., Cozzolino, D., *et al.* (2009) Discovering a chemical basis for differentiating wines made by fermentation with "wild" indigenous and inoculated yeasts: role of yeast volatile compounds. *Australian Journal of Grape and Wine Research*, **15** (3), 238–248.
17. Saberi, S., Cliff, M.A., van Vuuren, H.J.J. (2012) Impact of mixed *S. cerevisiae* strains on the production of volatiles and estimated sensory profiles of Chardonnay wines. *Food Research International*, **48** (2), 725–735.
18. Medina, K., Boido, E., Fariña, L., *et al.* (2013) Increased flavour diversity of Chardonnay wines by spontaneous fermentation and co-fermentation with *Hanseniaspora vineae*. *Food Chemistry*, **141** (3), 2513–2521.
19. Bellon, J.R., Schmid, F., Capone, D.L., *et al.* (2013) Introducing a new breed of wine yeast: interspecific hybridisation between a commercial *Saccharomyces cerevisiae* wine yeast and *Saccharomyces mikatae*. *PLoS One*, **8** (4), e62053.
20. Azzolini, M., Tosi, E., Lorenzini, M., *et al.* (2015) Contribution to the aroma of white wines by controlled *Torulaspora delbrueckii* cultures in association with *Saccharomyces cerevisiae*. *World Journal of Microbiology and Biotechnology*, **31** (2), 277–293.
21. Hyma, K.E., Saerens, S.M., Verstrepen, K.J., Fay, J.C. (2011) Divergence in wine characteristics produced by wild and domesticated strains of *Saccharomyces cerevisiae*. *FEMS Yeast Research*, **11** (7), 540–551.
22. Soden, A., Francis, I.L., Oakey, H., Henschke, P.A. (2000) Effects of co-fermentation with *Candida stellata* and *Saccharomyces cerevisiae* on the aroma and composition of Chardonnay wine. *Australian Journal of Grape and Wine Research*, **6** (1), 21–30.
23. Carrascosa, A.V., Bartolome, B., Robredo, S., *et al.* (2012) Influence of locally-selected yeast on the chemical and sensorial properties of Albariño white wines. *LWT – Food Science and Technology*, **46** (1), 319–325.
24. Sadoudi, M., Tourdot-Maréchal, R., Rousseaux, S., et al. (2012) Yeast–yeast interactions revealed by aromatic profile analysis of Sauvignon Blanc wine fermented by single or co-culture of non-*Saccharomyces* and *Saccharomyces* yeasts. *Food Microbiology*, **32** (2), 243–253.
25. King, E.S., Kievit, R.L., Curtin, C., *et al.* (2010) The effect of multiple yeasts co-inoculations on Sauvignon Blanc wine aroma composition, sensory properties and consumer preference. *Food Chemistry*, **122** (3), 618–626.

30

Appearance of Reduced Aromas during Bottle Storage

30.1 Introduction

Each year, over 10 000 wines are entered into the International Wine Challenge (IWC) in London, one of the largest and most prestigious such competitions in the world. Receiving a Gold Medal at the IWC can be a winery's ticket to exploding sales, but faulty wines are still routinely observed – typically, about 6% of entries [1]. The major faults deemed to be present in these wines are associated with cork taint (Chapter 18), *Brettanomyces* spoilage (Chapters 12 and 23.3), oxidation (Chapter 24), and reduced sulfur aromas (Chapters 10 and 22.4) (see Figure 30.1). Of these major faults, the factors responsible for the appearance of reductive aromas in bottle are arguably the most mysterious.

30.2 Potential latent sources of compounds responsible for reduced aromas

Reduced aromas are correlated with wines possessing suprathreshold concentrations of sulfurous compounds – namely, H_2S, CH_3SH, and dimethyl sulfide (DMS) (Chapter 10). Reductive aromas are more commonly observed in wines bottled in packaging with low oxygen ingress (i.e., under a tin-lined screw cap) [2], presumably because of the absence of oxidative reactions.[1] DMS formation is well established to arise from non-oxidative hydrolysis of *S*-methylmethionine (Chapter 23.3). However, the identity of the precursors capable of releasing H_2S and CH_3SH during storage is not as clear. Two early hypotheses are listed below:

- Classic hypothesis 1. Mercaptans can be oxidized to their corresponding disulfides during wine production; for example, CH_3SH will form dimethyl disulfide. Because disulfides have higher sensory thresholds than their corresponding mercaptans, they can enter the bottle unnoticed, but will reform mercaptans during anaerobic storage [3].

[1] Although many issues may be encountered in the winery, the closure or packaging type eventually plays a relatively large role in the presence of faults. Cork taint and oxidation can be associated with cork closures, and the use of screw cap and some technical closures to overcome this brings a new conundrum, that of reductive characters.

Understanding Wine Chemistry, First Edition. Andrew L. Waterhouse, Gavin L. Sacks, and David W. Jeffery.

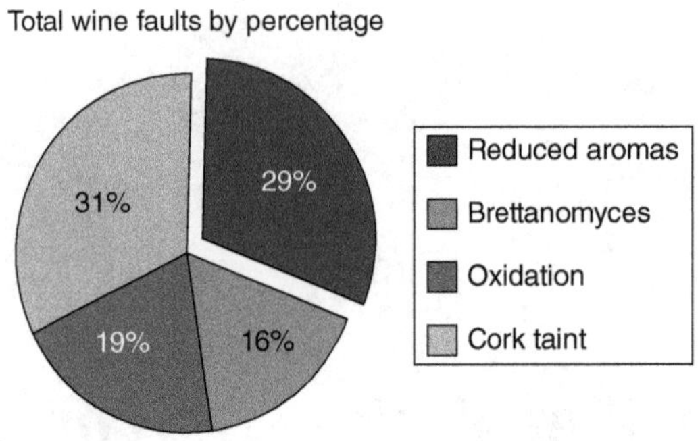

Figure 30.1 Distribution of faults in wines from 2008 International Wine Challenge in London, UK. Of the 10 000+ wines entered, approximately 6% were described as faulty [1]

- Classic hypothesis 2. Mercaptans arise during bottle storage from acid hydrolysis of thioacetates, for example, methyl thioacetate [4].

There is evidence that both disulfides and thioacetates can be formed during fermentation, and that they can release mercaptans early in a wine's life. However, as noted in a review by Ugliano, there is scant evidence that these hypotheses are valid for actual bottled wines [5]. For example, the appearance of CH_3SH is uncorrelated with either dimethyl disulfide or methyl thioacetate disappearance. On the other hand, oxidation of wines containing thiols results in their loss – but through coupling with quinones (Chapter 24), not disulfide formation. Finally, the classic hypotheses cannot explain the formation of H_2S during storage.

Searches for alternate precursors of these latent compounds are an active area of research. Metal-catalyzed hydrolysis of sulfur containing amino acids (cysteine, methionine) has been proposed because addition of transition metals will enhance the rate of H_2S and CH_3SH formation [5]. Other work has demonstrated that copper–thiol complexes (e.g. CuS) can stay dispersed in wine following copper additions, and do not necessarily fully precipitate – leading to the hypothesis that these complexes serve as latent thiol precursors of H_2S and other thiols during storage [6].

Still other latent sources of H_2S and CH_3SH may exist. For example, the potential contribution of non-volatile asymmetric disulfides or trisulfides (e.g., adducts with other thiol compounds) has not been well explored. Also, quinone–thiol adducts formed following wine oxidation (Chapter 24) are assumed to be stable – but this is not beyond doubt. Perhaps these adducts can subsequently release H_2S and CH_3SH, or other thiols, during prolonged storage. Finally, other potent (and common) contributors to reduced aromas may have been overlooked. If the standing hypotheses are not valid, then these questions are founded on unknown chemical processes; that is, what "reducing" agent/s or reactions are participating or occurring? Wine storage is a rare situation of very long lifetimes of reactive solutions where kinetically slow processes, perhaps those not commonly observed otherwise, have time to occur.

Determining the likely latent sources of H_2S, CH_3SH, or other causes of reduced aromas is more than interesting chemistry – it is also a crucial step in eliminating an increasingly vexing problem. Determining their identity should lead to better prevention, remediation, and detection strategies for reduced aroma formation in bottle. Currently, some winemakers will add ascorbic acid to predict the appearance of reduced aromas, but the rationale for this test is based on disulfides as latent precursors, and these tests have not been validated. In recent years, winemakers have benefited from accelerated aging tests, for example, contact tests to predict the likelihood of potassium bitartrate instability (Chapter 26.1), and it is expected that a validated accelerated reduction assay would be equally beneficial.

References

1. Goode, J. and Harrop, S. (2008) *Wine faults and their prevalence: data from the world's largest blind tasting,* 16èmes Entretiens Scientifiques Lallemand, Horsens.
2. Godden, P., Francis, L., Field, J., *et al.* (2001) Wine bottle closures: physical characteristics and effect on composition and sensory properties of a Semillon wine – 1. Performance up to 20 months post-bottling. *Australian Journal of Grape and Wine Research*, **7** (2), 62–105.
3. Limmer, A. (2005) Do corks breathe? Or the origin of SLO. *Australian and New Zealand Grapegrower and Winemaker*, **497**, 89–98.
4. Rauhut, D., Kurbel, H., MacNamara, K., Grossmann, M. (1998) Headspace GC-SCD monitoring of low volatile sulfur compounds during fermentation and in wine. *Analusis*, **26** (3), 142–145.
5. Ugliano, M. (2013) Oxygen contribution to wine aroma evolution during bottle aging. *Journal of Agricultural and Food Chemistry*, **61** (26), 6125–6136.
6. Franco-Luesma, E. and Ferreira, V. (2014) Quantitative analysis of free and bonded forms of volatile sulfur compoulds in wine. Basic methodologies and evidences showing the existence of reversible cation-complexed forms. *Journal of Chromatography A*, **1359**, 8–15.

31

Grape Genetics, Chemistry, and Breeding

31.1 Introduction

Vitis, the genus name of grapes, contains over 60 species [1, 2], although almost all commercially important varieties belong to a single domesticated species that originated in Eurasia – *V. vinifera*. Despite these grapes having preferred flavor characteristics for wine production, they tend to be susceptible to pests, diseases, and extreme temperatures; species native to North America and East Asia are generally better adapted to surviving these stressors. For example, *V. riparia* can tolerate winter temperatures down to –40 °C and *V. muscadinia* is resistant to several diseases capable of devastating *vinifera* (e.g., Pierce's disease caused by an insect-transmitted bacteria) [2]. However, these wild species tend to be low yielding and produce wines with undesirable sensory characteristics, including high acidity, low astringency, and excessive herbaceous aromas (Table 31.1).

31.2 Breeding new varieties

Although most of the major commercial grape cultivated varieties, or *cultivars* (Chardonnay, Riesling, Pinot Noir), are hundreds of years old, new grape varieties are periodically introduced. For example, Cabernet Sauvignon is believed to have been produced from a cross[1] of Cabernet Franc with Sauvignon Blanc centuries ago [6]. Grape breeders create new varieties by pairing two different species to produce a "hybrid" grape in an attempt to get advantageous traits from both parents (e.g., cold hardiness from wild species combined with desirable yield or flavor attributes from *V. vinifera*). For example, the popular juice grape cultivar "Concord," developed by Ephraim Bull in the nineteenth century, possesses pest resistance from its *V. labrusca* parent, and perfect flowers[2] and high yields from its *V. vinifera* parentage [2].

[1] In vineyards, grapes are usually *clonally* propagated – that is, a cutting from an existing grape vine will be planted, often after grafting on to a grape rootstock. The resulting grapevine will be genetically identical to the grape from which it was propagated. A cross refers to a grape grown from a seed, and will have genetic material from both male and female parents. These crosses can occur by chance or intentionally, and are the standard means to develop new grape varieties.

[2] Perfect, or hermaphroditic, flowers contain both male (stamen) and female (carpel) components. Perfect flowers are advantageous to growers because they can self-pollinate, and because all plants will produce fruit. Wild grapes are usually unisexual, and 50% of plants (males) will not produce fruit.

Understanding Wine Chemistry, First Edition. Andrew L. Waterhouse, Gavin L. Sacks, and David W. Jeffery.

Table 31.1 Representative concentrations of some key wine components that differentiate typical red vinifera varieties from *V. riparia* and *V. labruscana* (Concord)

Wine component[a]	*Vitis* species[b]			Notes	Chapter
	V	R [3]	L [4, 5]		
Titratable acidity (g/L as tartaric)	5–6.5	35	9.5	Correlates with sourness	3
Condensed tannin (mg/L catechin equivalents)	500–700	<50	<50	Correlates with astringency, color stabilization	14
3-Isobutyl-2-methoxypyrazine (ng/L)	3–17	56	ND[c]	"Green pepper," detection threshold = 2 ng/L	5
Methyl anthranilate (μg/L)	0.06–0.6	ND	600–3000	"Grape Kool-aid," detection threshold = 300 μg/L	5

[a] Grapes harvested at maturity (i.e., 20–24 °Brix for *vinifera* and *riparia*, 16 °Brix for *labruscana*), without deacidification.
[b] V = *V. vinifera*, R = *V. riparia*, L = *V. labruscana* hybrids of the American *labrusca* species and European *vinifera* species. Values for *vinifera* are found in previous chapters.
[c] Not detected.

In the past it could take grape breeders decades to develop and release a new grape variety. The majority of seedlings produced from *V. vinifera* × *V. riparia* would not possess all of the desirable traits of both parents, and it could take several years of field testing and small-scale winemaking to determine which offspring would produce viable new varieties. Because modern grape breeding often utilizes complex crosses of hybrids over several generations of grapes, the process of crossing and evaluation often needs to be repeated several times before commercial-scale field trials and release of a new variety to growers and the public.

Advances in grape genomics are expected to greatly accelerate this breeding process. The grape genome was sequenced by two separate consortia in 2007 [2] and was the first perennial fruit sequenced. Grapes have approximately 30 000 genes, or about 50% more than humans. While about half of these genes are strongly similar to genes found in other plants, many have unknown functions. Of the many approaches to determine the gene(s) responsible for a trait, the most common current approach is *mapping*. Typically, mapping involves associating physical points on the chromosomes (markers) from a group (*population*) of individual grapes and statistically comparing this to data on particular traits, such as disease resistance or production of flavor-related compounds (the *phenotype*).[3] Often, a population consists of siblings generated from the same parents (*linkage mapping*), but it is also possible to study a population of unrelated individuals (*association mapping*). Like many traits in grapes, most chemical components vary quantitatively – that is, there will be a range of values observed because their formation is under the control of multiple genes. Thus, most studies of grape chemistry involve *quantitative trait loci (QTL) mapping*, in which multiple genetic regions controlling a trait are identified. Once identified, the function of candidate gene(s) can be confirmed by expressing the gene in a different organism, or by selectively overexpressing or silencing the gene. Eventually, the specific forms (*alleles*) of the gene responsible for a given phenotype can be determined.

In comparison to studies on disease resistance, studies of genes controlling fruit chemistry traits were relatively sparse. Examples of chemistry-related genes that have been characterized include:

[3] The cost of genome sequencing has fallen dramatically in the period from 2005 to 2015 – for the 500 million base pair grape genome, the cost has decreased from over US$1 M to close to $1000 per individual genome. However, at the time of the writing of this book, it is still too expensive to fully sequence all individuals in a mapping population. For now, mapping studies in plants often rely on analysis of a small subset of the genome referred to as *markers*, which serve as mileposts in the genome. Depending on the analysis, the gaps between markers may include a few genes or hundreds of genes, which can complicate interpretation of which gene is actually responsible for a given phenotype. Alternatively, if likely candidate genes are known, these may be sequenced in full.

- *Total anthocyanins and anthocyanin 3,5-diglucosides.* Differences among red, white, and pink grapes can largely be explained by allelic variation in a transcription factor (*VvmybA1*) responsible for expression of a 3-*O*-glycosyltransferase (UDGT) involved in the final step of anthocyanin biosynthesis [7]. Similarly, anthocyanin 3,5-diglucoside production is not observed in *vinifera* due to a mutation in a 5-*O*-glycosyltransferase gene [8].
- *Methoxypyrazines.* Differences in methoxypyrazine (MP) formation between Cabernet Sauvignon (high MP) and Pinot Noir (low MP) could be explained by allelic variation in *VvOMT3*, a methyltransferase involved in the final step of MP biosynthesis [9].
- *Monoterpenes.* The higher monoterpene content of Muscat varieties could be explained by variations in the 1-deoxy-D-xylulose 5-phosphate synthase gene (*VvDXS*). Analogous genes are known to be responsible for the first steps of monoterpene biosynthesis in other plants [10].

Monoterpenes and anthocyanins – and to a lesser extent MPs – are under strong control by a single major QTL. Variation in primary metabolites – such as sugars and malic acid – are expected to be controlled by multiple genes since they are involved in multiple pathways within the grape berry.

31.3 Genetics and selection

Knowledge of the genes controlling fruit quality and other traits should assist grape breeders in their efforts to produce new varieties. In principle, breeders could use biotechnology techniques to selectively delete or add a gene of interest to an existing variety, for example, insert a disease resistance gene into Cabernet Sauvignon. Although genetically modified (GM) grapes have been tested in several countries – mostly for research involving disease-resistance genes – no immediate commercial use is expected for these varieties [11]. GM organisms are currently treated with suspicion by many consumers, so market acceptance is problematic, and approval of commercial GM grape varieties would likely be challenging in many countries.[4]

Alternatively, knowledge of key genes can be used as part of *marker assisted selection* (MAS). MAS uses traditional crossings to generate new varieties, but these varieties can be screened at an early stage in breeding to eliminate those that lack desirable traits. For example, a breeder interested in producing a Muscat-type grape with good powdery mildew resistance could genetically screen seeds produced from crossing Muscat of Alexandria and wild American species, and only retain those crosses that possess the correct *DXS* variant (for high monoterpenes) and genes associated with disease resistance [12]. Assuming the new cross is accepted by consumers, such varieties would be more sustainable since disease control through pesticides represents a major financial and environmental cost. Beyond this, developing new grape varieties offers potential for novel grape and wine flavors that could be very attractive. As should be evident from Table 31.1, genetic variation can yield three orders of magnitude (or more) differences in key flavor compounds. This variation is often much greater than what can be achieved through manipulating viticultural or winemaking practices.

Finally, the knowledge generated from grape genetics is valuable in interpreting the results of viticultural studies. Researchers have historically observed empirical effects of growing conditions on grape or wine chemistry, but their ability to interpret these data sets was limited. For example, well-drained soils and low water availability are common features of wine regions capable of producing darkly colored red wines.

[4] Similar statements can be made about GM wine microorganisms. For example, a GM yeast capable of malolactic conversion was approved for commercial use in the United States in 2003, and another strain capable of degrading urea (a precursor of ethyl carbamate) was approved in 2006. However, these yeasts have not achieved widespread use in part because of restrictions on export of resulting wines to regions that do not permit GM yeasts, that is, European Union countries. Nonetheless, GM research allows for targeted phenogenetic insight and can be used to guide traditional selection approaches.

A recent report shows that water restriction results in decreased expression of a *UDGT* responsible for producing anthocyanins from anthocyanidins [13]. These genetic advances will allow for a mechanistic interpretation of variation in grape chemistry arising from differences among cultivars or growing conditions – in other words, "molecular viticulture."

References

1. Young, P.R. and Vivier, M.A. (2010) Genetics and genomic approaches to improve wine quality, in *Managing wine quality*, Vol. 1, *Viticulture and wine quality* (ed. Reynolds, A.G.), Woodhead Publishing and CRC Press, Oxford and Boca Raton.
2. Reisch, B.I., Owens, C.L., Cousins, P.S. (2012) Grape, in *Fruit breeding* (eds Badenes, M.L. and Byrne, D.H.), Springer, New York, pp. 225–262.
3. Sun, Q., Gates, M.J., Lavin, E.H., *et al.* (2011) Comparison of odor-active compounds in grapes and wines from *Vitis vinifera* and non-foxy American grape species. *Journal of Agricultural and Food Chemistry*, **59** (19), 10657–10664.
4. Kluba, R.M. and Mattick, L.R. (1978) Changes in nonvolatile acids and other chemical constituents of New York State grapes and wines during maturation and fermentation. *Journal of Food Science*, **43** (3), 717–720.
5. Nelson, R.R., Acree, T.E., Lee, C.Y., Butts, R.M. (1977) Methyl anthranilate as an aroma constituent of American wine. *Journal of Food Science*, **42** (1), 57–59.
6. Bowers, J.E. and Meredith, C.P. (1997) The parentage of a classic wine grape, Cabernet Sauvignon. *Nature Genetics*, **16** (1), 84–87.
7. This, P., Lacombe, T., Cadle-Davidson, M., Owens, C.L. (2007) Wine grape (*Vitis vinifera* L.) color associates with allelic variation in the domestication gene *VvmybA1*. *Theoretical and Applied Genetics*, **114** (4), 723–730.
8. Jánváry, L., Hoffmann, T., Pfeiffer, J., *et al.* (2009) A double mutation in the anthocyanin 5-*O*-glucosyltransferase gene disrupts enzymatic activity in *Vitis vinifera* L. *Journal of Agricultural and Food Chemistry*, **57** (9), 3512–3518.
9. Dunlevy, J.D., Dennis, E.G., Soole, K.L., *et al.* (2013) A methyltransferase essential for the methoxypyrazine-derived flavour of wine. *The Plant Journal*, **75** (4), 606–617.
10. Emanuelli, F., Battilana, J., Costantini, L., *et al.* (2010) A candidate gene association study on muscat flavor in grapevine (*Vitis vinifera* L.). *BMC Plant Biology*, **10** (241).
11. Anonymous (2015) Grape Vine, GMO Compass. 2015 [updated March 26, 2015]. Available from: http://www.gmo-compass.org/eng/database/plants/73.grape_vine.html.
12. Emanuelli, F., Sordo, M., Lorenzi, S., *et al.* (2014) Development of user-friendly functional molecular markers for *VvDXS* gene conferring muscat flavor in grapevine. *Molecular Breeding*, **33** (1), 235–241.
13. Castellarin, S.D., Pfeiffer, A., Sivilotti, P., *et al.* (2007) Transcriptional regulation of anthocyanin biosynthesis in ripening fruits of grapevine under seasonal water deficit. *Plant Cell Environment*, **30** (11), 1381–99.

32

Analytical Innovations and Applications

32.1 Introduction

Wine is a chemically complex beverage with many hundreds of compounds known to significantly contribute to sensory properties, impact stability, or affect product safety (see preceding chapters). The ability to characterize these compounds has enabled winemaking operations to be better managed, resulting in improved consistency and quality of the final products [1]. In addition, analysis is important in terms of meeting regulatory requirements, detecting adulterants, discovering new components, and determining authenticity. Wine researchers continue to seek innovative approaches to the art of chemical analysis as analytical capabilities evolve and increasingly sophisticated instruments (and software) become available [e.g., 2–4]. Briefly, progress in the sophistication of available analytical methods can be summarized as follows:

- Single result. Most of the classical chemical methods, including precipitation, distillation, gravimetry, titrimetry, elemental analysis.
- Single result. Shift to physical (early instrumental) methods, such as colorimetry, conductivity, forms of optical spectroscopy, followed by more modern instrumental methods (nuclear magnetic resonance (NMR), mass spectrometry (MS), infrared (IR), and ultraviolet/visible (UV/vis) spectroscopy, atomic absorption and emission spectroscopy (AAS/AES)).
- Multiple results. Detection of multiple analytes with a single method, often involving the coupling of separation approaches to selective detectors, such as gas chromatography (GC) and high-performance liquid chromatography (HPLC) coupled to MS.
- Multiple data sets. Instrumental methods for chemical fingerprinting with chemometric analysis of data (perhaps from multiple instruments), specific analytes not necessarily identified, miniaturization and automation for routine monitoring and modeling, using techniques such as NMR, mid-IR (MIR), MS, UV/vis, and biosensors.
- Targeted and untargeted approaches. Attempt to determine multiple analytes/majority of components within samples (e.g., using GC-MS, HPLC-MS, and NMR) as outlined below.

Understanding Wine Chemistry, First Edition. Andrew L. Waterhouse, Gavin L. Sacks, and David W. Jeffery.

To fully understand the variables affecting wine chemical composition, production practices, sensory properties, and ultimately consumer preference, it would be ideal to target all known important compounds and measure them in every experiment (i.e., *metabolomics*).[1] Metabolomic studies in wine and elsewhere generally fall into one of three categories: informative, discriminative, and predictive.

1. Informative, *What compounds are typically present, and how much?* Metabolomics may be used to generate a database of typical compounds in wines and their concentrations.
2. Discriminative. *How does a particular variable affect chemical composition?* Metabolomics can be used to globally assess chemical differences correlated with changes in grape variety, growing region, yeast strain, or other parameters.
3. Predictive. *How do we model a hard-to-measure variable from easier-to-measure attributes?* For example, sensory data are tedious and expensive to collect with human panels, but most sensory attributes arise from multiple rather than singular chemical stimuli. Multivariate models based on chemical data can be used to model sensory data.

In contrast to other "omics" fields – notably genomics – there is not a universal approach to metabolomics, due to the heterogeneous chemical properties and wide range of concentrations observed for small molecules. *Targeted metabolomics* approaches intending to measure a subset of potentially important compounds, for example, all known wine aroma compounds, therefore often rely on multiple analytical approaches. For example, to comprehensively explore relationships between red wine aroma chemical composition and quality, 9 separate extraction and GC methods were used to quantify 110 volatile compounds [5] – many additional methods would have been necessary to evaluate key non-volatile components like polyphenols, polysaccharides, metals, etc.

On the other hand, *untargeted metabolomics* approaches have arisen to study wine chemical profiles (i.e., a large range of metabolites in wine). Although these approaches generally use minimally selective sample preparation [6], they will not be fully comprehensive; for example, trace level compounds may not be visible due to interferences. However, because of their less biased approach, untargeted metabolomics can identify correlations with vineyard and winemaking variables that might have been otherwise overlooked. This chapter provides examples of analytical innovations and applications to wine science, including rapid analytical methods, and targeted and untargeted analyses of wine that aid overall understanding of the links between grape and wine compositions.

32.2 Typical approaches to wine analysis

Many analyses in wine begin with trying to explain a specific phenomenon:

> *Why does this wine smell musty? How can I increase concentrations of GSH-3-MH, a likely precursor of the fruity smelling 3-mercaptohexanol? Why do some wines have problems with potassium bitartrate instabilities and others do not?*

The approaches used for quantitative or qualitative analysis of compounds in wine are usually borrowed from elsewhere in food research (e.g., see Reference [7]), and more broadly from studies of natural products.

---

[1] The term *metabolomics* comes from systems biology, where it has been defined as the "measurement of the dynamic multi-parametric metabolic response of living systems to pathophysiologic stimuli or genetic modification" [8]. The metabolomics term is used more ambiguously in investigations of wine and other foods, but could be rewritten as "measurement of many potentially important small molecules in one or more wines."

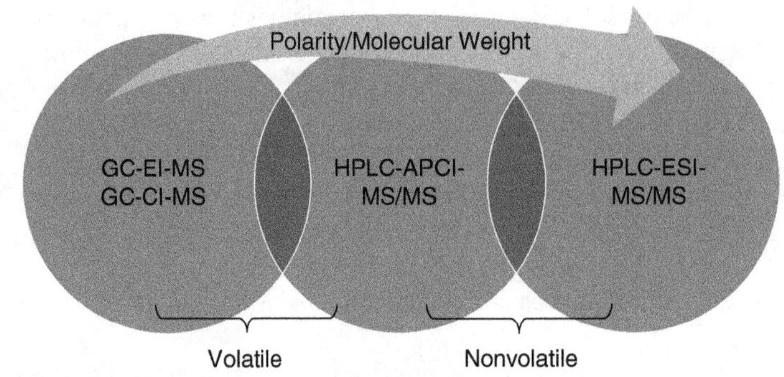

Figure 32.1 Common instrumentation and ionization modes used for gas or liquid phase analysis of compounds in wine depending on volatility, and as a function of increasing polarity and molecular weight

- In some cases likely analytical targets are already known from the literature, or can be rationalized from basic chemical knowledge. For example, the structure of the GSH-3-MH precursor and a knowledge of plant biochemistry suggested that (*E*)-2-hexenal and glutathione were likely precursors and important for further study [9].
- When key compounds responsible for a phenomenon are unknown, it is common to use techniques borrowed from natural products discovery (e.g., see Reference [10]). Samples are fractionated by chromatography or some other technique and evaluated for activity (e.g., flavor) by an appropriate assay.[2] Ideally, a control lacking activity is also available for comparison. Examples of assays used in the discovery of wine compounds are matched to the phenomenon of interest:
  - Identification of key aroma compounds typically uses the human nose as a selective and sensitive "bioassay." In practice, this is accomplished with an olfactometer ("sniff-port")[3] as a detector for GC (GC-O) [11–13].
  - Less routinely, taste/mouthfeel compounds may be detected by tasting fractions following HPLC separation (LC-taste) [14]. While online coupled GC-O instrumentation is relatively common for wine analysis, LC-taste is often conducted offline, with HPLC fractions being collected and assessed sensorially in a separate step (e.g., see References [15] to [18]).
  - Other assays include the use of spectrophotometers, colorimeters, or human subjects for detecting pigments; for example, potassium bitartrate model systems for detecting compounds capable of preventing tartrate instabilities [19].
- Components of active fractions can then be identified by conventional analytical approaches, such as GC or HPLC coupled to various detectors (especially an MS, Figure 32.1) [20, 21] or NMR to identify compounds (Table 32.1). Ideally, identity confirmation is based on comparison to authentic standards.
- Once target analytes are identified, GC- and HPLC-based approaches are often used for routine quantification of targeted compounds. Typically, some form of extraction and separation of analytes is required prior to their instrumental analysis, both to remove interferences and to pre-concentrate the analytes (Table 32.1). Higher concentration compounds (>10 mg/L) may be amenable to spectroscopic means of detection, as described below.

---

[2] In drug discovery, this assay will often take the form of a biological model, e.g. testing the toxicity of the fractions on a pathogenic organism or cancer cells in studies to detect novel antibiotics or chemotherapeutics.

[3] The first reported coupling of GC to an olfactometer dates to 1964, or about a decade after the initial report on GC, and describes its use in the fragrance industry. Routine reports of the application of GC-O to research questions were sparse until the 1990s, in part because of lack of commercial sniffing ports.

Table 32.1 Routine techniques and instrumentation used for chemical analysis of wine

Procedure	Comments	Typical for Volatiles (V) or nonvolatiles (N)
Isolation		
Solid-phase extraction (SPE)	Passage of wine down a cartridge packed with adsorbent – selection of phases availabe, elute with different solvents	V, N
Liquid-liquid extraction (LLE)	Extraction of wine with organic solvent followed by concentration of organic extracts – range of solvents can be used	V, N
Solid-phase microextraction (SPME)	Fiber coated with solid-phase adsorbent exposed to wine headspace (or directly immersed into wine) – selection of phases available including combinations	V
Stir bar sorptive extraction (SBSE)	Stirrer bar coated with solid-phase adsorbent stirred in wine – several phases available	V
Purge and trap (PT)	Analytes purged from wine by inert gas flow and trapped by adsorbent – selection of phases available	V
Distillation	Solvent-assisted flavor evaporation (SAFE) or simultaneous distillation extraction (SDE) – range of solvents can be used	V
Separation		
Gas chromatography (GC)	Components separated in gas phase with a temperature gradient due to differences in boiling point and polarity of analytes – selection of stationary phases available	V
High-performance liquid chromatography (HPLC)	Components separated in liquid phase as a function of mobile phase composition due to differences in polarity, size and charge of analytes – selection of stationary phases available	N
Capillary zone electrophoresis (CZE)	Components separated in the liquid phase as a result of an applied electrical potential due to electroosmotic flow of buffer mobile phase, and charge on analyte – range of buffer compositions used (pH, ionic strength, additives)	N
Detection		
Mass spectrometer (e.g. single or triple quadrupole, time-of-flight (TOF), ion trap (IT))	Detects ionized molecules based on mass-to-charge ratio (*m/z*); tandem mass spectrometry (MS/MS) enhances selectivity, sensitivity and compound identification capability – excellent detector for quantitative and qualitative analysis, can be coupled to HPLC or GC	V, N
Dioade array dector (DAD)	Detects molecules that absorb light in the UV/visible region (190-600 nm) – good for many wine components containing aromatic rings, e.g., polyphenols	N
Fluorescence detector (FLD)	Detects molecules that fluoresce based on excitation and emission wavelengths – good for analysis of some polyphenols	N

(*Continued*)

Table 32.1 (*continued*)

Procedure	Comments	Typical for Volatiles (V) or nonvolatiles (N)
Refractive index detector (RID)	Detects changes in refractive index of eluent as compounds elute – good for compounds lacking a chromophore	N
Flame ionization detector (FID)	Detects current produced by compounds ionized in a flame – universal detector for organic analytes	V
Sulfur chemiluminescence detector (SCD)	Detects chemiluminescence of SO (produced from combustion of sulfur compounds) upon reaction with ozone – selective for sulfur compounds	V
Pulsed-flame photometric detector (PFPD)	Detects characteristic emissions of elements excited through combustion – selective for sulfur and phosphorous compounds	V
Atomic emission detector (AED)	Detects characteristic emissions of elements excited by a plasma light source – sensitive to a range of elements common to wine compounds	V
Electron capture detector (ECD)	Detects changes in current due to capture of electrons by analytes – selective for electronegative elements, especially halogenated compounds	V
Ionization modes for MS		
Electron ionization (EI)	Used with GC-MS, positive ion mode, high energy ionization leads to extensive fragmentation	V
Chemical ionization (CI)	Used with GC-MS, positive or negative ion mode, useful for more polar compounds, less fragmentation than EI	V
Electrospray ionization (ESI)	Used with HPLC-MS, positive or negative ion mode, can generate multiply-charged analytes (allowing analysis of high molecular weight samples), minimal fragmentation (although in-source fragmentation is possible), works better for polar analytes, used with MS/MS	N
Atomospheric pressure chemical ionization (APCI)	Used with HPLC-MS, positive or negative ion mode, minimal fragmentation (although in-source fragmentation is possible), useful for volatile/lower polarity analytes, used with MS/MS	N

Recent innovations to this standard approach involve two-dimensional capabilities (e.g., “heart-cut” GC and fully comprehensive GC or HPLC) to enhance separation for complex samples such as wine [22–25]. Capillary zone electrophoresis (CZE) is another innovative separation technique that has been used for analysis of polyphenols in wine [26, 27], although far less commonly than HPLC.

32.3 Multivariate data analysis and chemometrics

Typical analytical procedures are often univariate (one variable), meaning individual calibrations are constructed for the determination of single components. This approach necessitates that the components of interest can be separated from matrix interferences or selectively detected in some way (e.g., GC-MS analysis of aroma volatiles) [28]. Multivariate (multiple variable) calibrations are used when interferences cannot be eliminated, but rather are compensated for through an appropriate mathematical model (Figure 32.2). The basic steps for developing multivariate calibrations are as follows:

- *Calibrate.* Measure a parameter of interest using a selective, established technique (e.g., malic acid by HPLC with UV detection) in a wide range of wines. Then measure a large number of variables (e.g., IR absorbances at various wavelengths) for the same wines.
- *Model.* Using statistical software, develop a multivariate model for the parameter of interest.
- *Validate.* Check the calibration, preferably using samples that were not used to generate the model.

Multivariate models are especially common for techniques that produce spectral data such as IR and UV/vis. Modeling is often performed through partial least squares (PLS) regression in which numerous variables (e.g., whole spectra) are compressed into several latent variables.[4] An alternative modeling approach, multiple linear regression (MLR), uses a selected number of variables (e.g., specific wavelengths) for calibration, but there is a loss of chemical (e.g., functional group) and diagnostic (outlier detection) information in this approach, among other shortcomings [28]. Similar statistical approaches may be used to develop calibrations based on chromatographic data sets to predict parameters like flavor attributes.

Beyond generating predictive models, multivariate statistical methods may also be used in informative and discriminative analyses involving large data sets to identify key distinguishing features [29]. *Unsupervised* tools such as principal component analysis (PCA), hierarchical cluster analysis (HCA) and factor analysis (FA) are often used as part of initial exploration of data sets to determine variables that best explain variation within a data set, or to detect groupings of samples. *Supervised* tools like PLS, MLR, linear discriminant analysis (LDA), and canonical variate analysis (CVA) are used to determine the measurements that best differentiate samples based on known classes. Applications of multivariate statistics to analytical problems are outlined below, but further details of the statistical methods are beyond the scope of this chapter and can be found elsewhere (e.g., see References [30] to [35]).

32.4 Chemometrics in practice – rapid methods for wine analysis

Beyond chromatographic and mass spectrometric based methods, spectroscopic techniques utilizing different wavelengths of light (UV/vis, IR) and NMR are also applied to the analysis of wine. These approaches are often less selective, in that they are unable to obtain an isolated signal for most analytes, but when combined with chemometrics they can offer rapid, non-destructive measurement of compositional parameters for decision-making or classification purposes [34–37]. In addition, relatively non-specific chemical sensor arrays (e.g., metal oxide semiconductor, optical, amperometric/potentiometric, colorimetric) can be employed to investigate components in wine. These form the basis of electronic nose and tongue (e-nose, e-tongue) devices that can be utilized to artificially evaluate wine sensory properties, differentiate samples, or monitor processes [38–44]. Although these chemometric approaches are less appropriate for research labs, these systems offer great potential for implementation in process control in wineries (as well as other food processing environments) because they are simple to use, provide rapid results, are reasonably inexpensive and require minimal sample preparation [45].

Assessment of wine astringency using an e-tongue or MIR spectroscopy has been evaluated as a rapid alternative to the usual, time-consuming sensory testing undertaken by a trained panel [42]. As with other rapid methods, a reference method is used to prepare calibration models that relate the parameter of interest

[4] Latent variables, so-called because they are "hidden" in the original data, are new variables (known as scores) calculated from the measured variables (which are merely sums of the latent variables). The scores can be plotted to visualize the structure of the observations making up the data set. Note that PLS can also stand for projection to latent structures (by means of partial least squares regression).

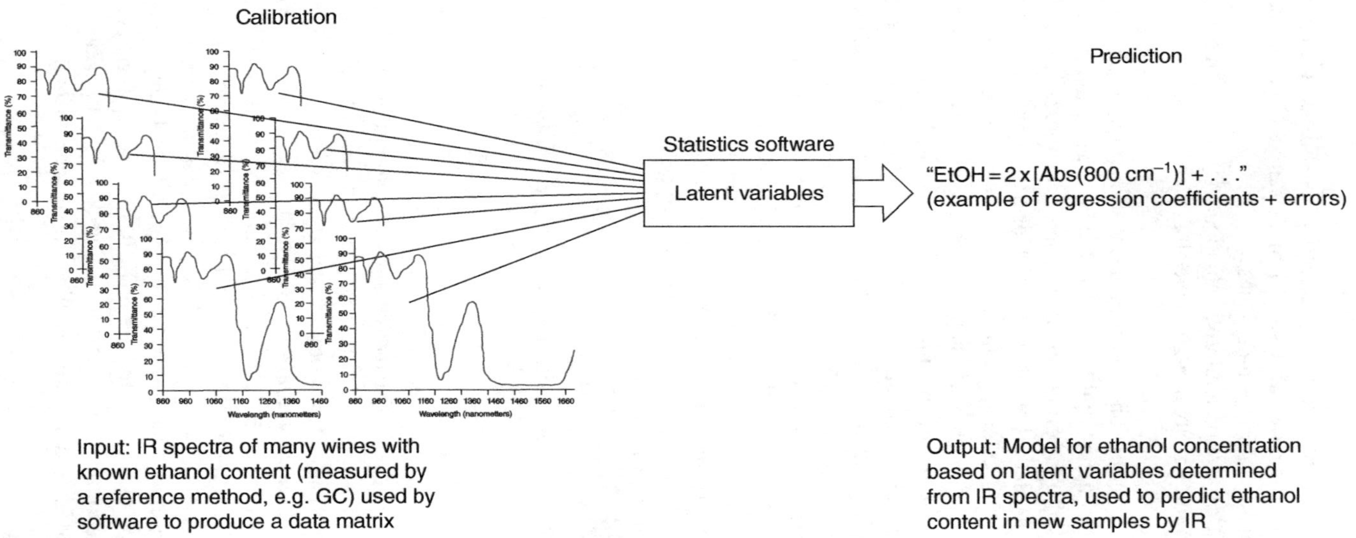

Calibration
Prediction
Statistics software
Latent variables
"EtOH = 2 x [Abs(800 cm⁻¹)] + . . ."
(example of regression coefficients + errors)
Transmittance (%)
Wavelength (nanometers)
Input: IR spectra of many wines with known ethanol content (measured by a reference method, e.g. GC) used by software to produce a data matrix
Output: Model for ethanol concentration based on latent variables determined from IR spectra, used to predict ethanol content in new samples by IR

Figure 32.2 *Example of how raw spectral data (IR in this case) can be used in conjunction with a quantitative reference method (e.g., GC for ethanol content) to build a multivariate calibration for predicting the concentration of components in wine. Commercial instruments with inbuilt calibrations based on this approach are available for simple measurement of wine compositional parameters such as ethanol*

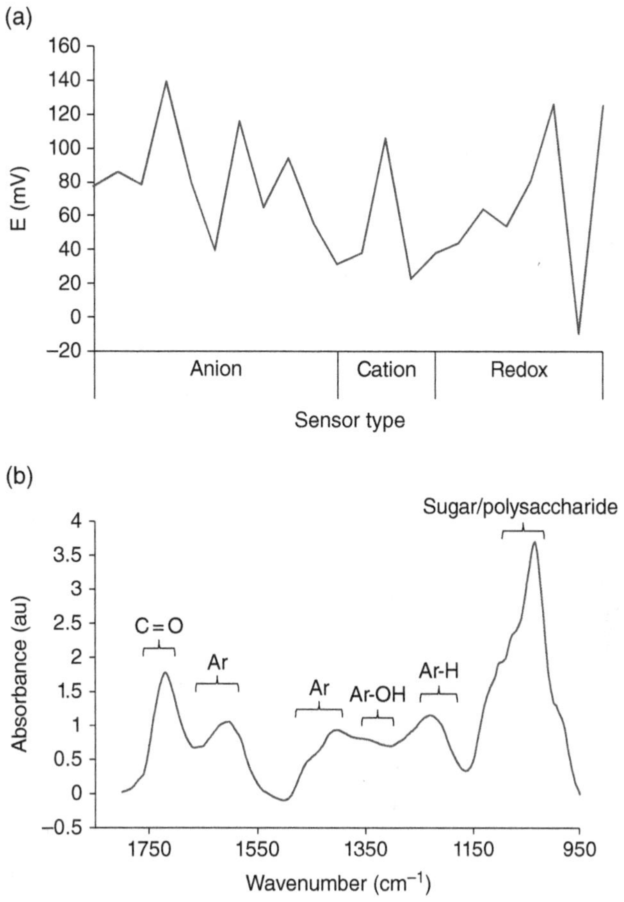

Figure 32.3 Examples of outputs used to develop prediction models for gelatin index as a measure of astringency from the analysis of a wine by (a) e-tongue with a range of potentiometric sensors and (b) MIR in the fingerprint region, showing tentative functional group assignments (Ar = aromatic). Data from Reference [42]

(in this case, gelatin index as determined by precipitation of tannins and spectral measurements, Chapter 33) to the output from the new technique or sensor. The e-tongue consisted of potentiometric sensors that were sensitive to organic ions and redox reactions, and the MIR fingerprint region (1800–950 cm^{-1}) was selected as it contains absorbances related to phenolic compounds [42] (Figure 32.3). As often occurs with multivariate approaches, the detection limits are above that of primary methods – the PLS models for both e-tongue and MIR approaches were suitable for red and rosé wines, but not for white wines.

Volatiles may also be determined in a similar manner by using an e-nose consisting of an array of metal oxide semiconductors. This approach has been applied to automatically monitor wine evolution during storage, and could be envisaged as a tool for winemakers to detect the development of problem aromas, thereby allowing for earlier intervention [43]. In this case, GC-MS was used as the reference method to determine volatile compounds, including a number of fatty acids, and ethyl and acetate esters, in one white wine and one red wine. PLS regression models were developed to predict the results of GC-MS analysis, with some compounds such as hexyl acetate and ethyl octanoate being quite well modeled (Figure 32.4). As with other approaches based on chemometric models, expanding the range of wines would further "train" the e-nose to recognize patterns and predict volatile concentrations in novel samples.

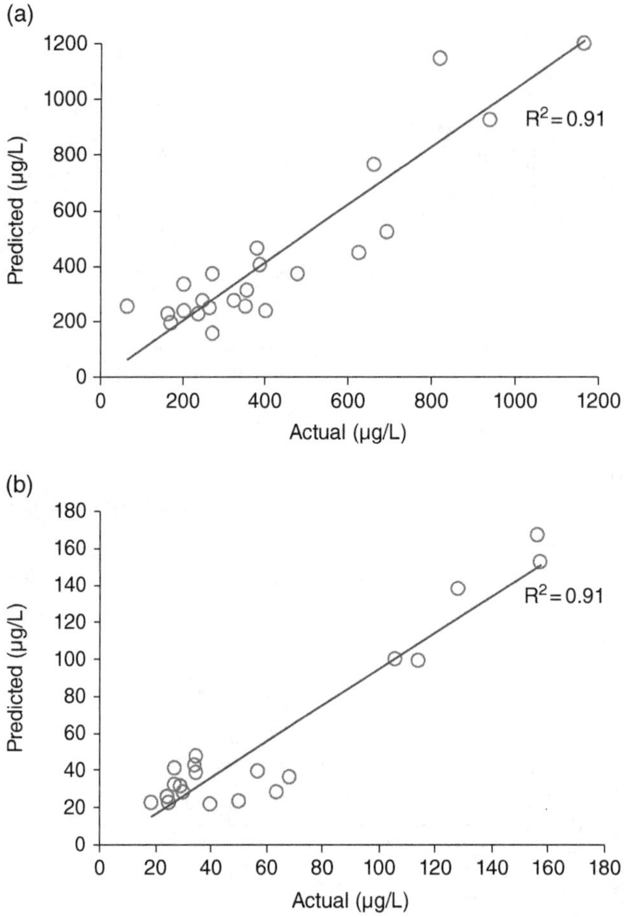

Figure 32.4 PLS model of GC-MS (actual) and e-nose (predicted) data for (a) hexyl acetate and (b) ethyl octanoate in wine. Data from Reference [43]

Rapid analytical methods do not necessarily require specialized equipment, and simple and cheap devices are more likely to be widely adopted for routine use. Analyses that utilize light are especially attractive, as in the case of instruments based on IR, but UV/vis light sources are also exploited in novel ways. One such example involved the construction of an in-line sensor employing a short path-length flow cell and miniaturized arrangement of light-emitting diodes (LEDs) and photodiodes to measure phenolics and color throughout red wine fermentation [46]. This work demonstrated the possibility of using such a sensor in place of a UV/vis spectrophotometer.

Another unique approach to rapid analysis involved a smartphone camera, paper microfluidics, and eight dyes (Figure 32.5) to assess "taste" upon addition of a drop of red wine [47]. The RGB color intensities (red = 680 nm, green = 540 nm, blue = 470 nm) of the dyes, analyzed from images taken using a digital smartphone camera, were found to correlate with the concentrations of different wine taste components, such as organic acids, salts, and sugars. RGB color intensities for different wine samples could be classified according to grape variety and wine style (body and sweetness) by principal component analysis (PCA). In particular, the method may be able to discriminate oxidized from non-oxidized wine samples, and could potentially act as a simple and effective quality control tool during wine production [47].

Figure 32.5 Simple approach to rapidly classifying red wine based on (a) pre-loading and drying dyes in wells of a paper microfluidic chip, (b) adding red wine to the center of the chip, and (c) photographing the chip with a smartphone camera to enable analysis of RGB images. Source: San Park 2014 [47]. Reproduced with permission of Royal Society of Chemistry

32.5 Targeted and untargeted metabolomics of wine

The chemometric examples of the previous section used relatively simple instrumentation incapable of isolating individual compounds. While useful for analyses of known targets, the lack of molecular identification makes these techniques poorly suited for researchers interested in developing novel hypotheses, or identifying unknown chemical factors that discriminate samples. True metabolomics studies based on targeted or untargeted analyses typically require the selectivity of MS coupled to one- or two-dimensional chromatography[5] (and, less frequently, NMR).

Due to their diversity and relative ease of measurement, a common target of metabolomics studies are phenolic compounds. These are often quantified by HPLC-DAD-MS, and targets include anthocyanins (in red wines), flavonols, and hydroxycinnamates. Untargeted analysis of non-volatiles by HPLC-MS/MS may also be used for phenolic compounds, and metabolomic approaches often utilize HPLC with high-resolution

[5] One-dimensional separations involve one column and a single phase (e.g., HPLC with a reversed-phase C18 (ocatadecyl silica) column or GC with a polar wax (polyethylene glycol) column). In contrast, two-dimensional (orthogonal) separations involve differing selectivity between two column stationary phases (e.g., polar wax and non-polar dimethylpolysiloxane GC columns) for improved separation of analytes. This necessitates more complicated instrumentation that is able to transfer analytes between first and second dimensions.

MS (in positive and negative ion modes) to allow determination of empirical formulae [35]. Data processing is often performed without initially identifying (or only tentatively identifying) individual compounds. In this way, several thousand *features* (putative metabolites) can be detected, with laborious efforts of identification being reserved only for statistically interesting compounds. For instance, over 400 compounds were tentatively identified (acids, sugars, phenols, and other compounds) in one study, and 15 of those could then be used to discriminate Graciano and Tempranillo wines [48].

Untargeted GC-MS methods have also been developed for informative, predictive, and discriminative analyses. One example of discriminative analyses involved fully comprehensive 2-D GC (GC×GC) and detection by TOF-MS[6] combined with software that can automatically detect peaks and deconvolute spectra [49]. A total of 375 compounds was identified by headspace SPME of five commercial Cabernet Sauvignon wines differing in region and vintage, and >85% of the compounds were significantly different between the wines. For predictive analyses (modeling of flavor attributes) multivariate data analysis enabled the use of ordinary 1-D GC-MS instruments for detection of several hundred compounds, as shown for Hunter Valley (HV) Semillon [50]. Automated analysis of chromatographic profiles followed by PCA revealed the compounds best correlated with wine age and style. PLS regression was then used to develop predictive models of HV Semillon sensory attributes (characters such as honey, toast, and orange marmalade) based on compounds identified by GC-MS profiling.

A key question looking forward is, "*Will untargeted metabolomic approaches replace classic targeted approaches*?" For research inquiries that do not need to establish causation – particularly authentication studies and related discrimination studies (Chapter 28) – untargeted approaches appear to have particular utility. However, untargeted approaches have been less effective for detecting causative agents, for example, determining specific compounds responsible for sensory attributes like aroma. Of the 10 000 volatiles detected in foodstuffs, very few (<3%) appear to contribute to aroma of the food [51], which means that many artifactual correlations are expected in untargeted analyses. For example, untargeted analyses of compounds responsible for the black pepper aroma of Shiraz wines found a strong correlation with a compound with no black pepper aroma (i.e., a *marker compound*, in this case α-ylangene). Further studies using the traditional bioassay-based approach (GC-O) were necessary to identify rotundone as the causative agent [52]. For sensory properties that emerge from multiple compounds (e.g., many fruity aromas), the hybrid approach of targeted metabolomics may turn out to be most appropriate, as highlighted by the study mentioned earlier that related the concentration of more than 100 volatile compounds to wine quality attributes [5]. While these comprehensive approaches currently require a large number of separate analyses, they should be facilitated by advances in both databases and instrumentation, as seen in metabolomics of other biological systems [53].

References

1. Buglass, A.J. and Caven-Quantrill, D.J. (2011) Analytical methods, in *Handbook of alcoholic beverages: technical, analytical and nutritional aspects* (ed. Buglass, A.J.), John Wiley & Sons, Ltd, Chichester, UK, pp. 629–932.
2. Vershinin, V.I. and Zolotov, Y.A. (2009) Periodization of the history of chemical analysis and analytical chemistry as a branch of science. *Journal of Analytical Chemistry*, **64** (8), 859–867.
3. Karayannis, M.I. and Efstathiou, C.E. (2012) Significant steps in the evolution of analytical chemistry – Is the today's analytical chemistry only chemistry? *Talanta*, **102**, 7–15.
4. Pérez-Bustamante, J.A. (1997) A schematic overview of the historical evolution of analytical chemistry. *Fresenius' Journal of Analytical Chemistry*, **357** (2), 151–161.
5. San Juan, F., Cacho, J., Ferreira, V., Escudero, A. (2012) Aroma chemical composition of red wines from different price categories and its relationship to quality. *Journal of Agricultural and Food Chemistry*, **60** (20), 5045–5056.

[6] TOF-MS is often used with GC×GC because the second dimension separation occurs in a matter of seconds. This means rapid mass spectral acquisition rates (e.g., up to 500 Hz) are required to provide enough data points to define extremely narrow peaks and identify compounds.

6. Cevallos-Cevallos, J.M., Reyes-De-Corcuera, J.I., Etxeberria, E., *et al.* (2009) Metabolomic analysis in food science: a review. *Trends in Food Science and Technology*, **20** (11–12), 557–566.
7. Marsili, R. (ed.) (2012) *Flavor, fragrance, and odor analysis*, 2nd edn, CRC Press, Boca Raton, FL.
8. Nicholson, J.K. and Lindon, J.C. (2008) Systems biology: Metabonomics. *Nature*, **455** (7216), 1054–1056.
9. Capone, D.L. and Jeffery, D.W. (2011) Effects of transporting and processing Sauvignon blanc grapes on 3-mercaptohexan-1-ol precursor concentrations. *Journal of Agricultural and Food Chemistry*, **59** (9), 4659–4667.
10. Sarker, S.D. and Nahar, L. (eds) (2012) *Natural products isolation*, 3rd edn, Humana Press, New York.
11. Acree, T.E. (1997) GC/Olfactometry: GC with a sense of smell. *Analytical Chemistry*, **69** (5), 170A–175A.
12. Chin, S.-T., Eyres, G.T., Marriott, P.J. (2015) Application of integrated comprehensive/multidimensional gas chromatography with mass spectrometry and olfactometry for aroma analysis in wine and coffee. *Food Chemistry*, **185**, 355–361.
13. Delahunty, C.M., Eyres, G., Dufour, J.-P. (2006) Gas chromatography–olfactometry. *Journal of Separation Science*, **29** (14), 2107–2125.
14. Reichelt, K.V., Peter, R., Ley, J.P., *et al.* (2010) LC Taste® as a novel tool for the identification of flavour modifying compounds, in *Proceedings of the 12th Weurman Symposium* (eds Blank, I., Wust, M., Yeretzian, C.), 2008, Interlaken, Switzerland, Zürcher Hochschule für Angewandte Wissenschaften, Winterthur.
15. Pineau, B., Barbe, J.-C., Van Leeuwen, C., Dubourdieu, D. (2009) Examples of perceptive interactions involved in specific "red-" and "black-berry" aromas in red wines. *Journal of Agricultural and Food Chemistry*, **57** (9), 3702–3708.
16. Sáenz-Navajas, M.-P., Ferreira, V., Dizy, M., Fernández-Zurbano, P. (2010) Characterization of taste-active fractions in red wine combining HPLC fractionation, sensory analysis and ultra performance liquid chromatography coupled with mass spectrometry detection. *Analytica Chimica Acta*, **673** (2), 151–159.
17. Falcao, L.D., Lytra, G., Darriet, P., Barbe, J.C. (2012) Identification of ethyl 2-hydroxy-4-methylpentanoate in red wines, a compound involved in blackberry aroma. *Food Chemistry*, **132** (1), 230–236.
18. Lytra, G., Tempere, S., Revel, G.d., Barbe, J.-C. (2012) Impact of perceptive interactions on red wine fruity aroma. *Journal of Agricultural and Food Chemistry*, **60** (50), 12260–12269.
19. Balakian, S. and Berg, H.W. (1968) The role of polyphenols in the behavior of potassium bitartrate in red wines. *American Journal of Enology and Viticulture*, **19** (2), 91–100.
20. Flamini, R. (ed.) (2008) *Hyphenated techniques in grape and wine chemistry*, John Wiley & Sons, Ltd, Chichester, UK.
21. Flamini, R. and Traldi, P. (2010) *Mass spectrometry in grape and wine chemistry*, John Wiley & Sons, Inc., Hoboken, NJ.
22. Marriott, P.J., Eyres, G.T., Dufour, J.-P. (2009) Emerging opportunities for flavor analysis through hyphenated gas chromatography. *Journal of Agricultural and Food Chemistry*, **57** (21), 9962–9971.
23. Marriott, P.J., Chin, S.-T., Maikhunthod, B., *et al.* (2012) Multidimensional gas chromatography. *TrAC Trends in Analytical Chemistry*, **34**, 1–21.
24. Kalili, K.M., Vestner, J., Stander, M.A., de Villiers, A. (2013) Toward unraveling grape tannin composition: application of online hydrophilic interaction chromatography × reversed-phase liquid chromatography–time-of-flight mass spectrometry for grape seed analysis. *Analytical Chemistry*, **85** (19), 9107–9115.
25. Willemse, C.M., Stander, M.A., Tredoux, A.G.J., de Villiers, A. (2014) Comprehensive two-dimensional liquid chromatographic analysis of anthocyanins. *Journal of Chromatography A*, **1359**, 189–201.
26. Sáenz-López, R., Fernández-Zurbano, P., Tena, M.T. (2004) Analysis of aged red wine pigments by capillary zone electrophoresis. *Journal of Chromatography A*, **1052** (1–2), 191–197.
27. Moreno, M., Arribas, A.S., Bermejo, E., *et al.* (2011) Analysis of polyphenols in white wine by CZE with amperometric detection using carbon nanotube-modified electrodes. *Electrophoresis*, **32** (8), 877–883.
28. Olivieri, A.C. (2015) Practical guidelines for reporting results in single- and multi-component analytical calibration: a tutorial. *Analytica Chimica Acta*, **868**, 10–22.
29. Eriksson, L., Johansson, E., Kettaneh-Wold, N., Wold, S. (2013) *Multi-and megavariate data analysis: basic principles and applications*, 3rd edn, MKS Umetrics AB, Malmö, Sweden.
30. Gishen, M., Dambergs, R.G., Cozzolino, D. (2005) Grape and wine analysis – enhancing the power of spectroscopy with chemometrics. *Australian Journal of Grape and Wine Research*, **11** (3), 296–305.
31. Lavine, B. and Workman, J. (2006) Chemometrics. *Analytical Chemistry*, **78** (12), 4137–4145.
32. Lawless, H.T. and Heymann, H. (2010) *Sensory evaluation of food*, Springer, New York.
33. Hong, Y.-S. (2011) NMR-based metabolomics in wine science. *Magnetic Resonance in Chemistry*, **49**, S13–S21.

34. Cozzolino, D. and Smyth, H. (2013) Analytical and chemometric-based methods to monitor and evaluate wine protected designation, in *Comprehensive analytical chemistry: food protected designation of origin – methodologies and applications* (eds Miguel de la, G. and Ana, G.), Elsevier, Oxford, UK, pp. 385–408.
35. Versari, A., Laurie, V.F., Ricci, A., *et al.* (2014) Progress in authentication, typification and traceability of grapes and wines by chemometric approaches. *Food Research International*, **60**, 2–18.
36. Cozzolino, D., Cynkar, W., Shah, N., Smith, P. (2011) Technical solutions for analysis of grape juice, must, and wine: the role of infrared spectroscopy and chemometrics. *Analytical and Bioanalytical Chemistry*, **401** (5), 1475–1484.
37. Godelmann, R., Fang, F., Humpfer, E., *et al.* (2013) Targeted and nontargeted wine analysis by ^{1}H NMR spectroscopy combined with multivariate statistical analysis. Differentiation of important parameters: grape variety, geographical origin, year of vintage. *Journal of Agricultural and Food Chemistry*, **61** (23), 5610–5619.
38. Lozano, J., Arroyo, T., Santos, J.P., *et al.* (2008) Electronic nose for wine ageing detection. *Sensors and Actuators B: Chemical*, **133** (1), 180–186.
39. Zeravik, J., Hlavacek, A., Lacina, K., Skladal, P. (2009) State of the art in the field of electronic and bioelectronic tongues – towards the analysis of wines. *Electroanalysis*, **21** (23), 2509–2520.
40. Buratti, S., Ballabio, D., Benedetti, S., Cosio, M.S. (2007) Prediction of Italian red wine sensorial descriptors from electronic nose, electronic tongue and spectrophotometric measurements by means of genetic algorithm regression models. *Food Chemistry*, **100** (1), 211–218.
41. Peris, M. and Escuder-Gilabert, L. (2013) On-line monitoring of food fermentation processes using electronic noses and electronic tongues: a review. *Analytica Chimica Acta*, **804**, 29–36.
42. Simoes Costa, A.M., Costa Sobral, M.M., Delgadillo, I., *et al.* (2015) Astringency quantification in wine: comparison of the electronic tongue and FT-MIR spectroscopy. *Sensors and Actuators B: Chemical*, **207** (0), 1095–1103.
43. Lozano, J., Santos, J.P., Suárez, J.I., *et al.* (2015) Automatic sensor system for the continuous analysis of the evolution of wine. *American Journal of Enology and Viticulture*, **66** (2), 148–155.
44. Ghanem, E., Hopfer, H., Navarro, A., *et al.* (2015) Predicting the composition of red wine blends using an array of multicomponent peptide-based sensors. *Molecules*, **20** (5), 9170.
45. Smyth, H. and Cozzolino, D. (2013) Instrumental methods (spectroscopy, electronic nose, and tongue) as tools to predict taste and aroma in beverages: advantages and limitations. *Chemical Reviews*, **113** (3), 1429–1440.
46. Shrake, N.L., Amirtharajah, R., Brenneman, C., *et al.* (2014) In-line measurement of color and total phenolics during red wine fermentations using a light-emitting diode sensor. *American Journal of Enology and Viticulture*, **65** (4), 463–470.
47. San Park, T., Baynes, C., Cho, S.-I., Yoon, J.-Y. (2014) Paper microfluidics for red wine tasting. *RSC Advances*, **4** (46), 24356–24362.
48. Arbulu, M., Sampedro, M.C., Gomez-Caballero, A., *et al.* (2015) Untargeted metabolomic analysis using liquid chromatography quadrupole time-of-flight mass spectrometry for non-volatile profiling of wines. *Analytica Chimica Acta*, **858**, 32–41.
49. Robinson, A.L., Boss, P.K., Heymann, H., *et al.* (2011) Development of a sensitive non-targeted method for characterizing the wine volatile profile using headspace solid-phase microextraction comprehensive two-dimensional gas chromatography time-of-flight mass spectrometry. *Journal of Chromatography A*, **1218** (3), 504–517.
50. Schmidtke, L.M., Blackman, J.W., Clark, A.C., Grant-Preece, P. (2013) Wine metabolomics: objective measures of sensory properties of Semillon from GC-MS profiles. *Journal of Agricultural and Food Chemistry*, **61** (49), 11957–11967.
51. Dunkel, A., Steinhaus, M., Kotthoff, M., *et al.* (2014) Nature's chemical signatures in human olfaction: a foodborne perspective for future biotechnology. *Angewandte Chemie International Edition*, **53** (28), 7124–7143.
52. Wood, C., Siebert, T.E., Parker, M., *et al.* (2008) From wine to pepper: rotundone, an obscure sesquiterpene, is a potent spicy aroma compound. *Journal of Agricultural and Food Chemistry*, **56** (10), 3738–3744.
53. Fuhrer, T. and Zamboni, N. (2015) High-throughput discovery metabolomics. *Current Opinion in Biotechnology*, **31**, 73–78.

33

New Approaches to Tannin Characterization

33.1 Introduction

The perception of astringency ("drying, rough, puckering") in wine is largely credited to condensed tannins (proanthocyanidins). As described in Chapter 14, condensed tannins in grapes and other plants are polymers of flavanoid monomers, and possess innumerable structures due to variations in subunit number, subunit structure/stereochemistry, and connection points. Even assuming only two subunit choices, no branching, and a single initiator, a tannin 30-mer could have $2^{30} \rightarrow 1$ billion different structures. This chemistry is further complicated in wines where proanthocyanidin linkages undergo hydrolysis and subsequently rearrange or else react with a large number of wine components, for example, anthocyanins or aldehydes (Chapters 24 and 25). Consequently, methods to characterize tannins typically fall into one of three categories:

- Characterization of the tannin monomers, such as through the use of acid phloroglucinolysis followed by HPLC. These approaches can determine the mean degree of polymerization and identify constituent flavan-3-ol subunits of the average proanthocyanidin in the sample [1]. However, these approaches will typically not degrade (or measure) non-standard interflavan linkages formed during aging.
- Operational characterization to measure the concentration of intact tannin based on a common chemical feature, for example, measurement of the late eluting broad "hump" at 280 nm in HPLC separation [2, 3].
- Functional characterization of astringency, for example, protein precipitation methods.

Of existing analytical methods, those based on precipitation of tannins by proteins or other macromolecules have been particularly successful in modeling astringency in wine, as determined by trained sensory panels (Chapter 14). Harbertson and Adams adapted the 1978 method of Hagerman and Butler [4] for wine, using bovine serum albumin as the binding substrate [5]. Sarneckis *et al.* utilized methyl cellulose as a precipitant to develop a comparable method [6]. Following precipitation, the phenolic content of the pellet or supernatant can be measured. These precipitation methods can achieve very high correlations between "astringency intensity" and "tannin" ($r^2 > 0.8$), likely because they mimic reactions that would occur between tannins and lubricating salivary proteins in the mouth [7].

Understanding Wine Chemistry, First Edition. Andrew L. Waterhouse, Gavin L. Sacks, and David W. Jeffery.

33.2 The challenge of astringency subclasses

While precipitation based assays are useful, they only model the overall intensity of perceived astringency. However, winemakers and others commonly use terms to discriminate types of astringency, and several reports have shown that the wines can be distinguished by trained panels based on either temporal behavior (e.g., how astringency intensity decays once the wine is expectorated) or subterms ("grainy," "velvety," "puckering," and more) [8]. At this point, it is unclear if these subterms arise from structural differences among tannins or the presence or absence of other wine components [9]. A challenge for chemists today is to look for analytical methods that could produce results that relate to these variations in perception.

Traditional protein-precipitation methods use an excess of protein, so that strong and moderate binding tannins are fully precipitated and consequently not well distinguished. McRae *et al.* have hypothesized that the strength of tannin–protein binding, as measured by isothermal titration calorimetry (ITC) [10], could be related to the qualities of astringency perception, particularly its time-intensity properties [11]. The enthalpy of binding of isolated wine tannin to polyproline, a synthetic model protein, drops by over 30% in young versus old wines [11]. While ITC methods are difficult to execute, comparable data can be achieved through HPLC analyses at different temperatures [12].

Relating the astringent subterms to specific chemical components is also an ongoing area of research, in wine and other fields. For example, some flavonol glycosides (Chapter 13) are reported to have "velvety astringent" sensory properties [13]. This astringent characteristic may have a different mechanism than that described for the "puckering astringency" of condensed tannins. Instead of affecting astringency indirectly through protein precipitation, these compounds may act directly on tactile receptors [14]. While intriguing, this hypothesis has not been validated with sensory studies on multiple wines. Other subterms may arise from the combined presence of other flavor compounds and condensed tannins; for example, high acid and high tannin may be responsible for the perception of "green tannin" [15]. Finally, future advances in relating tannin chemistry to sensory properties may benefit from using novel analytical chemistry approaches, such as detailed mass spectral data on tannin composition [16].

References

1. Kennedy, J.A. and Jones, G.P. (2001) Analysis of proanthocyanidin cleavage products following acid-catalysis in the presence of excess phloroglucinol. *Journal of Agricultural and Food Chemistry*, **49** (4), 1740–1746.
2. Waterhouse, A.L., Ignelzi, S., Shirley, J.R. (2000) A comparison of methods for quantifying oligomeric proanthocyanidins from grape seed extracts. *American Journal of Enology and Viticulture*, **51** (4), i383–389.
3. Mouls, L., Hugouvieux, V., Mazauric, J.P., *et al.* (2014) How to gain insight into the polydispersity of tannins: a combined MS and LC study. *Food Chemistry*, **165**, 348–353.
4. Hagerman, A.E. and Butler, L.G. (1978) Protein precipitation method for the quantitative determination of tannins. *Journal of Agricultural and Food Chemistry*, **26** (4), 809–812.
5. Adams, D.O. and Harbertson, J.F. (1999) Use of alkaline phosphatase for the analysis of tannins in grapes and red wines. *American Journal of Enology and Viticulture*, **50** (3), 247–252.
6. Sarneckis, C.J., Dambergs, R.G., Jones, P., *et al.* (2006) Quantification of condensed tannins by precipitation with methyl cellulose: development and validation of an optimised tool for grape and wine analysis. *Australian Journal of Grape and Wine Research*, **12** (1), 39–49.
7. Caceres-Mella, A., Pena-Neira, A., Narvaez-Bastias, J., *et al.* (2013) Comparison of analytical methods for measuring proanthocyanidins in wines and their relationship with perceived astringency. *International Journal of Food Science and Technology*, **48** (12), 2588–2594.
8. Gawel, R., Oberholster, A., Francis, I.L. (2000) A "mouth-feel wheel": terminology for communicating the mouth-feel characteristics of red wine. *Australian Journal of Grape and Wine Research*, **6** (3), 203–207.

9. Bajec, M.R. and Pickering, G.J. (2008) Astringency: mechanisms and perception. *Critical Reviews in Food Science and Nutrition*, **48** (9), 858–875.
10. Poncet-Legrand, C., Gautier, C., Cheynier, V., Imberty, A. (2007) Interactions between flavan-3-ols and poly(l-proline) studied by isothermal titration calorimetry: effect of the tannin structure. *Journal of Agricultural and Food Chemistry*, **55** (22), 9235–9240.
11. McRae, J.M., Falconer, R.J., Kennedy, J.A. (2010) Thermodynamics of grape and wine tannin interaction with polyproline: implications for red wine astringency. *Journal of Agricultural and Food Chemistry*, **58** (23), 12510–12518.
12. Revelette, M.R., Barak, J.A., Kennedy, J.A. (2014) High-performance liquid chromatography determination of red wine tannin stickiness. *Journal of Agricultural and Food Chemistry*, **62** (28), 6626–6631.
13. Hufnagel, J.C. and Hofmann, T. (2008) Orosensory-directed identification of astringent mouthfeel and bitter-tasting compounds in red wine. *Journal of Agricultural and Food Chemistry*, **56** (4), 1376–1386.
14. Schöbel, N., Radtke, D., Kyereme, J., *et al.* (2014) Astringency is a trigeminal sensation that involves the activation of G protein–coupled signaling by phenolic compounds. *Chemical Senses*, **39** (6), 471–487.
15. Jones, P.R., Gawel, R., Francis, I.L., Waters, E.J. (2008) The influence of interactions between major white wine components on the aroma, flavour and texture of model white wine. *Food Quality and Preference*, **19** (6), 596–607.
16. Kuhnert, N., Drynan, J.W., Obuchowicz, J., *et al.* (2010) Mass spectrometric characterization of black tea thearubigins leading to an oxidative cascade hypothesis for thearubigin formation. *Rapid Communications in Mass Spectrometry*, **24** (23), 3387–3404.

Index

Understanding Wine Chemistry, First Edition. Andrew L. Waterhouse, Gavin L. Sacks, and David W. Jeffery.